Jesús Muñoz Morcillo, Caroline Y. Robertson-von Trotha (eds.)
Genealogy of Popular Science

History of Science and Technology | Volume 1

Jesús Muñoz Morcillo, classicist (PhD) and art historian (PhD), is research fellow at the ZAK | Centre for Cultural and General Studies and the Institute of Art and Architecture History (IKB) at the Karlsruhe Institute of Technology (KIT).
Caroline Y. Robertson-von Trotha,
sociologist (PhD), is founding director of the ZAK | Centre for Cultural and General Studies at Karlsruhe Institute of Technology (KIT), coordinator of the German network of the Anna Lindh Foundation, member of the Culture Committee of the German UNESCO Commission, and chairlady of the Academic Council for Culture and Foreign Policy (WIKA) at the Institute for Foreign Cultural Relations (ifa).

Jesús Muñoz Morcillo, Caroline Y. Robertson-von Trotha (eds.)

Genealogy of Popular Science

From Ancient Ecphrasis to Virtual Reality

[transcript]

ZAK | Centre for Cultural and General Studies in cooperation with the Institut für Kunst- und Baugeschichte/FG Kunstgeschichte at Karlsruhe Institute of Technology (KIT) This project was funded by the Schleicher Foundation in cooperation with the KIT Foundation. The editors thank the Schleicher Foundation, the Commerzbank, and the KIT Foundation for the great support that made possible the conference and the edited volume.

Bibliographic information published by the Deutsche Nationalbibliothek
The Deutsche Nationalbibliothek lists this publication in the Deutsche Nationalbibliografie; detailed bibliographic data are available in the Internet at http://dnb.d-nb.de

Cover layout: Maria Arndt, Bielefeld
Cover illustration based on "Scuola di Atene" (School of Athens), Raphael (1483-1520), 1509, Fresco, Vatican Museums.
Layout and Typeset by Jesús Muñoz Morcillo, Ph.D.
Proofread by Kareem James Abu-Zeid, Ph.D.
Printed by Majuskel Medienproduktion GmbH, Wetzlar
Print-ISBN 978-3-8376-4835-5
PDF-ISBN 978-3-8394-4835-9
https://doi.org/10.14361/9783839448359

Printed on permanent acid-free text paper.

Contents

List of Figures

INTRODUCTORY ARTICLES

ON THE TRAIL OF POPULAR SCIENCE IN ANTIQUITY

MODERN TIMES: ARTS AND SCIENCES AND MEDIA

Preface

Popular science is a complex phenomenon with various sociopolitical, economic, and aesthetic dimensions. Because of this multilayered reality, research into popular science cannot be entirely contained within any single scholarly discipline. In interdisciplinary research fields such as Science Studies, the Sociology of Science, or Science and Technology Studies (STS), the approach to the phenomenon of popular science seems to be filtered through issues of the present day. Here, historical approaches are missing. The History of Science, on the other hand, lacks aesthetic criteria to explain the poly-medial dimension of popular science. And philological and iconographic disciplines, such as Classical Philology or Art History, seldom bring into focus the dissemination of scientific and technological knowledge as a cultural practice.

This book aims to initiate an interdisciplinary, genealogically reflected debate about popular science as a recurrent cultural technique. In this volume, the authors elucidate the polyvalent category of popular science in a multidisciplinary and diachronic way. The focus lies on both its cultural construction and the formal and functional techniques which characterize the dissemination of cultural and scientific knowledge in its contexts of production and reception.

Understanding the production of popular science is crucial because experts rely on much more than just scientific facts. Indeed, science often builds on societal values, rhetorical means, or popular traditions. Understanding the reception contexts of popular science is likewise relevant since previous knowledge, moral peculiarities, and concerns of a specific target group constitute the agency that indirectly coins what kind of (scientific) knowledge becomes popular. Furthermore, the modern awareness of a societal dimension of science doesn't imply the nonexistence of similar practices in previous periods. Even non-programmatic, random, or unconscious actions that popularize knowledge are part of the same family tree.

The genealogical approach proposed in this volume requires a loosening up of very specific contemporary terms such as 'science communication' or 'public understanding of science' in favor of more general ones. The terms 'science popularizer' and 'popular science' already appear in specialized literature for phenomena related to the period from the 19th century onwards (cf. e.g., Marsak 1959; Daum 2002, 2009).[1] Oliver Hochadel uses the similar German term "öffentliche Wissenschaft" (i.e., "public [therefore also popular] science") for the German Enlightenment (2003). Robertson-von Trotha (2007, 2012) coined that term ("Öffentliche Wissenschaft") in the 1990s, and used it to denote the contemporary practice of dialog- and process-based science communication in a broader sense with significant societal implications. In the present essay, we build on these experiences and theoretical approaches, opting for the versatile term 'popular science,' which unifies the notions of 'science communication' and 'public science' in a more general, diachronic sense. Thanks to its lack of terminological specificity, the term 'popular science' can indeed mark both new and past phenomena of science popularization as forms of agency and as cultural products at the same time.

As in the case of changing cultural techniques, there is no linear history of the phenomenon of 'popular science' that would begin at a certain point, e.g., with the ancient, moralizing didactic poem, and end at another point, e.g., with the use of Virtual Reality (VR) apps representing the ultimate embodiment of science and society in a unique immersive experience. This idea of continuity belongs to the realm of metaphysics and teleological thinking. But there is a way to understand contextual and historical similarities in popular science not as an unbroken continuity but as a series of accidents, each with their own unique manifestations. On the one hand, the historical contingency of the applicable terms for public science events leads to a dissociation of likely related phenomena. But, on the other hand, if we can get to the bottom of each action, then we can unveil specific relationships of forces inherent to popular science experiences. For example, the question about the intention behind the popularization of science is as legitimate today as it was in previous eras.

At the beginning of his well-known article entitled "Nietzsche, Genealogy, History" Foucault (1977: 139) puts it in a nutshell: "Genealogy is gray," i.e., it has no easily definable contours; it is "meticulous," because one needs a lot of clues and erudition to contruct its "cyclopean monuments"; it is "patiently documentary," since patience is the only way to obtain the evidence you are striving for; it deals with documents that can be reused several times—namely, with "parchments" in a figurative sense: that precious paper made from processed animal skin heavily used for copying manuscripts during the middle ages. Indeed, parchment was suitable for writing and overwriting several times by repeatedly scratching the

1 All bibliographical references quoted in the preface are listed at the end of the first introductory chapter by Jesús Muñoz Morcillo.

surface. And we can also add that genealogy has ramifications or offshoots of different lengths.

This volume follows the genealogical idea in both general and mimetic terms, offering a great variety of research topics from different periods, diverse methodological approaches, and even different degrees of 'growth' and 'blossoming'if the reader allows us the comparison with a living (genealogical) tree and its crown.

Despite all imagery, this definition applies to the fact that forgotten or even unconscious action models often precede those cultural techniques and programmatic agencies that we, otherwise, would only identify with modern contexts. We would like to put forward the claim that this cultural and historical amnesia probably concerns the phenomenon of popular science as well.

A genealogical approach requires interdisciplinary methods, humility, and also the patience needed to develop those methods. In the present volume, we are taking the first step in juxtaposing and bringing together the relevant disciplines in language that is comprehensible and accessible. Philological and aesthetic analyses unfold alongside specific disciplines such as the Sociology or History of Science. We offer a multilayered examination of the cultural technique of popular science; this should throw some light on a research field that has hardly been investigated so far, even if it does not cover every important aspect related to it.

Indeed, this book does not pretend to be a perfect handbook that conveys all possible manifestations of popular science. It does not even come close to Alfred Korzybski's idea of the map of a territory as the model of reality necessary for understanding what would otherwise remain inaccessible (Korzybski 1933). Our attempt only represents some minimal geographical details on the surface of Korzybski's ideal abstraction. It has a valuable self-reflexiveness, indeed, but in an almost living way, like the branches of a tree that still has to flourish. Therefore, even if this is a useful approach to the history and aesthetics of popular science and a starting point for more accurate contextualizing and conceptualizing attempts, we cannot specify any clear boundaries yet—perhaps because there are no real boundaries between nature, science, and culture. Indeed, we might even ask ourselves if those defining boundaries exist at all, those limits that 'kill' the object, releasing the magic force that transforms the map into the territory itself. The reader will judge if our abstraction of the map as a growing tree which might never render all variables of this phenomenon is a useful one.

One of the most important contributions of this book is to consider the popularization of the humanities as an essential part of these processes, as important as the natural sciences—or as even more relevant than them—since the key of communication also relies on cultural techniques related to the humanities. In times of ecocritical positions, there is no point in perpetuating any artificial divide between culture, nature, and science. In this sense, and even if not all authors in this book directly quote him, the genealogy of popular science also relies on Bruno Latour's ideas on modernity as the production of hybrids (Latour 1993). Science, nature, art, language, and mathematical models belong together. In this very sense, we attach importance to art, as a dissemination technique for science and culture, and to art history, as the subject of popular science, at the same time. Together with classical studies, art history is the best-represented discipline in this volume, and for good reason: Rhetoric and visuality are the pillars of popular science, even if today's discourses on science communication tend to disregard both the contributions and the misleading dangers of rhetorical means. Coded in words and images, the intention behind popular science, therefore, becomes a key for understanding societies and their moral values.

The book opens with an introductory chapter on the origins of popular science (Muñoz Morcillo). This text informs new and ancient notions and practices of science communication, re-contextualizing ancient and pre-modern rhetorical and societal practices that lead to the question of the adequacy of speaking about a genealogy of science popularization. Oliver Hochadel's chapter offers a longue durée approach with a contribution on the continuities in the history of science popularization from the Enlightenment to our days. These two texts are crucial for a proper orientation and a fruitful reading of the book, which comprises twenty-one different case studies connected, above all, by societal implications and the persistence of verbal and visual rhetoric. Each contribution resembles a bough, or a section of a bough, capable of growing further and bearing many more fruits.

Case studies are essential since this book makes an evidence-based claim about the intrinsic connections between popular science and society, regardless of historical time and geographical region. Still, this thesis is formulated as an invitation to look for further ramifications and entanglements in the vast research field of what we could call 'popular science studies.' We rely, therefore, on the curiosity of the readers, who may be very varied in their interests: from philologists, archaeologists, and art historians to sociologists, philosophers, and historians of science.

Each one of these readers should be able to find something valuable here, from the relevance of rhetorical devices and material culture to the specific logic structure of doxographical texts;[2] from the societal uses of general knowledge in antiquity to the creation of knowledge-spreading proto-encyclopedias and epitomes in pre-modern times; from the popular interactions of art and science during the Enlightenment to modern media coverage on topics ranging from atomic energy to

2 Doxography is a term coined by the German classicist Hermann A. Diels and refers to the works of ancient historians collecting popular opinions (*dóxai*) attributed to past philosophers and scientists.

neuroscience, not to mention the sociological analysis of science communication in popular new media such as the web video or the science slam.

The reader will find a huge selection of case studies that prove the diachronic, polyhedral, and multilayered nature of popular science: worthy reflexions about the use of amphorae for disseminating central elements of Greek culture in the Mediterranean area (Streicher); the relevance of preliminary exercises in ancient Greek schools for the popularization of Greek mythology and literature as a living practice (Fernández Delgado); new insights into the role of Aristotelian rhetoric for doxographers—the ancient knowledge popularizers (Martín-Velasco)—or into the importance of description (*ékphrasis*) as a knowledge device for a broad audience in Greek dramas (Matías Pérez); the analysis of argument schemes including simplified scientific explanations and shortened quotations in authors from the Second Sophistic (Parussel); reflections on ancient zoological descriptions and representations that shed light on specific societal practices in Rome (Engster); rhetorical means for making invisible natural phenomena understandable (Rossetti); Hero of Alexandria's rhetorical techniques for bringing the mechanics of his artefacts alive for a broader audience and their impact in the Renaissance (Roby); the order of knowledge in pre-modern encyclopedias, their popularity, and similarities to contemporary internet-based practices (Herweg); a revealing case study about pre-modern ephemeral literature on comets and related public scientific discussions (Gindhart); the relevance of short versions (epitomes) of larger texts in late antiquity and pre-modern times for condensing and popularizing knowledge (especially history books) among a broad readership (Sehlmeyer); the revaluation of visual perception as part of modern reasoning in Enlightenment art (Jehle); the popularity (despite anti-popularizations attempts) of optical devices thanks to their use by artists in the first half of the 19th century (Fiorentini); the emergence of popular writings on art history (Axtmann); the agency of ornamental prints and pattern sheets in the popularization of aesthetic ideas in the second half of the 19th century (Manker); the relationship between popular science and esoteric literature in the avant-garde (Immelmann); the visual metaphors of cybernetics that disseminated a new understanding of nature as an ecosystem (Trüper); the controversial public discussion on nuclear power in US popular journals of the Cold War era (Kloppen); the use of everyday images and stereotypes for directing public attention to neuroscience in the German magazine *Gerhirn & Geist*, which planted the seeds for new visual preferences (Hommrich); the role of fiction in popular media such as movies and graphic novels or comics for translating and visualizing scientific agency (Klohs); or the perpetuation of gender issues in innovative formats of science communication such as the science slam (Hill).

Considering such a variety of approaches from different disciplines, the editors have tried to foster accessible language across the book, which has chiefly entailed the use of translations for all quotations from classical texts and the use of punctual explanations for specialized terminology. However, interdisciplinarity also means finding compromise solutions, and as such the reader will at times be compelled to make some efforts, which will hopefully be compensated with fruitful insights and new inspiration.

This volume is the result of the conference "Genealogy of Popular Science. From Ancient Ecphrasis to Virtual Reality"[3] held on June 15-16, 2018 at the Karlsruhe Institute of Technology (KIT), Germany.[4] The event was organized by the ZAK I Centre for Cultural and General Studies in cooperation with the Institut für Kunst- und Baugeschichte/Fachgebiet Kunstgeschichte. The ZAK's research addresses, among many things, topics within the fields of globalization, identity, cultural and technological change, intercultural communication, and the mediation of interdisciplinary competencies and their relevance in an international context. Thus, the ZAK contributes to the scientific reflection and shaping of social and cultural practice. One of the main research areas of the ZAK concerns Public Science and New Media, which not only includes studies on current practices of science communication, such as the production of online videos or the use of VR technologies in museums, but also research about the origins of popular science and its development through the ages.

The conference and the publication of this book were made possible thanks to a generous grant by the Schleicher Foundation in cooperation with the KIT Foundation to promote Dr. Muñoz Morcillo's postdoctoral research. The editors of this volume want to again thank the Schleicher Foundation, the Commerzbank, and the KIT Foundation for their great support, which made this project possible. Since its foundation in 2013, the KIT Foundation has been promoting research, teaching, innovation, and academic life at KIT. We thank the team of the KIT Foundation, especially foundation manager director Kathrin Krause and manager Dagmar Seelig, for their supportive, kind, and close cooperation. A special thank goes to Prof. Dr. Thomas Hirth, Vice-President for Innovation and International Affairs at KIT for taking the time to introduce the KIT to our international guests with a welcome speech. We highly appreciate this recognition of how advance studies in the humanities shape the KIT research landscape.

3 To differentiate the ancient mode of ecphrasis from the modern one, we use the transliteration of the Greek term ἔκφρασις (that is, *ékphrasis*) throughout the book.

4 The conference took place at the Karlsruhe Institute of Technology (KIT), Fritz-Haller-Hörsaal, Building 20.40, Englerstr. 7, KIT Campus South Karlsruhe.

We also owe our thanks to our project partners Prof. Dr. Oliver Jehle, Dr. Alexandra Axtmann, and Prof. Dr. Mathias Herweg, who contributed suggestions, ideas, and their own scholarly writings that enrich this volume in significant ways. For the selection of the papers and the peer review process, we want to thank all members of the Selection Committee, who spent their precious time reading and evaluating all submitted abstracts. These are, by name: Dr. Joachim Allgaier, Prof. Dr. José Antonio Fernández Delgado, PhD candidate Andrea Geipel, Prof. Dr. Mathias Herweg, Dr. Oliver Hochadel, Prof. Dr. Oliver Jehle, Dr. Christine Mielke, Dr. Laura Miguélez Cavero, Dr. Eva Noller, Prof. Dr. Courtney Ann Roby, and Dr. Ralf Schneider. We also thank all contributing authors for their insightful texts. And last but not least, we also want to thank the ZAK team and all helpers that made the conference and this book possible, especially Stephanie Rothe and Klemens Czurda for assuming great responsibility during the conference, Jens Görisch for all financial aspects and many helpful tips, and Dr. Kareem James Abu-Zeid for his meticulous proofreading of the entire volume.

Jesús Muñoz Morcillo
Caroline Y. Robertson-von Trotha

Introductory Articles

The Origins of Popular Science as a Rhetorical and Protreptical Practice

Jesús Muñoz Morcillo[*]

ABSTRACT

Despite modern scholars' efforts to explain the origins of science communication, their approaches to this phenomenon are often a response to current affairs and thereby ignore its historical roots in the classical world, where the phenomenon first emerged as knowledge popularization in general terms. Most modern authors confine their research to a modern concept of science and society that excludes the much broader idea of ancient episteme. In this introductory chapter, I propose a general approach to science communication as popular science—i.e., to the popularization of knowledge in antiquity—that focuses on dissemination formats, rhetorical strategies, and cultural techniques, all of which signal certain diachronic continuities and transformations.

* For creating the appropriate conditions for researching the genealogy of popular science from an aesthetical and historical perspective, I want to thank Prof. Dr. Caroline Y. Robertson-von Trotha and Prof. Dr. Oliver Jehle. For their useful remarks on the function of rhetoric for disseminating knowledge in antiquity, I want to thank Prof. Dr. José Antonio Fernández Delgado and Prof. Dr. Francisca Pordomingo Pardo. The final version of this paper has also benefited from a research stay at the Getty Research Institute (GRI) in Los Angeles thanks to a grant awarded by the Vokswagen Foundation. And for corrections, interesting suggestions, and formatting help, I owe my gratitude to Kareem James Abu-Zeid and Stephanie Rothe.

INTRODUCTION

Science communication[1] is a modern cultural practice that scholars mostly associate either with postwar research and education politics (e.g., Robertson-von Trotha 2007, 2012) or with the positivistic turn of the 19th century (e.g., Fyfe/Lightman 2007). Science communication researchers have recently stressed that the history of science communication is, for the most part, a desideratum of research.[2] Despite efforts to explain the origins of this intricate phenomenon through its sociopolitical, economic, and aesthetic dimensions, most of the relevant research fields—such as the History of Science, the Sociology of Science, or Science and Technology Studies (STS)—approach the popularization of science from the perspective of current affairs, ignoring to a great extent its historical origins and typology in ancient literature and culture, not to mention its diachronic traditions and transformations. Even historians of science have written little about this dimension of science communication. This is probably due to the difficulty of historicizing the notion of science communication and to certain methodological shortcomings inherent in explanations of the poly-medial dimensions of the science popularization. Indeed, most writings in this discipline overlook the philological and iconographic nature of this issue.

In this introductory chapter, I propose a general approach to the subject. For that, I focus on the origins of science communication as the popularization of knowledge, i.e., popular science, in the Greco-Roman world. An overview of these origins can lead to a deeper analysis of recurrent structures, (dis)continuities, and transformations. The working hypothesis here is that knowledge popularization has the ability to both adapt to and shape the dominant societal and political reality. To date, scholars have primarily been concerned with finding precursors of science communication that match either Enlightenment ideals (Hochadel 2003; Wolfschmidt 2017) or modern theories on institutionalized science (Rob-

---

1 There are many different terms related to the general idea of 'science communication' as communication with laypeople, such as: the public understanding of science; the public communication of science and technology; public science; *divulgation scientifique*; dissemination; etc. The popular expression 'science communication' is a 20th-century coinage. But broadly speaking, the meaning of 'science communication,' when applied to antiquity, can be said to be the 'popularization of knowledge with a purpose' or simply 'popular science.' As explained in the preface, 'popular science' is the preferred term for a diachronic use throughout this book.

2 See, e.g., the special issue of the Journal of Science Communication 16/3 (2017), on the "History of Science Communication."

ertson-von Trotha 2012). More recent efforts that try to set the course for a programmatic history of science communication, recognizing rhetoric as an essential influence, seem to espouse the view of science communication as a modern, professionalized phenomenon that began after the positivistic revolution of the 19th century (Bauer 2017). In contrast to this, my approach transcends the current historical and conceptual boundaries that constrain popular science within the bounds of the history of modern science. I accomplish this by considering every verbal or visual rhetorical action with a specific purpose to be an act of science communication, thereby enhancing the scope of study to include the ancient and pre-modern epistemic understanding of science. Whenever ancient philosophers, scholars of the Middle Ages, or modern scientists communicate science, they are doing much more than simply passing down knowledge. The cultural interdependence between communicational intent and audience reception has led to different strategies of persuasion over time. Rhetorical skills and the intentions behind that rhetoric are as important as the scientific content itself. Accordingly, the history of popular science is also a history of ancient rhetorical persuasion.

Rather than merely re-dating the origins of popular science in antiquity and thereby repackaging it as 'old wine' that has survived in 'new bottles,' my overview shows that the ancient pragmatism behind this practice frequently outweighed modern idealistic presumptions of individual liberation or the harmonic co-evolution of science and society through scientific literacy. In this sense, I also make a plea for the revision of some general assumptions, e.g., those related to modern issues of trust and legitimation (Kohring 2016a, 2016b; Weingart/Guenther 2016). It seems that each period's practices of science popularization are mirrors of its corresponding historical context.

In this introductory chapter, I focus on an argument supporting the idea that knowledge popularization was a common and diversified communication practice in antiquity. I begin with a brief introduction to the notion of popular science, surveying the current state of scholarship and outlining the conceptual and terminological differences between modern and ancient forms of science communication. This leads to a revision of the widespread opinion that science popularization originated during the Enlightenment without any reference to older archetypes. In the second section, I explore the variety of scientific writings intended for a non-expert public in the Greco-Roman world, offering an exemplary selection from different genres such as didactic poetry, dialogues, and epistles, which reveal information about the target audience and the purposes behind the dissemination of knowledge. In the third and fourth sections, I complete the picture by drawing particular attention to literary and visual rhetorical strategies for popularizing knowledge, such as *ékphrasis* (ἔκφρασις) and the philosophical legitimation of artistic representation as a medium of knowledge and wisdom. Finally, in a short conclusion, I summarize the results and provide a brief comparison with modern conceptions of popular science. It seems that ancient knowledge popularization has some similarities to current practices of science communication, which raises the question about whether it is possible to formulate a genealogy of public science.

NOTIONS OF POPULAR SCIENCE

For many researchers, the practice of science popularization began as a counterpoint to the high degree of specialization of modern scientific disciplines during the 19th century (MacLeod 1996; Morus 1998; Auerbach 1999; Fyfe/Lightman 2007; Huang 2015; Weingart/Guenther 2016). Other scholars argue that the historical roots of today's science communication, understood as the public communication of scientific knowledge, go back to a time before the definition of contemporary science and professional scientists. Those authors believe that this practice did not originate with the specialization of the sciences and the institutional funding of research activities, but rather with the work of explainers of science during the Enlightenment (Stewart 1992; Bensaude-Vincent 2001; Hochadel 2003; Gregory 2016: 2). In any case, recent literature focuses on the sociopolitical changes after the Second World War, which first led to the appearance of professionalized and institutionalized science communication. Examples of this are the 'scientific literacy' programs in the United States of America in the 1960s and, after the neoliberal turn of the 1980s in the United Kingdom, the concept of the 'public understanding of science' formulated in the Bodmer Report (1985, "Public Understanding of Science," see, e.g., Knight 2006; Robertson-von Trotha 2007, 2012; Short 2013). The Bodmer Report identified British society's disaffection with science and technology as a crucial problem with regard to economic development and progress. Specialized literature considers these two institutional measures as forerunners of modern science communication.

The meaning of popular science under the label of 'science communication' is manifold, even today. We can identify three basic meanings of this term: 1. The popularization of science, in line with the information deficit model[3] in the

3 This term from the field of Science and Technology Studies (STS) was coined in the 1980s; the core assumption of the deficit model is that public hostility to science and technology is due to a lack of information and a missunderstanding of science.

mass media, i.e., through the reduction of complexity, which includes not only informal teaching strategies but also critical journalism (cf., e.g., Bauernschmidt 2018). 2. Public science as dialogue-oriented science communication, whether institutionalized or not (Robertson-von Trotha 2012). And 3. PR work as institutionalized science communication at universities, research centers, and other institutions, which constitutes a kind of obsequious journalism (Kohring 2004, 2012). Since no scholars to date have attempted a reconstruction of all the practices of 'science communication' in antiquity and their sociopolitical contexts, we cannot compare the different qualities of this cultural technique with ancient counterparts, but we can start by working with the most general notion, namely, the first one. Indeed, the roots of popular science in a broad sense, i.e., understood as the popularization of knowledge with a purpose, extend far into the past. Scientific and philosophical knowledge, i.e., technology and the humanities, were communicated to the public long before the popularization of scientific demonstrations during the Enlightenment, the specialization of the sciences in the 19th century, or the neoliberal turn of the 20th century.

Despite the apparent existence of different practices of popular science in different historical periods, the research on this topic is scarce. Science and Technology Studies (STS) in particular seem to focus on recent decades, from the 1960s to the present day. There are some remarkable exceptions, however. These include: the abovementioned analysis of public science practices in the Enlightenment (Stewart 1992; Bensaude-Vincent 2001; Hochadel 2003); some approaches to communication cultures and the role of popular speech in the Middle Ages (Faulstich 1996; Wandhoff 2003), including the production of universal chronicles in vernacular languages (Herweg 2016) and the use of myths as a form of knowledge in the epic literature of the 13th century (Gebert 2013); the relation between public science and power in the Victorian era in Great Britain (MacLeod 1996); some works on general ancient communication techniques (Wilke 2012) and on ancient technical writings from a rhetoric perspective (Doody/Föllinger/Taub 2012: 233-236); and a recent interesting monograph on specific communication strategies for "advertising" ancient machines through rhetorical descriptions, which could be called technical *ékphrasis* (ἔκφρασις, cf. Roby 2016).

Even if some authors occasionally point out that certain genres of scientific writing, such as the didactic poem, were suitable for a broad educated audience in antiquity (cf. e.g., Effe 1977: 195-199, 2005: 37; Christmann 2003), a comprehensive survey on the popularization of scientific knowledge in antiquity—or even an introduction to this topic—still seems to be a desideratum of research. Against this background, it is also understandable that researchers are still a long way off from attempting a diachronic approach to popular science as a recurrent and changing cultural technique.

In the present text, I focus on the origins of science communication, understood as the popularization of knowledge (episteme) with a concrete purpose, in the Greco-Roman world. This also provides the foundation for the first research question, namely: Which communication formats, rhetorical strategies, and cultural techniques from antiquity can be said to constitute the origins of popular science, i.e., 'science communication' as the popularization of knowledge with a purpose? The main problem here is that our knowledge of the public in antiquity is still subject to debate, since it is extremely difficult to keep different phenomena apart. For example, teaching and the popularization of knowledge are related but not identical things, and the boundaries are often blurred here. Indeed, we can consider the secondary school standards of Hellenistic times (i.e., those of the grammar schools, with their practical exercises in composition, cf. Marrou 1956: 170-175, 274-283) an exceptional case of knowledge popularization. Furthermore, 'science' and 'technology' were also much broader concepts back then than they are today, and meant something closer to 'knowledge' (*epistḗmē*, ἐπιστήμη) and 'skill' (*téchnē*, τέχνη) respectively.

Nevertheless, in general terms we can assess whether a text was meant to be for initiates and experts (esoteric texts), an educated audience, i.e., the interested public (exoteric texts), or even for a broader audience (as in the incidental or essential depictions in popular genres such as drama or the epigram). Finally, I hypothesize about the possible continuity and transformation of some ancient communication strategies in modern science communication, in particular regarding the adaptation of genres such as the didactic poem, the letter, the dialogue, or protreptic speech, as well as the adaptation of certain structural features and persuasion strategies such as *ékphrasis* (ἔκφρασις). Because of the complexity and magnitude of a diachronic approach, we cannot satisfyingly answer the larger research question, i.e., to what extent ancient rhetorical and cultural qualities of 'public science' may have survived, even if formats, terminology, and epistemic categories have changed through the ages. However, we can put the first piece of the puzzle in place by making the effort to synthesize the Western origins of science communication in some remarkable 'scientific' and popular ancient writings.

ANCIENT POPULAR WRITINGS FOR A NON-EXPERT PUBLIC

The Didactic Poem as a Social Event and a Long-Term Popularizer of Knowledge

The proverb "Observe due measure: and proportion is best in all things"[4] is a central thought from the earliest didactic poem of the ancient world, Hesiod's *Works and Days* (Ἔργα καὶ ἡμέρα, ca. VIII BC). Despite our temporal distance to Hesiod, this ancient phrase—applying measure and proportion in everyday life situations—still has a very practical use today. Many ancient didactic poems have transported popular and technical knowledge to the present day. Didactic poems are probably the first example of a systematic popular communication of knowledge. According to scholarly tradition, Hesiod wrote *Works and Days* because of a dispute with his brother Perses. He presents the lifestyle of the good peasant and the cosmic order of those days in dactylic hexameters. In *Works and Days*, Hesiod not only explains to his brother how to work the land for a good harvest and how to build his house, but also passes down some popular knowledge. A good example of this is the "Ages of Man" section (verses 106-201, γένος [...] ἄνθροπων), which recounts the stages of human existence since the creation of the world, from the paradisiac Golden Age to the painful and arduous Iron Age. This popular myth, which, according to the structuralist Jean-Pierre Vernant (1965), actually focused on the benefits of justice (*dýkē*, δύκη) in contrast to man's arrogance (*hýbris*, ὕβρις), somehow became part of the ancient general education and remained unquestioned for centuries. We find almost the same interpretation of the Ages of Man in the Roman Empire many centuries after Hesiod's lifetime (e.g., Ovid, *Metamorphoses* 1.89-150). It seems that this description of the divine origin of human civilization was an excellent starting point for certain common values and beliefs, including the idea that a better and fairer age was still to come (cf. Virgil, *Eclogues* 4). The first Church Fathers merged the ancient mythical concept of human ages with the biblical Jewish tradition. In the 4th century AD, Saint Jerome tried to fit the Hesiodic ages into a historical chronology. Saint Augustine probably followed pagan traditions when he formulated his influential Christian periodization of human history as six ages of the world, from the creation of Adam to the Revelation (cf. Archambault 1966). For centuries, the homily was the primary medium for teaching people

4 Μέτρα φυλάσσεσθαι© καιρὸς δ' ἐπὶ πᾶσιν ἄριστος. (Hesiod, *Works and Days*, 694, translation by Hugh G. Evelyn-White).

about the new synthesis of pagan philosophy and Christian theology. Through the popularization of universal chronicles in the medieval period, not just in Latin but also in vernacular languages, the Ages of Man became common knowledge among the literate population. Furthermore, the names of two Hesiodic ancient ages still correspond to the current archaeological classification of two prehistoric periods: the Bronze Age and the Iron Age.

Since the 'publication' of *Works and Days*, the Hesiodic poem has enjoyed great popularity. If we believe the imperial rhetoric tradition, Hesiod once competed against Homer in a public poetry contest (Περὶ Ὁμήρου καὶ Ἡσιόδου καὶ γένους καὶ ἀγῶνος αὐτῶν, i.e., "On Homer and Hesiod, their Ancestry and their Contest," cf. West 1967). The incident seems to be inspired by Hesiod's own reference to a poetic competition in *Works and Days* (vv. 650-659). An anonymous compiler described the supposed poetic competition between Homer and Hesiod as taking place during the funeral games for King Amphidamas at Chalcis, in Euboea. Hesiod defeated Homer, "a rhapsodist wandering from city to city" (περιέρχεσθαι κατὰ πόλιν ῥαψῳδοῦντα), because, in the opinion of Panedes, brother of the deceased King Amphidamas, there is a moral and didactic superiority in Hesiod's poetry, which inspires both agriculture and freedom, while Homer's verses describe wars and murders. Even if this contest was merely a fictional rhetorical exercise, the fact that a general interest in useful knowledge undeniably persisted long after Hesiod's lifetime is evidence for the importance of poetry as a popular medium capable of passing down moral values, practical knowledge, and wisdom. The reported contest between Hesiod and Homer, along with textual evidence about the existence of this kind of poetic competition (e.g., *Work and Days* 656-659, Aristophanes, *Freedom* 1282-83), demonstrates that epic poetry and theater were not the only genres that enjoyed a broad audience. Poets also declaimed didactic poetry in solemn literary competitions (*agōnes*, ἀγῶνες), and some beautiful excerpts of didactic poetry were even considered suitable for reciting and commenting on during feasts (*sympósia*, συμπόσια) or for memorizing at school (*gymnásion*, γυμνάσιον). In this sense, Fernández Delgado (1976) pointed out several decades ago that gnomic poetry, to which Hesiod's work belongs, reached a different public than epic poetry.

Due to the variety of public events in ancient times, as well as other paths of cultural dissemination such as public readings and symposia, different target groups were reached, both in Greece and Rome (Momigliano 1978; Wiseman 2015). The common elements of the production and communication of knowledge in antiquity seem to be orality and the visual culture. Plato considered writing a mnemonic support without validity *per se* for the production of truth (cf. *Phaedrus* and *Seventh Letter*). The superiority of the oral tradition, visual experience, and dialectical methods over a simple written production of knowledge lasted many centuries. Indeed, ancient knowledge was the result of vivid dialectical methods and the calculated dissemination of didactic poems, letters, or lectures to different target groups. In any case, some kind of public performance was necessary for achieving the desired communication goal. Plato believed that knowledge was only possible through the combination of verbal description and sense perception (cf. Plato, *Seventh Letter*, 341be). However, even Plato, who criticized the poetic arts as flawed imitations of nature that were inferior to philosophy, was aware that *philopoiētaí* (φιλοποιηταί), i.e., "the lovers of poetry," would advocate not only for poetic pleasure but also for the societal benefits of poetry (*Rep.* 10.607d-608a). Indeed, some ancient authors considered Homer the inventor of geography as a discipline, to cite just one example. The Stoic historian Strabo (64 BC to circa 24 AD) praised Homer's descriptive realism (Strabo, *Geography* 1.1.1-11). The Epicurean philosopher Philodemus of Gadara (circa 110-35 BC) had a more cautious opinion on the educational function of poetry—one that, on the other hand, excuses Homer's mistakes. Indeed, Philodemus believed that there is no need for a mimetic link when making geographic descriptions, since the Homeric poem and poetry in general are about fictional situations (Jensen 1923, IV n. 2; here, following Lombardo 2002, n. 19). As we can see, there were different assumptions about poetry's ability to produce and disseminate true knowledge. However, the diachronic production of didactic (and also epic) poetry for the dissemination of general and specific knowledge, as well as archaeological discoveries related to Homer's descriptions and geographical information, support the argument that ancient poetry played an active role in the communication of traditions, cultural values, religious beliefs, and new knowledge among the population.

Two crucial categories of communication experts existed within this context: the rhapsodists, for broader audiences; and the grammarians, for the first stages of education. Plato considered the rhapsodist to be an interpreter or mediator (*hermēneús*, ἑρμηνεύς) of the poet, who was also, for his part, the interpreter of the gods (Plato, *Ion* 535a). The grammarians (*grammatistaí*, γραμματισταί, *Teeteto* 163c) were also interpreters and propagators of knowledge that was encapsulated in poetry, and made important contributions to the general education of Greek and Roman citizens.[5]

The popularization of poetic texts through lyrical competitions, public lectures, and private events (e.g., symposia) shows just how important the arts of rhetoric

5 For more on education in the different periods of Greek antiquity, cf. Marrou 1956.

and performance were in the ancient world for spreading ideas, cultural values, and new knowledge to a public audience. This tradition, most likely initiated by Hesiod, even applies to the pre-Socratic philosophers of the sixth century BC, who were the first "natural scientists." Parmenides formulated his monistic theory in dactylic hexameters, according to which everything can be attributed to a principle. The works *Purifications* and *On Nature* by Empedocles are also in hexameters. According to Diogenes, Aristotle considered Empedocles the forerunner of rhetoric (D.L. 8.57 and 9.25). Even in Hellenistic times, long after Plato's criticism on poetry, some Stoic philosophers, including Zeno's follower Cleanthes (331-231 BC), preferred poetry as a visually descriptive medium for the dissemination of their teachings to a broad audience (cf. Pohlenz 1939). The grammarian Diomedes asserted that "the poet himself speaks" in didactic poetry (i.e., in contrast to the characters in epics, tragedies, or comedies; Diomedes 482.30-483.3 Keil). This characteristic is typical for the creation of a bond between the 'poet-scientist' and the audience, and is present in every didactic poem from Hesiod to Lucretius and beyond. Still in the Hellenistic period, Epicurus wrote letters as epitomes of his teachings for a broad audience, and sometimes even for advanced students (Muñoz Morcillo 2016). However, these texts of his are often still quite difficult to understand; from today's point of view, Epicurus seems to have been a neglectful writer.

Figure 1: Frontispiece of Pope Sixtus IV's De rerum natura *Manuscript, Detail*

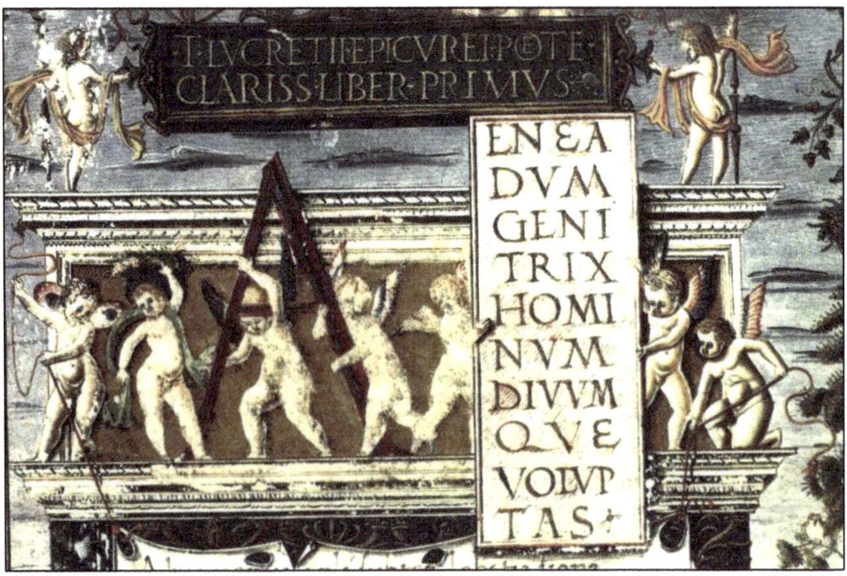

Source: Biblioteca Apostolica Vaticana, Vat. lat. 1569, fol. 1r; Photo: Public Domain

Nevertheless, this was probably due to the circumstances surrounding the passing-down of Epicurean teachings that were extremely discriminated against in Christian times. A few epistles and some fragments are all that survived. However, Epicurus advocated for a clear and plain writing style for philosophy, even if his extant texts do not always match this position. Similar to the Stoic philosophers, he defended the natural meaning of language and a philosophy for every man based on common sense, as we can see from his *Epistle to Herodotus* (*Ep. ad Hdt.* 75-76). His teachings gained great prominence thanks to one of his later Roman successors, Lucretius, who wrote *De Rerum Natura* (*DRN*), an extremely popular and influential didactic poem about the atomistic theory, which included plenty of visual metaphors and mythological examples. After its 'rediscovery' in the 15th century, this didactic poem again would gain some degree of popularity among rulers, patrons, and humanists of the Renaissance (see *Fig. 1*).[6] Despite the impact of the formal, refined, and sophisticated poetry from the Alexandrian School, and in contrast to his contemporaries the Neoterics, Lucretius put the elegance of words in the service of educational goals. We can infer that his poem was conceived for an educated audience that misunderstood Epicurus' hedonism. The dedication to the reluctant Epicurean student Gaius Memnius at the beginning of the text and further exhortations to him in other parts of the poem make this quite clear.[7]

In the proem to the fourth book (*DRN* 4.11-22), Lucretius presents his text as "sweet honey" (cf. *DRN* 2.398-407, 3.176-207) made up of smooth atoms. His didactic poetry thus helps audiences swallow the Epicurean medicine (cf. *DRN* 1.931-950, 4.8-25, cf. Gruber 2008). The metaphor of the right philosophy as medicine is similar to comparisons made by Cynics and Stoics, as we can read in the *Tabula Cebetis* (1st-2nd Century AD, cf. *Tabula Cebetis* 3, Musonius Rufus, 1.1.9). As late as the 12th century, the Byzantine polymath John Tzetzes evokes the topos of "speeches sweet as honey" (cf. Λογούς εκίνει μελιχρούς in *Chiliades* 8 *Hist.* 253, v. 1017 referring to Cleopatra's eloquence) and reminds us that a "sect" of Epicurean philosophers said that "honey was the tenth part of ambrosia" (Το μέλι, μέρος δέκατον ἐκάλουν ἀμβροσίας in *Chiliades* 8 *Hist.* 251, v. 984). Horace, "a pig from Epicurus' sty" (Epicuri de grege porcum, *Epist.*, 1.4.10), compared himself to a bee: "I, like Matine bee, / in act and guise, / that culls its sweets through toilsome hours, / am roaming Tibur's banks along, and

6 About the rediscovery of *De Rerum Natura* and its possible impact on the arts and the sciences of the Renaissance see Greenblatt 2011. Cf. also Sehlmeyer's article in this book, p. 336 n. 89.

7 Cf. Lucr. *DRN* 1.26, 42, 411, 1052; 2.143, 182; 5.8, 93, 164, 867, 1282.

fashioning with puny powers /a laboured song."[8] In fact, Lucretius and the convinced Epicurean Horace compare themselves to diligent bees.[9] Lucretius claims that he would, like the bees in flowery meadows, collect the raw material for his honey from the papyrus scrolls, i.e., from the books, and from Epicurus' "golden words" (*aurea dicta*). Horace, in turn, uses the bee metaphor to differentiate himself from old-fashioned, lofty epic poets who rely on quantity and showmanship. Hardworking and humble like the bee of the dawn (the *apis matinae*), he diligently gathers only the best of various sources to compose small but elaborate songs. In the tradition of Hellenistic poetry, Horace makes the πόνος, i.e., his research 'efforts,' to create something elegant and pleasant. Similarly, Lucretius does not want to write great quantities of verse (*non multis versibus*) either, but rather in lovely and elegant verse (*potius suavidicis*) (*DRN* 4, 180-182).

If the didactic poem is considered honey, then the ancient 'science communicator' can be said to be as industrious as a bee, collecting the best thoughts, experiments, and stories of 'science' in a sweet and salutary poetic composition.

There are also popular poems on scientific topics that followed an original prose text. Aratus's famous astronomical poem *Phainomena*, which was translated into Latin and commented on many times,[10] is actually based on an earlier, lost prose work by the astronomer and mathematician Eudoxus of Cnidus (cf. Taub 2004: 47; 2017: 29). Aratus' didactic poem *Diosemeia* (on weather and astronomical phenomena) was also very popular in the Greek and Roman world. Hellenistic poets, Latin poets, rhetoricians, politicians, and even the author of *Acts of the Apostles* were familiar with Aratus' poems, as we can infer from the record of Paul's speech in Athens in 51 AD: "As some of your own poets have said, 'We are his offspring too,'" quoting Aratus' hymn to Zeus at the beginning of *Phaenomena*.[11] The presence of such a broad audience—even if it was an educated one—is a clear argument in favor of the idea that the popularization of

---

8 Ego apis Matinae / more modoque / grata carpentis thyma per laborem / plurimum circa nemus uvidique Tiburis ripas operosa parvos / carmina fingo (Latin text in Shorey/Laing 1919; translation by Conington 1882).

9 Horace, *Carmina* 4.2.27-32; for a similar comparison in Lucretius cf. *DRN*. 3.9-13, cf. also Waszink 1974: 20-23; Morgan 1998: 262-270; Markovic 2008: 18-19.

10 *Phaenomena*, translated and commented on, e.g., by Cicero, Ovid, Germanicus, and Avienus, the last as late as the 4th century AD.

11 Cf. Acts 17.28: ἐν αὐτῷ γὰρ ζῶμεν καὶ κινούμεθα καὶ ἐσμεν, ὡς καί τινες τῶν καθ' ὑμᾶς ποιητῶν εἰρήκασιν· τοῦ γὰρ καὶ γένος ἐσμέν; cf. Arat. *Phaenomena* 5; see also Faber 1993.

knowledge was a common practice in antiquity, and was primarily supported by didactic poetry.

The tradition of didactic writings on physics, philosophy, and manifold applied techniques in poems was indeed prolific in the late Roman Republic and the Roman Empire, and lasted through the end of antiquity and beyond.[12] Greek and Latin didactic poetry embraced a large variety of topics, including cosmology (e.g., Hesiod's *Theogony*), natural philosophy (Parmenides, Xenophanes, Empedocles[13]), astronomy (Aratus' *Phaenomena*, Manilius' *Astronomica*, Avienus' *Aratea*), agriculture (e.g., Hesiod's *Works and Days*, Virgil's *Georgics*), fishing (e.g., poems on halieutics), ethnographic and landscape descriptions (e.g., Ausonius' *Mosella*), cosmetics (e.g., Ovid's *Medicamina*), relationship tips for men and women (e.g., Ovid's *Ars amatoria*), and many other things.

We also find interesting uses of didactic poetry in the Middle Ages. In mid-12th-century Constantinople, the polymath and polygraph John Tzetzes wrote a miscellaneous didactic work in political verse (i.e., decapentasyllabic verse) known as *Chiliades* (actually *Book of Histories*), where he preserves precious information on a great variety of literary, historical, philosophical, and technical topics. Tzetzes also wrote the didactic poem *Allegories of the Iliad*, books I-XV, for the purpose of introducing Empress Irene of the Comnenian Court (born Bertha von Sulzbach) to the foundational text of Greek literature. However, as Cesaretti has stated (1991: 188), the subsequent books, i.e., XVI-XXIV, which are dedicated to Constantine Cotertzes, are different from the books (I-XV) that were dedicated to the Empress. The main differences are that books XVI-XXIV are longer and introduce the allegorical explanation of προσωποποιία, i.e., "personification," which Tzetzes had already used in his *Exegesis Iliadis*, but not in the first fifteen books of his *Allegoriae Iliadis*. Cesaretti also argues that the personification was probably "troppo difficile per l'augusta," i.e., "too complex" for Empress Irene (Cesaretti 1991: 190).

Even at the first universities, the didactic poem was considered a great medium for popularizing and memorizing scholarly knowledge. In the Schola Medica Salernitana, for example, the French physician and poet Gilles de Corbeil composed two brief poems on uroscopy and pulsology, *De Urinis* and *De Pulsibus*, as mne-

---

12 For an analysis of didactic poetry in the Middle Ages, cf. Haye 1997.

13 Farrington (1939: 184-185, 214 n.9) proposed that these three pre-Socratic philosophers probably chose to write in verse in order to reach a wider audience. Despite recent criticism of this position (Penwill 1995: n.10), the fact remains that Empedocles dedicated one of his poems, *Purifications*, to the people of the city of Acragas, thereby explicitly addressing the general public (see Fr. 112, Diogenes 8.62).

monic and pedagogic aids for his students. Thanks to the rhetorical quality, the popularity, and the repeatability of didactic poetry, the ideas and knowledge passed down by this kind of literature have been preserved to the present day.

After the Middle Ages, poetry and science stopped 'playing' together at this level, even though there were certain cyclical returns to the ancient literary and cultural techniques in Western Europe. During the Midlands Enlightenment, i.e., in the late-1700s, some scientists wrote book-length scientific treatises in verse. For example, the astronomer and philosopher Roger Joseph Boscovich (1711-1787) wrote a poem based on his theory of solar and lunar eclipses in Latin verse, titled *De Solis ac Lunae defectibus* (1760, translated into French in 1779). Boscovich had his academic colleagues in mind as the target audience. A few decades later, the British physician and natural philosopher Erasmus Darwin (1731-1802) wrote the poem *The Botanic Garden* (1791), in which he illustrated and updated the classifications of the father of modern taxonomy, the Swedish physician Carl Linnaeus. In contrast to Boscovich, Erasmus Darwin wrote in the vernacular language with the intention of popularizing botanic principles among the interested public.

In both cases, the primary goal of this poetry was not to go back to the *archēgétai* (ἀρχηγέται, i.e., the inventors of the genres, even if a "Botanic Muse" guides the reader in Darwin's poem), but to render homage to scientific achievements popularizing science within and beyond academic circles as a matter of reputation. We can, therefore, venture that modern didactic poetry is a matter of intellectual needs, societal status, and the adaptation of traditions, rather than a revival of ancient forms of knowledge popularization. Erasmus Darwin popularized science beyond poetic skills—his impressive experiments on galvanism with small animals such as eels and vorticella were some of the inspirations behind Mary Shelley's *Frankenstein*.

At the end of the 18th century, it was still possible to gain some publicity with a didactic poem, as the French poet Jacques Delille did with his *Jardins, ou l'art d'embellir les paysages* (*Gardens, or the Art of Embellishing Landscaptes*, 1782). The same author published another important didactic poem at the beginning of the 19th century: *Les Trois Règnes de la Nature* (*The Three Realms of Nature*, 1808; cf. Phillips 2010). However, by 1900 this kind of poetry had almost disappeared. This was probably due to several factors. Among those, we should mention the upswing of positivism, the specialization of scientific disciplines—which made it almost impossible to be a physicist and a poet or philosopher at the same time—, the speed of scientific discoveries, the growing independence of the arts and literature, a certain literary self-awareness, and a general refusal to write on behalf of the natural sciences (cf. Phillips 2010).

The transition from verse to prose in popularizations of science was already underway in late antiquity (e.g., with the *poikilographía*) and especially in the Early Middle Ages with the emergence of universal chronicles that were influenced by the Bible as well as by Greek and Roman historians. Today, the vast majority of popular science books are published in prose. Our times even have a notorious, modern precedent: one of Erasmus Darwin's grandchildren wrote the most famous popular science book of the modern era, a book that was written, like most modern bestsellers, in prose: *The Origin of the Species*, by Charles Darwin.

Even if science and poetry no longer work together for instructional purposes for broad audiences as in ancient times, there has recently been a renaissance of this symbiotic combination of scientific knowledge and poetic expertise for the popularization of scientific knowledge among children and young people (cf. Phillips 2010; Crawford 2006). Nevertheless, concerning the cultural functions and the performativity of didactic poetry, we can find similarities in today's popular media, such as scientific documentaries (especially since Carl Sagan's and Ann Druyan's series *Cosmos* came out in 1980), popular science web videos (cf. Muñoz Morcillo et al. 2016), live science presentations (e.g., science slams, cf. Muñoz Morcillo 2020b), and even online music videos on scientific topics (Allgaier 2012). Taking into account that ancient poetry was mostly accompanied by music and frequently also by performances (i.e., dance), we can see that the 'new' science music video format has a lot in common with ancient didactic poems. Indeed, both have 'lyrics' and a didactical purpose. Of course, there are historical contingencies and contextual factors that make these practices quite different. It is not even known if today's YouTubers are aware of such ancient traditions. However, Horace's doctrine of "benefit and amusement" (*Ars poetica* 333) as the main goals of poetry still has validity in this context: *aut prodesse volunt aut delectare* YouTubers.

Dialogue and Drama: Between Dialectic Knowledge and Fictional Stereotypes

In addition to poetry, other genres were used in antiquity for communicating natural science, cognitive theories, technology, politics, ethics, and other subjects. Plato and Xenophon wrote dialogues for the dialectical discussion and presentation of complex moral and philosophical questions. The dialogue was an ideal format for reproducing controversial questions, critical opinions, and different ideologies. On the basis of the societal scandal triggered by Socrates' death, we can infer that intellectual life in antiquity did not take place in an ivory tower. Indeed, philosophers often taught in public and open spaces such as the *stoá poikílē* (στοά ποικίλη, the painted porch) or Epicurus' *kḗpos* (κῆπος, garden), where even women and slaves were welcomed (cf. Gordon 2012: 72-108). Free citizens could participate in the philosophical and political life of the city for centuries, but women and slaves often could not take part in public affairs. The first Sophists were rhetorical experts and popular public speakers, but they were more interested in the human essence and the power of words than in understanding natural phenomena, in contrast to their pre-Socratic predecessors. Plato criticized them by asserting that their rhetorical skills, i.e., *téchnē rhētorikḗ* (τέχνη ῥητορική), were as deceptive as the *téchnē graphikḗ* (τέχνη γραφική), the skills of painters and sculptors (*Plt.* 598d, 602d, *Sph.* 235a-b), thereby implying that painters and sculptors took part in the formation of public opinion through their visual works. There is a kind of Platonic mimesis at play here that is based on the true knowledge of nature: the *mímesis historikḗ* (μίμησις ἱστορική), i.e., imitation according to scientific knowledge, in contrast to *mímesis doxomimētikḗ* (μίμησις δοξομιμητική), imitation according to opinions (or opinion-imitation, cf. *Sph.* 267e). The latter could be considered an ancient forerunner of modern 'citizen science' (cf. Irwin 1995; Bonney et al. 2009: 7), insofar as knowledge is sometimes based on laypersons' opinions or testimonies. Indeed, many zoological descriptions provided by Aristotle are based on experiences by laypeople; for example, he took notes on cetaceans reported by fishermen and recorded their testimonies as valuable knowledge in his *Historia Animalium*.[14]

As for the Sophists, the comedian Aristophanes parodied their bizarre rhetorical and speculative art in one of his most remarkable and popular works: *The Clouds* (cf. especially vv. 156-164). Fiction, including fictional works of drama, often portrays the prevailing stereotypes of the time. In this case, the 'intellectual' of the second half of the 5th century BC is depicted in *The Clouds* (a comedy, and therefore also a kind of 'dialogue') as a bold and greedy cheater who is a master of rhetorical skills and does not hesitate to use them for his benefit. Intellectuals were caricatured not only in literature. Painted ceramics also served to amuse the public by portraying stereotypical features. In a *kýlix* (a wine-drinking cup) from the Vatican Museums (*Fig.* 2), we see a depiction of Aesop, the Greek fabulist, as a weird thinker: With a thin body, an oversized head, a forehead furrowed by lines,[15] and his mouth open, he listens to the teachings of a fox sitting

14 Cf. HA 6.12 on Dolphins, Whales, and Seals.

15 According to Giorgio Vasari, Michelangelo's forehead was "furrowed by seven streight lines" (Vasari's *Lives*, translation by Hinds 1980: vol. 4, 178-179). This seven lines are also present in da Volterra's broze bust of Michelangelo. Britton (2009: 17) identifies this detail with ps-Aristotle's mention of the "cloudy brow", which signifies

in front of him. Like the figure of Socrates, from Aristophanes's comedy, he is unattractive, looking like a Silenus (i.e., a companion of Dionysus), and he obviously neglects his appearance. Indeed, he has uncombed long hair, an incipient bald spot, and an unkempt beard. This picture of ancient eloquent and persuasive intellectuals is not very different from the modern, popular iconography of the mad scientist that we all know from Hollywood films (cf. Weingart et al. 2003; Kirby 2008a, 2008b). These depictions are influenced by many factors related to the corresponding societal and political context, but one thing seems not to have changed much: the most effective way to disseminate scientific stereotypes has always been based on fiction, satire, or comedy. Therefore, the cultural practice of describing prototypical intellectuals can be considered a manifestation of a homeostatic, self-regulated system for disseminating scientific knowledge.

Philosophical and theatrical dialogues became important instruments for passing down knowledge, debates, assumptions, and stereotypes that sometimes lasted until modern times. The Romans adopted the Greek tradition of the philosophical dialogue and exported it to the rest of the empire. Cicero's *Tusculana Disputationes* or Tacitus' *Dialogus de oratoribus* are great examples of the late Roman Republic and the Roman Empire (1st century), even if these works were not oriented to a broad audience in modern terms.

In late antiquity, we again find that the dialogue served as a popular medium for passing down cultural, technological, and antiquarian knowledge for an educated public. An interesting example is the *Deipnosophistae* (*The Banquet of the Learned*) by the Greek rhetorician and grammarian Athenaeus of Naucratis (whose floruit came around the end of the 2nd and the beginning of the 3rd century AD). Among many literary, historical, and antiquarian references contained in this lengthy dialogue, Athenaeus narrates the construction of the largest ship of antiquity, the Syracusia, by Archimedes, who was the 'scientific advisor' of Hieron II, the tyrant of Syracuse (*Deipnosophistae* 5.40). Following the description of a Greek physician named Moschion, Athenaeus gives some specifications: It weighs up to 1800 tons, can carry 200 soldiers and 1942 passengers, and has enough space for a catapult. As the 'Titanic' of that era, it was meant to impress the public even beyond its home in Magna Graecia. In addition to the application of a principle of buoyancy that was probably based on that of Archimedes (*On Floating Bodies*, Περὶ τῶν ἐπιπλεόντων σωμάτων) for making the ship floatable and the invention

"self-will as in the lion or the bull" (cf. ps-Aristotles, *Physiognomica* 1244-1247). There is some continuity in the representation of intellectual stereotypes.

of something similar to a windlass[16] for drawing the enormous vessel down into the sea, there is an extremely interesting explanation of how to protect the keel from biofouling by "applying to the wood leaden tiles and putting below pieces of cloth impregnated with pitch."[17] The Syracusia episode and some of Archimedes' military inventions in the service of the same king, such as the heat ray machine, i.e., a sort of six-angled mirror or parabolic reflector for burning enemy ships with bundled rays of sunlight (cf. Anthemius, Περί παραδόξων μηχανημάτων 153; Tz., *H.* 2.119, Ἑξάγωνόν τι κάτοπτρον), are probably the first documented cases of both scientific advising and a trust in scientific knowledge for clear political and economic purposes. Indeed, Heron II of Syracuse sent the biggest ship ever made in antiquity[18] to Alexandria as a gift for the Ptolemaic Kingdom, but also as a strategy for obtaining technical and political credibility beyond his own borders through the mediatization of a singular achievement—in this case, the construction of the biggest ship ever known at that time. We can consider this episode one of the first cases of 'science communication' on behalf of political propaganda.

The encyclopedic length of the dialogue *Deipnosophistae* is a sign of a new development in the production and passing-down of knowledge, namely, a move away from evolving dialectics and toward the conservational function of literature. Athenaeus, like most scholars of late antiquity, was concerned with the idea that the golden age of Greek culture and philosophy had come to an end. The dialogue as a communication format was no longer a dedicated form for producing knowledge in a Platonic, dialectical sense, but rather a literary genre that had become suitable for conserving miscellaneous knowledge, sometimes even critical opinions on officially forbidden topics. Even the oldest printed encyclopedia—'encyclopedia' itself being a Greek compound word meaning 'general education'—, the *Margarita Philosophica*, mostly compiled by the German Carthusian humanist writer Gregor Reisch, was written as a dialogue between student and teacher.

The classical dialogue's development into encyclopedic writing is a compelling case of authorial intentionalism. One of these dialogues became a notable 'bestseller' in late antiquity and the Middle Ages: Boethius' *Consolatio Philosophiae*, which the Roman scholar wrote while he was in prison around 524 AD. This dialogue was intensively copied before the invention of the printing press

16 The Greek word is *helíkē* (ἑλίκη), which is probably from *hélix* (ἕλιξ), meaning twisted or curved, or referring to anything that assumes a spiral shape, according to the Liddell-Scott-Jones Greek-English Lexicon.

17 *Deipnosophistae* 5.40: μολυβδίναις δὲ κεραμίσιν ἐπεστεγνοῦντο πρὸς τὸ ξύλον, ὑποτιθεμένων ὀθονίων μετὰ πίττης.

18 For more information on ancient ships and seamanship, see Casson (1971).

and translated into many European languages (Bastert 2013). Even if we are no longer familiar with the philosophical dialogue as a format for science dissemination, there is a similar modern format that exhibits a large degree of continuity with the past, although it is largely lacking in depth and dialectical momentum: the interview. However, probably the most original use of the dialogue format for the sake of science popularization in modern times is Jane Marcet's popular 'conversations' on chemistry, botany, religion, and economics at the beginning of the 19th century (Bahar 2001). Since then, a dialogue with the public has become a common programmatic element of modern science communication, at least in theory.[19] The dialogue in general seems to have some universal value for communicating (and producing) knowledge.

Epistles as Exoteric, Popularizing Texts

So-called exoteric texts (from *exōterikós* [ἐξωτερικός], 'outwards'), which probably addressed a target group similar to what we would today call the 'interested public,' testify to the popularity of ancient schools of philosophy. These texts consisted of clearly formulated letters that were presented in the course of introductory public 'courses' (e.g., as a forerunner of later propaedeutics at the first universities and Studia Generalia of the Late Middle Ages)—a method that is very similar to the later dissemination of the Gospels by the Apostles. Without the popularity of knowledge through epistles and speeches, Christianity wouldn't have been able to put its stamp on Western civilization. Many of the letters handed down by Diogenes Laertius can be regarded as examples of ancient philosophy popularization. In contrast, the more difficult, esoteric texts (in the ancient sense, from *esōterikós* [ἐσωτερικός], 'inwards') were usually much longer and more complex, and were therefore not suitable for laypeople.

In antiquity, epistolography would also become an important medium and a common format for spreading knowledge in certain circles. The letters of Seneca are one good example of this. Similar to Lucretius in his doctrinal poem, in his letters Seneca advises his friend Lucilius on how to become a good Stoic.

The epistle as a popularizing medium reached a notable climax with the writings of the Apostles. Their letters are outstanding and successful examples of the aggressive popularization of a theological conception of the world.

Furthermore, in combination with knowledge dissemination, the 12th-century polymath and Byzantine scholar John Tzetzes produced a large number of letters

19 Cf. Roberson von Trotha 2012; cf. European initiatives such as 'Science with and for Society,' or 'Responsible Research and Innovation.'

and explanatory poems in which he displayed his self-awareness as a great communicator and a man of knowledge (Pizzone 2017). Epistolography became so prominent that the Latin schools of the Middle Ages introduced the so-called *Ars dictaminis* (the art of prose composition, mostly for writing letters), whose aim was to enable all educated people to express themselves in the epistolary format (Black 1987; Grendler 1989).

Today, letters for a broad audience are still important carriers of popular science knowledge. A comparable modern example with a substantially broader range than ancient epistles would be the blog post, but we are also familiar with 'open letters' and similar formats within newspapers.

Protreptics on Behalf of Popular Science

For educated people in some ancient Greek πόλεις (*póleis*), it was not particularly difficult to gain access to knowledge. Plato's Academy, Aristotle's Lyceum, and Epicurus' Garden were relatively open places of teaching and learning. Sophists and philosophers even developed special communication formats to attract students. An interesting example is that of protreptic speeches—i.e., exhortations—which served to attract new students to schools of rhetoric or philosophy.[20] Famous ancient protreptic speeches are the "advertising speeches" written by Aristotle and Iamblichus, which exhort people to study philosophy. Before them, the first Sophists were already publicly advertising their rhetoric courses in similar speeches, as we know, e.g., from Plato's *Euthydemus* (*Euthd.* 278c). We can distinguish between sophistic and philosophical protreptics, i.e., between speeches for the commodification of rhetorical skills and speeches for the promotion of a particular philosophical school, i.e., between persuasion and 'real' knowledge. The philosophical protreptic has a large and varied tradition in antiquity. We have protreptic texts from Hellenistic times to late antiquity and beyond. A few prominent examples are Aristotle's *Protreptikós* on philosophy, Galen's *Protreptikós* on medicine as the best of the arts, Iamblichus' fragmentary *Protrepticus*, the *Tabula Cebetis* on moral purification, and Boethius' extremely popular *Consolatio philosophiae*—the last two texts were written in dialogue form.

Protreptic speech was very similar in its function to the later public disputations of the Middle Ages, to homilies, and to epideictic speeches in general, three genres that were intended, respectively, to attract new students, to disseminate theological ideas among the population, and to impress upon people's imaginations.

---

20 Cf. Liddell-Scott-Jones' translation of προτρεπτικός as "hortatory."

The Church Fathers also adopted classical rhetorics as a strategy for the communication of patristic knowledge, and later for the communication of scholastic knowledge as well. In late antiquity, the Christianization of the Roman Empire dissolved the syncretic pantheon of the Romans and took over the pantheon's ideological and social functions. The old stories about the origins of the world had to be adapted. The Church Fathers began to "contribute" to societal life and knowledge. There was an urgent need for knowledge transfer to support the new world order in all cultural and scientific spheres. The doctrines of Christianity were linked to Neoplatonism and Ptolemy's geocentric model. Augustine (354-430 AD) wrote the first didactic sermons in *De doctrina Christiana*. He continued the protreptic rhetorical tradition in the service of the Church. In his exegetical homilies, Basil the Great also described the new image of nature, which was connected with antiquity. Homilies and sermons became the main communication media of the early Church. In this way, the Archbishop of Constantinople, Dio Chrysostom (i.e., "the golden-mouthed," ca. 349-407 AD), gained great fame as an inspiring speaker, even long after his death.

In the Middle Ages, some literary forms of philosophy had a highly protreptic style, e.g., the *Summa Contra Gentiles* by Thomas Aquinas (Jordan 1986), and public disputations in general. The latter were popular events that replaced regular classes at the universities of the 13th and 14th centuries, and can be understood as formalizations of philosophical dialogue (Novikoff 2012) for an interdisciplinary academic audience. For a broader public, including foreign students and masters as well as religious and civil authorities, there was the *quodlibetal* disputation, from the Latin word *quodlibet* ('whatever'). In this type of disputation, any member of the audience could propose a question or a set of questions. As Novikoff (2012) points out, the popularity of the *quodlibetal* disputation shows the massive cultural influence that the scholastic tradition had on the society of that time.

Since protreptic speeches were originally performative discourses, it is not surprising that visual argumentation eventually became a crucial factor in them too. Later protreptic methods would build on a combination of authorial qualities and persuasive images linked to experiments' visual evidence, especially during the scientific revolution of the Renaissance, which inaugurated a new and flourishing visual culture.[21] Scientists and artists worked together, staging the pictures of science as a new popular medium, especially in botany, mineralogy, anatomy, astronomy, and zoology.

21 Cf., e.g., Vesalius' famous drawing of an anatomically prepared arm in the title page of *De humani corporis fabrica* (1543).

During the Enlightenment, natural scientists and demonstrators supported the legitimation of their research by adopting the idea of the *theatrum anatomicum* of the Renaissance to perform striking scientific experiments (Hochadel 2003: 27; cf. Klestinec 2011). In the *theatrum anatomicum*, new students and the interested public could attend an autopsy or be informed about scientific discoveries. The intention behind this kind of demonstration and other public experiments, such as those on magnetism and electricity, in salons, coffee houses, or fairground booths was to consolidate and increase the prestige and authority of the lecturer through a kind of 'show of force' (Jütte 1998: 244; Hochadel 2003: 21-25). Providing a demonstration of one's experiments was the new audiovisual and protreptic way of attracting new students by communicating a simple message: scientists are the new masters of natural forces. Hands-on experiments, illusions, automata, and intriguing games became an integral part of science popularization in the 18th century (Stafford 1996).

Protreptic speech and visual demonstrations live on in some communication activities that aim to stimulate the growth of technology- and information-based economies, e.g., STEM (Science, Technology, Engineering, and Math) activities for secondary schools and specific gender- or family-related events such as Girls Days or Open-Door Days at modern universities. Even YouTubers practicing science communication, such as Derek Muller (Veritasium, Element of Truth), have developed sophistic and philosophical protreptic strategies for maintaining and growing their community of followers. Some of them use their channels as showrooms to display filmmaking skills that interested institutions and individuals can buy (e.g., Kurzgesagt), while other YouTubers develop a narrative of science as the ultimate truth (Veritasium), or promote knowledge and curiosity as edifying attitudes (Brainscoop)—approaches that bring to mind Lucretius' metaphor of philosophy as a safe harbor (*DRN* 5.7-12).[22] In contrast to older times, no one today would expect an institutional protreptic speech in favor of philosophy or the humanities in general—the most one might expect is a speech in defense of them.[23] However, not so long ago, the disciplines that were being put under pressure by society and that were being forced to legitimate their societal benefits were precisely those that modern societies now wholeheartedly support (Hochadel 2003: 15-28). In summary, the protreptic tradition, with its focus on the promotion of specific disciplines or certain attitudes toward life, has contin-

22 On protreptics in YouTube science videos, especially regarding German YouTuber Mai Thi Nguyen-Kim, cf. Muñoz Morcillo 2020b: 215-219.

23 An interesting exception are protreptic science slam speeches in favor of the humanities (Muñoz Morcillo 2020b).

ued with a similar structure all the way to the present day, albeit with changes to its subject matter and motivations.

RHETORICAL STRATEGIES FOR POPULARIZING KNOWLEDGE

The short Hippocratic treatise *On the Right Way to Behave* (Περὶ εὐσχημοσύνης, 8-12) contains interesting advice on the way a physician should appear and how a physician should behave when visiting a patient or discussing matters with other colleagues. The primary purpose of this treatise is to give physicians some orientation in terms of pedagogical and rhetorical measures, thereby helping them gain prestige and confidence. Furthermore, a physician's reputation is the topic of several epigrams and epitaphs from the period. In some satiric epigrams in particular, it seems that there was a popular perception of Hippocratic doctors as being obsessed with the benefits of the administration of contraries, either by dietary prescriptions (*Aten* 499 d) or by surgical intervention (e.g., the coction or draining of a humor). In the most hilarious example, the physician's mere presence could kill the patient (*AP* XI 123). As for the epitaphs and reliefs on sarcophagi, the ἰατρός (*iatrós*, physician) does indeed seem to be a reputable person (GV2040) who saved many lives (GV244), although sometimes with the comment that he was unable to save himself (GV1934).

In some funerary reliefs we also find the elegant and reputable figure of the doctor as described in *De habitu decenti* (Περὶ εὐσχημοσύνης, *On the Right Way to Behave*), sometimes in the company of Asclepius' daughter Hygeia, treating or healing children (*Fig.* 3): flowing drapery or elegant sitting postures are the norm here. However, none of the physicians depicted in the *Anthologia Palatina* or in the epitaphs seem to explicitly mention the controversial theory of similars (akin to today's homeopathy), probably because Hippocratic physicians used to combine 'homeopathic' and 'allopathic' treatments depending on the concrete situation and the needs of the patient. However, in the *Hippocratic Corpus* we find evidence of some controversy between the two ancient medical theories, especially in *De locis in homine* (Περὶ τόπων τῶν κατὰ ἄνθρωπον, *On Places in Man* 42). The justification of the treatment with similars when other methods don't make sense resembles the structure of the *progymnásmata anaskeuḗ* (ἀνασκευή, confirmation) and *kataskeuḗ* (κατασκευή, refutation). This supports the notion that ancient physicians often received elaborate rhetorical training, which they viewed as being necessary for their public activity. Therefore, the visual and rhetorical strategies of Hippocratic physicians concern their authoritative appearance, the short description (ἔκφρασις) of treatments for laypeople, and their rhetorical training in the 'counterargument' (*antílexis*, ἀντίλεξις), especially regarding explanatory statements about the theory of similars.

Figures 2 and 3: Aesop, with a Fox, from the Central Medallion of a Kýlix, ca. 470 BC; Physician Amphiaraos Treating the Shoulder of a Young Man, from a Votive Relief, Offering of Archinos to Amphiaraos (detail), ca. 400-380 BC

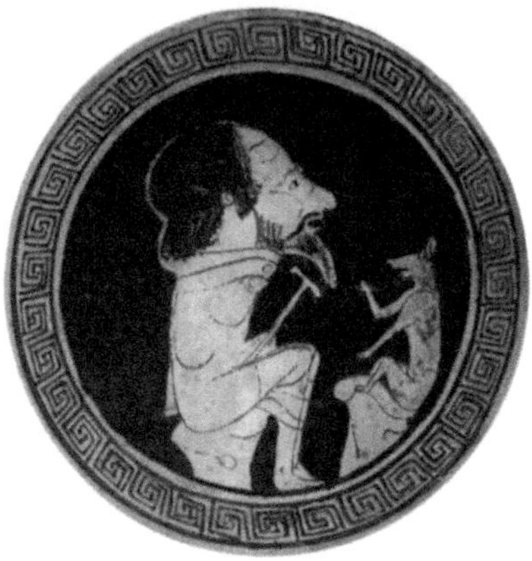

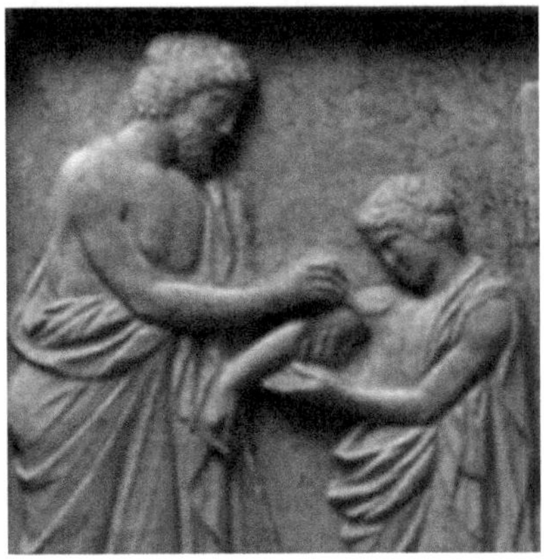

Source: Gregorian Etruscan Museum, Vatican City; Photo: Free Art Licence; Athens, National Archaeological Museum; Photo: Wellcome Collection. Attribution 4.0 International (CC BY 4.0)

The famous phrase "Life is short, art is long" (Ὁ βίος βραχύς, / ἡ δὲ τέχνη μακρή, *Aph.* 1.1) also belongs to the Hippocratic aphorisms. Here, 'art' is used in the sense of *téchnē* (τέχνη), i.e., craftsmanship or technique in general (Jones 1931: 97-98). This aphorism does not end here, however, but goes on to say that "the opportunity is fleeting, experimentation is dangerous, and judgment is difficult" (ὁ δὲ καιρὸς ὀξύς, / ἡ δὲ πεῖρα σφαλερή, / ἡ δὲ κρίσις χαλεπή), and then adds the following: "The physician must not only be prepared to do what is right himself, but also to make the patient, the attendants, and externals cooperate."[24] What is meant, therefore, is not that life is too short for the acquisition of both theoretical and practical knowledge. Rather, Hippocrates considers teaching and explaining medical

24 δεῖ δὲ οὐ μόνον ἑωυτὸν παρέχειν τὰ δέοντα ποιεῦντα, ἀλλὰ καὶ τὸν νοσέοντα, καὶ τοὺς παρεόντας, καὶ τὰ ἔξωθεν. Translation by Charles Darwin Adams (1868).

knowledge and facts to laymen and students just as much an 'art' as medicine itself.[25]

The stylistic comparison of popular and professional texts reveals the most important difference between them: popular texts are much more rhetorically polished. Rhetorical skills make the good communicator. This is no different today. Some authors even consider modern science a kind of skilled rhetorical autobiography of science, or an obsequious form of journalism (cf. Baudouin 1993; Gross 1994, 2003; Condit et al. 2012; Kohring 2004, 2012). Rhetoric was so important and successful that even the theoretical formulation of the preliminary rhetorical exercises, the *progymnásmata*, i.e., the rhetoric schoolbooks of the Roman Empire, were passed down through the centuries without significant changes.[26] The most interesting preliminary exercise with regard to ancient popular science was probably the *ékphrasis* (ἔκφρασις),[27] the art of making vivid and clear descriptions of things, persons, animals, seasons, and situations, among other things.[28] Many of the most memorable and enjoyable textual passages—from didactic texts, epic poetry, drama, historiography, or novels—belong to the ecphrastic literary tradition, which probably began with the famous description of the Shield of Achilles in *The Iliad* (18.478-608). The *ékphrasis* is probably the most complex and influential exercise of the *progymnásmata*. A good *ékphrasis* should be able to put the described object right before the eyes of the listener (Theon, 118.7-120.11 Spengel; Patillon-Bolognesi 1997, 66-69), similar to the

25 Communication between experts and laypeoplersons was already an issue in antiquity. The awareness about the complexity of the art (τέχνη) of public communication was testified to by the father of medicine Hippocrates when he attested "life is short, but art is long" (Hippocrates, Aphorismi, 1, 1). Seneca translated the aphorism of Hippocrates—Ὁ βίος βραχύς, Ἡ δὲ τέχνη μακρή—with the Latin sentence "vitam brevem esse, longam artem," formulated in indirect speech as part of a larger phrase. In Western literature, the Latinized aphorism is more familiar in reverse order: "ars longa, vita brevis." Hippocrates was referring to the limited time available to the physician to master both the medical art itself and the art of communication with laypeople.

26 These exercises were known in Western Europe thanks to Priscian' translation of ps-Hermogenes' manual at the beginning of the 6th century CE. During the Renaissance, the complete *progymnásmata* tradition reached Western Europe thanks, above all, to the translation of Aphthonius' manual (cf. Kraus 2008).

27 For an introduction to the *ékphrasis* as a rhetorical device, cf. Webb 1999, 2009. For an exposition of the ecphrastic canons from antiquity to Byzantium, cf. Muñoz Morcillo 2019, 2020c.

28 "A speech which leads one around (περιηγηματικός) bringing the subject matter vividly (ἐναργῶς) before the eyes." (Theon's *Progymnasmata*, here after Webb 2009: 11)

Latin expression "subiecto sub oculos." (Quintilian, *Institutio Oratoria* 9.2,40-42; see also Cicero, *De Oratore* 3.53; 202; *Orator* 139) According to Fernández Delgado and Pordomingo Pardo, the educational stage for teaching the *progymnásmata* could be between the end of the grammarian class (comparable to primary school today) and the beginning of the rhetoric school. This level seems to have been attained by most of the citizens with basic school education (cf. Fernández Delgado 2007: 273-306; Fernández Delgado/Pordomingo Pardo 2016). In explaining *ékphrasis*, the rhetoricians of the Roman Empire surprisingly quoted prose writers—above all the historian Thucydides—more often than poets when citing best-practice examples. Thucydides' texts were suitable for the vivid representation of crucial moments of history. Thucydides himself said that he wanted to build a "monument for all time,"[29] a monument that would also appeal to future listeners and readers, with his *History of the Peloponnesian War*. However, not every citizen had access to the pleasure of private listening. The 'audiobook' of classical antiquity mostly consisted of a slave reading a papyrus roll for his patron (cf., e.g., Petronius' *Satyricon* 74, 4). Accordingly, only wealthy citizens could enjoy this sumptuous pleasure. The common people depended on public events. In any case, *ékphrasis* was a very useful technique for spreading knowledge about new discoveries and achievements, as Courtney Ann Roby (2016: 14) explains:

> "The power of technical ecphrasis to delineate and propagate technological knowledge does not confine authors to their native technological cultures; indeed, a vital role in self-reflective scientific development is played by texts that celebrate (explicitly or implicitly) the technological prowess of others."

The author even makes an interesting comparison with modern "journalistic coverage of developments like Sputnik that helped fuel the space race between the US and USSR" (Roby 2016: 14 n. 51).

29 For the meaning of this sentence, see Momigliano (1978). Thucydides may be trying to keep a certain distance to Herodotus, who is said to have made a small fortune with his public lectures. Instead, Thucydides would have presented his work as being the result of solid research that probably wasn't suitable for winning a lecturing award, but that would endure forever. Indeed, Herodotus used to include fables in his *Histories*, which had a high degree of public appeal. Nevertheless, Thucydides also provided many memorable narrative passages and descriptions that were often quoted by Greek rhetoricians, even several centuries after his death.

The importance of rhetoric for science communication endured until modern times. Considering that schools of rhetoric were the primary education of most free citizens in the Roman Empire, and that the *progymnásmata* were translated into Latin,[30] and were learned, practiced, and recommended by influential European Humanists such as Erasmus of Rotterdam, Ramon Llull, and Georg Henisch, we can assume that the use of description in the sense of *subiecto sub oculos* and *evidentia* (i.e., vividness, from the Greek word *enárgeia*, ἐνάργεια) was common for many centuries. For example, the rhetorical awareness of artists and scholars of the Renaissance in the tradition of *ékphrasis* and other *progymnásmata* was crucial for the development of a critical history of art in modern languages: Giorgio Vasari's training in Arezzo's Latin School (cf. Black 1987) and his long friendship with his Latin teacher Pollastra are evidence that the origins of art history as a literary practice have much to do with ancient rhetoric (Muñoz Morcillo 2020a). As Svetlana Alpers has shown (Alpers 1960, 1995), Vasaris' descriptions of art follow ecphrastic standards such as amplification, psychological focus, and the topos of realistic representation, regardless of whether he is describing the work of Giotto, Masaccio, or Leonardo da Vinci.

As late as the 17th century, we can still find rhetoric manuals in Latin with the 14 practical exercises, including the *descriptio* (*ékphrasis*, ἔκφρασις).[31] In this sense, modern rhetorical skills are the result of applying the ancient rhetorical practice to modern languages. This is not least due to the prior efforts of Stoic philosophers (especially Cleanthes) and Epicurean philosophers to vindicate the usefulness of poetry and rhetoric based on a "glottological materialism" (i.e., the natural representational value of language) and a new conception of the Aristotelian *phantasía* (φαντασία, mental representation) for the dissemination of knowledge and the achievement of *sophía* (σοφία, wisdom, cf. Schönberger 1995: 171). Overcoming Plato's criticism of the veracity of poetry and rhetoric was essential for the further development of popular science. Indeed, Aristotle's notion of 'fantasy' (*phan-*

30 Cf. Johnson 1943; Arcos Pereira/García de Paso 2015; Cuyás de Torres 2016a, 2016b.

31 See, e.g., Rodolphus Agricola's translation of Aphthonius' *progymnásmata* with annotations by Reinhard Lorich (Amsterdam 1649), where *ékphrasis* is defined as follows: "De Descriptione, quae Graece Έκφρασις dicta. Descriptio est oratio expositiva, quae narratione id, quod propositum est, diligenter velut, oculis subjicit." On the influence of the *progymnásmata* in the Middle Ages, the Renaissance, and subsequent periods, cf. Awianowicz 2007; Arcos Pereira/García de Paso 2015; Pérez Custodio 2015). On the multiple editions of Aphthonius's manual, cf. Kraus 2008.

tasía)[32] can be equated to intuition and creative imagination in a pre-modern sense. Cicero (*Orator* 8-11), Pseudo-Longinus (*On the Sublime* 15), Philostratus (*Life of Apollonius of Tyana* 6.19), and above all Plotinus (cf., e.g., *The Enneads* 5.9, 11) laid the foundations of artistic creation as a result of *phantasía*, which enabled the representation of the world in all its earthly and divine contexts, and which gradually led to the consideration of the visual arts as being close to wisdom—in a pre-modern sense similar to the notion of the 'fine arts' or the *Bildende Künste*. This privilege was usually reserved for poetry in antiquity. The basis of artistic creation stopped being the servile mimesis or recreation of a physical reality perceived through the senses and instead became the imagination that produces images more efficiently by directly addressing what has been conceived of in the mind (Philostratus, *Life of Apollonius of Tyana* 6.19).

VISUAL RHETORIC'S LEGITIMATION FOR COMMUNICATING KNOWLEDGE

Greek history, as well as official myths and scientific achievements that shaped the identity of the Greeks, was expressed not only in words but also in pictures, sculptures, and the dramatic arts—i.e., by means of visual rhetoric. The fastest way to spread ideas and knowledge, as well as prejudices, among the population was with theatrical performances, and in particular the performance of comedies, in which most of the arts (music, lyric, dance, and even sculpture) were brought together. The contribution of the visual and performing arts to the dissemination of scientific and technical knowledge in antiquity and in later times is a crucial issue that has been inadequately studied to date.[33]

Antiquity's visual rhetoric was probably an "impact-oriented *dynamis*," as argued by Nadia J. Koch (2005), i.e., a vivid connection between production and reception with a focus on visual eloquence and persuasion, similar to the traditional rhetorical arts. The reason for these similarities is to be found in the intellectual efforts of the first Sophists. The Sophist philosophers tried to create universal terms for the productive arts, such as medicine, rhetoric, painting, and music, thereby promoting the transfer of technical terms between disciplines and

32 For a short introduction to the Aristotelian *phantasía*, see Lombardo 2002, chapter III, "La phantasía y el pathos;" on the gnoseological origins of *phantasía* in Aristotle, cf. Περὶ μνήμης καὶ ἀναμνήσεως, 449b-31, 450a4-5; on *phantasía* in Aristotle in general, see also Hagemeier 2008.

33 For a study of this issue in modern times, cf., e.g., Krause 2016.

allowing for a better public understanding (Koch 2005: 8). As an example from later times, the abovementioned Greek orator Dio Chrysostom considered that Homeric verses describing Zeus and a statue made by Pheidias mould and give expression "to man's conception of the deity" (*The Olympic Discourse* 26).[34] Indeed, painters, sculptors, orators, and today's filmmakers have played decisive roles in shaping their respective societies with the knowledge of their time. Several centuries before Dio Chrysostom, Plato argued that beauty (*kalón*, καλόν) was difficult (*Hippias Major*, 304e) since there were many factors involved—not only the aesthetic pleasure (*hedonḗ*, ἡδονή) one takes in a visual work of art, but also the utility and purpose, i.e., the practical value (*agathós*, ἀγαθός), of the object. Furthermore, the term καλόν (beauty) encompasses not only physical objects and qualities, such as forms, colors, sounds, but also broader contexts, such as those of customs, laws, or even lessons (*Hippias Major*, 286c-293d; *Gorgias*, 474d-475a). Plato did not deny the beauty of works of art as long as they produce aesthetic pleasure (almost in a modern sense), but this kind of pleasure should be differentiated from the intellectual pleasure that Plato considered the 'right' kind. The beauty based on physical phenomena can lead us to wrong conclusions (*Phaedo*, 72e-78b). Overcoming Plato's criticism of the moral representational validity of the visual arts seems to have been more difficult than vindicating the role of rhetoric in education and the dissemination of knowledge. Nevertheless, the Aristotelian concept of *phantasía*, and its successive interpretations through Stoicism, Epicureanism, Cicero (*Orator* 8-11), Pseudo-Longinus (*On the Sublime* 15), Philostratus (*Life of Apollonius of Tyana* 6.19), and above all Plotinus (cf., e.g., *The Enneads* 5.9,11), paved the way for a gradually approach of the visual arts into the domain of knowledge and *sophía* (σοφία, wisdom), which not only enabled the emergence of strong artistic awareness in the Renaissance, but also the notion of art as a medium of cultural values and scientific knowledge. However, beyond philosophical considerations, there was a *de facto* visual culture for popularizing knowledge in ancient times: mosaics and ceramics, but also comedies and dramas, were suitable popular media, as we have seen, for spreading established knowledge, historical episodes, and even assumptions and stereotypes.

The epistemological awareness of the image that originated in antiquity and that had fully evolved by the Renaissance would later enable the introduction of 'cabinets of wonder,' together with their cultural dimensions, as a new tradition in the 15th-16th centuries, where science and art are exhibited side-by-side in the sense of Leibniz' idea of a *theatrum naturae et artis* (Bredekamp 2007: 77-78).

34 Translation by Cohoon 1932. See also Arcos Pereira/García de Paso 2015.

Most of the 'cabinets of wonders' would have to wait until the French Revolution to become part of a broad public communication of science, but art exhibits in churches, basilicas, and public spaces would also inform people's perception of biblical episodes and general knowledge, such as the ideas of proportion, light, and beauty, as well as our place in history according to Christian interpretations.

Figures 4 and 5: Historiated Letter with Galen's Vivisection Scene, from De humanis corporis fabrica, *Andreas Vesalius (1543); Emblem on the Natural Marriage of the Vine and the Elm,* Emblematum libellus *(1534), Andrea Alciati*

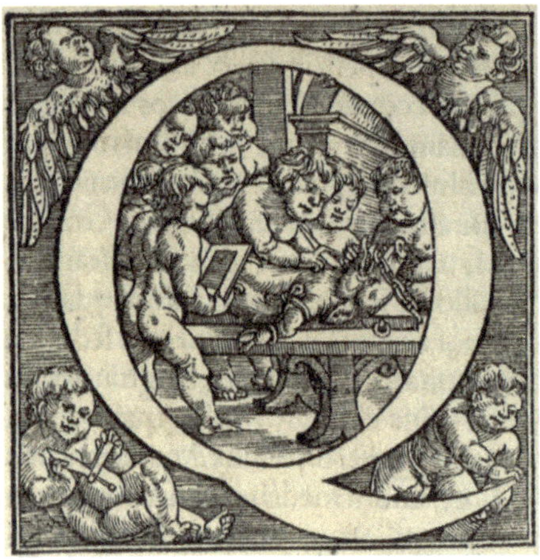

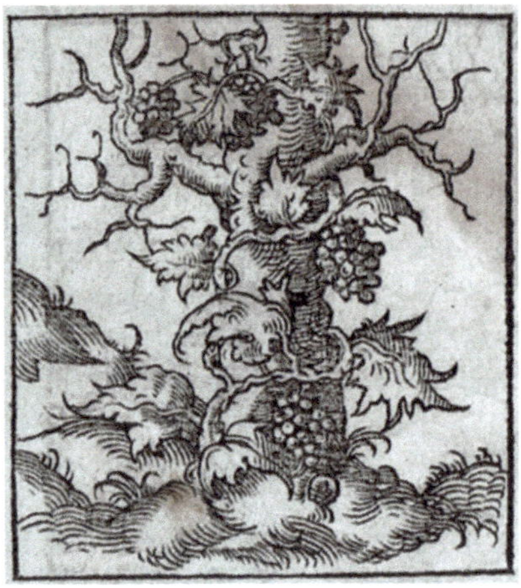

Source: Getty Research Institute; Photos: Jesús Muñoz Morcillo

Certain Renaissance humanists provide outstanding examples of the merging of artistic and scientific investigations to communicate knowledge, including Leonardo da Vinci (see, e.g., his "Sketch of Uterus with Foetus," ca. 1511-13, cf. Eskridge 2003) and especially Andreas Vesalius (see his *Tabulae anatomicae sex* from 1538 and his *De humani corporis fabrica* from 1543). We even find examples of the popularization of ancient ecphrastic texts via images in different knowledge areas. Although Vesalius corrected ancient anatomical misconceptions thank to his painstaking dissections of human corpses, in *De humani corporis fabrica* he illustrates, for example, Galen's *ékphrasis* of a vivisection to explain the recurrent laryngeal nerve in the scene decorating the opening letter Q (*Fig. 4*). This is an exceptional case of ecphrastic tradition. Vesalius illuminated the first letter of his book with the vivisection of a pig carried out by a group of cupids. This scene corresponds with the famous anatomical experiment, as vividly described by Roman physician Galen (*De pracognitione* 5.9-21). With this experiment, Galen wanted to convince his detractors of the crucial role of vivisection in understanding how the body works—in this case, how the recurrent laryngeal nerve contributes to creating voice.

Johann Wonnecke von Kaub (Johannes de Cuba) followed Dioscorides and Galen, among other authors, for the descriptions and illustrations of his popular late-15th-century book *Gart der Gesuntheit* (*The Garden of Health*). The botanists Brunfels, Manardo, and Cordus identified classical descriptions with living plants, which led to corrections (and confusions) of image-text correlations. Georgius Agricola acknowledged his debt to ancient authors such as Pliny, Vitruvius, Straton, and Theophrastus, and produced woodcuts by following ancient motifs—e.g., soda exploitation at the Nile following an *ékphrasis* by Pliny (*Naturalis Historia* 31.46), or gold panning by the Argonauts (*De re metallica*, book 8, p. 330), which is actually the visual rationale for the 'golden fleece' myth. Indeed, right after explaining how the Thuringians and others separate gold particles from sand and water using skins of oxen or horses, Agricola rationalizes the story of the 'golden fleece' explaining that the Colchians placed "the skins of animals in the pools of springs; and since many particles of gold had clung to them when they were removed, the poets invented the 'golden fleece' of the Colchians." (Translation by Hoover/Hoover 1950: 331-332) The physician Conrad Gessner published his encyclopedic *Historia animalium* (1551-58) with many illustrations. His zoological descriptions drew not only upon observation but also upon ancient, medieval, and contemporary authors, with Aristotle, Pliny the Elder, and Herodotus being his primary sources. Finally, Renaissance collections of emblems, starting with Alciati's *Libellus emblematum* (1534), also contain manifold classical ecphrastic references to the natural world, and are laden with allegorical and symbolic meaning. A good example is "the marriage of the vine and the elm," (*Fig.* 5) an image borrowed from a famous visual metaphor that Horace and Virgil previously introduced as the idea of eternal friendship. The vine plant embraces the elm even after its death. Being its motto "friendship lasting even beyond death" (amicitia etiam post mortem durans, Alciati 1934: 16). This image corresponds with an agricultural practice based on the observation of the natural behavior of the vine plant. But the observation of climbing plants had a powerful impact on the production of emblems with different meanings. Needless to say, a ubiquitous visual culture also dominates today's science communication.

CONCLUSIONS: ANCIENT AND MODERN CONCEPTIONS OF SCIENCE

The present text has provided an overview of science popularization's techniques in antiquity, which we can consider the 'origins' of Western popular science in general terms. I estimate that some ancient communication structures have operated over long periods of time, with their contents and intentions changing depending on the dominant Weltanschauung of each successive time period. Therefore, the examples discussed in this chapter may open up new perspectives on the analysis of popular science from a genealogical point of view. A requirement for the genealogical approach in Foucault's terms (1971) is painstaking interdisciplinary research accompanied by a conceptual relaxation that avoids any transcendental interpretation of history. Therefore, it is necessary to expand the narrow modern definition of science so as to include all historical knowledge areas, especially those that don't fit into modern positivistic definitions. In other words, the epistemic or practical constitution and dissemination of knowledge in general, rather than the more specific modern idea of science, should be the subject of diachronic analysis.

I have examined many different approaches to the popularization of knowledge: didactic poems on different topics, the public exposure of philosophers, the protreptical tradition for attracting new students, the awareness of communication between experts and lay people, the role of fiction and humor in stereotypical depictions of ancient 'scientists,' and the importance of vivid descriptions (ἐκφράσεις) and narratives for successful communication in public lectures and written texts. These are some of the most important characteristics that define the popularization of knowledge in antiquity. Many general ideas and practices have a homeostatic nature and are also still prevalent today, such as the prominent link between science communication and rhetorical skills, the awareness of different target groups, the idea of science popularization as a means for dispensing actual philosophical medicine (i.e., similar to today's 'deficit model'), the protreptic nature of programmatic science communication, and the persistent use of stereotypes.

A reference in the *Hippocratic Corpus* (*On the Right Way to Behave*, Περὶ εὐσχημοσύνης 8-12) reveals that both communication between experts and laypeople and communication among experts were very common practical topics among physicians. A rhetorical education was recommended to them to convey their arguments convincingly. Moreover, Greek thinkers also discovered the rhetorical power of images and their ability to both express and shape notions and preconceptions about the world and its rules.

It seems that the popularization of knowledge in antiquity has more similarities to current practices of science communication than one would typically assume. Ancient authors, too, had programmatic means of communicating science, and could almost be said to have practiced philosophical or rhetorical lobbying. They worked with a great variety of media and genres, from the didactic poem to the broadly accessible medium of Greek drama, from 'written machines' (i.e., artificats vividly described through *ékphrasis*) to stereotypical representations of intellectuals. There was also great intellectual competitiveness and awareness of one's authorship, even in the Middle Ages. John Tzetzes in particular was proud of understanding and explaining Homer's poems better than his predecessors and contemporaries. We find intriguing precursors of modern science communication concepts such as the 'information deficit model' and 'citizen science.' The latter, for example, is present in the zoological reports written by Aristotle. These findings raise the legitimate question about whether a genealogy of science popularization through the Middle Ages and the Renaissance and on to modern times can be constructed, together with its continuities and transformations. I think that such a genealogy can be sketched out, though it must include noteworthy instances of discontinuity.

However, some crucial questions of our time related to political aspects of science communication, especially those concerning the institutional funding of science and technology, e.g., the need for legitimation or the issue of citizens' trust in science, cannot be traced back to antiquity without additional research. In other words, we do not know how Hellenistic kings or Roman emperors would have profited from philosophy in material and economic terms, or if citizens' trust in science and in technological progress was a subject of discussion in democratic Athens. From today's perspective, we can venture that trust or mistrust in science is always somehow part of this process, since modern science communication has a clear function as an instrument of power exposed to calculated mediatization. However, can we apply the general idea of public science as part of modern power structures to ancient times? Regarding the totality of this phenomenon, this remains an unanswered question. In general terms, we can say that political and religious changes had an impact on the preferred forms and topics of communication during each time period. This phenomenon points to the existence of an interdependency between societal transformations and popular science strategies. According to this, scientific knowledge and its popularization have always been part of their respective societies. In other words, every age has the popular science it deserves.[35]

35 This phrase is based on Beat Wyss' statement "Jede Zeit hat die Bilder, die sie verdient," i.e., "Every age has the images that it deserves" (Wyss 2013: 178), which probably builds on Aby Warburg's remark "Jede Zeit hat die Renaissance der Antike,

REFERENCES

Alciati, Andrea (1934): Emblematum libellus, Paris: Christianus Wechelus.

Allgaier, Joachim (2012): "On the Shoulders of YouTube: Science in Music Videos." In: Science Communication 35/2, pp. 266-275.

Alpers, Svetlana (1960): "Ekphrasis and aesthetic attitudes in Vasari's Lives." Journal of the Warburg and Courtauld Institutes 23, pp. 190-215.

Alpers, Svetlana (1995): "Ekphrasis und Kunstanschauung in Vasaris Viten." In: Gottfried Boehm/Helmut Pfotenhauer (eds.): Beschreibungskunst, Kunstbeschreibung. München: Wilhelm Fink, pp. 217-253.

Archambault, Paul (1966): "The Ages of Man and the Ages of the World. A Study of Two Traditions." In: Revue d'Études Augustiniennes et Patristiques 12/3-4, pp. 193-228.

Arcos Pereira, Trinidad/García de Paso, María Dolores (2015): "La traducción al latín de algunos términos y su repercusión en los comentarios a los Aphthonii Progymnasmata." In: Cuadernos de Filología Clásica. Estudios Latinos 35/2, pp. 319-336.

Auerbach, Jeffrey A. (1999): The Great Exhibition of 1851: A Nation on Display, New Haven: Yale University Press.

Awianowicz, Bartosz Bogusław (2007): Progymnasmata in the Theory and Practice of the Humanistic School from the Late 15th to the Mid-18th Century, Nicolaus Copernicus University, Poland. (PhD Thesis)

Bahar, Saba (2001): "Jane Marcet and the Limits to Public Science." In: The British Journal for the History of Science 34/1, pp. 29-49.

Bastert, Bernd (2013): "Kontinuitäten eines »Klassikers«. Zur spätmittelalterlichen deutschen Rezeption der ›Consolatio Philosophiae‹ des Boethius." In: M Eikelmann/U. Friedrich (eds.), Praktiken europäischer Traditionsbildung im Mittelalter: Wissen – Literatur – Mythos, Berlin: Akademie Verlag, pp. 117-140.

Baudouin, Jurdant (1993): "Popularization of Science as the Autobiography of Science." In: Public Understanding of Science 2/4, pp. 365-373.

Bauer, Martin W. (2017): "Kritische Beobachtungen zur Geschichte der Wissenschaftskommunikation." In: H. Bonfadelli/B. Fähnrich/C. Lüthje/J. Milde/M. Rhomberg/M. S. Schäfer (eds.), Forschungsfeld Wissenschaftskommunikation: Springer Verlag, pp. 17-40.

die sie verdient" ("Every age has the Renaissance of antiquity that it deserves"), Gombrich (1986: 238). However, this phrase is probably based on an older and much more famous quote by Joseph de Maistre from 1811, "Toute nation a le gouvernement qu'elle mérite" ("Every nation gets the government it deserves," Maistre/Blanc 1860: 196).

Bensaude-Vincent, Bernadette (2001): "A Genealogy of the Increasing Gap between Science and the Public." In: Public Understanding of Science 10/1, pp. 99-113. DOI: 10.1088/0963-6625/10/1/307

Black, Robert (1987): "Humanism and Education in Renaissance Arezzo." In: I Tatti Studies: Essays in the Renaissance 2, pp. 171-237.

Bonney, Rick/Ballard, Heidi/Jordan, Rebecca/McCallie, Ellen/Phillips, Tina/Shirk, Jennifer/Wilderman, Candie C. (2009): "Public participation in scientific research: defining the field and assessing its potential for informal science education." Washington, DC: Centre for Advancement of Informal Science Education (CAISE).

Bredekamp, Horst (2007 [1993]): Antikensehnsucht und Maschinenglauben. Die Geschichte der Kunstkammer und die Zukunft der Kunstgeschichte, Berlin: Wagenbach.

Britton, Piers D. (2009): "Michelangelo and the rerum corporis in Vasari's *Vite*: The Limits of Physiognomy." In: Notes in the History of Art 28/3, pp. 13-18.

Casson, Lionel (1971): Ships and Seamanship in the Ancient World, Princeton: Princeton University Press.

Cesaretti, Paolo (1991): Allegoristi di Omero a Bisanzio. Ricerche ermeneutiche (XI-XII secolo), Milano: Guerini Studio.

Christmann, Eckhard (2003): "Zum Verhältnis von Autor und Leser in der römischen Agrarliteratur: Bücher und Schriften für Herren und Sklaven." In: Marietta Horster/Christiane Reitz (eds), Antike Fachschriftsteller, Literarischer Diskurs und sozialer Kontext, Stuttgart, pp. 21-152.

Clark, Donald L. (1952): "The Rise and Fall of Progymnasmata in Sixteenth and Seventeenth Century Grammar Schools." In: Speech Monographs 19/4, pp. 259-263.

Cohoon, J. W. (ed.) (1932): Dio Chrysostom. Discourses 1-11, Loeb Classical Library, Cambridge, MA: Harvard University Press.

Condit, Celeste M./Lynch, John / Winderman, Emily (2012): "Recent Rhetorical Studies in Public Understanding of Science: Multiple Purposes and Strengths." In: Public Understanding of Science 21/4, pp. 386-400.

Conington, John (ed.) (1882): Horace. The Odes and Carmen Saeculare of Horace, London. George Bell and Sons.

Crawford, Robert (ed.) (2006): Contemporary Poetry and Contemporary Science, Oxford: Oxford University Press.

Cuyás de Torres, María Elisa (2016a): "Las ediciones de los Progymnasmata Rhetorica de Antonio Llull." In: Cuadernos de Filología Clásica. Estudios Latinos 36/2, pp. 255-277.

Cuyás de Torres, María Elisa (2016b): "Las variantes de lectura en las ediciones de los Progymnasmata Rhetorica de Antonio Llull." In: Revista de Estudios Latinos 16, pp. 155-177.

Doody, Aude/Föllinger, Sabine/Taub, Liba (eds.) (2012): Structures and Strategies in Ancient Greek and Roman Technical Writing. Special Issue of Studies in History and Philosophy of Science 43.

Effe, Bernd (1977): Dichtung und Lehre. Untersuchungen zur Typologie des antiken Lehrgedichts, München: C. H. Beck.

Effe, Bernd (2005): "Typologie und literarhistorischer Kontext: Zur Gattungsgeschichte des griechischen Lehrgedichts." In: Marietta Horster/Christiane Reitz (eds), Wissensvermittlung in dichterischer Gestalt, Stuttgart: Franz Steiner, pp. 27-44.

Eskridge, Robert (2003): The Enduring Relationship of Science and Art. Adaptation of a Lecture titled "Exploration and the Cosmos: The Consilience of Science and Art." Accessed March 3, 2020. http://www.artic.edu/aic/education/sciarttech/2a1.html.

Faber, Riemer A. (1993): "The Apostle and the Poet: Paul and Aratus." In: Clarion 42/13. Accessed March 3, 2020. http://spindleworks.com/library/rfaber/aratus.htm.

Farrington, Benjamin (1939): Science and Politics in the Ancient World, London: George Allen and Unwin.

Faulstich, Werner (1996): "Medien und Öffentlichkeiten im Mittelalter: 800-1400." In: Werner Faulstich (ed.), Die Geschichte der Medien 2. Göttingen: Vandenhoeck & Ruprecht, pp. 269-272.

Faulstich, Peter (2006): Öffentliche Wissenschaft. Neue Perspektiven der Vermittlung in der wissenschaftlichen Weiterbildung, Bielefeld: transcript.

Fernández Delgado, José Antonio (1976): La poesía gnómica griega, una forma de expresión de la antigua poesía de composición oral, Salamanca: Universidad de Salamanca.

Fernández Delgado, José Antonio (2007): "Influencia literaria de los progymnásmata." In: J. A. Fernández Delgado/F. Pordomingo Pardo/A. Stramaglia (eds.), Escuela y literatura en Grecia Antigua. Actas del Simposio Internacional, Universidad de Salamanca, 17-19 de Noviembre de 2004, Cassino: Università degli Studi di Cassino, pp. 273-306.

Fernández Delgado, José Antonio / Pordomingo Pardo, F. (2016): "Musical Ekphrasis and Diegema in Longus' Daphnis and Chloe." In: Greek, Roman, and Byzantine Studies 56: 696-708.

Foucault, Michel (1971): "Nietzsche, la genealogie, l'histoire." In: Suzanne Bachelard/Georges Canguilhem/François Dagognet/Michel Foucault/Martial Gueroult/Michel Henry/Jean Laplanche/Jean-Claude Pariente/Michel Serres (eds.), Hommage à Jean Hyppolite, Paris: Presses Universitaire de France, pp. 145-72.

Foucault, Michel (1977 [1971]): "Nietzsche, Genealogy, History." In: Donald F. Bouchard (ed.), Language, Counter-memory, Practice, Trans. D. F. Bouchard, Sherry Simon, Ithaca: Cornell University Press, pp. 139-64.

Fyfe, Aileen/Lightman, Bernard (eds.) (2007): Science in the Marketplace: Nineteenth-Century Sites and Experiences, Chicago: University of Chicago Press.

Gebert, Bent (2013): Mythos als Wissensform: Epistemik und Poetik des Trojanerkriegs Konrads von Würzburg, Berlin: DeGruyter.

Gombrich, Ernst H. (1986): Aby Warburg: An Intellectual Biography, with a Memoir on the History of the Library by F. Saxl, Chicago: University of Chicago Press.

Gordon, Pamela (2012): The Invention and Gendering of Epicurus, Ann Arbor: University of Michigan Press.

Greenblatt, Steven (2011): The Swerve. How the World Became Modern, New York: W.W. Norton.

Grendler, Paul F. (1989): Schooling in Renaissance Italy: Literacy and Learning, 1300-1600, Baltimore: Johns Hopkins University Press.

Gross, Alan G. (1994): "The Roles of Rhetoric in the Public Understanding of Science." In: Public Understanding of Science 3/1, pp. 3-23.

Gross, Alan G. (2003): "Book Review: Shaping Science with Rhetoric: The Cases of Dobzhansky, Schrödinger, and Wilson." In: Public Understanding of Science 12/1, pp. 109-110.

Gruber, Gwendolyn M. (2008): "Honey for the Soul: Lucretius' Sweet Poetry." 104th Annual Meeting of The Classical Association of the Middle West and South, Tucson, Arizona, April 16-19, 2008. (Abstract)

Hagemeier, Martin (2008): Phantasia bei Aristoteles. Eine Untersuchung über die philosophische Konzeption der phantasia bei Aristoteles, Norderstedt: Grin.

Haye, Thomas (1997): Das Lateinische Lehrgedicht im Mittelalter: Analyse einer Gattung, Leiden, New York and Cologne: Brill.

Herweg, Mathias (2016): "Erzählen unter Wahrheitsgarantie – Deutsche Weltchroniken des 13. Jahrhunderts." In: G. Wolf/N. Ott (eds.), Handbuch Chroniken des Mittelalters, Berlin: De Gruyter, pp. 145-179.

Hinds, Allen B. (1980 [=1963]): The Lives of the Painters, Sculptors and Architects. 1-4 Vols. Translated by Allen B. Hinds, ed. by William Gaunt, London-Melbourne-Toronto: Dent & Sons.

Hochadel, Oliver (2003): Öffentliche Wissenschaft. Elektrizität in der deutschen Aufklärung, Göttingen: Wallstein Verlag.

Hoover, Herbert C./Hoover Lou H. (1950 [1912]): Georgius Agricola. De Re Metallica, New York: Dover Publications.

Huang, Hsiang-Fu (2015): Commercial and Sublime: Popular Astronomy Lectures in Nineteenth Century Britain, London: University College London.

Irwin, Alan (1995): Citizen Science: A Study of People, Expertise and Sustainable Development, London: Routledge.

Jensen, Christian Cornelius (ed.) (1923): Philodemos, Über die Gedichte. Fünftes Buch, Berlin: Weidmann.

Jones, W.H.S. (ed.) (1931): Hippocrates, Volume IV: Nature of Man. Regimen in Health. Humours. Aphorisms. Regimen 1-3. Dreams, Harvard University Press.

Johnson, Francis R. (1943): "Two Renaissance Textbooks of Rhetoric: Aphthonius' *Progymnasmata* and Rainolde's *A Booke Called the Foundacion of Rhetorike*." In: The Huntington Library Quarterly 6/4, pp. 427-444.

Jordan, Mark D. (1986): "The Protreptic Structure of the Summa Contra Gentiles." In: The Thomist 50/2, pp. 173-209.

Jütte, Robert (1998): "Die Entdeckung des „inneren" Menschen 1500-1800." In: Van Dülmen, R. (ed.), Erfindung des Menschen. Schöpfungsträume und Körperbilder 1500-2000, Vienna: Böhlau Verlag.

Kirby, David A. (2008a): "Hollywood Knowledge: Communication between Scientific and Entertainment Cultures." In: D. Cheng/M. Claessens/N. R. J. Gascoigne/J. Metcalfe/B. Schiele/S. Shi (eds.), Communicating Science in Social Contexts, New York: Springer, pp. 165-181.

Kirby, David A. (2008b): "Cinematic Science: The Public Communication of Science and Technology in Popular Film." In: Massimiano Bucchi/Brian Trench (eds.), Routledge Handbook of Public Communication of Science and Technology, London and New York: Routledge, pp. 67-94.

Klestinec, Cynthia (2011): Theaters of Anatomy. Students, Teachers, and Traditions of Dissection in Renaissance Venice, Baltimore: The Johns Hopkins University Press.

Knight, David (2006): Public Understanding of Science. A History of Communicating Scientific Ideas, New York: Routledge.

Koch, Nadia J. (2005): "Bildrhetorische Aspekte der antiken Kunsttheorie." In: Rhetorik. Ein internationales Jahrbuch 24, pp. 1-13.

Kohring, Matthias (2016a): Science Communication and Trust in Science: A Critique of the Traditional Discourse on Science Communication. *JCOM* 15(05). Accessed March 3, 2020. https://jcom.sissa.it/archive/15/05/JCOM_1505_2016_C00/JCOM_1505_2016_C04.

Kohring, Matthias (2016b): Misunderstanding Trust in Science: A Critique of the Traditional Discourse on Science Communication. *JCOM* 15(05). Accessed March 3, 2020. https://jcom.sissa.it/archive/15/05/JCOM_1505_2016_C00/JCOM_1505_2016_C04.

Kohring, Matthias (2012): "Die Wissenschaft des Wissenschaftsjournalismus. Eine Forschungskritik und ein Alternativvorschlag." In: C. Y. Robertson-von Trotha/J. Muñoz Morcillo (eds.), Öffentliche Wissenschaft und Neue Medien: die Rolle der Web 2.0-Kultur in der Wissenschaftsvermittlung, Karlsruhe: KIT Scientific Publishing.

Kohring, Matthias (2004): "Die Wissenschaft des Wissenschaftsjournalismus. Eine Forschungskritik und ein Alternativvorschlag." In: C. Müller (ed.), SciencePop. Wissenschaftsjournalismus zwischen PR und Forschungskritik, Graz and Vienna: Nausner & Nausner, pp. 161-183.

Korzybski, Alfred (1933): Science and Sanity. An Introduction to Non-Aristotelian Systems and General Semantics. The International Non-Aristotelian Library Pubblishing Company, pp. 747-761.

Kraus, Manfred (2008): "Aphthonius and the Progymnasmata in Rhetorical Theory and Practice". In: David Zarefsky/Elizabeth Benacka (eds.): Sizing Up Rhetoric, Long Grove, IL: Waveland Press, pp. 52-67.

Krause, Kelly (2016): "A Framework for Visual Communication at *Nature*." In: Public Understanding of Science 26/1, pp. 15-24.

Latour, Bruno (1993 [=1991]): We Have Never Been Modern, Cambridge, MA: Harvard University Press.

Lombardo, Giovanni (2002): La estética antigua. Madrid: La balsa de la Medusa. (Tr. Francisco Campillo)

MacLeod, Roy M. (1996): Public Science and Public Policy in Victorian England, Aldershot: Brookfield.

Maistre, Joseph de/Blanc, Albert (eds.) (1860) Correspondance diplomatique, tome 2. Paris: Michel Lévy frères libraires éditeurs.

Markovic, Daniel (2008): The Rhetoric of Explanation in Lucretius' *De rerum natura*, Boston: Brill.

Marrou, Henri I. (1956): A History of Education in Antiquity, Madison: the University of Wisconsin Press.

Momigliano, Arnaldo (1978): "The Historians of the Classical World and their Audiences: Some Suggestions." In: Annali della Scuola Normale Superiore di Pisa, Serie III, Vol. VIII, Fasc. I, pp. 59-75.

Morgan, Teresa J. (1998): Literate Education in the Hellenistic and Roman Worlds, Cambridge, UK: Cambridge Classical Studies.

Morus, Iwan Rhys (1998): Frankenstein's Children: Electricity, Exhibition and Experiment in Early Nineteenth-Century London, Princeton: Princeton University Press.

Muñoz Morcillo, Jesús/Czurda, Klemens/Robertson-von Trotha, Caroline Y. (2016): "Typologies of the Popular Science Web Video." In: Journal of Science Communication 15 (04) A02, pp. 1-32. Accessed July 20, 2020. http://jcom.sissa.it/sites/default/files/documents/JCOM_1504_2016_A02.pdf

Muñoz Morcillo, Jesús (2016): "Epicurus and the Recipient's Question in the Letter to Herodotus." Revista de Filología Clásica Myrtia 31, pp. 339-372.

Muñoz Morcillo, Jesús (2019): "Aproximación a los cánones de la écfrasis, entre tradición literaria e influencia escolar," In: Rivista di Cultura Classica e Medioevale, RCCM LXI/2, pp. 475-495.
Muñoz Morcillo, Jesús (2020a): "Progymnasmatic Ekphrasis at the Latin School of Arezzo and Vasari's 'Memory Images.'" Session 77.3: Constructing a Classical Tradition: East and West. Annual Meeting Society of Classical Studies, January 2-5, 2020, Washington DC. (Paper)
Muñoz Morcillo, Jesús (2020b): Ursprünge und Protreptik des Science Slams. In: Philipp Niemann, Laura Bittner, Christiane Hauser, Philip Schrögel (eds.), Science-Slam. Multidisziplinäre Perspektiven auf eine populäre Form der Wissenschaftskommunikation, Wiesbaden: Springer VS.
Muñoz Morcillo, Jesús (2020c): La ékphrasis griega, de la Antigüedad a Bizancio, Frankfurt am Main and New York: Peter Lang. (forthcoming)
Novikoff, Alex J. (2012): "Toward a Cultural History of Scholastic Disputation." In: American Historical Review 117/2, pp. 331-364.
Pearce, Fred (2009): "Climategate: Anatomy of a Public Relations Disaster." In: Yale Environment 360. Accessed March 3, 2020. http://e360.yale.edu/features/climategate_anatomy_of_a_public_relations_disaster.
Penwill, John L. (1995): "Image, Ideology and Action in Cicero and Lucretius." In: Ramus 23/1-2, pp. 68-91.
Pérez Custodio, Violeta (2015): "La recepción de los ejercicios retóricos del Pseudo-Hermógenes en la España del Renacimiento." In: José María Maestre Maestre /Sandra Inés Ramos Maldonado/Manuel Antonio Díaz Gito/María Violeta Pérez Custodio/Bartolomé Pozuelo Calero/Antonio Serrano Cueto (eds.), Humanismo y pervivencia del mundo clásico. V. Homenaje a Juan Gil, Madrid-Alcañiz: Instituto de Estudios Humanísticos-CSIC, pp. 1351-1364.
Phillips, Anna Lena (2010): "Epic Science. At an International Colloquium this September, Researchers will Tell the Story of Science Poetry's Evolution." In: American Scientist, 98/5, pp. 383. Accessed March 3, 2020. http://www.americanscientist.org/issues/pub/epic-science.
Pizzone, Aglae (2017): "The *Historiai* of John Tzetzes: a Byzantine 'Book of Memory'?" In: Byzantine and Modern Greek Studies 41, pp. 182-207.
Pohlenz, Max (1939): "Die Begründung der abendländischen Sprachlehre durch die Stoa." In: Nachrichten der Akademie der Wissenschaften zu Göttingen. Philologisch-Historische Klasse, I, 3.6, pp. 155-198.
Robertson-von Trotha, Caroline Y. (2007): "Öffentliche Wissenschaft – ein notwendiger Dialog." In: J. Klaus/H. Vogt (eds.), Wissensmanagement und wissenschaftliche Weiterbildung. Dokumentation der Jahrestagung der Deutschen Ge-

sellschaft für wissenschaftliche Weiterbildung und Fernstudium an der Universität Karlsruhe (TH), Hamburg: DGWF, pp. 97-134.

Robertson-von Trotha, Caroline Y. (2012): "Öffentliche Wissenschaft im Spiegel der Web 2.0-Kultur." In: C. Y. Robertson-von Trotha/J. Muñoz Morcillo (eds.), Öffentliche Wissenschaft und Neue Medien. Die Rolle der Web 2.0-Kultur in der Wissenschaftsvermittlung, Karlsruhe: KIT Publishing.

Roby, Courtney A. (2016): Technical Ekphrasis in Greek and Roman Science and Literature. The Written Machine between Alexandria and Rome. Cambridge, UK: Cambridge University Press.

Schönberger, Otto (1995): „Die Bilder des Philostratos". In: Boehm/Pfotenhauer (eds.): Beschreibungskunst – Kunstbeschreibung. München: Wilhelm Fink, pp. 143-176.

Shorey, Paul/Laing, Gordon J. (1919): Horace, Odes and Epodes, Chicago: Benj. H. Sanborn & Co.

Short, Dan Brian (2013): "The Public Understanding of Science: 30 Years of the Bodmer Report." In: the School Science Review 95, pp. 39-43.

Stafford, Barbara M. (1996): Artful Science. Enlightenment Entertainment and the Eclipse of Visual Education, Cambridge, MA: MIT Press.

Stewart, Larry (1992): The Rise of Public Science. Rhetoric, Technology, and Natural Philosophy in Newtonian Britain, 1660–1750, Cambridge, UK: Cambridge University Press.

Taub, Liba (2004): Ancient Meteorology, London-New York: Routledge.

Taub, Liba (2017): Science Writing in Greco-Roman Antiquity, Cambridge, UK: Cambridge University Press.

Vernant, Jean-Pierre (1965): Mythe et pensée chez les Grecs: Études de psychologie historique, Paris: Maspero.

Wandhoff, Haiko (2003): Ekphrasis. Kunstbeschreibungen und virtuelle Räume in der Literatur des Mittelalters, Berlin: De Gruyter.

Waszink, Jan Hendrik (1974): Biene und Honig als Symbol des Dichters und der Dichtung in der griechisch-römischen Antike, Opladen: Westdeutscher Verlag.

Webb, Ruth (1999): "Ekphrasis Ancient and Modern. The Invention of a Genre." In: Word & Image 15, 1, pp. 7-18.

Webb, Ruth (2009): Ekphrasis, Imagination and Persuasion in Ancient Rhetorical Theory and Practice, Farnham/Burlington: Ashgate.

Weingart, Peter/Muhl, Claudia/Pansegrau, Petra (2003): "Of Power Maniacs and Unethical Geniuses: Science and Scientists in Fiction Film." In: Public Understanding of Science, 12/3, pp. 279-287.

Weingart, Peter/Guenther, Lars (2016): Science Communication and the Issue of Trust. *JCOM 15(05)*. Accessed March 4 2020. https://jcom.sissa.it/archive/15/05/JCOM_1505_2016_C00/JCOM_1505_2016_C01.

West, Martin L. (1967): 'The Contest of Homer and Hesiod,' In: The Classical Quarterly 17, pp. 433-450.

Wilke, Jürgen (2012): "Die Kommunikationswissenschaft und die Antike." In: Christina Kuhn (ed.), Politische Kommunikation und öffentliche Meinung in der antiken Welt, Stuttgart: Franz Steiner Verlag, pp. 31-55.

Wiseman, Timothy P. (2015): The Roman Audience: Classical Literature as Social History, Oxford: Oxford University Press.

Wolfschmidt, Gudrun (ed.) (2017 [2002]): Popularisierung der Naturwissenschaften, Diepholz: GNT-Verlag.

Wyss, Beat (2013): Renaissance als Kulturtechnik, Hamburg: Philo & Philo Fine Arts.

From Rational Recreation to Fun with Science. Continuities in the History of Science Popularization since the Enlightenment

Oliver Hochadel[*]

ABSTRACT

This article proposes a *longue durée* approach for the history of science popularization. Only by looking at the last three centuries as one large unit can some revealing continuities emerge. This perspective might enrich our current discussions about how to popularize science. Drawing on a large amount of secondary literature, the article will focus on three of these continuities: 1) the constant tension between entertainment and education; 2) the rhetoric about science being "in," alternating with the topic of science having become incomprehensible; and 3) the fierce competition of scientific practitioners for audiences using the media available at the time, as well as the flipside of this, i.e., the highly diverse ideologies and value systems that underpin any kind of science popularization. In sum, historical actors from the Enlightenment to the present day constantly negotiate, police, and dispute who is entitled to popularize science and in what way.

* I would like to thank Agustí Nieto-Galan and Klaus Taschwer for many fruitful discussions on the continuities in the history of science popularization over the last two decades. And I am very grateful to Jesús Muñoz Morcillo for inviting me to be part of this venture. Research for this article was supported by the Grup de recerca consolidat i finançat (2017 SGR 1138, AGAUR-Generalitat de Catalunya).

A CONFESSION

How do historians of science find their topics? We are usually very eloquent in putting forward an explanation that stresses the intrinsic value of the topic we are studying. We highlight the general relevance of the issue and emphasize that we have spotted an important gap in existing knowledge (a *Forschungslücke*) that we are eager to fill. But if we were to probe deeper, quite a few of us would admit that personal predilections (which may even include a fun factor), chance (e.g., stumbling on new sources), or—*horribile dictu*—even opportunism (what qualifies us for the job market) play an important role in our thematic choices. Looking at my own research biography, I would plead guilty to all of these charges. My major research topics of the last 25 years are scattered over three centuries and deal with quite diverse fields of knowledge: electricity in the German Enlightenment, the history of zoological gardens in the 19th and early 20th centuries, and the recent history of human origins research (mostly paleoanthropology). Regardless of whether I am explaining my work on a night out in a bar or in a serious job interview, my strategy to integrate these three apparently entirely unrelated topics into a meaningful research agenda goes like this: All three cases form part of the history of science popularization. Hence, I am interested in how the relationship between science and its publics has evolved over time.

So even if my choice of research topics was considerably indebted to circumstance, they—and this was hardly planned—add up to a little more than just three case studies on the history of science popularization. Dabbling with different centuries and topics led me to think that the historical study of science popularization suffers from "temporal myopia." Most studies, even collected volumes and large research projects, generally focus only on a given period of time—a century at most, and more often only a few decades. There are good practical reasons for this limitation (even a group of researchers can only cover so much ground) as well as "content reasons": Historical periods differ in many respects from each other, therefore generalizations are tricky if not utterly erroneous. Nevertheless, and in my case this is a circumstance born out of circumstance: I stumbled time and again on structurally identical or at least similar arguments, tropes, and rhetorical strategies in my hodgepodge of case studies. This made me aware of a number of continuities in the history of science popularization that I would like to discuss in this article.

AN AVALANCHE—THE SCHOLARSHIP ON SCIENCE POPULARIZATION

In the last three decades the history of science popularization has become a huge topic in the history of science. It is now a commonplace that we may learn a lot about the inner workings of science itself if we analyze how it interacts with the media and the general public. And more than this: In line with the communicative turn in the history of science (Secord, 2004a), a neat separation between the "strictly scientific" and the "merely popular" seems increasingly difficult, if not utterly arbitrary (Hilgartner, 1994; and to cite one case study: Egmond/Mason, 1996). Numerous studies have shown that these two spheres were and are intertwined, constantly feeding off each other.

Yet this large amount of scholarship focuses mostly on a specific period.[1] To stress this once more, the titles just listed represent only a fraction of the literature on popular science. There is one clear peak though: A substantial share of the literature deals with science popularization in Victorian Britain (and this article reflects this as well). We might have to be careful not to turn Victorian Britain into a paradigm from which to judge all other cases.[2]

This avalanche of literature on science popularization has left historians of science uneasy. And so they do what historians do: They historicize. Topham (2009a, 2009b; see also Topham 2007: 135-138 and *passim*) and others (Schirrmacher 2008; Nieto-Galan 2016: xiii; and see the early work by Cooter/Pumfrey 1994: 239) have called for a strict contextualization of "popular science." The meanings of terms such as "popularization" and "popular science" vary substantially over time but also between different social and political settings. I fully subscribe to this demand of historicizing.

Nevertheless, this call for contextualization in a sense aggravates the problem of temporal myopia. The ever-growing number of often excellent case studies on the history of science popularization atomizes the field. Topham himself, and also Bernadette Bensaude-Vincent (2009: 361, 366), clearly sense the lack

1 For the 18th century, cf. Stewart 1992; Golinski 1992; Hochadel 2003; Lynn 2006; Bensaude-Vincent/Blondel 2008; for the 19th century, cf. Bensaude-Vincent/Rasmussen 1997; Daum [1998] 2002; Morus 1998; Schwarz 1999; Goschler 2000; Secord 2003 [2000]; Fyfe/Lightman 2007; Lightman 2007a; Samida 2011a; Kember/Plunkett/Sullivan 2012a; and for the 20th century, cf. Ash/Stifter 2002; Nikolow/Schirrmacher 2007; Bowler 2009; Schirrmacher 2008 and 2013.

2 See the already-present criticism of this kind of "parochialism" by Secord 2004a: 669; Daum 2009: 322 and 324.

of a *longue durée* approach. Comprehensive overviews of the history of science popularizations are rare (see Nieto-Galan 2016)[3] and clearly pose an enormous challenge.

Obviously, this article will not be able to draw this much-coveted big picture. It will instead try to identify some striking continuities in science popularization between the Enlightenment and today. To cover three centuries in a single article seems overly ambitious if not outright ludicrous. Therefore, the "essay" in the title should be understood as a disclaimer. It will not be possible to adequately historicize all the examples I shall cite hereafter. I will therefore not even attempt to do so. Rather, I will try to turn a selective overview of the existing scholarship into a research agenda.

One may argue that in a sense everything has changed in the history of science popularization since the 18th century: science itself (from a rather obscure beginning to a practice that completely and irreversibly changed our world); the scientists (specialization, professionalization, and multiplication); the communicators and also the spaces and media that are deployed to popularize science; and of course the audiences as well. For example, the 19th century fundamentally differs in many aspects from the 18th century. To name but a few of these differences: the printing revolution, the increasingly literate masses, the new venues to present science to growing audiences, and the professionalization of the popularizers.

My point is that the numerous changes are all-too-obvious. More surprising are the continuities, i.e., the recurring arguments, tropes, and patterns. In this article I shall focus on three pairs of continuities. 1) The eternal tension between instruction and entertainment might be the most obvious one. 2) Yet the slogan that "science is in" (i.e., fashionable) is also recurrent—as is its opposite, "science is out" (i.e., incomprehensible). 3) The fierce competition for audiences, their attention, their money, and the legitimation they convey represents another continuity. This feature has its flipside too: Popular science has always been policed: Who is allowed to popularize what and in what way to whom?

There are some good reasons for starting this search for continuities in the Enlightenment (and thus for limiting it to the last three centuries): the formation of the public sphere, including the first public lectures on natural philosophy and thus the emergence of the figure of the popularizer and the natural philosophers' determination "to make themselves heard" (Stewart 1992: xxviii); the first journals dedicated to natural philosophy and, a little later, specialized books popularizing natural philosophy to selected audiences such as women (e.g., Sutton 1995; Terrall 1995) or children (Secord 1985; for the 19th century, see: Myers 1989; Fyfe 2008). But in

3 For a different approach, see Kretschmann 2003.

no way would I like to suggest a clear-cut break for the early 18th century. Mutatis mutandis, some of the continuities certainly "extend backwards," and this volume provides ample material in favor of a much longer *longue durée* perspective.

INSTRUCTION THROUGH ENTERTAINMENT

In 1764, Johann Christoph Thenn proposed to the board of the St. Anna School in Augsburg in southern Germany to introduce experimental physics into the school curriculum. Thenn was a pastor but also a vocal and active popularizer of natural philosophy. In his memorandum he argued that natural philosophy should not only be taught "through thorough proofs, but mainly through a large number of experiments." This would "not only foster the most rational entertainment but also maintain unflagging attention among the listeners." We may call this idea to capture the attention of the students the "incentive theory" (Hochadel 2003: 114-131, quote 122).[4]

As this volume shows, the coupling of entertainment and instruction goes much further back than the 18th century. It can already be found in antiquity. As scientific knowledge tends to have the reputation of being rather dry and difficult, the popularizer needs to lure his audience with entertainment. With the help of a humorous story, an impressive explosion, or the promise of science's utility, the popularizer tries to arouse the interest of his audience. While the incentives may vary greatly, the argument of the incentive theory has changed little over the course of time. In antiquity the trope was: In order to make the public drink from the bitter vessel of knowledge, you need to put sweet honey on the rim (see Muñoz Morcillo in this volume). In the 18th century popularizers of all sorts, whether an itinerant lecturer of experimental physics, a reform-oriented school teacher, or an author of a popular science book, always insisted that you had to capture the attention of your audience first before being able to teach them anything. And nothing is more useful than an optical illusion, an electric shock, a ghost appearing on a screen, or a strange exotic creature. Hence, appealing to the senses—first and foremost sight, but also hearing, smell, and touch—is a prerequisite of any attempt to popularize.

A vocal advocate of the incentive theory is Georg Christoph Lichtenberg (1744-1799), professor of experimental physics at the University of Göttingen in the later decades of the 18th century. From his experience as a lecturer, he knew very well that a "physical experiment that bangs is always worth more than a

4 Unless otherwise noted, all translations in this essay are my own.

quiet one" (Lichtenberg 1967-1972: I, 624), or that "the more fashionable part of the world does not begin to learn until you show them, through proof—which at the same time must be fun—that you can also learn science for fun, or in order to play with it" (Lichtenberg 1967-72: III, 315).

Exotic animals in princely menageries evoked an analogous discourse among their visitors. The French courtier and naturalist Emmanuel de Croÿ (1718-1784) pondered in his private journal about what the Versailles menagerie actually represented: "Was it entertainment or education?" (cited in Robbins 2002: 64). Later on, in the 19th century, the zoological gardens "viewed themselves very much within the rational recreation unit [...] entertainment could be, and was, synthesised with education" (Cowie 2014: 124).

Science popularization comes in many forms and types of media. This article emphasizes the performance of science popularization, in particular the public demonstration of natural phenomena aided by instruments and other devices. Yet the tension between entertainment and instruction is also omnipresent in popular science books, magazines, and lectures. Bernard Le Bovier de Fontenelle (1657-1757), often considered one of the first popularizers of science, writes in his long standing bestseller *Entretiens sur la pluralité des mondes* (1686): "it has been my design to amuse and instruct them [the readers] at the same time" (quoted from Nieto-Galan 2016: 32).

The "spectacularization" of science has been assigned by many historians of science to the 18th century or even earlier (Schaffer 1983; Bensaude-Vincent/Blondel 2008; Werrett 2010; Thébaud-Sorger 2015), while other scholars point to the 19th century (Morus 2007; Samida 2011a; Kember/Plunkett/Sullivan 2012a). The reconstruction and display of dinosaurs in museums and other spaces as a practice of spectacular science in the first third of the 20th century has been at the center of a number of recent books (Heumann et al. 2018; Nieuwland 2019; Rieppel 2019; see also the article by Manias 2016, and the earlier article by Secord 2004b).

From the late 20th century up to the present day, science centers have become the emblematic institutions for this blending of fun and instruction, seconded by terms such as curiosity, exploration, amazement, trying out, and discovery. Science centers promise to provide science and guarantee fun.[5] School reformer Thenn would have subscribed to this argument. In the word "edutainment" the two concepts of entertainment and education have been fused.

---

5 E.g., "Wissenschaft mit Spaßgarantie" (Science with a Guarantee of Fun), Bremen https://universum-bremen.de; "Vom Staunen zum Denken" (From Wonder to Insight), Phaeno Science Center, Wolfsburg https://www.phaeno.de/angebote/schulen.

ENTERTAINMENT VERSUS INSTRUCTION

In 1764, the members of the St. Anna School board voiced their doubts concerning Thenn's proposal to introduce experimental natural philosophy into the curriculum. Would it not be necessary to teach the pupils some basics in mathematics first? And is there not a danger that they would watch the experiments "without any reason or concept" ("ohne allen Verstand und Begriff") and walk out of class with as little understanding as before? (Hochadel 2003: 125f).

The skeptical reaction of the school board is the classical counter-argument against the incentive theory: Time and again natural philosophers scolded the popularizers (or other natural philosophers) for being too amusing, for providing too much fun instead of explanations and truth, for appealing to the senses instead of reason, for being in love of the spectacular instead of instructing their audience (examples: Franklin/Wilcke 1983 [1758]: XXXIV; Suckow 1761: prologue and 451; Kästner 1783: 361). If the subject matter is presented in an overly colorful guise, the actual goal will be missed and the audience will learn nothing. Those who only appeal to the senses will not reach the mind. Give the audience pleasure, certainly, but please be reasonable as well, the Enlightenment had already warned. And the 19th century followed these lines. Seeing is not enough was the criticism launched at the reconstructions of dinosaurs on display in the Crystal Palace in London in the 1850s. The models bore no explanations and only very few visitors bothered to buy a guide (Secord 2004b: 159). Zoological gardens tried from about the 1850s onwards to pass on some basic knowledge to their visitors through guides and labels put on the cages, only to routinely lament that these were being ignored. Zoo-goers simply preferred looking exclusively at the wild animals (Hochadel 2008; Cowie 2014: 115).

Where rational instruction ends and the mere spectacle begins is an open question for the 18th as well as the 19th centuries.[6] Morus (2010: 778) writes about the public culture of physics in Victorian London: "Scientific performances were and are calculated to appeal to the senses rather than the intellect, however they might be glossed otherwise." Simplification, trivialization, and—even worse—distortion are the accusations regularly levelled against such attempts to popularize. In the 20th century popular science programs on television seemed to embody, for the critics, all these evils in their most condensed form.

"The need to attract the largest possible audience pushed television's version of science, whether intended as education or fiction, even more toward sensationalism, politics, celeb-

---

6 Withers (2011) discusses this in the context of Victorian Britain.

rities, and representation and away from the discussion of ideas, away from the real, away from attention to the thought and reasoning behind scientific conclusions and recommendations." (LaFollette 2012: 229)[7]

Entertainment and education function like two poles forming a connection onto which the act of popularization can be placed. But where exactly? According to Lichtenberg the teacher needs to find the proper "middle way" (Lichtenberg 1967-72: III, 315; cf. Kempf 1991: 215). Lichtenberg even uses the metaphor of the sweet lure we know from antiquity.[8] In one of his early texts, the introduction to the *Göttingischen Beyträgen zum Nutzen und Vergnügen*, he analyzes if and how such a journal would be able to divulge useful knowledge. Lichtenberg says that it is a great art to blend the right amount of "sweets" into this kind of instruction. The sweet element should not be too miniscule, because then it would be too obvious that it is about instruction. But neither can it be too sizable, or the public will only go for the sugary part and discard the rest (Lichtenberg 1768; cf. Kempf 1991: 257; Hochadel 2003: 266).

French anthropologist Jean Louis Armand de Quatrefages (1810-1892) wrote a host of popular articles for the journal *Revue des deux mondes*. He reflects about his style: "In general, I have discarded overly technical details and dealt almost exclusively with general issues. I often tried to imitate the physician who uses honey to disguise the medicine whose taste would otherwise disgust the patient..." (Quatrefages 1854: xiii-xiv; also quoted in Drouin/Bensaude-Vincent 1996: 415). A timeless recipe, it seems.

The debate on how entertaining science has to be in order to be appealing to the general public is fought with practically identical arguments in different centuries. Michael Lynn (2006: 150) writes about popular science in the French Enlightenment: "The greatest problems arose when the balance between usefulness and entertainment tipped too strongly in favor of the latter." Bernard Lightman (2007b: 124) writes about the highly successful Victorian popularizer, Professor Pepper: "The combination of instruction and entertainment in public science proved during this period to be volatile and unstable. At what point did science lecturers like Pepper go too far in incorporating entertainment?" Or with respect to the multiplication of venues for popular science at the time: "The organizers of these sites were always playing with the balance between instruction and en-

7 For this specific tension between format and content inherent in science on TV, see Clack and Brittain (2007: 21) and the special issue edited by Florensa, Hochadel and Tabernero (2014).

8 See the introduction by Muñoz Morcillo in this volume.

tertainment in their efforts to produce a form of rational recreation..." (Fyfe/Lightman, 2007: 11). Iwan Morus Rhys wrote in a number of publications about the culture of the electrical spectacle in 19th-century Britain, in particular in London, highlighting the very same tension. One of his articles (Morus 2007) is even titled: "The Two Cultures of Electricity: Between Entertainment and Edification in Victorian Science."

In August 1900 a Danish journalist criticized the Copenhagen Zoo, which also put exotic human beings on display. "He recognised that the gardens combined education and entertainment but argued that a proper balance between the two elements had not been struck in this case" (Andersen/Hjermitslev 2009: 160). Similar criticisms had already been launched in the mid-19th century against the Regent's Park Zoo in London (Jones 1997: 13-15; Ito 2014: 133, 167). In 1880 the Viennese zoologist Friedrich Knauer asserted that "A modern zoo needs to combine instruction and entertainment"—yet in practice the latter took over. Ethnographic shows, highly popular in the later part of the 19th century, have also been described as hovering between science and entertainment (Goodall 2002: 83; Bruckner 2003).

The British Egyptologist Margaret Murray (1863-1963) tried hard to find the right balance between crowd-drawing spectacle and teaching when she unrolled mummies in museums in the early 20th century (Sheppard 2012). Peter Bowler analyzed the rhetoric of the British popular science book market in the first half of the 20th century. He writes:

"One of the most difficult problems facing authors and publishers was achieving the right balance between education and entertainment. If too serious, the material looked like a textbook and would put off those unwilling to tackle really serious study. If too populist in tone, it looked ephemeral and seemed to offer no prospect of genuine self-improvement." (Bowler 2009: 85)

Science centers have been criticized along similar lines, i.e., as only providing entertaining spectacle and very much focusing on the senses without imparting any profound, lasting knowledge. The debate remains open. In fact, little is truly known about what visitors actually learn (Unterstell 2012).

The edutainment of the 21st century is the descendant of the rational recreation of the Enlightenment. The German expression at the time was "Vernünftiges Vergnügen" (reasonable/rational fun). A contradictio in adjecto? This question might be misleading because there is no simple answer to it. In a sense, the argument of how to popularize science has not changed since the Enlightenment.

The tension between entertainment and instruction seems inherent in any kind of science popularization (see Nieto-Galan 2016: xv, 18, 32, 73).

The criticism that science popularization was too spectacular was directed not only against the actual practice but obviously also against the persona of the popularizer. Their credibility and motives were put into question, regardless of whether we look at the itinerant lecturers of the Enlightenment, the entrepreneurs of scientific spectacles in the 19th century (with a spectrum that includes figures such as Professor Pepper as well as P.T. Barnum), the producers of stunning underwater documentaries Hans Hass and Lotte Baierl in the 1950s, or the anatomist Gunther von Hagens and his travelling exhibition Body Worlds from our own time. These professional popularizers were routinely lambasted for their overtly commercial interest. What's worse, their critics maintained, instead of instructing their audiences, they supposedly spread erroneous ideas, distorted knowledge, ignorance, or even superstition. It is revealing to pay close attention to this policing of science popularizers and their precarious status (see Hochadel 2003, for the 18th century; and Lightman 2007a, for Victorian Britain). For example, in 1804 the scientific performer Étienne-Gaspard Robertson (1763-1837) was described as a "conjuror" ("Taschenspieler"), the exact opposite of a "physicist" (Hochadel 2006: 14-16). This may be understood as a boundary work (for this concept, see Gieryn 1983) that constructs the scientific showman as the contrast foil for the ideal scientist and the values he represents: disinterest in material gain, the pursuit of new knowledge and valid explanations, honesty, and modesty.

SCIENCE IS IN!

In 1789, the German physician and writer Christoph Wilhelm Hufeland summarizes the excitement and interest of the general public for the sciences. They "really have made themselves indispensable. And is there still a well-mannered circle in which one does not talk about elementary fire, magnetism, electricity, *Principe oxygène*, the ultimate causes of things...?" (emphasis in original). Even "the most serious sciences have now taken off their pedantic expression. Dressed in pleasing fashionable robes, they are not denied access to any ladies' gathering" (quoted in Hochadel 2003: 16f).

In France it seems that this was already the case in the 1730s. One of the most successful popularizers of his time was Abbé Nollet. Outside his *cabinet* [workshop] in Paris, where he taught natural philosophy through spectacular experiments, "the carriages of duchesses, couples, and pretty women" would clog the street, as Émilie du Châtelet joked in a letter to Francesco Algarotti in April

1736 (Besterman 1953-1977: V, 139, and see also IL, 139; Sutton 1995: 225; Hochadel 2003: 124f; Lynn 2006: 22). A Polish visitor to Paris by the name of Moszynskí wrote in 1785: "Physics and chemistry have become a sort of epidemic infecting everybody, including women" (Thébaud-Sorger 2015: 142).

To proceed to the 19th century: In his presidential address William Harcourt claimed at the annual meeting of the British Association for the Advancement of Science in 1839 that "the very amusements, as well as the conveniences of life, have taken a scientific colour" (quoted from Kember/Plunkett/Sullivan, 2012: 2). Just five years later, in 1844, the *London Polytechnic Magazine* observed with great satisfaction "that taste for physical philosophy... is becoming so general" (quoted from Morus 1998: 70).

Just like in the Enlightenment, "fashionable" is a term widely used to describe the interest of the general public for specific, often new fields in science. On 17 July 1874, the *Irish Times* even called Darwin's theory of evolution a "fashionable doctrine" (quoted from De Courcy 2010: 59). On 31 March 1877 the *Kölnische Zeitung* refers to the emerging field of anthropology as a "fashionable science" ("Mode Wissenschaft," quoted from Zimmerman 2001: 118).

To be sure, each of these quotes stem from entirely different contexts. Yet there is one interesting commonality: Many of these citations are meant to be tongue-in-cheek, ironic, critical, or even outright dismissive. The undertone says: It's *only* a fashion and will thus disappear as quickly as a new fashion trend in clothes. In its extreme variant, the etiquette "fashionable" is used to discredit a new scientific theory. German physicist Ernst Gehrcke, convinced that Einstein's rise to international fame in the public sphere in the early 1920s was based on mass delusion, called the relativization of space and time a temporary fashion ("Modeerscheinung," te Heesen 2006: 150). What is fashionable and appeals to the masses (and even to women!—this discourse is clearly gendered) immediately raises suspicion. What can a field of knowledge, a theory, an experiment, or a book be worth if it is so popular? Being "in" is always a double-edged sword. It calls for scientific policemen who patrol the boundaries of true science, protecting it from epistemologically and morally dubious intruders from the popular sphere. We shall come back to this "boundary work" later on.

SCIENCE IS OUT!

The trope that science has become fashionable among the general public has its flipside, too: the increasing incomprehensibility of science and the alleged retreat of the scientist into the ivory tower. This growing gap between science and society is another continuity in the history of science popularization.[9]

The argument goes like this: For a while it was fashionable to talk about electricity, comets, or even mathematics at social gatherings (e.g., Terrall 2002: 32-34). Yet the specialization of the disciplines that went hand-in-hand with the professionalization of the practitioners of science made these topics increasingly incompatible with the demands of polite conversation. These topics required specialist knowledge and created an uncomfortable gap between experts and non-experts; they were thus banned from the salon.

In the preceding sentences I have paraphrased the argument of Steven Shapin and James Secord, two eminent scholars who made structurally similar arguments for two different centuries.

> "The divorce between 'the two cultures' was by no means irrevocable by the end of the eighteenth century... but the withdrawal of the man of science from the general conversation was then well under way. One could not have a polite conversation with an author one could not understand; you could only be lectured at." (Shapin 2003: 173)

Secord (2007: 52) makes the same case for the late 19th century:

> "To discuss the electron theory or fossil botany at ordinary dinners and soirees, other than when these subjects were in the news, was (with rare exceptions) seen to be talking shop."

These are topical arguments, and Secord (2007: 51) himself reflects on their recurrence: "The world of the generalist is always the world we have lost."

In the 20th century this argument evolves into the two-culture debate. A German author (Anonymous 1904) laments in the journal *Kosmos* that everybody is supposed to know about literature and music but that this is not the case for the functioning of a thermometer or an electrical tram (quoted from Schirrmacher 2007: 53). As is well known, the reference point of this debate is the 1959 Rede Lecture. In "The Two Cultures" C.P. Snow describes a deep rift between scientific and literary culture. It is curious that some of Snow's examples are—once more—from dinner parties. One of the anecdotes is from a dinner at one of the colleges in Cambridge in the 1890s and attributed to the historian A.L. Smith (1850-1924). It ends with the phrase: "Oh, those are mathematicians! We never talk to *them*" (Snow 1959: 2; emphasis in original). With respect to his own time, Snow laments that "highly educated members of the non-scientific

9 On the ideology of term "gap" and the attempts to bridge it, see Felt 2000: 26-30.

culture couldn't cope with the simplest concepts of pure science.... What is a machine-tool? I once asked a literary party; and they looked shifty." (Snow 1959: 15)

We can still encounter an analogous argument in the late 20th and early 21st centuries. The cognitive psychologist and well-known popular science writer Steven Pinker lamented in the 1990s:

> "In a gathering of today's elite, it is perfectly acceptable to laugh that you barely passed Physics for Poets... and have remained ignorant of science ever since... But saying that you have never heard of James Joyce... is... shocking..." (quoted from Adams 2007: 140)

The renowned Austrian quantum physicist Anton Zeilinger put forward an Austrian variation of virtually the same argument: "Not to know who Schrödinger is might even be cool. Not to know who Mozart is is considered embarrassing." (Zeilinger 2000: 9; see also Zeilinger 2005: 239)

The *longue durée* perspective seems to indicate that the supposed gulf between the general public (often represented by its bourgeois elites who love to mingle at dinner parties) and scientists was already opened up in the Enlightenment. This abyss has been "rediscovered" time and again by historians of science (in a more analytic tone) as well as by popularizers—scientists such as Anton Zeilinger and scientifically trained writers such as C.P. Snow—in a more exhortative tone. To build bridges over this "gulf" has become a major argument in the legitimation of science popularization.

These two discourses—"science is in" and "science is out"—appear and reappear time and again. They lead an antagonistic and dialectical relationship. They point to a similar pair of opposed concepts: comprehensibility and incomprehensibility. Yet the pairs in/out and intelligible/unintelligible do not always correspond: Newton's highly mathematical theory of the *Principia* was successfully popularized by mechanical demonstrations (Stewart 1992), yet Einstein's highly mathematical theory of relativity owed much of its pubic appeal to its alleged incomprehensibility.

COMPETING FOR AUDIENCES—SCIENCE IN THE MARKETPLACE

The historiography on the popularization of science has highlighted the central role of the audience, in a number of respects. Scientific popularization often serves a legitimizing function (see Shinn/Cloître 1985: 59; Weingart 2001: Ch. 6; and the following case studies: Hochadel 2013; Carandell 2013; Florensa 2016). Appealing directly to the public has become an important strategy since the 18th century in order to defend or attack a controversial theory,[10] or to construct a new academic discipline—e.g., ethology in the 1930s (Taschwer/Föger 2003) or evolutionary psychology in the 1990s (Cassidy 2006).

Historians of science have time and again highlighted the diversity of publics in terms of class, academic training, gender, age, and other parameters (Cooter/Pumfrey 1994; case studies by Terrall 2002: 32; and Adelman 2009). In the Enlightenment polite society was crucial in granting science the status of a respectable activity (e.g., Sutton 1995; Walters 1997; Terrall 2002).

There is a broad consensus among historians of science that these publics are not mere passive receivers of knowledge that is being handed down to them. Rather, they exert a strong influence on how science popularization is practiced, as judges of the validity of scientific claims but also as clients paying for demonstrations, books and magazines, instruments, toys, and other commodities.

Scholars of Victorian popular science have coined the term of "Science in the marketplace" (Fyfe/Lightman, 2007). They convincingly show that popularizers are competing for both the attention of their audiences and their money. The concept of the marketplace helps us understand Victorian popular science much better. "In the nineteenth century, perhaps especially, science had to compete as a form of cultural good in a competitive marketplace as perhaps never before. The result was, often, that 'popular' science was more 'popular' than scientific." (Withers, 2011: 102)

The British case did not differ much from the German one. Nyhart (2009: 219) said the following about natural history in 19th-century Germany: "Anyone seeking to broaden the audience for natural history museums had to acknowledge stiff, even overwhelming competition for the attention of the casual visitor." Yet when searching for continuities in the history of science popularization, we might say the same about the previous century. The fierce competition of lecturers for paying audiences in the mid- and late 18th century has been described with regards to Paris (Lynn 2006: 51-55), to London and England as a whole (Stewart 1992; Schaffer 1993; Money 1993), to the Netherlands (Roberts 1999: 696-699), and to the German lands (Hochadel 2003: 210-214). The enormous public interest in balloons from 1783 onwards is another instructive example of popular science becoming a booming business (Keen, 2006; Lynn, 2010). For these scholars, advertising, consumption, commodification, and the marketplace are key terms for describing the dynamics of science popularization in the Enlightenment. Larry Stewart, for example, writes: "Experimental natural phi-

10 Cosmology seems to be a point in case: see, e.g., Wessely 2013, on Hanns Hörbiger's World Ice Theory; Nieto-Galan 2016: 177-178, on James Lovelock's Gaia theory; Edford 2007, on Brian Greene and superstring theory.

losophy would become a commodity to be traded in the marketplace." (1992: 97; see also xxiv-xxv, 128) While acknowledging the substantial differences with respect to the emerging mass society of the second half of the 19th century in particular, there is sufficient evidence to talk about a market for science popularization as early as the 18th century.

THE MEDIALIZATION OF SCIENCE

Capturing the attention of the public is crucial for popularizers, as we have seen. Therefore, it is important to analyze the role of the media. The sociologist of science Peter Weingart has coined the term "medialization of science." He is referring to the "increasing importance of the media for scientific communication" and the "instrumentalization of the media through science in order to deal with priority conflicts and the mobilization of public support" (Weingart 2001: 244). Weingart asserts "a new relationship between science and the media" (2001: 232; also see Weingart 2012). More recently, he and his colleagues used the term "the sciences' media connection" (Rödder/Franzen/Weingart 2012).

Instructive and well-known cases of the medialization of science include: the debate about the extinction of the dinosaurs supposedly caused by the impact of an asteroid, which began in the 1980s (Clemens 1986); the cold fusion saga of 1989 (Lewenstein 1995); and the discovery of the "hobbit" in paleoanthropology in 2004 (Goulden 2013). The highly medialized research on ancient DNA is also hard to understand without this concept (Dobson Jones 2017).

Yet some of the features of the "science-media connection" seem to be a far less recent phenomenon. Again, I would argue that the medialization of science goes way back into the 19th century. The debate about the theory of evolution is the most obvious example (Secord 2003 [2000]). Fields that captured the public imagination, such as prehistory, were debated widely in the press. Researchers very consciously entered the public arena in order to make their claims (Hochadel/Carandell/Florensa 2016). For the very same reason, classical archaeology (Samida 2011b), earthquake research (Coen 2013: ch. 3), polar research (Schimanski and Spring 2015), and experimental physiology (Nieto-Galan 2015) provide further examples for the medialization of science *avant la lettre*.

Moreover, in the early and mid-20th century scientists deliberately used the media to put forward their often-controversial ideas, seeking legitimization for those ideas. Instructive examples are astronomy (Halley's comet of 1910; Simões et al. 2013), paleoanthropology once more (the Steinau hoax of 1911; Hochadel 2016), neurology (how Oscar Vogt made the brain a public subject in the 1920s; Hagner 2002), and oceanography (the science-media cooperation on the Danish Galathea expedition of the early 1950s; Nielsen 2009).

One might even go out on a limb and search for cases of the medialization of science in the 18th century. True, there were no mass media at the time and literacy was low, but scientific controversies often spilled out into the open. During the Enlightenment scholars deliberately entered the public sphere and tried to win over public opinion for their specific claims. They strategically used media (broadly conceived) such as newspapers, weekly journals, brochures, and pamphlets.

The introduction of lightning rods into the public arena provides some instructive examples. The Purfleet controversy in London in the 1760s and 1770s about the best shape for the top of the lightning rod (Home 2009) and the European-wide debate on the source of the dry fog of the summer of 1783 (Hochadel 2009) were fought out largely in the media *sensu lato* of the time. The debate about mesmerism in Paris in the 1780s might be another example (see Darnton 1968). Although this issue needs more investigation, it seems that the medialization of science—trying to win over the support of the public by addressing them through print—does have a historical dimension dating back at least to the Enlightenment.

MORALIZING THE PUBLICS

Historians of science agree: The popularization of science is never ideologically neutral. More or less explicitly, there is always some kind of social, political, or moral agenda associated with it. It is not only about the mere content, the bare scientific facts. In myriad ways, popularization is steeped in ideologies and value systems.

To cite a typical example from the Enlightenment, the German writer C.F.D. Schubart hails in his *Deutsche Chronik* (3. St., 7.4.1774) the soap-boiler Karas: "Instead of filling his free hours with gambling, drinking, and vulgar babbling (as do most artisans), he occupied himself with mathematics, physics, language, music" (reprinted in Schubart, 1989: 8f; also see Hochadel, 2003: 108). It seems that what kind of knowledge the common man imbibes is only of secondary importance. What really matters is spending one's time properly—dedicating oneself to personal improvement and not succumbing to the omnipresent vices.

The very same argument is put forward countless times in the 19th century. It is articulated with a greater sense of urgency given the rapidly growing working class and its endangered morals. An important part of the *raison d'être* of a number of new urban institutions is to provide sound and rational recreation for the workers, a mixture of leisure and instruction: This is the case for public parks (Conway 1991: 29-35 on Great Britain; Taylor 1999 on the United States), zoological gardens (Wessely 2008: 66-73 on Berlin and Vienna; Ito 2014 on London; Hochadel/Valls 2016 on Barcelona; Vasta/Buzzi 2013 and Vasta 2018: 169 on Buenos Aires), natural history museums (Bennett 1995; Hill 2005, on museums in Great Britain; Köstering 2003: 259-263 on German natural history museums; Rieppel 2012: 461 and Rader/Cain, 2014 on US natural history museums), and even itinerant menageries (Cowie 2014: 66). The rich culture of popular natural history in the 19th century needs to be understood as a civic education for the masses, often still with religious overtones. Observing nature and its creatures attentively is considered to sharpen the mind of the budding naturalist and to protect against superstition (Nyhart 2009: 185). The enormously successful Viennese Volksbildung (university extension) movement around 1900 wanted to use the popularization of science to teach its thousands of students how to think critically and make rational judgments (Taschwer 2002).

It is thus important to note that the values transmitted through science popularization are not necessarily bourgeois (or, in Gramscian terms, hegemonic; see Nieto-Galan 2011). The ideology in which the knowledge is suffused may differ widely, varying between patronizing and emancipatory, authoritarian and democratic. For example, in the early 20th century, the anarchist movement in Barcelona had its own agenda of popularizing science and medicine, denying any kind of authority of the academically trained (Girón/Molero-Mesa 2016). In the Weimar Republic, German Socialists and Communists had developed their own ideas and practices for popularizing science (Hopwood 1996). Observing the stars with a telescope might provide comfort in troubled times of war and authoritarian rule, as recent studies on amateur astronomy during the Franco dictatorship suggest (Ruiz-Castell 2016). Not to mention the more obvious uses of science popularization in the processes of nation building and the formation of national identities in the 19th and 20th centuries. To repeat: The driving ideology or value system may vary, yet science popularization has clearly been conceived of as a cure for social ailments and thus as a mean of social control from at least the 18th century onwards (Drouin/Bensaude-Vincent 1996: 420).

CONCLUSIONS

The *longue durée* perspective we have taken in this article entails, to repeat this point once more, its own kind of problems, especially: a lack of historicization and a somewhat indiscriminate lumping together of cases. Yet the insights this approach promises might outweigh its shortcomings, both on a practical level, i.e., with respect to public debates about the importance of science popularization, as well as on a more scholarly level, i.e., with respect to the numerous but somehow unconnected works of scholarship on the history of science popularization.

On a practical level, the continuities we have identified in the last three centuries might help deconstruct a number of recurrent tropes in our current debates on science popularization. As we have seen, the tension between entertainment and instruction dates back to at least the Enlightenment. When we discuss the pros and cons of edutainment (for an example in archaeology, see Clack/Brittain 2007) and what form it should take, we might benefit from the historical dimension that this article has tried to provide. The diagnosis of a "boom" of popular science diagnosed since the early 2000s (fancy magazines, new television formats including reality TV, the "reinvention" of the science museum, the use of social media for science communication, etc.) should be met with a healthy dose of skepticism. We should rather ask: When did scientists *not* turn to the public in order to communicate their findings and plead for public support (including money for research!) and the social legitimacy of their research agenda?

A conceptual evergreen in the public discourse is the scientist in the ivory tower, cooped up in his or her research and aloof from all worldly matters. This image has a long history too. Peter Bowler (2009: 4), looking at Britain in the first half of the 20th century, named it the myth of the isolated professional. I would venture to say that scientists never lived in an ivory tower. This image is rather a rhetorical device that intends to mobilize them: Please come down from your lofty towers! Take a stand on key issues that concern society as a whole! Explain your own work! One recent example is the public debate on GMOs (genetically modified organisms). Since the turn of the millennium and under the threat of having their research budgets cut, biologists had to speak out and defend their research, facing a public worried about the possibly damaging effects of GMOs on the environment and on our food (see, e.g., the Swiss case: Dubochet/Narby/Kiefer 1997).

The scholar in the ivory tower finds her/his opposite in the media-savvy or even media-hungry scientist who knows how to use or manipulate public opinion to foster his or her aims and claims. We should of course keep a critical eye on this kind of self-promotion (see e.g., Kjærgaard 2011), yet at the same time it seems helpful to put the figure of the attention-craving scientist in historical perspective—this is by no means an entirely new phenomenon.

And to end with some remarks addressed to historians of science: Through its structure, this article has tried to group the vast material into opposing tropes: education versus entertainment, "science is in" versus "science is out," and enticing and policing the audiences at the same time. Yet I would like to insist that these opposites are more than just a narrative device to organize the material. The history of popularizing science seems to be characterized by a constant tension between form and content, between the opening and closing of the scientific community towards the public, between comprehensibility and incomprehensibility, between offering attractively packaged knowledge and controlling the message, between promising fun and trying to instill virtues. The active search for the epistemological legitimation and social acceptance of science (and specific claims) through popularization is always accompanied by a boundary work that tries to define and enforce what may count as "proper" science popularization and who should be entitled to popularize.

This essay has merely tried to overcome a temporal myopia and distill some revealing continuities in the history of science popularization. Obviously, much more work is needed in order to draw out, color, and refine the lines sketched here. Further research, I would argue, would have to keep in mind the big picture in a principled way. Only then, by looking at a large period of time, be it three centuries or an even greater span, as well as by covering the largest possible range of regions and cultures (admittedly, this article has been quite limited in its geographical scope), will the continuities in the history of science popularization emerge further. And these—shall we say, underexplored—continuities should be as relevant to the historian of science as the rather well-known differences among the seemingly infinite number of case studies.

REFERENCES

Adams, Jon (2007): Interference Patterns: Literary Study, Scientific Knowledge, and Disciplinary Autonomy, Lewisburg: Bucknell University Press.

Adelman, Juliana (2009): "Animal Knowledge. Zoology and Classification in Nineteenth-Century Dublin." In: Field Day Review 5, pp. 109-121.

Andersen, Casper/Hjermitslev, Hans H. (2009): "Directing Public Interest: Danish Newspaper Science 1900-1903." In: Centaurus 51/2, pp. 143-167.

Anonymous (1904): "Moderne Bildung." In: Kosmos 1, pp. 1-2.

Ash, Mitchell G./Stifter, Christian H. (eds.) (2002): Wissenschaft, Politik und Öffentlichkeit. Von der Wiener Moderne bis zur Gegenwart, Vienna: WUV.

Bennett, Tony (1995): The Birth of the Museum: History, Theory, Politics, London and New York: Routledge.

Bensaude-Vincent, Bernadette (2009): "A Historical Perspective on Science and its 'Others.'" In: Isis 100/2, pp. 359-368.

Bensaude-Vincent, Bernadette/Rasmussen, Anne (eds.) (1997): La science populaire dans la presse et l'édition. XIXe et XXe siècles, CNRS Histoire, 1251-4357, Paris: CNRS Editions.

Besterman, Theodore (ed.) (1953-1977): Voltaire's Correspondence, 107 Vols., Geneva: Institut et Musée Voltaire.

Bishop, Rebecca (2017): "Evolution im Zoo: Die Veranschaulichung einer Theorie am lebenden Objekt." In: A. Schwarz (ed.), Streitfall Evolution. Eine Kulturgeschichte, Cologne, Vienna and Weimar: Böhlau, pp. 103-124.

Bowler, Peter J. (2009): Science for All: The Popularization of Science in Early Twentieth-Century Britain, Chicago: The University of Chicago Press.

Broks, Peter (2006): Understanding Popular Science. In: Issues in Cultural and Media Studies, New York: Open University Press.

Bruckner, Sierra Ann (2003): "Spectacles of Human Nature: Commercial Ethnography between Leisure, Learning, and *Schaulust*." In: H. G. Perry/Matti Bunzl (eds.), Worldly Provincialism. German Anthropology in the Age of Empire, Ann Arbor: The University of Michigan Press, pp. 127-155.

Carandell, Miquel (2013): "Newspapers as the Arena of Scientific Controversy: The Debate about the 'Orce Man' in Spanish Mass-Media." In: Lorraine Locke/Simon Locke (eds.), Knowledges in Publics: Beyond Deficit, Engagement and Transfer, Cambridge, UK: Cambridge Scholars Publishing, pp. 150-170.

Cassidy, Angela (2006): "Evolutionary psychology as public science and boundary work." In: Public Understanding of Science 15/2, pp. 175-205.

Clack, Tim/Brittain, Marcus (2007): "1. Introduction: Archaeology and the Media." In: T. Clack/M. Brittain (eds.), Archaeology and the Media, Walnut Creek: Left Coast Press, pp. 11-65.

Clemens, Elisabeth S. (1986): "Of asteroids and dinosaurs. The role of the press in the shaping of scientific debate." In: Social Studies of Science 16/3, pp. 421-456.

Coen, Deborah R. (2013): The Earthquake Observers: Disaster Science from Lisbon to Richter, Chicago and London: The University of Chicago Press.

Conway, Hazel (1991): People's Parks: The Design and Development of Victorian Parks in Britain, Cambridge, UK: Cambridge University Press.

Cooter, Roger/Pumfrey, Stephen (1994): "Separate spheres and public places. Reflections on the history of science popularization and science in popular culture." In: History of Science 32, pp. 237-267.

Courcy, Catherine de (2010): Dublin Zoo: An Illustrated History, Cork: Collins.

Cowie, Helen (2014): Exhibiting Animals in Nineteenth-Century Britain: Empathy, Education, Entertainment, Basingstoke: Palgrave Macmillan.

Darnton, Robert (1968): Mesmerism and the End of the Enlightenment in France, Cambridge, MA: Harvard University Press.

Daum, Andreas W. (2002 [1998]): Wissenschaftspopularisierung im 19. Jahrhundert. Bürgerliche Kultur, naturwissenschaftliche Bildung und die deutsche Öffentlichkeit 1848-1914, 2nd ed., Munich: Oldenbourg.

Daum, Andreas W. (2009): "Varieties of popular science and the transformations of public knowledge. Some historical reflections." In: Isis 100/2, pp. 319-332.

Dobson Jones, Elizabeth (2017): The Search for Ancient DNA in the Media Limelight: A Case Study of Celebrity Science, PhD thesis, University College London.

Drouin, Jean-Marc/Bensaude-Vincent, Bernadette (1996): "Nature for the People." In: Nicholas Jardine/James A. Secord/Emma C. Spary (eds.), Cultures of Natural History, Chicago: Chicago University Press, pp. 408-425.

Dubochet, Jacques/Narby, Jeremy/Kiefer, Bertra (1997): L'ADN devant le souverain. Science, démocratie et génie génétique, Geneva: Editions Médecine et Hygiène, Département Livre Georg.

Edford, Rarchel (2007): "The Elegance of The Elegant Universe: Unity, Beauty, and Harmony in Brian Greene's Popularization of Superstring Theory." In: Public Understanding of Science 16/4, pp. 441-454.

Egmond, Florike/Mason, Peter (1996): "Skeletons on Show: Learned Entertainment and Popular Knowledge." In: History Workshop Journal 41/Spring, pp. 92-116.

Felt, Ulrike (2000): "Why Should the Public 'Understand' Science? Some Aspects of Public Understanding of Science from a Historical Perspective." In: Meinolf Dierkes and Claudia von Grote (eds.), Between Understanding and Trust. The Public, Science and Technology, Amsterdam: Harwood Academic Publishers, pp. 7-38.

Florensa, Clara/Hochadel, Oliver/Tabernero, Carlos (eds.) (2014): Science on Television, Actes d'Història de la Ciència i de la Tècnica (Nova Època) 7 (Special Issue).

Franklin, Benjamin (1983 [1758]): Briefe von der Elektricität, transl. Johann Carl Wilcke, Braunschweig: Vieweg.

Fyfe, Aileen (2008): "Tracts, classics and brands: science for children in the Nineteenth Century." In: Julia Briggs, Dennis Butts, and M. Grenby (eds.), Popular Children's Literature in Britain, 1700 1900, Aldershot: Ashgate, pp. 209-228.

Fyfe, Aileen/Lightman, Bernard (eds.) (2007): Science in the Marketplace. Nineteenth-Century Sites and Experiences, Chicago: The University of Chicago Press.

Gavroglu, Kostas (2012): "Science popularization, hegemonic ideology and commercialized science." In: Journal of History of Science and Technology 6, pp. 85-99.

Gieryn, Thomas F. (1983): "Boundary-Work and the Demarcation of Science from Non-Science: Strains and Interest Professional Ideologies of Scientists." In: American Sociological Review 48/6, pp. 781-795.

Girón, Álavaro/Molero-Mesa, Jorge (2016): "La Rosa de Foc: Anarchist culture, urban spaces, and the management of scientific knowledge in a divided city." In: O. Hochadel/A. Nieto-Galan (eds.), Barcelona: An Urban History of Science and Modernity, 1888-1929, London: Routledge, pp. 115-135.

Golinski, Jan V. (1992): Science as Public Culture: Chemistry and Enlightenment in Britain, 1760-1820, Cambridge and New York: Cambridge University Press.

González-Silva, Matiana/Herran, Néstor (2009): "Ideology, elitism and social commitment: Alternative images of science in two *fin de siècle* Barcelona newspapers." In: Centaurus 51/2, pp. 97-115.

Goodall, Jane R. (2002): Performance and Evolution in the Age of Darwin: Out of the Natural Order, London: Routledge.

Goulden, Murray (2013): "Hobbits, Hunters and Hydrology: Images of a 'Missing Link', and its Scientific Communication." In: Public Understanding of Science 22/5, pp. 575-589.

Hagner, Michael (2002): "'Mit der Tagespresse sich zu befassen, ist nicht jedermanns Geschmack.' Oskar Vogt und die Öffentlichkeit des Gehirns." In: A. te Heesen (ed.), Cut and paste um 1900. Der Zeitungsausschnitt in den Wissenschaften, Berlin: Vice Versa, pp. 116-130.

Heumann, Ina/Stoecker, Holger/Tamborini, Marco/Vennen, Mareike (eds.) (2018): Dinosaurierfragmente: Zur Geschichte der Tendaguru-Expedition und ihrer Objekte, 1906-2017, Göttingen: Wallstein.

Hilgartner, Stephen (1990): "The dominant view of popularization. Conceptual problems, political uses." In: Social Studies of Science 20/3, pp. 519-539.

Hill, Kate (2005): Culture and Class in English Public Museums, 1850-1914, Burlington, VT: Routledge.

Hochadel, Oliver (2003): Öffentliche Wissenschaft. Elektrizität in der deutschen Aufklärung, Göttingen: Wallstein.

Hochadel, Oliver (2006): "Zauberhafte Aufklärung. Etienne-Gaspard Robertson zwischen Schaustellerei und Wissenschaft." In: Brigitte Felderer/Ernst Strouhal (eds.), Rare Künste. Zur Kultur- und Mediengeschichte der Zauberkunst; Vienna: Springer, pp. 3-19.

Hochadel, Oliver (2008): "Der Tiger als Hyäne. Tierschilder im Zoo." In: Thomas Brandstetter/Dirk Rupnow/Christina Wessely (eds.), Sachunterricht. Fundstücke aus der Wissenschaftsgeschichte. Festschrift für Mitchell Ash, Vienna: Löcker, pp. 248-254.

Hochadel, Oliver (2009): "'In nebula nebulorum'. The Foggy Summer of 1783 and the Introduction of Lightning Rods in the German Empire." In: P. Heering/O. Hochadel/D. Rhees (eds.), Playing with Fire. Histories of the Lightning Rod; Philadelphia: American Philosophical Society, pp. 45-70.

Hochadel, Oliver (2013): El mito de Atapuerca. Orígenes, ciencia, divulgación, Bellaterra: Edicions UAB.

Hochadel, Oliver/Carandell, M./Florensa, C. (eds.) (2016): Scoops, Scams and Scuffles. The Construction of Prehistoric Knowledge in Newspapers, Centaurus (Special Issue), 58.

Hochadel, Oliver/Valls, Laura (2016): "Civic Nature. The transformation of the Parc de la Ciutadella into a space for popular science (1872-1917)." In: O.

Hochadel/A. Nieto-Galan (eds.), Barcelona: An Urban History of Science and Modernity, 1888-1929, London/New York: Routledge, pp. 25-45.

Home, R.W. (2009): "Points or Knobs: Lightning Rods and the Basis of Decision Making in Late Eighteenth Century British Science." In: P. Heering/O. Hochadel/D. Rhees (eds.), Playing with Fire. Histories of the Lightning Rod, Philadelphia: American Philosophical Society, pp. 98-120.

Hopwood, Nick (1996): "Producing a socialist popular science in the Weimar Republic." In: History Workshop Journal 41/1, pp. 117-153.

Ito, Takashi (2014): London Zoo and the Victorians, 1828-1859, Woodbridge: Boydell.

Jones, Robert W. (1997): "'The sight of creatures strange to our clime': London Zoo and the consumption of the exotic." In: Journal of Victorian Culture 2/1, pp. 1-26.

Kästner, Abraham Gotthelf (1783): Vermischte Schriften, 2nd ed., Altenburg.

Keen, Paul (2006): "The 'Balloonomania': Science and Spectacle in 1780s England." In: Eighteenth-Century Studies 39/4, pp. 507-535.

Kember, Joe/Plunkett, John/Sullivan, Jill A. (eds.) (2012a): "Introduction." In: Kember/Plunkett/Sullivan 2012b.

Kember, Joe/Plunkett, John/Sullivan, Jill A. (eds.) (2012b): Popular Exhibitions, Science and Showmanship, 1840-1910, London: Pickering & Chatto.

Kempf, Thomas (1991): Aufklärung als Disziplinierung. Studien zum Diskurs des Wissens in Intelligenzblättern und gelehrten Beilagen der zweiten Hälfte des 18. Jahrhunderts, Munich: Iudicium.

Kjærgaard, Peter C. (2011): "Ida and Ardi: The fossil cover girls of 2009." In: The Evolutionary Review 2, pp. 1-9.

Knauer, Friedrich (1880): "Ueber Thiergärten I. Der Thiergarten einst und heute," Deutsche Zeitung, 23.10., p. 4.

Köstering, Susanne (2003): Natur zum Anschauen. Das Naturkundemuseum des deutschen Kaiserreichs 1871-1914, Cologne: Böhlau.

Kretschmann, Carsten (ed.) (2003), Wissenspopularisierung. Konzepte der Wissensverbreitung im Wandel, Berlin: Akademie Verlag.

Kretschmann, Carsten (ed.) (2006): Räume öffnen sich. Naturhistorische Museen im Deutschland des 19. Jahrhunderts, Berlin: Akademia-Verlag.

LaFollette, Marcel Chotkowski (2012): Science on American Television. A History, Chicago: The University of Chicago Press.

Lewenstein, Bruce V. (1995): "From Fax to Facts: Science Communication in the Cold Fusion Saga." In: Social Studies of Science 25/3, pp. 403-436.

Lichtenberg, Georg Christoph (1768): "Ueber die Wochenblätter überhaupt, als eine Einleitung zu dem Göttingischen." In: Göttingische Gelehrte Beyträge zum Nutzen und Vergnügen, 1, April 6.

Lichtenberg, Georg Christoph (1967-72): Schriften und Briefe, ed. Wolfgang Promies, 4 Vols., Munich: Hanser.

Lightman, Bernard (2007a): Victorian Popularizers of Science. Designing Nature for New Audiences, Chicago and London: The University of Chicago Press.

Lightman, Bernard (2007b): "Lecturing in the Spatial Economy of Science." In: Fyfe/Lightman 2007, pp. 97-132.

Lynn, Michael R. (2006): Popular Science and Public Opinion in Eighteenth-Century France, Studies in Early Modern European History, Manchester: Manchester University Press.

Lynn, Michael R. (2010): The Sublime Invention: Ballooning in Europe, 1783-1820, London: Pickering & Chatto.

Manias, Chris (2016): "The Lost Worlds of Messmore & Damon: Science, Spectacle & Prehistoric Monsters in early-twentieth century America." In: Endeavour 40/3, pp. 163-177.

Money, John (1993): "Teaching in the Market Place, or 'Caesar adsum jam forte: Pompey aderat': The retailing of knowledge in Provincial England during the Eighteenth Century." In: John Brewer/Roy Porter (eds.), Consumption and the World of Goods, London: Routledge, pp. 335-380.

Morus, Iwan Rhys (1998): Frankenstein's Children: Electricity, Exhibition, and Experiment in Early-Nineteenth-Century London, Princeton: Princeton University Press.

Morus, Iwan Rhys (2007): "'More the aspect of magic than anything natural': The philosophy of demonstration." In: Fyfe/Lightman 2007, pp. 336-370.

Morus, Iwan Rhys (2010): "Placing Performance." In: Isis 101/4, pp. 775-778.

Myers, Greg (1989): "Science for Women and Children: The Dialogue of Popular Science in the Nineteenth Century." In: John Christie/Sally Shuttleworth (eds.), Nature Transfigured. Science and Literature, 1700-1900, Manchester and New York: Manchester University Press, pp. 171-200.

Nielsen, Kristian Hvidtfelt (2009): "In quest of publicity: the science-media partnership of the Galathea Deep Sea Expedition from 1950 to 1952." In: Public Understanding of Science 18/4, pp. 464-480.

Nieto-Galan, Agusti (2011): "Antonio Gramsci Revisited: Historians of Science, Intellectuals, and the Struggle for Hegemony." In: History of Science 49/4, pp. 453-478.

Nieto-Galan, Agusti (2015): "Mr Giovanni Succi Meets Dr Luigi Luciani in Florence. Hunger Artists and Experimental Physiology in the Late Nineteenth Century." In: Social History of Medicine 28/1, pp. 64-81.

Nieto-Galan, Agusti (2016): Science in the Public Sphere: A History of Lay Knowledge and Expertise, New York: Routledge.

Nieuwland, Ilja (2019): American Dinosaur Abroad: A Cultural History of Carnegie's Plaster Diplodocus, Pittsburgh: University of Pittsburgh Press.

Nikolow, Sybilla/Schirrmacher, Arne (eds.) (2007): Wissenschaft und Öffentlichkeit als Ressourcen füreinander. Studien zur Wissenschaftsgeschichte im 20. Jahrhundert, Frankfurt a.M.: Campus.

Nyhart, Lynn K. (2009): Modern Nature: The Rise of the Biological Perspective in Germany, Chicago and London: The University of Chicago Press.

Quatrefages, Armand de (1854): Souvenirs d'un naturaliste, I, Paris: Charpentier.

Rader, Karen A./Cain, Victoria E. M. (2014): Life on Display. Revolutionizing U.S. Museums of Science and Natural History in the Twentieth Century, Chicago: The University of Chicago Press.

Rieppel, Lukas (2012): "Bringing Dinosaurs Back to Life: Exhibiting Prehistory at the American Museum of Natural History." In: Isis 103/3, pp. 460-490.

Rieppel, Lukas (2019): Assembling the Dinosaur. Fossil Hunters, Tycoons, and the Making of a Spectacle, Cambridge, Mass.: Harvard University Press.

Robbins, Louise E. (2002): Elephant Slaves and Pampered Parrots: Exotic Animals in Eighteenth-Century Paris, ed. Harriet Ritvo, Animals, History, Culture, Baltimore/London: Johns Hopkins University Press.

Roberts, Lissa L. (1999): "Science becomes Electric: Dutch Interaction with the Electrical Machine during the Eighteenth Century." In: Isis 90/4, pp. 680-714.

Rödder, Simone/Franzen, Martina/Weingart, Peter (eds.) (2012): The Sciences' Media Connection – Public Communication and its Repercussions, Dordrecht: Springer.

Ruiz-Castell, Pedro (2016): "Making telescopes and partying with the stars: Amateur astronomy in Spain during Franco's dictatorship." In: Journal for the History of Astronomy 47/2, pp. 194-209.

Samida, Stefanie (ed.) (2011a): Inszenierte Wissenschaft. Zur Popularisierung von Wissen im 19. Jahrhundert, Bielefeld: transcript.

Samida, Stefanie (2011b): "Archäologische Berichterstattung in der Presse des 19. Jahrhunderts. Ein Plädoyer für medienhistorische Forschungen." In: Saeculum 61/2, pp. 283-304.

Schaffer, Simon (1983): "Natural Philosophy and Public Spectacle in the Eighteenth Century." In: History of Science 21, pp. 1-43.

Schaffer, Simon (1993): "The consuming flame: Electrical showmen and Tory mystics in the world of goods." In: John Brewer/Roy Porter (eds.), Consumption and the World of Goods, London: Routledge, pp. 488-526.

Schimanski, Johan/Spring, Ulrike (2015): Passagiere des Eises. Polarhelden und arktische Diskurse 1874, Vienna and Cologne: Böhlau.

Schirrmacher, Arne (2007): "Der lange Weg zum neuen Bild des Atoms." In: S. Nikolow/A. Schirrmacher (eds.), Wissenschaft und Öffentlichkeit als Ressourcen füreinander. Studien zur Wissenschaftsgeschichte im 20. Jahrhundert, Frankfurt a.M.: Campus, pp. 39-73.

Schirrmacher, Arne (2008): "Nach der Popularisierung. Zur Relation von Wissenschaft und Öffentlichkeit im 20. Jahrhundert." In: Geschichte und Gesellschaft 34, pp. 73-95.

Schirrmacher, Arne (2013): "Communicating Science: National Approaches in Twentieth-Century Europe." In: Science in Context 26/3.

Schwarz, Angela (1999): Der Schlüssel zur Modernen Welt. Wissenschaftspopularisierung in Großbritannien und Deutschland im Übergang zur Moderne (ca. 1870-1914), Stuttgart: Franz Steiner.

Secord, James A. (1985): "Newton in the Nursery: Tom Telescope and the Philosophy of Tops and Balls, 1761–1838." In: History of Science 23/2, pp. 127-151.

Secord, James A. (2003 [2000]): Victorian Sensation. The Extraordinary Publication, Reception, and Secret Authorship of Vestiges of the Natural History of Creation, Chicago/London: The University of Chicago Press.

Secord, James A. (2004a): "Knowledge in transit." In: Isis 95/4, pp. 654-672.

Secord, James A. (2004b): "Monsters at Crystal Palace." In: S. de Chadarevian/ N. Hopwood (eds.), Models. The Third Dimension of Science, Stanford: Stanford University Press, pp. 138-169.

Secord, James A. (2007): "How scientific conversation became shop talk." In: Aileen Fyfe/Bernard Lightman (eds.), Science in the Marketplace. Nineteenth-Century Sites and Experiences, Chicago: The University of Chicago Press, pp. 23-59.

Shapin, Steven (2003): "The Image of the Man of Science." In: Roy Porter (ed.), Eighteenth-Century Science, The Cambridge History of Science, 4, Cambridge, UK: Cambridge University Press, pp. 159-183.

Sheppard, Kathleen L. (2012): "Between Spectacle and Science: Margaret Murray and the Tomb of the Two Brothers." In: Science in Context 25/4, pp. 525-549.

Shinn, Terry/Cloître, Michel (1985): "Expository Practice: Social, Cognitive and Epistemological Linkage." In: T. Shinn/R. Whitley (eds.), Expository Science. Forms and Functions of Popularization, Dordrecht: Kluwer, pp. 31-60.

Simões, Ana/Zilhão, Isabel/Diogo, Maria Paula/Carneiro, Ana (2013): "Halley Turns Republican: How the Portuguese Press Presented the 1910 Return of Halley's Comet." In: History of Science 51/2, pp. 199-219.

Snow, Charles Percy (2001 [1959]): The Two Cultures and the Scientific Revolution, London: Cambridge University Press.

Stewart, Larry (1992): The Rise of Public Science. Rhetoric, Technology and Natural Philosophy in Newtonian Britain, 1660-1750, Cambridge, UK: Cambridge University Press.

Suckow, Laurenz Johann Daniel (1761): Entwurf einer Naturlehre, Jena: Cuno.

Taschwer, Klaus (2002): Wissenschaft für viele. Zur Wissensvermittlung in der Wiener Volksbildungsbewegung um 1900, PhD thesis, Universität Wien.

Taschwer, Klaus/Föger, Benedikt (2003): Konrad Lorenz. Biographie, Vienna: Zsolnay.

Taylor, Dorceta E. (1999): "Central Park as a Model for Social Control: Urban Parks, Social Class and Leisure Behavior in Nineteenth-Century America." In: Journal of Leisure Research 31/4, pp. 420-477.

Terrall, Mary (1995): "Gendered Spaces, Gendered Audiences: Inside and Outside the Paris Academy of Sciences." In: Configurations 2, pp. 207-232.

Terrall, Mary (2002): The Man who Flattened the Earth: Maupertuis and the Sciences in the Enlightenment, Chicago: The University of Chicago Press.

Thébaud-Sorger, Marie (2015): "Spectacles de Sciences." In: S. van Damme (ed.), Histoire des Sciences et des Savoirs. Vol. 1, De la Renaissance aux Lumières, Paris: Seuil, pp. 133-153.

Topham, Jonathan R. (2007): "Publishing 'Popular Science' in Early Nineteenth-Century Britain." In: Fyfe/Lightman 2007, pp. 135-168.

Topham, Jonathan R. (2009a): "Introduction: Historicing 'popular science.'" In: Isis 100/2, pp. 310-318.

Topham, Jonathan R. (2009b): "Rethinking the history of science popularization / popular science." In: Faidra Papanelopoulou/Agustí Nieto-Galan/Enrique Perdiguero (eds.), Popularizing Science and Technology in the European Periphery, 1800-2000, Ashgate: Aldershot, pp. 1-20.

Unterstell, Rembert (2012): "Wissen to go? Science Center und 'Centermania.' Das 'Erlebnis Wissenschaft' in der Unterhaltungsgesellschaft." In: Gegenworte 28, pp. 64-66.

Vasta, Marina (2018): Viaje pintoresco y excursión científica. El Jardín Zoológico de Buenos Aires, 1888-1924, Buenos Aires: Serie Tesis del IAA.

Vasta, Marina/Buzzi, Dino (2013): "Habitar el zoológico, Habitar la ciudad. Representaciones Urbanas en las Guías del Jardín Zoológico de Buenos Aires (1904-1934)," 3er Congreso Iberoamericano de Teoría del Habitar (ALTEHA).

Walters, Alice N. (1997): "Conversation Pieces: Science and Politeness in Eighteenth-Century England." In: History of Science 35/2, pp. 121-154.

Weingart, Peter (2001): Die Stunde der Wahrheit? Vom Verhältnis der Wissenschaft zu Politik, Wirtschaft und Medien in der Wissensgesellschaft, Weilerswist: Velbrück Wissenschaft.

Weingart, Peter (2012): "The Lure of the Mass Media and its Repercussions on Science." In: Simone Rödder, Martina Franzen, and Peter Weingart (eds.), The Sciences' Media Connection – Public Communication and its Repercussions, Dordrecht: Springer, pp. 17-32.

Werrett, Simon (2010): Fireworks: Pyrotechnic Arts and Sciences in European History, Chicago: The University of Chicago Press.

Wessely, Christina (2008): Künstliche Tiere. Zoologische Gärten und urbane Moderne, Berlin: Kadmos.

Wessely, Christina (2013): Welteis. Eine wahre Geschichte, Berlin: Matthes & Seitz.

Withers, Charles W. J. (2011): "Scale and the geographies of civic science. Practice and experience in the meetings of the British Association for the Advancement of Science in Britain and in Ireland, ca. 1845-1900." In: David N. Livingstone/Charles W. J. Withers (eds.), Geographies of Nineteenth-Century Science, Chicago/London: The University of Chicago Press, pp. 99-122.

Zeilinger, Anton (2000): "Zum Nutzen der Forschung," HEUREKA, Das Wissenschaftsmagazin im Falter, 4, pp. 8-9.

Zeilinger, Anton (2005): Einsteins Spuk. Teleportation und andere Mysterien der Quantenphysik, trans. Friedrich Giese, Munich: C. Bertelsmann.

Zimmerman, Andrew (2000): "Science and 'Schaulust' in the Berlin Museum of Ethnology." In: Constantin Goschler (ed.), Wissenschaft und Öffentlichkeit in Berlin, 1870-1930, Stuttgart: Steiner, pp. 65-88.

Zimmerman, Andrew (2001): Anthropology and Antihumanism in Imperial Germany, Chicago: The University of Chicago Press.

On the Trail of
Popular Science in Antiquity

Mythology and Rhetoric Exercises at the Greek School

José Antonio Fernández Delgado

ABSTRACT

The sources we have at our disposal for the study of school mythology in Greece are of two kinds. The first sources are papyri and other writing media (wood tablets, ostraca, parchments), which were generally used at school or in the school sphere. The second source is the rhetoricians' writings, which are of two kinds: the theoretical progymnasmatic texts and the broad series of rhetoric compositions destined to serve as models for school practice. While my earlier research on the use of mythology in Greek schools focused on both aspects of rhetoricians' writtings, the current essay will be centred on the school exercises on papyri and other materials, the richness of whose mythological contents is no less abundant than that attested in the theoretical part. The testimonies studied do not cover all the school exercises, a huge part of which has hardly any mythological content. Rather, the focus shall be on the *progymnásmata*, or school exercises used as the introduction to the study of rhetoric at the end of the *grammatikós* class. To this end, we use the only one existing corpus of *progymnásmata* on papyrus and other writing materials that has been recently assembled and presented as a doctoral thesis at the University of Navarra under the supervision of the author. In this way, my contribution to the present book mainly consists of filling a gap in the existing bibliography on Greco-Roman school studies—however strange it might be to foreground the mythological thematic element, which is so inherent in classical literature and culture in general.

INTRODUCTION

In a previous study, we addressed mythology's impact on the Greek school, considering its treatment by the rhetors that wrote the progymnasmatic treatises—i.e., the manuals by Theon of Alexandria (1st c.), Hermogenes or Pseudo-Hermogenes (2nd-3rd c),[1] and Aphthonius (4th c.), the best known of the three[2]—and its presence in the repertoire of model-exercises of *progymnásmata* for their use by teachers and pupils, attributed without much evidence to the great rhetor Libanius of Antioch (cf. Gibson 2008: XXIII-XXV; Ureña Bracero 2007), teacher of Aphthonius (cf. Fernández Delgado 2017). In the previous study we revealed, both in the theory and in the practice of progymnasmatic doctrine, the significant presence of mythology: from its broad casuistry of details and data, depending on the nature of the argumentative topoi and the types of exercise, through to the methods of their propaedeutics, the literary sources and main branches of the Greek myths, even including the rationalist critique of the myth and the figure of one of its greatest exponents, namely, Palaephatus, who wrote texts on the rationalisation of myths in the 4th century BC, and whose work has in turn been recently considered a progymnasmatic product (cf. Ramón García 2001). All of which has yet more merit when we consider that the teaching of mythology, which was so essential in a culture such as the Greco-Roman one, whose literature and art were largely informed by mythological subject-matter, did not have an actual subject in the schematic curriculum of studies in the Greek school, but instead had to be taught within the context of the exegesis of various texts (cf. Marrou 1970, 204f). At the same time, this serves to show, within the framework of this book on the genealogy of popular science, the broad diversity of the means of accessing knowledge and its popularization.

On this occasion, our research about the impact of mythology in the Greek school will focus on the other side of the coin, i.e., on the school exercises themselves. Produced by the actual pupils, or by the teacher to serve as examples, they are the direct products of the school setting, and accurately mirror the level of the students' learning. We are not going to be dealing here with the difficulties often entailed in the identification of classroom material among the papyri, or something so seemingly straightforward as the distinction between an exercise produced by a pupil and the work of a teacher. For this miracle of recovery to take place, even

1 The authorship of this progymnasmatic doctrine attributed to Hermogenes has been questioned, cf. Patillon 2008: 165-168.

2 There is a new edition, with an introduction and notes, by Patillon & Bolognesi 1997 and Patillon 2008 (cf. review of the latter by Fernández Delgado 2010), and a translation, with notes and an introduction, into Spanish by Reche Martínez 1991 and into English by Kennedy 2003.

two thousand years after they were produced, we have had to pass through two centuries of exploring, finding, and publishing papyri and other forms of written media (small wood tablets, ostraca, and parchments) preserved by the dry sands of the Egyptian desert, a cultural milieu in which Greco-Roman teaching had been introduced, as in the metropolis, since its conquest by Alexander the Great. It has also been extremely useful, we might even say essential, for this study to have been able to refer to a repertoire, now theoretically complete (although in the field of papyri one never knows what the last word is), of the myriad progymnasmatic exercises preserved on papyri and other media. We refer to the one that is forthcoming, with a translation, historical bibliography, and a study of its corresponding palaeographic, formal, and content features, by Cristina Iturralde inher PhD dissertation under the author's supervision.[3] This material belongs to a stage of schooling transitioning between the class of the *grammatikós* (akin to our secondary education, for example) and the higher education of the *rhetorician,*[4] and is by far the more expressive from the perspective of mythology. We shall be focusing our study on it, ignoring the school exercises from previous stages (primarily reading-writing and grammar), not because of any shortage of good compilatory repertoires, but instead because the few mythological references in them tend to be limited to isolated names or brief allusions.[5]

---

3 Cristina Iturralde Mauleón, *Progymnásmata en papiro y otros materiales*, PhD dissertation undertaken initially at Salamanca University involving a contract linked to my research project funded by the MEC (Spanish Ministry of Education and Culture) and then to the University of Navarra under the generous joint supervision of José Bernardino Torres Guerra by virtue of a grant linked to his research project, also funded by the MEC, as well as prolonged research stays at Oxford University; PhD awarded by the University of Navarra 2016, the dissertation is currently in press.

4 The exact timing of the teaching of *progymnásmata* (prior to the *gymnásmata* or the rhetorical exercises themselves, *declamationes* or *melétai*, in any case) is something that has been questioned, as the tendency in Rome was, according to the testimony of Quintilian (*Inst.* I 9. 6), for them to be assumed by the *grammatici* while the Greek rhetors were reluctant to give way in what they considered to be their domain. Perhaps the first exercises in the series, from the fable to the *gnṓmē*, which are the most elementary as well as the most versatile and frequent, both on classroom papyri (except for the narration) as in their use by literary authors, would be the first to be delegated to the *grammatikoí*; cf. Fernández Delgado 2007: 278-281.

5 Our previous study did not consider the six models of *melétē/declamatio* on a mythological theme—*Embassy Speech of Menelaus to the Trojans, Embassy Speech of Ulysses to the Trojans, Achilles' Reply to Ulysses' Embassy Speech, Orestes' Apology, Poseidon's Speech against Ares the Murderer, Ares' Apology*—included among the 51 models of *melétē* attributed to Libanius (Foerster VIII) and corresponding to the later

We introduce and comment on the mythological data organizing it not as is usually done in the respective treatises by gods, heroes, cycles, or specific divinities, which would lead to a highly non-systematic outcome given the often scarce and random representativeness of the surviving papyrus material, but instead by following the order the *progymnásmata* tend to adopt in the rhetoricians' treatises. Furthermore, this will allow for distinguishing the best and worst represented exercises accordingly and explaining the reason for those differences, as well as the numerous coincidences and divergences regarding the list of *progymnásmata* attributed to Libanius, which despite some gaps is the most complete surviving one to this day (cf. Gibson 2008).

A simplified list of the *progymnásmata* postulated by the rhetoricians consists of the following: fable (*mýthos*), narration (*diḗgēma*), anecdote (*chreía*), maxim (*gnṓmē*), refutation/confirmation (*anaskeuḗ/kataskeuḗ*), commonplace (*koinòs tópos*), encomium (*enkṓmion*), invective (*psógos*), comparison (*sýnkrisis*), personification (*ēthopoiía*), description (*ékphrasis*), argument (*thésis*), and proposal of law (*nómou eisphorá*). Of these, the exercises identified in the classroom materials—whether as isolated exercises or as the most common anthologies of exercises, sometimes corresponding to schoolbooks—are as follows: *mýthos*, *chreía*, *gnṓmē*, *enkṓmion*, *sýncrisis*, *ēthopoiía*, *ékphrasis* (with doubts), and *thésis* (with doubts); of these, the *mýthos*, *chreía*, and *gnṓmē* are by far the best represented, followed at some distance by the *enkṓmion* and *ēthopoiía*.[6]

---

stage of the teaching of rhetoric, and I shall not be considering now whether any papyrus fragments of mythological *melétē* may have been identified.

6 Cf. Iturralde, in press. The fact that fable, anecdote, and maxim are the best represented *progymnásmata* is explained by their versatility to fit into any kind of discourse, as shown by their frequent appearance in the literature. The relative employability of the encomium and *ēthopoiía* explains their frequency in the more complex exercises. The scarce presence of *sýncrisis* may be because it is in fact a comparison of two encomiums/invectives; that of the *ékphrasis* of the better-known type, which is the *ékphrasis* of the artistic object, was the one least favored by progymnasmatic theory, and other types (e.g., the *ékphrasis* of events) may be confused with other kinds of discourse, which may also explain the absence of the narration among the classroom *progymnásmata* papyri. The scarce presence of the *thésis* may be because it is restricted to a philosophical-scientific type of discourse, while the complete absence of the commonplace and the proposal of law may be because of the legal nature of their discourse. The absence of the *anaskeuḗ/kataskeuḗ* may be because it can be part of other *progymnásmata* and, in fact, Theon does not consider it a separate exercise. In contrast, the narration is by far the one best represented in the repertoire of progymnasmatic models attributed to Libanius, followed some way off by the encomium/invective and the *ēthopoiía*, and then further back still by *ékphrasis* and a handful of examples of fable,

THE TESTIMONIES

The papyrus *progymnásmata* that, to a greater or lesser extent, contain mythological references (and not historical or other ones) are: four *mýthoi*, three *chreíai*, six *gnomae*, eight *enkṓmia* or sets of *enkṓmia*, twenty-three *ēthopoiíai* or sets of *ēthopoiíai* (a couple of them are doubtful), a possible *ékphrasis*, out of a total of 22 exercises of *mýthos*, two of *chreía*, five of *gnṓmē*, nine of *enkṓmion*, one of *sýncrisis*, four of *ēthopoiía*, and one of *ékphrasis*, plus 54 partial anthologies of some of these exercises.[7] The myths alluded to may involve the Olympian gods and other divinities, heroes of the Trojan cycle, specifically heroes and heroines from the *Iliad* and pre- or post-*Iliad* texts, and to a far lesser degree from the *Odyssey*, some metamorphoses, as well as elements from Hesiodic or Argive, Attic, and Thessalian mythology (Triops and Erysichthon), as well as some Biblical stories.[8] Some of these myths coincide with the content in the

---

chreía, commonplace, anaskeué/kataskeué, sýncrisis, thésis, and (fewer still) proposal of law. Regarding fable and *chreía* (as well as *gnṓmē* in its case), the author may not have considered it necessary to include more examples; commonplace, *thésis*, and above all proposal of law are more restricted to the type of discourse, of a legal nature in the latter case and of a philosophical-scientific nature in the former; *anaskeué/kataskeué* and *sýnkrisis* are barely present in the repertoire, the former, as I have said, possibly because it can be part of other *progymnásmata* (and, in fact, Theon's list does not include it as such), and the latter because in reality it is a comparison between two encomiums/invectives.

7 The progymnasmatic models of the repertoire attributed to Libanius that contain mythological references are as follows: 39 of the 41 narrations, the six refutationes/confutationes, six of the 17 encomiums/invectives, two of the five *sýnkrisis*, 21 of the 27 *ēthopoiíai*, and 13 of the 30 *ekphráseis*.

8 *Ēthopoiía* of Cain after killing Abel and *eidōlopoiía* (a subtype of the *ēthopoiía*) pronounced by Abel after having been killed by Cain (*P. Bodmer* XXX-XXXVII). Here, too, there are differences with the mythical content of the progymnasmatic models, in which, because there are more of them, besides the type of gods and heroes present in the school exercises, and in a similar proportion to those from the *Iliad* and the *Odyssey*, there are others belonging to the Theban and Corinthian cycles, to those of the Argonauts (e.g., Medea), foundation myths, and sundry metamorphoses, albeit missing Hesiodic and Thessalian references or Biblical stories. Specifically, these latter ones, which have been identified in a miscellaneous papyrus codex, the *Codex Visionum*, from the early 5th c. (cf. Fournet 1992, Miguélez-Cavero 2013), constitute a good example of how it appears that school practice kept a step ahead of the corresponding theoretical treatises in the assumption of contemporary subject matter, Christian in this case. Cf. Ureña Bracero 2005.

school models attributed to Libanius (or the few adduced by the treatise written by his disciple Aphthonius), although the majority do not coincide, and those that do so do not occur in the same type of exercise, as we shall indicate in due course. As in Libanius' models, nonetheless, here too the models often have a moral sense or at least an exemplifying one. We shall now provide a list of them, dedicating more time to those cases of greater import from a mythological perspective or those in the best condition.

Out of the four fable exercises with mythological references found on papyrus—the few that can be included in this exercise, such as in the case of the *chreía* or the *gnṓmē* —one (*P. Köln* VI 250, B II; 2nd-3rd c.) contains a reference to the Fates; another (*Tabulae Ceratae Assendelftianae* (86) = *T. Leiden Univ. Libr.* 109 III *verso*; 3rd century AD; a "folder" of small wooden tablets from Palmira) to the "mother of the gods"; another (*P. Amh.* II 26, col. II; 3rd-4th c.; a Greek version of a Latin fable) to Demeter/Ceres; and the most interesting one (*Tabulae Ceratae Assendelftianae* (86) = *T. Leiden Univ. Libr.* 109 IV *verso*) places its cautionary tale in the mouth of Hermes.[9]

Out of the three papyrus *chreíai* with mythological references, all three are adjudicated to Diogenes, while one includes a comparison with Pelops and Oenomaus (*P. Vindob.* G 29946, col. VI; 2nd-3rd c.), and another a reference to Autolychus and Hermes in the *Autolychus* by Euripides (*P. Rain.* III 32 = *P. Vindob.* G 19766; 2nd c.);[10] the third *chreía*, an ostracon from Thebes in Egypt, contains a highly didactic reference to the Muses (*SB* I 5730 = *O. Thompson* II 2; 3rd-4th c.). According to this, when the muses asked Diogenes where they lived, he replied: in the soul of educated people (*pepaideuménoi*).

Out of the six mythological *gnomae* found in school material, one (*P. Cair.* inv. 56226; 1st-3rd c.; from Oxyrhynchus) asks "Why did Prometheus, who is said to be the one that gave form...?", a question that, thanks to its complete transmission by Stobaeus (III 2, 26 Hense & Wachsmuth), we know continues as follows: "... to us and all other living creatures, give animals a single nature, to each one according to their species?" (whereas in contrast to the animals, it is understood, human beings are extremely complex).[11] Two *gnomae* refer to Zeus (*P. Cair.* inv. 56227, 1st-3rd c., from Oxyrhynchus, and *O. Berol.* inv. 12319, dating from the 3rd-2nd c. BC no less, from Philadelphia), one as "Zeus is the one that sends our daily bread," the other paraphrasing the saying "The rain Zeus sends never pleases everyone." Two are adjudicated to Menander, from among the many maxims attributed to him; one on slander refers to the Attic myth of

9 Cf. Babr. *fab.* 117 Luzzatto & La Penna.

10 Cf. Luppe 1991; Pechstein 1998; Pordomingo 2007.

11 Prometheus is the object not of a *gnṓmē* but instead of *ékphrasis* in the repertoire of progymnasmatic models attributed to Libanius (no. 19).

Hippolytus and Theseus (*P. Vindob.* G 19999 A and B; 1st c. AD; from Soknopaiou Nesos), another urges the young man to flee from Dionysus and further so from Aphrodite (*P. Giss.* 348 = *P. Iand.* 5. 77; 2nd-3rd c.; from Arsinoe). A third *gnṓmē* predicates that "Eros is the oldest of all the gods" (*P. Bour.* 1. 141; s. IV), following *Theogony*'s declaration (v. 120).

Out of the eight exercises or, as appropriate, the series of exercises of encomium with a mythological content identified on papyrus, one (*P. Köln* VII 286; 2nd-3rd c.) is in praise of Dionysus, another (*P. Oxy.* VII 1015; 3rd c.) of Hermes, and another (*P. Oxy.* XVII 2084; 3rd c.) is in praise of the fig, containing references to Hermes and Dionysus in addition to the sweet voice of the ancient sage Nestor from the *Iliad.*[12] A papyrus dating back to such an early time as the 3rd c. BC (*P. Mil. Vogl.* III 123) has passed down fragments of an anthology of texts whose schematic and formulaic nature have led to their identification as sundry progymnasmatic encomiums, which are assumed to be examples proposed by the teacher for their subsequent completion by the pupils. They are dedicated to Minos (presented as the mythical founder of the Thalassocracy and the Greek islands and, upon his death, a fair judge of Avernus), Rhadamanthus (like the former, a son of Zeus, with the special circumstance of the god's predilection for Crete, the hero's place of birth, where Zeus was also born and to where he took his beloved Europa), and Tydeus (as a member of the expedition of the Seven against Thebes, son-in-law of the Argive King Adrastus, deliverer from the threat arising from fifty cities, and, in personal terms, friend to his companions—a particularly exemplary fact for schoolchildren). He is also characterized with the no less exemplary values of good diplomatic attributes (a clever compensation, no doubt, for the educational shortcomings he was attributed by mythical tradition), loved by the gods and in particular by Athena, himself, and his offspring (Diomedes). This exercise may have been followed by another dedicated to Diomedes himself.[13]

Another encomiastic exercise (to the so-called Flower of Antinous) is part of what has been classified as an anthology of school sketches, from a later period (2nd-3rd century), in this case not only of *enkṓmia*, as in the previous anthology, but of sundry *progymnásmata* (*P. Mil. Vogl.* I 20; from Tebtunis). It contains similar references to a series of plant divinities known for being the object of infatuation of heroes or gods—Heracles, Dionysus, the Nymphs, and Apollo—,

12 Cf. Fernández Delgado/Pordomingo (2020), with bibliography. Pordomingo considers these *enkṓmia* with references to Hermes to be school compositions dedicated to a festival of this god.

13 Cf. Fernández Delgado 2012. An encomium to Diomedes is no. 1 among those in the repertoire attributed to Libanius.

dying young and metamorphosing as deities into plants, namely, Hyllus, Crocus, Cypress, Narcissus, and Hyacinthus.[14]

An encomium to Antinous (*P. Oxy.* L 3537; 3rd-4th c., *Fig. 1*), the beloved of the emperor Hadrian, in this case a contemporary character and yet the subject of several poetic texts, contains references to Hermes, the Muses, Zeus, and Leto.[15] An encomium to Achilles, with mandatory references to Zeus, Eachus, Peleus, Thetis, and Chiron, among others, alternates with an *ēthopoiía* of Clytemnestra addressing Orestes at the moment the latter is going to kill her, in a papyrus prose anthology (*P. Vindob.* G 29789; 3rd-4th c.; from Soknopaiou Nesos), as opposed to the usual use of the hexameter in school *ēthopoiía* on papyrus and according to common practice in the models of the repertoire of *progymnásmata* attributed to Libanius or in the samples provided by Aphthonius.[16] A later encomium to Colluthus refers to a hero in the *Odyssey*, Alcinous, in another papyrus anthology (*P. Cair. Masp.* II 67187v; 6th c., found in the archives of the notary and teacher Dioscorus of Aphrodito and written in Antinopolis) in which it alternates with an *ēthopoiía* on the relationship between Achilles and Polyxena, likewise the subject of another papyrus *ēthopoiía* from the same archive (*P. Cair. Masp.* II 67186v).[17] Finally, we also have an encomium to the judge Domninus with an acrostic, a very typical school game from that same archive belonging to Dioscorus (*P. Cair. Masp.* III 67316v); it includes references to Orpheus, son of Calliope, and to Nestor, and alternates with an *ēthopoiía* of Achilles following his death caused by Polyxena; so among the school papyri of Dioscorus, we encounter the application of the enco-

14 Of these, only Hyacinthus is the object of a *progýmnasma*, narration no. 2, in the repertoire attributed to Libanius.

15 Cf. Ureña Bracero 1999; Miguélez-Cavero 2008, no. 19. Of these gods, only Leto is the object of a *progýmnasma*, the *diḗgēma* or narration no. 29, in the repertoire attributed to Libanius.

16 Cf. Carvounis 2004; Pordomingo 2007, with bibliography. It is to be assumed that the use of hexameter instead of prose in many school exercises (not in progymnasmatic repertoires or treatises), a use that tends to be accompanied at the same time by a largely Homeric language, constitutes a merit and a higher degree of suitability as it involves a mainly Homeric or comparable subject matter and a language whose mastery already posed serious difficulties for the pupils of progymnasmatic schooling, in *koinḗ*. Cf. Ureña Bracero 2005. Achilles is the object of both an encomium (no. 3) and an invective (no. 1) (which shows just how far the challenges posed in rhetorical practice went at school), as well as of *sýnkrisis* with Diomedes (no. 1) and with Ajax (no. 2), in the repertoire attributed to Libanius, as well as *sýnkrisis* with Hector in the example the rhetor Aphthonius provided in his progymnasmatic treatise.

17 An *ēthopoiía* of Polyxena on Achilles and an *ékphrasis* of Polyxena sacrificed by Neoptolemus constitute, respectively, nos. 16 and 18 in the repertoire of *progymnásmata* attributed to Libanius.

mium to contemporary personages and, on the other hand, an insistence on the subject of Polyxena as the object of an *ēthopoiía*.[18]

Figures 1 and 2: Encomium, P. Oxy. L 3537; Sulpicius Maximus' Gravestone

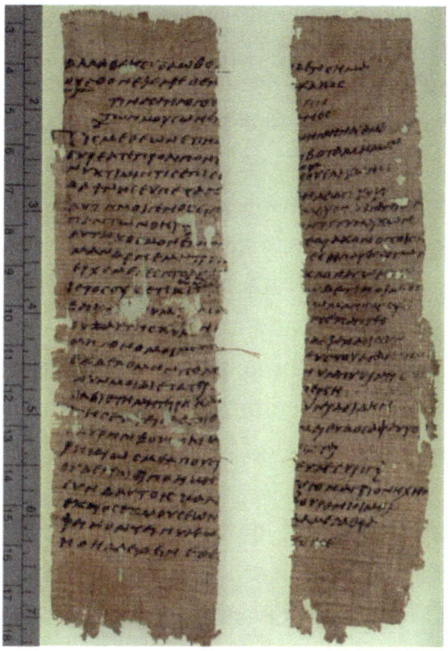

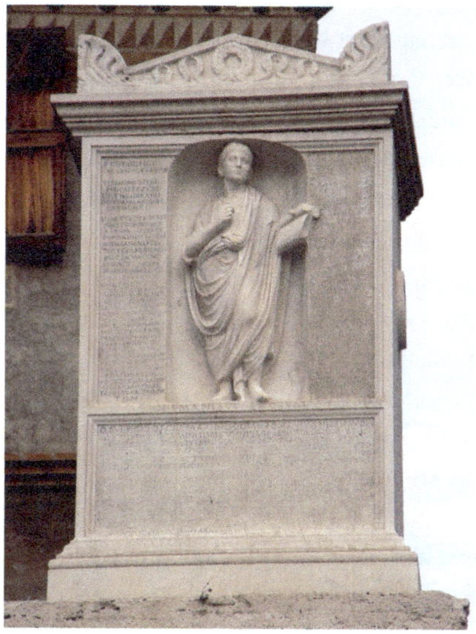

Sources: POxy Oxirrinchus Online; Musei Capitolini, Rome

A similar number of cases to those cited corresponds to the exercises of *ēthopoiía* with mythological content identified on papyrus or other kinds of media, with the advantage that among them there are several examples in an acceptable state of preservation. Following the same chronological order of the casuistry used with the other exercises, we shall begin with the oldest example that has survived to this day, which is at the same time the best preserved, because in contrast to all the other cases it is precisely the only one we have on stone (*IG* XIV 2012, *Fig. 2*) and not on papyrus or any other of the writing materials used at school. This is so because the composition was written by an eleven-year-old Roman boy, Quintus Sulpicius Maximus, for a school competition on extempore Greek verse, against the backdrop of the great Capitoline Games held in Rome in AD 94 (reign of Domitian). The child won the prize and then, following his premature death, his parents had the poem engraved on the niche containing in the middle a statue of the boy on the funerary stele they dedicated to him, which is now to be found in the Centrale Montemartini, an exhibition

18 Cf. Fournet 1999: no. 42; Jarcho 1999.

hall of the Capitoline Museum in the Piazza del Campidoglio, with a replica of the monument reconstructed on a Roman wall piece from Porta Salaria.[19]

The poem is of considerable interest for many reasons, none of which are far from the spirit of the present book. It is an example of the existence of competitions that, even from the classroom, fostered not only the literary application of progymnasmatic teaching but also the oral nature, if not of the composition itself, at least of its performance, reception, and literary dissemination at such an advanced date as the end of the 1st c.; it also testifies to the preponderance, still at that time, of Greek over Latin as the language of schooling and culture in Rome, an issue that the great rhetor Quintilian (*Inst.* I 1. 12-13) felt the need to oppose in favor of teaching Latin. Apart from being the oldest surviving example, it is the only case of a school *ēthopoiía* outside Egypt, and not precisely in Greece but instead in Rome, the great heir of its culture and system of teaching. This shows, if there were any need to do so, that progymnasmatic practice is a genuine part of Greek teaching and not something linked explicitly to colonized Egypt as an instrument of acculturation, as has sometimes been suggested by ignoring the *ēthopoiía* of Sulpicius (cf. Fournet 1992; Fernández Delgado 1994).

From the viewpoint of interest to us here, namely, its mythological content, the exercise, with the engraving containing 43 hexameters, refers to "the words assumedly spoken"[20] by Zeus to Helios after the latter had let his son Phaeton drive his chariot around the earth and through his inexperience had set it on fire. The subject, although well known,[21] neither belongs to the type of mythology most commonly found in the classroom (which is Homeric) nor has it been addressed in any other of the *progymnásmata*—whether those of pupils or rhetoricians—that we know of, whereby it has a certain originality within the context of these testimonies. An essential feature of its treatment by Sulpicius in terms of didactic exemplarity is the monumental reprimand that the furious Zeus gives Helios for surrendering to his young son's foolish whim and for the devastating effects the stupid prank had for the entire cosmos, as well as the threat Zeus makes to have Helios take up the reins and the straight path of his chariot, which is in fact what the text is really all about.

Another long and very well-preserved example, a full text of 26 hexameters on papyrus (*P. Oxy.* XLII 3002), nonetheless poses the difficulty of adscription between *ēthopoiía* and paraphrase, with the latter being a type of exercise that Theon (62-64 Sp.) considers a true *progýmnasma,* but which is not included as such by the rhetoricians in their successive lists of these exercises (cf. Fernández

19 Cf. Fernández Delgado/Ureña Bracero 1991 (with photograph).

20 This or certain variations thereof is the title that characterizes the exercise of *ēthopoiía* or "characterization"; cf. Fournet 1992.

21 Among other reasons, it provides the plot for a tragedy by Euripides called Phaethon, of which large fragments have been recovered thanks in part to a papyrus, as well as to a poem by Ovid.

Delgado 2012). The difficulty lies in the fact that, lacking a title like so many other papyrus with *ēthopoiíai*, the words Athena and Hera speak to Achilles (with allusions to his parents, Thetis and Peleus) urging him to put aside his anger toward Agamemnon and fight alongside him against the Trojans (with references to Priam and Hector), which make up the exercise, albeit sufficiently different to the text of the *Iliad* to avoid being classified as a simple paraphrase, do not adopt an original approach, but instead follow a model present in *Iliad* I 207-214 (where the one providing the advice is Athena sent by Hera), contrary to regular practice in the exercise of *ēthopoiía*.[22] It does include some of the inventiveness often found in the genre when Athena warns Achilles that if he does not obey her, someone in the future may sing of Achilles' anger, clearly referring to Homer's *Iliad post litteram*. What's more, the advice Athena and Hera give to Achilles not to fight his friends but instead to cooperate with them against their enemies is also of easy application in schooling, and more so considering the heavily Greek moral principle of loving one's friends and hating one's enemies.[23]

Another school *ēthopoiía*, belonging to the same anthology of progymnasmatic sketches as the aforementioned Flower of Antinous (*P. Mil. Vogl.* I 20), is likewise of interest from the perspective of its mythical content, not so much because of the hero it refers to, Heracles, who is well represented in the rhetoricians' progymnasmatic casuistry, from Libanius[24] to Aphthonius (34 Rabe);[25] and indeed, among the school papyri themselves there is one (*P. Oxy.* L 3537, *Fig. 1*) with a reference to Heracles. Instead, this *ēthopoiía* sketch is noteworthy above all because of the particular, and at the same time ingenious, version that it provides of the words Heracles would say (in a semi-preserved title) to the El-

22 As we have already noted, the repertoire of progymnasmatic models attributed to Libanius contains an encomium (no. 3) and also, as a show of inventiveness typical of the genre, an invective (no. 1) toward Achilles, as well as an *ékphrasis* of Hera (no. 16) and another of Pallas Athena (no. 22).

23 A well-known maxim at least since as far back as Theognis (cf. 1032f., 1318a-b): ...ἐγὼ τοῖσιν μὲν ἐπαρκέσω οἵ με φιλεῦσιν,/ τοῖς δ' ἐχθροῖσ' ἀνίη καὶ μέγα πῆμ' ἔσομαι; "May the great wide brazen sky fall upon me [...] if I aid not such as love me, and become not a pain and great grief unto such as hate" (v. 869-872, transl. by Edmonds 1931); a similar thought can be found in the earlier Hes. *Op.* 342; Archil. fr. 23 W.; Sapph. fr. 5 V., v. 6s; Sol. fr. 13 W., 5s; Pi. *P.* 2, 83-84; and Call. *Ep.* 52, 1-2; cf. Fernández Delgado 1984.

24 In his repertoire, Heracles is the protagonist of two *diēgēmata* or narrations (23 and 24) and four *ékphrasis* (13, 14 Heracles and Antaeus, 15 Heracles and the Lion's Skin, and 26 Heracles and the Erymanthian Boar).

25 As an example of the *ēthopoiía progýmnasma*, Aphthonius provides the words spoken by the hero in response to the orders given by Eurystheus (orders that according to mythographic tradition, by contrast, Heracles obeyed without a murmur).

eusinian officer (*dadouchos*) who allegedly tried to impede his initiation into the Mysteries. Faced with this refusal, Heracles haughtily replies that he has been initiated into more important mysteries, and has enjoyed a night in company of Core (Persephone, the goddess of the underworld). This implies upsetting the usual order of events of the Heraclean myth, which places his descent into the underworld at the end of everything, preceding his fit of madness, in order to provide an argument that thoroughly undermines the attempt to stop him from taking part in the Eleusinian Mysteries (possibly because he was a foreigner, cf. Colomo 2004; Fernández Delgado/Pordomingo 2008).

One progymnasmatic testimony (*P. Ryl.* III 487; 3rd-4th c.) is remarkable for something that is relatively scarce in school papyri and, to a lesser extent, in the rhetoricians' models, namely, references to the subject matter of the *Odyssey*, compared to the preponderance of testimonies involving the *Iliad*.[26] It contains an identified exercise with allusions to characters from the *Odyssey* (Circe, Polyphemus, and Amalthea) and fragments apparently of an *ēthopoiía* (the whole papyrus contains fragments of 37 hexameters) with an original approach supposedly spoken by Odysseus on his return to Ithaca revealing himself to his servant Philoetius and calling for help to punish Eurymachus and the suitors, in close connection with the theme of Book XXI in the *Odyssey*. More original still, and in fact unique, is the *ēthopoiía* in the *recto* of a papyrus already mentioned because of the encomium to Antinous contained in its *verso* (*P. Oxy.* L 3537; 3rd-4th c.), following another unidentified exercise with a reference to Heracles; for this *ēthopoiía* the title has been reconstructed, "Hesiod's words after being inspired by the Muses," as well as the beginning and end of its 24 hexameters. It has subsequently been discovered how the beginning of these concealed something no less appetizing for scholars, as we have already seen, namely, an acrostic, consisting of the Homeric hemistich τὸν δ' ἀπαμειβόμενος προσέφη ("and he said to him in reply," cf. Agosti 1997; Carvounis 2004). The text contains references to the mythology of the *Theogony* (Helicon, divine lineages, wars between giants), the *Eeas* (lineages of heroes and women), and *Works and Days* (Zeus) as the subject of an original canto in which the author announced he has passed not from a shepherd to an *aoidós* (oral epic poet), as described by the preface to the Theogonic narrative, but from pastoral poet to urban poet, according to the particular aesthetics of the Hellenistic era.

26 The repertoire of *progymnásmata* attributed to Libanius contains an *ēthopoiía* (no. 25) spoken by Odysseus after killing the suitors and another two Odysseus addresses to the Cyclops (nos. 23, 24), as well as an encomium to Odysseus (no. 2). The example of *gnṓmē* provided by the progymnasmatic treatise of Aphthonius concerns poverty and refers to Odysseus as a beggar.

A certain anthology of *ēthopoiíai* (*P. Graves* = *TCD pap.* inv. D 6; s. IV-VI; perhaps from Thebes) attracts our attention because of the originality of the topics addressed in its titles (cf. Ureña Bracero 1999; Carvounis 2004)—"What would Calliope say to console Thetis?";[27] "What would Odysseus say when ordering Menelaus not to bury Ajax's corpse?";[28] "What would Triops say when Erysichthon spent his entire fortune and was still not satisfied?"; "Apollo's words following the massacre of Niobe's children"[29]—and also for the mythical allusions contained on the papyrus's surviving fragments, of these *ēthopoiíai* (Memnon and Orpheus in the first, Demeter and Erinia in the third) and of others (Aphrodite, Erinias, and Hermes; Helen, Leda, Castor, and Polydeuces; Aphrodite and Priam; Patroclus, Hippolytus, and Niobe).

Another *ēthopoiíai* anthology on papyrus (*P. Heid.* inv. 1271; 5th-6th c.) is noteworthy precisely because of the ingenuity of its six compositions within the Iliadic or the pre- and post-Iliadic subject matter, with such brevity and charm that to some extent they remind us of epigrams, a genre with which the hexametric *ēthopoiíai* was in fact confused until its identification as *progýmnasma.*[30] Their titles, in this case expressed in shortened form (as if they were actually short poems), are: "Phoenix during the embassy seeking to convince Achilles to set aside his anger"[31] (based on asking him irately whether the centaur Chiron did not teach him to fight so that now he has learned to express anger over a girl), "A Greek woman encountering Helen in Greece" (challenging her indignantly, and with some insistence over how she has the gall to present herself there after causing the death of so many Greeks), "A Greek after Hector had killed Patroclus and bearing his arms" (reminding him that he will have to face Achilles on the morrow),[32] "Zeus when Aphrodite comes to him railing against

27 Subject addressed by Quintus of Smyrna III, 633ss. Cf. Fernández Delgado 1994; Ureña Bracero 1999.

28 Ajax is the object of a refutation (no. 2)/confutation (no. 3) (new rhetorical challenge like the one mentioned of Achilles), a *sýncrisis* with Achilles (no. 2), three *ēthopoiíai* (no. 5 before committing suicide, no. 6 after his madness, no. 7 on being stripped of Achilles' weapons), and an *ékphrasis* (23), in the repertoire of progymnasmatic models attributed to Libanius.

29 *ēthopoiía* no. 8 in the Libanius repertoire is "What words would Niobe speak before the corpses of her children?"

30 In fact, these exercises have in turn been related to a series of epigrams transmitted in Book IX (451-480 and 449, 126) of the Greek Anthology, cf. Fernández Delgado/Ureña Bracero 1991; Fernández Delgado 1994.

31 The Libanius progymnasmatic repertoire contains an *ēthopoiía* (no. 14) on Chiron when he hears that Achilles is living in the girls' quarters (i.e., dressed as a woman in Deidamia's court).

32 *ēthopoiía* no. 3 in the Libanius repertoire, along with an example of a mixed *ēthopoiía* of *ēthos* and *páthos* adduced by his disciple Aphthonius (35 Rabe), contains the words spoken

Diomedes because he injured her hand" (reproaching her for concerning herself with matters of war instead of her injuring herself with her darts), "Aphrodite when Diomedes is injured by Aegiale" (a post-Homeric subject, although clearly connected to the previous one), and "[...] when Agamemnon fails to aid Orestes" (with references to Menelaus and Iphigenia).[33]

A diptych (*Heitsch* 38; s. V-VI; from Cairo) made of wood—a common material used as a writing medium at school—contains, alternating with a *gnṓmē* by Menander, seven verses of an *eidolopoeia* (*eidōlopoiía*), a subtype of the *ēthopoiía* pronounced by imaginary beings (Hermog. 20 Rabe; Aphth. 34 Rabe), in which the specter of Achilles, following the end of the Trojan War, addresses his fellow countrymen as they embark for Hellas, reproaching them for hastening away without giving him his part of the spoils, and alluding to "Pluto's nocturnal Fate."[34] Out of a further three *ēthopoiíai*, one of them (*P. Cair. Masp.* II 67186v) is doubtful and concerns another post-Iliadic myth involving the romance between Achilles and Polyxena; another (*P. Cair. Masp.* II 67187v) alternates, as on other occasions, with an encomium, in this case that of Colluthus already mentioned; and another (*P. Cair. Masp.* III 67316v) contains the words pronounced by the hero upon the death of Polyxena.[35] Both myths are connected, among other testimonies, in Euripides' *Hecuba* (vv. 109-115), where the chorus tells how Hecuba's daughter, Polyxena, is to be sacrificed at the tomb of Achil-

---

by Achilles over the corpse of Patroclus; a *sýncrisis* of Achilles and Hector is the one adduced by Aphthonius as an example of this *progýmnasma*, and Hector, as strange as it may seem (but cf. the aforementioned case of Achilles), is the object of *psógos* no. 3 in Libanius; the most difficult yet is a genuine feature of the *progymnásmata* (cf. Lucian's *Eulogy of the Fly*, with a clear rhetorical influence, to cite just one example).

33 Narration no. 5 in the Libanius repertoire concerns Agamemnon. If we look closely, all the cases in this papyrus, apart from some allusions of a didactic nature, such as the teachings of Chiron (the subject, by the way, of a work attributed to Hesiod), share the same attitude of reproach and indignant questioning on the part of the one asking for explanations over different criticizable attitudes, which despite being a frequent feature in the *ēthopoiía*, by being exclusively grouped together, here can only suggest a particular didactic purpose of such an important, we might hazard, subgenre of the *ēthopoiía*.

34 Cf. Fernández Delgado/Ureña Bracero 1991, 59; Fournet 1992; Jarcho 1999, 88, who comments on the link between the subject and *Hecuba* by Euripides. *Ēthopoiía* no. 4 in Libanius addresses the subject "What did Achilles say when the Greeks were being defeated?"

35 As we have seen so far, the type of *ēthopoiía* expressed in the sight of someone's corpse also truly constitutes a frequent subgenre of the same. *Ēthopoiía* no.16 in the Libanius repertoire is "What would Polyxena have said upon receiving the order to be taken by the Greeks because she has been told she is to be Achilles' bride?"; and *ékphrasis* no. 18 is "Polyxena being murdered by Neoptolemus."

les after he had appeared and claimed his part of the spoils.[36] Also concerning the Homeric hero par excellence, the school champion Achilles, there are a further two surviving titles of *ēthopoiía* (*P. Cair. Masp.* III 67353v), in one of which Achilles speaks to his mother Thetis about his armor, as well as an exercise that is difficult to define (*PSI* XIV 1399; 6th-7th c.) regarding Neoptolemus and Achilles.

Finally, among the surviving school papyri of Dioscorus of Aphrodito there is likewise an example (*P. Cair. Masp.* II 67188v) of a progymnasmatic subject that is also very enlightening, even though it is not Homeric and is double in this case, being an *ēthopoiía* by Apollo to his two beloveds, Hyacinthus and Daphne, at the same time.[37]

CONCLUSIONS

To conclude, the mythological content of the progymnasmatic exercises preserved on papyri and other writing materials constitute a highly important complement, both in quantity and in variety, to the mythical wealth provided by the repertoire of progymnasmatic models attributed to Libanius, the respective progymnasmatic examples of his disciple Aphthonius, and the sundry data handled by the rhetoricians, with the particularity that the papyrus testimonies and others provide a direct testimony of school practice in the handling of mythical material. At the same time, all this school material shows how mythical knowledge and its dissemination in Antiquity did not depend, as it does today, on the existence of mythological repertoires or of a specific subject in the school curriculum, but rather, first, on the reading and hearing of the many literary authors who dealt with mythology and, second, on the practice carried out in school learning.

The progymnasmatic exercises identified in school material on papyrus, whether isolated exercises or, more often, anthologies of exercises, sometimes corresponding to schoolbooks, are the fable, *chreía*, *gnṓmē*, encomium, comparison, *ēthopoiía*, *ékphrasis* (with doubts), and the *thésis* (with doubts); of these, the fable, *chreía*, and *gnṓmē* are by far the most numerous, followed at some distance by the encomium and *ēthopoiía*. The papyrus *progymnásmata* that, to a greater or lesser extent, contain mythological references (and not historical or

36 Regarding Euripides' influence on school rhetoric, cf. the recent study by Lentano 2018, which cites another well-known one by Raffaella Cribiore 2001.

37 Cf. Fournet 1999: no. 41; Carvounis 2004: 81. Narration no. 2 in the progymnasmatic repertoire attributed to Libanius deals with Hyacinthus and no. 17 with Daphne; *ēthopoiía* no. 11 is on "What would a painter say if, when seeking to paint a portrait of Apollo on laurel wood (*dáphnē* in Greek), the wood failed to absorb the paint?"

other ones) are as follows: four fables, three *chreía*, six *gnomae*, eight encomiums or sets of encomiums, and twenty-three *ēthopoiía* or sets of *ēthopoiíai* (a couple of them doubtful), as well as one possible *ékphrasis*.

The myths alluded to may refer to the Olympian gods and other divinities, heroes of the Trojan cycle, specifically heroes and heroines of the *Iliad*, pre- and post-Iliadic texts, and to a much lesser extent of the *Odyssey*, some metamorphoses, as well as, in contrast to what is seen in the theoretical texts and the progymnasmatic models, some Hesiodic mythology, and Argive, Attic, and Thessalian mythology—and even some Biblical stories, with what that implies in terms of the school exercises' ability to assume contemporary subject matter. Some of the myths do indeed coincide with the content in school models attributed to Libanius or Aphthonius, although the majority do not, and those that do coincide do not share the same type of exercise. Nevertheless, as in the Libanius models, here too the myths often have a moral sense or at least an exemplifying one. The treatment of the topics, however, although stylistically poorer, tends to better the models insofar as the particularity and inventiveness of their versions are concerned.

Some *ēthopoiía* exercises share the same attitude of reproach and indignant questioning on the part of the person asking for criticizable attitudes to be explained, which although a common feature in *ēthopoiía*, by sometimes appearing grouped, can only suggest an exclusively didactic purpose in such an important—we might say—subgenre of the *ēthopoiía*. As we have seen, the type of *ēthopoiía* spoken over someone's corpse constitutes a numerous subgenre of the same. An extraordinary case of school *ēthopoiía* is that addressed by Zeus to Helios for lending his chariot to his son Phaeton. It is unique for a number of reasons: because of the specific myth involved and its didactic use; because it is proudly engraved on the stone funeral stele that the parents dedicated to the author, an eleven-year-old boy; because its provenance is not Egypt, but Rome; and because it is the oldest surviving one. There are other points of interest in the text, none of which are in any way far from the spirit of this book: This *ēthopoiía* is proof of the existence of school competitions in the major public festivals, with what that implied for encouraging learning; it is evidence of the prevalence, still at that time (end of the 1st c. AD), of Greek over Latin as the language of learning and culture in Rome, and also that this type of teaching was used from one end of the Hellenized world to the other, and therein lies its continued presence in Europe down through the centuries (cf. Kraus 2020).

REFERENCES

Agosti, Gianfranco (1997): "P.Oxy. 3573r: etopea acrostica su Esiodo." In: Zeitschrift für Papyrologie und Epigraphik 119, pp. 1-5.

Carvounis, Aikaterini Nina (2004): Transformation of Epic. Reading Quintus of Smyrna Posthomerica XIV, Oxford: Oxford University.

Colomo, Daniela (2004): "Heracles and the Eleusinian Mysteries: *P.Mil.Vogl.* I 20, 18-32 revisited." In: Zeitschrift für Papyrologie und Epigraphik 148, pp. 87-98.

Cribiore, Raffaella (2001): "The Grammarian's Choice: The Popularity of Euripides' Phoenissae in Hellenistic and Roman Education." In: Yun Lee Too (ed.), Education in Greek and Roman Antiquity, Leiden: Brill, pp. 241-59.

Edmonds, John Maxwell (ed.) (1931): Elegy and Iambus, Vol. I, Cambridge, MA: Harvard University Press.

Fernández Delgado, José Antonio (1984): "Sabiduría popular y epos sapiencial en los *Idilios* de Teócrito," In: *Apophoreta Philologica* Emmanueli Fernández-Galiano a sodalibus oblata, Madrid 1984 (*Estudios Clásicos* 87), pp. 325-332.

Fernández Delgado, José Antonio (1994): "Hexametrische *Ethopoiiai* auf Papyrus und anderen Materialien." In: A. Bülow-Jacobsen (ed.), Proceedings of the XX International Congress of Papyrology, Copenhagen: Museum Tusculanum Press, pp. 299-305.

Fernández Delgado, José Antonio (2007): "Influencia literaria de los *progymnásmata.*" In: J. A. Fernández Delgado/F. Pordomingo/A. Stramaglia (eds.), Escuela y Literatura en Grecia Antigua, Università degli Studi di Cassino, pp. 273-306.

Fernández Delgado, José Antonio (2010): Rev. *Anonyme, Rhétorique. Aphthonios, Progymnasmata.* Texte établi et traduit par Michel Patillon, Paris: Les Belles Lettres, 2007. LVIII+363 pp.: Gnomon 82, pp. 405-411.

Fernández Delgado, José Antonio (2012): "Modèles progymnasmatiques de l'époque hellenistique: *P.Mil.Vogl.* III 123." In: P. H. Schubert (ed.), Proceedings XXVI International Congress of Papyrology (Geneva 16-21 August 2010), Genève, pp. 117-125.

Fernández Delgado, José Antonio (2012): "La parafrasi omerica nei papiri scolastici." In G. Bastianini/A. Casanova (eds.), *Omero. Cent'anni di papiri.* Atti del convegno internazionale di studi (Firenze, giugno 2011), Firenze, pp. 159-176.

Fernández Delgado, José Antonio (2017): "El mito en la escuela griega: el testimonio de los rétores." In: M. Alganza/P. Papadopoulou (eds.), La mitología griega en la tradición literaria: de la Antigüedad a la Grecia contemporánea (Simposio Científico Internacional, XIV Encuentro sobre Grecia, Universidad de Granada, 2016), Universidad de Granada, pp. 51-65.

Fernández Delgado, José Antonio/Pordomingo, Francisca (2008): "P.M.Vogl. I 20: bocetos de *progymnasmata.*" In: ZPE 167, pp. 167-192.

Fernández Delgado, José Antonio/Pordomingo, Francisca (2020): "Niveaux de réalisation des *progymnasmata*, de l'école à la littérature: *L'enkômion.*" In: Pierre Chiron/Benoît Sans (eds.), Les progymnasmata en pratique, de l'Antiquité à nos jours, Paris: Éditions Rue d'Ulm, pp. 17-41.

Fernández Delgado, José Antonio/Ureña Bracero, Jesús (1991): Un testimonio de la educación literaria griega en época romana: IG XIV 2012=Kaibel, EG 618, Universidad de Extremadura.

Fournet, Jean-Luc (1992): "Une éthopée de Caïn dans le Codex des Visions de la Fondation Bodmer." In: Zeitschrift für Papyrologie und Epigraphik 92, pp. 253-266.

Fournet, Jean-Luc (1999): Hellénisme dans l'Égypte du VIe siècle: la bibliothèque et l'oeuvre de Dioscore d'Aphrodité, Cairo: Institut Français d'Archéologie Orientale.

Gibson, Craig A. (2008): Libanius' Progymnasmata: Model Exercises in Greek Prose Composition and Rhetoric, Translated with an Introduction and Notes, Atlanta: Society of Biblical Literature.

Iturralde Mauleón, Cristina (in press): Progymnásmata en papiro y otros materiales, Universidad de Navarra (PhD thesis).

Jarcho, V. (1999): "P. Oxy. 35437: A True Ethopoea?" In: Eikasmós 10, pp. 185-199.

Kennedy, G. A. (2003): Progymnasmata: Greek Textbooks of Prose Composition and Rhetoric, Leiden-Boston: Brill.

Kraus, Manfred (2020): "La pratique des Progymnasmata dans les écoles du XVe au XVIIIe siècle à travers des traductions latines d'Aphthonios." In: Pierre Chiron/Benoît Sans (eds.), Les Progymnasmata en pratique de l'antiquité à nos jours, Paris: Éditions Rue d'Ulm, pp. 267-282.

Lentano, Maria (2018): "'Onde si immolino tre vergini o più'. Un motivo mitologico nella declamazione latina." In: Maia 70/1, pp. 10-27.

Luppe, W. (1991): "Literarische Texte. Drama." In: Archiv für Papyrusforchung 37, pp. 82-83.

Marrou, Henry-Irenee (1970): Historia de la educación en la antigüedad, 2nd ed., Buenos Aires: Eudeba (Translation of the 3rd French edition, Paris, 1955).

Miguélez-Cavero, Laura (2008): Poems in Context. Greek Poetry in the Egyptian Thebaid 200-600 AD, Berlin: De Gruyter.

Miguélez-Cavero, Laura (2013): "Rhetoric for a Christian Community: The Poems of the Codex Visionum." In: A. J. Quiroga Puertas (ed.), The Purpose of Rhetoric in Late Antiquity. From Performance to Exegesis, Tübingen: Mohr Siebeck.

Patillon, M. (2008): Corpus Rhetoricum. En annexe: Pseudo-Hermogène, Progymnasmata, Textes établis et traduits par M. Patillon, Paris: Belles Lettres.

Patillon, Michel/Bolognesi, Giancarlo (1997): Aelius Theon, *Progymnnasmata*, Texte établi et traduit, Paris: Belles Lettres.

Pechstein, Nikolaus (1998): Euripides Satyrographos, Stuttgart: B. G. Teubner.

Pordomingo Pardo, Francisca (2007): "La recepciónn de Eurípides en la escuela: el testimonio de los papiros." In: J. B. Bañuls/F. De Martino/C. Morenilla (eds.), El teatro Greco-latino y su recepción en la tradición occidental 2, Bari: Levante Editori, pp. 255-296.

Ramón García, Daniel (2001): "Mitógrafos y progymnasmata: Paléfato, Heráclito y el anónimo vaticano como ejercicios de retórica." In: Euphrosyne: Revista de Filología Clássica 39, pp. 277-284.

Reche Martínez, M. D. (1991): Teón, Hermógenes, Aftonio, Ejercicios de retórica, Madrid: Biblioteca Clásica Gredos.

Ureña Bracero, Jesús (1999): "Homero en la formación retórico-escolar griega: etopeyas con tema del ciclo troyano." In: Emerita 67/2, pp. 315-339.

Ureña Bracero, Jesús (2005): "El uso de fuentes literarias, recursos retóricos y técnicas de composición en etopeyas sobre un mismo tema." In: E. Amato and J. Schamp (eds.), 'ΗΘΟΠΟΙΙΑ. La representation de caractères entre fiction scolaire et réalité vivante à l'époque impériale et tardive, Salerno: Hylluss, pp. 93-111.

Ureña Bracero, Jesús (2007): "Algunas consideraciones sobre la autoría de los *progymnasmata* atribuidos a Libanio." In: José A. Fernández Delgado/Francisca Pordomingo/Antonio Stramaglia (eds.), *Escuela y literatura en Grecia Antigua*, Università degli Studi di Cassino, pp. 645-689.

The Panathenaic Prize-Amphorae as Communication Media

Martin Streicher

ABSTRACT

The Panathenaic amphorae, which were produced on behalf of the polis (city-state) and distributed to the winners of sport competitions (agons) at the Great Panathenaia,[1] are typically decorated with depictions of Athena and the specific agons. Beyond their primary use as awards, they traveled to various places throughout the Mediterranean world via economic and social networks, where they were introduced into sepulchral, sacred, and profane contexts. The adoption of the vase form and its iconography in different media proves its popularity, although its reception strongly depended on the respective contexts of its use and on the cultural influence of the recipients, among other things. This vase type is a medium through which central elements of Greek culture were communicated to a large number of recipients in generally understandable language for over six centuries.

INTRODUCTION

Within popular science, the aspects of medium, message/topic, language, addressee, and transfer process are of central importance. The choice of medium and language (in this case, antique ceramics and the images on the vases) were decisive factors for conveying the desired subjects to a large group of addressees.

1 The Great Panathenaia was the largest festival in ancient Athens. It was celebrated every four years from the 6th century BC to the 4th century AD in honor of Athena, the patron goddess of Athens. In addition to its religious components, the sports competitions were a great attraction for the international audience and participants.

The ceramics examined in this article are Panathenaic amphorae, which were produced on behalf of the city-state and distributed to the winners of hippic, gymnic, and, in rare cases, musical agons during the Great Panathenaia—the city festival of the polis Athens—which was celebrated every four years.[2]

Figures 1 and 2: Panathenaic Amphorae, Copenhagen, Ny Carlsberg Glyptotek 3606 (Side A and B)

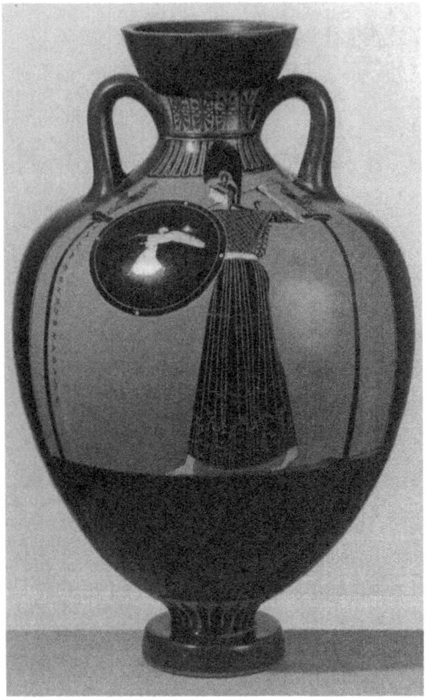
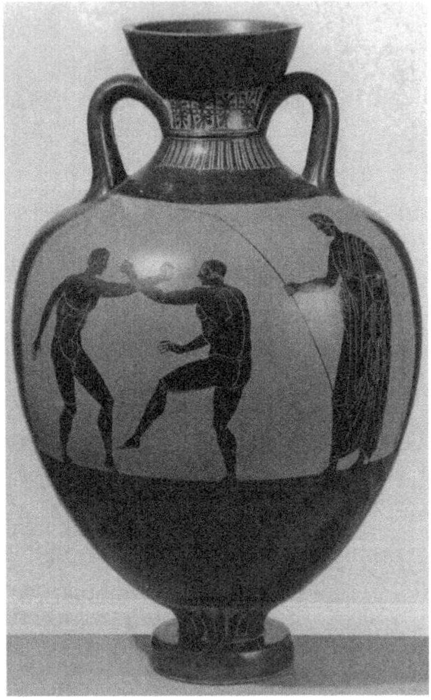

Source: Bentz, 1998, pl. 77, 5 172

The iconography of the vase (i.e., its themes) is marked by the goddess Athena depicted on the display side (*Fig. 1*) and the reproduction of the competition on the back side (e.g., chariot racing, running, boxing; see *Fig. 2*). Optional picture elements complete the scheme (e.g., *Níkai*, column figures, judges). Starting in the 4th century B.C., the inscriptions on the vases refer to the officials who were responsible for organizing the festival.

This article will focus on mythological and athletic topics, since, in contrast to historical aspects, these were communicated pictorially on the prize amphorae—and not in written form, as was the case with the names of the officials. Besides the myths, which are virtually omnipresent on antique vases, the second

2 For more on the Panathenaeic prize amphorae, see Bentz's post-doctoral thesis (1998), and also Eschbach (2017), who significantly expands the material basis of study.

subject treated in this article, athletics, was also frequently depicted (competition, training, award ceremonies, preparation, and follow-up). While on the prize amphorae Athena (myth) is represented on the display side and the competition (sport) on the back side, on other vase types there are also depictions of mythological sports scenes like those involving Patroklos and Pelops (Lewis 2009). Frequently, vessels with myth and sports representations were exported (ibid: 137), since they represented omnipresent themes in Greek culture. Their popularity makes them ideal test cases because "culture exists in internalized cultural knowledge in the members of the culture" (van de Put 2017: 76).[3] This is particularly true for the Panathenaic prize amphorae, which are very well suited for this type of examination due to their standardized painting scheme, our awareness of the producer and the occasion of production, the comparatively large number of preserved objects in various contexts in the Mediterranean region,[4] and their long production period. Nevertheless, the focus of this article lies not on existing knowledge of myths and sports, but on their pictorial dissemination through Greek amphorae.

MANUFACTURE

Knowledge of the consumer (the Athenian citizen), the production event (the Great Panathenaia), and the primary function of this central Athenian vase type (the awarding of a prize) allows a direct understanding of the scenes depicted on the amphorae. In addition to the representation of Athena, the deity of the polis, pictures of the competitions taking place there can also be found on the vases. The other (in part, mythical) persons represented on the vases (Nike as the goddess of victory, the competition judge, the officials responsible for the games) are also related to the primary reason for the production of the vases, i.e., the awarding of prizes at competitions. The intention of the producer, and thus the primary function of the pictures, is not to be found in the expansion of the consumer/viewer's knowledge, but rather in the domain of marketing for organizing games of panhellenic significance. The questions of whether this is the only possible reading of the pictures and what significance the ceramics would obtain in their further use (van de Put 2017: 75) will be taken up again later in this essay.

In addition to these content-related dimensions, the choice of medium and technology is also of interest for an understanding of the communication and reception processes, since both were already determined during the production of the

3 Concerning Pierre Bourdieu and Anthony Giddens.

4 The distribution area includes present-day Greece, Italy, Cyprus, Egypt, Libya, Turkey, the Levant, and the Black Sea region.

vases. The prize amphorae are the only type of vases that were produced using the black-figure technique, and this for a period of about six centuries—this will again be important for their reception and imitation. By choosing a painted vase as a prize, Athens clearly distinguished itself from other competition venues with prizes such as crowns and money, thus enabling greater communication potential.

DISTRIBUTION

Within the Panathenaic festival, the amphorae were awarded to successful athletes—up to 140 copies per winner, which differs from most other venues, where as generally painted ceramics were not awarded as prizes. After their primary use as a prize, the vessels reached various locations throughout the Mediterranean via economic and social networks (Kotsidu 2001). The distribution process of ceramics was subject to varying conditions in ancient times and cannot always be safely reconstructed by us today. Influencing factors include the place of production, the political situation, the purpose of use, the contents of the container, the place of discovery, and the distribution actors, who were not necessarily traders (Lund 2014: 301). In addition to the trade in (luxury) goods, the exchange of gifts and private property as well as, in some cases, the spoils of war can be seen as possible factors for the spread of objects in antiquity (Bol 2008).[5] Just how important the images were for the buyers has been already demonstrated elsewhere with reference to general ceramic production (Lewis 2009: 140), so this can also be safely assumed for the Panathenaic amphorae.

In contrast to the majority of antique ceramics, the type of vase in focus here was not primarily produced for a sales market; it was therefore not primarily involved in the cycle of supply and demand (Lewis 2009: 146). But trademarks on the amphorae prove that the vases were sometimes traded multiple times (Bentz 1998: 92-94). This shows that their distribution in the Mediterranean region did not take place exclusively through the victors themselves—who consecrated them in shrines, donated them for public occasions, or used them for private purposes—but also through the well-known trade routes for Athenian ceramics (Bentz 1998: 116-119). Thus it can be said that the individual amphorae did not necessarily remain in one place, and that their object-biographies often included several different stations.

5 While it is known that sculptures were occasionally repositioned, for ceramics this can only be assumed to a very small extent.

CONTEXTS

The distribution of the vases following the act of prize-giving took place in connection with their secondary uses in various sepulchral, sacral, and profane contexts, which were not always directly tied to Athens or the festival itself (Schierup 2017: 198).[6]

Ancient sanctuaries, with their frequent political, social, and economic functions, served as contact zones for "Händler, Gesandte, Söldner, Teilnehmer an überregionalen Kultfeiern" ("traders, envoys, mercenaries, participants in superregional cult celebrations," Schollmeyer 2008: 207). The concrete functions of the amphorae cannot always be precisely determined, so that even the purely practical uses of the vase or the oil contained within it cannot be excluded. The prize amphorae introduced in this context as votives or, respectively, as oil donations were not used solely as a religious communication medium between the consecrator and the deity, e.g., as a sign of gratitude for victory at the competition. The public positioning of a prestigious object also acts as a form of communication with the cult community, which shows the success and victory of an individual/family (Frevel 2007: 234-235). The fact that this applies not only in the case of representative building projects of the aristocratic upper class (Rohn 2008: 182), but also in the cases of the winners of gymnic and hippic agons, can be seen not only regarding the prize amphorae, but also regarding the statues of the winners and other forms of consecration.

Besides finds in residential houses (Bentz 1998: 106) and graves (Bentz 1998: 95-102), prize amphorae were also used in the public sector in the context of communal meals during or after festivals (Bentz 1998: 108). Panathenic prize amphorae were also found in areas where their oil was used for cooking, to cite another example. These vases were reserved for a more exclusive 'private' target group, which naturally implies influence on the respective target group. The communication possibilities and the effect of the objects vary considerably in different contexts (Schmidt 2009: 9). Thus, the reception of the images on the vases depended fundamentally on the contexts they were viewed in.

Under the circumstances outlined here, aspects such as quantity, visibility, accessibility, exhibition duration, and frequency are of central importance for the communication framework. In addition, the reading of the images and therefore the process of their reception was also heavily influenced by the context of their installation (van de Put 2017: 77). Thus, "Vasen [sind] schließlich immer auch Teil des kulturellen Kontextes an dem jeweiligen Fundplatz" ("vases [are] always part of the cultural context at the respective site of discovery," Fless 2003: 241), whereby both (i.e., the vases and the context) are subject to an interrelation: It will still be necessary within the framework of the reception process to discuss the fact that the contexts do not all belong to the Greek cultural sphere. The multitude of contexts

6 On the theory of context integration in the fields of sociology and anthropology, cf. van de Put 2017.

and their individual, sometimes clearly differing circumstances as well as the limited scope of the work do not allow for a definite analysis of the individual factors influencing the vases' reception.[7] It should be sufficient to point out that the prize amphorae were presented to a large public in various areas of life.

ADDRESSEES

The vases' different uses in various contexts already demonstrate that the group of recipients was not a homogenous one, but that it varied depending on the context of use. At the prize-giving ceremony, 1472 to 1567 amphorae per festival were publicly awarded to victorious athletes in archaic-classical times (Bentz 1998: 17). With such a large number of vases coming into circulation, it can be assumed that these were viewed by a festive community from all over Greece. Many of the vessels did not leave Athens, but were placed in local sanctuaries; in other cases, the oil in them was used for communal meals during or after the festival. The vases were thus already exhibited to a larger public immediately after they were handed over in Athens.

The post-award distribution mechanisms could hardly be controlled by the polis. However, the large number of amphorae filled with oil, which were handed over to individual persons, prove that the winners served as multipliers for the distribution of the amphorae, especially since they could hardly consume such large quantities of oil themselves. It can, therefore, be assumed—with all due caution—that the winners (and probably also the festive community[8]) have to be regarded as the primary recipients of the prize amphorae, but that the persons of the other contexts outlined here must at least be considered as secondary addressees. The presence of different, changing groups of recipients[9] of the prize amphorae requires their consideration in their respective contexts when investigating the phenomena of their communication (Gehrke 2016: 5). One problem is that in Southern Italy, for example, it is not always possible to clearly distinguish between indigenous and Greek groups, which makes it difficult to differentiate between intra- and intercultural communication and interaction (Schweizer 2012: 15).

While the victors, the consecrators, and the grave-owners were usually socially higher-placed persons, this is not to be assumed in principle for the partic-

7 A context with which archaeologists are often confronted is not the installation context, but the disposal point of the objects; this often makes further conclusions about the use and installation of the vases very difficult.

8 Competitors, cult personnel, Athenian citizens, guests, and officials, among others.

9 Public: festival participants, athletes, winners, cult personnel, visitors to shrines, merchants, mourners, and participants at communal meals—private: friends, guests, and family members.

ipants of the celebrations, the people partaking in public meals, and the visitors to the sanctuaries. The prize amphorae thus reached many different sectors of ancient society.

RECEPTION

So far, it has been demonstrated that prize amphorae were made accessible to different groups of recipients in their respective contexts. The next logical question is how these groups of people received the images on the vases.

One must distinguish between the recipients who belonged to the Greek *Koinḗ* group (or "community") and those who did not. While with the former a fundamental understanding of the subjects on the vases may be assumed because they belong to the same cultural group,[10] with the latter this cannot simply be taken for granted. While for the Greeks in Southern Italy it is presumed that the cultural imprint of the Greeks living in the motherland was largely the same (Paleothodoros 2003: 220), for the indigenous groups from Southern Italy "spezifisch regionale [...] Sehgewohnheiten" ("specific regional [...] viewing habits," Langner 2005: 64-65) have to be included, which offer alternatives to the specific Greek interpretations of the images. According to Martin Langner (Langner 2005: 62), the interpretation of the images was not necessarily imported as well. The prize amphorae presented two very central cultural elements of Greek culture to groups of non-Greek people, namely, the themes of myth/religion and sport. They thus served as an essential factor in the cultural exchange process. While the vases had a rather low economic value, they and their repertoire of images were valuable mediators of cultural knowledge. Here "Vorstellungen, Werte, Überzeugungen, Mentalitäten und Lebensstile" ("perceptions, values, convictions, mentalities, and lifestyles," Schmidt & Stähli 2012: 9) were presented. But were they always recognized?

Patrick Schollmeyer was able to show with reference to small-scale Cypriot sculptures that processes of reception often took place in the form of a synthesis of local and foreign elements (Schollmeyer 2008: 212). With reference to the images on the amphorae, it can be seen that they were not simply accepted in their original interpretation, but were reinterpreted and adapted when necessary. The fact that the images on the vases were transformed at reception is shown, for example, by Graeco-Scythian satyr portraits in the Bosporan Kingdom, i.e., a Greek image convention with adapted hair and beard (Langner 2005: 60). The same can be seen with an

10 This is, admittedly, an over-simplification of the reality of ancient life, in which there was great internal and external heterogeneity in Greek culture; this is particularly evident in the Hellenistic period, but is not exclusively valid for this epoch.

Etruscan name being included on a clearly Greek mythic depiction on an olpe from 7th-century (BC) Cerveteri (Mus. Naz. 110976, Schweizer 2012: 17-20 *Fig.* 2). Moreover, in Etruria the Amazons, for example, were reinterpreted so that positive qualities such as bravery and beauty were attributed to them (Puritani 2012: 109). Nevertheless, there was great demand in Etruria for painted Greek ceramics, including prize amphorae. The Perizoma Group, by depicting athletes dressed in a *perizoma* (loincloth) in almost 50 vessels from the Nikosthenes workshop —and thereby differing from the Greek depictions of nudity in sports—created a niche product in Etruria (Lewis 2009: 136). The acceptance of Athenian reproductions is also evident in the nude depictions of athletes on vases and mural paintings, such as in those from the Tomba degli Auguri and Tomba delle Bighe in Tarquinia (Steingräber 1985: nos. 42 and 47). However, there is no evidence that the Etruscans adopted Greek nudity, whether iconographically or in the real world (Wehgartner 2012: 63).

Even these few examples show that Greek and indigenous aspects were partly mixed up with—and thus that they influenced—cultural identity (Langner 2005: 60). We find this "semantischen Eklektizismus" ("semantic eclecticism," Pirson 2008: 319) in various domains, for example in architectural innovations.[11] The prize amphorae demonstrated particularly clearly that objects could also be used and understood contrary to their original purpose (that of the award ceremony), e.g., as consecrations, funeral gifts, and commercial objects.[12] It is not always possible to decide if the special actions (sports) or ideas (myths) depicted on them were always transmitted to the viewer in their original meaning, as the above example of nudity in Greek sport shows (Bumke 2008).[13] Thus, "die Nutzung von Griechischem als soziales Distinktiv" ("the use of Greek as a social distinction," Gehrke 2016: 10) often took place, so that an identification of the persons cannot be assumed with the explicitly depicted contents of the images but rather with their implicit meaning. The images, therefore, offer points of contact with the viewers (Langner 2005: 66), so that Athenian images could be linked to their own cultural background. The representations can be read literally (e.g., a boxing scene or chariot race) as well as metaphorically (e.g., Greekness or victoriousness), as has also been demonstrated for special anointing vessels (Heinemann 2009: 170).

The depictions on prize amphorae are also fictional images that provide insight into certain collective ideas/themes (Schmidt 2009: 13). To what extent the number and presence of the vases in the different contexts can be linked to ancient everyday

11 For a review of this topic, see, e.g., Pirson 2008, who discusses the tomb facades of Petra, the caryatids of the Heroon of Perikle in Limyra, or the Red Basilica in Pergamon, where the material remains clearly document the combination of indigenous and foreign elements.

12 An aspect that has also been proven for other ceramics; see Langner 2005: 55.

13 On the images of drinking customs found on ceramic finds as possible factors in the transfer of ideas, see Krausse 2003: 210.

reality or whether the vases provide concrete reference to the values of the Athenian elite or to Greekness as a whole cannot be answered in general terms (Lewis 2009: 135). The intercultural transfer process depended fundamentally on local conditions—a circumstance that applied not only to the reception of images, but also, for example, to the adoption of technical and architectural elements (Seeher 2008: 12). The open-ended quality of the images allowed viewers to read their own content into them (Mannack 2012: 51). They conveyed generally understandable ideas, made values and conventions visible, and offered "diskursive Strukturen" (discursive structures, Schmidt 2009: 12). The images were therefore not arbitrary, but rather deliberately chosen to communicate complex topics such as the socially constitutive elements represented by sport, religion, competition, and democracy (Carpenter 2009: 158). The paintings on prize amphorae are what Luca Giuliani calls descriptive pictures (Giuliani 2003), i.e., representations of generally valid processes. This makes them particularly suitable for reading (Lissarrague 2009) by different, heterogeneous viewers in various contexts. This also shows that the exclusive attribution of sports and myth images to Greek culture is too generalized (van de Put 2017: 76). Nevertheless, sport is often used as a sign of Greek identity. Philip II of Macedonia, for example, appears on coins as a rider on horseback with a taenia and a palm branch—a clear reference to the victory at the Olympic horse race in 356 BC (Langner 2005: 59). Myth images served as mediators of values, ideas, and behaviors and were thus identity-forming (Meyer 2009: 30). They were also an important means of communication. Thus, by means of myths represented in Greek iconography, for which there was great demand—e.g., in Apulia—ideas could be communicated to a non-Greek (in this case, Italic) audience (Carpenter 2009).

IMITATIONS AND ADAPTATIONS

The adoption of the vase form and its iconography in different media—such as coins (Thompson 1961), vase paintings (Frielinghaus 2001, see *Fig. 3*), mosaics (Valavanis 2001; see *Fig. 4*), wall paintings, and marble objects (Valavanis 2001)—and the production of local imitations (Musco et al. 2011) prove the popularity of prize amphorae and the reception of the themes distributed with them. It is striking that the subjects of the images were adopted, but not the inscriptions themselves, i.e., the references to the historical personalities and the year of competition. The reception was limited to abstract, easily adaptable elements of iconography, whereas concrete historical references played a negligible role. Werner Petermandl also points to the great appeal of sport, and the "Lust am Schauen" (desire to watch, Petermandl 2005), an indication of why the prize amphorae received such a great response throughout the Mediterranean region. However, not only the pictorial decoration itself but also its technique was adopted for the vase imitations. The reproductions not only made reference to deep-rooted and age-old elements of Greek culture, but also reflected the traditional technique of the amphorae, even at a time when it was otherwise no longer used. The choice of this traditional technique, perfected in Athens, was not arbitrary, but closely linked to the content communicated in the form of the vase and through its iconography. Technology was therefore not just a means to an end, but an integral part of communication.

Figures 3 and 4: Kýlix with Depiction of Athena and Panathenaic Amphora, Athens, Kerameikos Inv. 1941; Mosaic with Panathenaic Amphora, Delos, House of the Masks

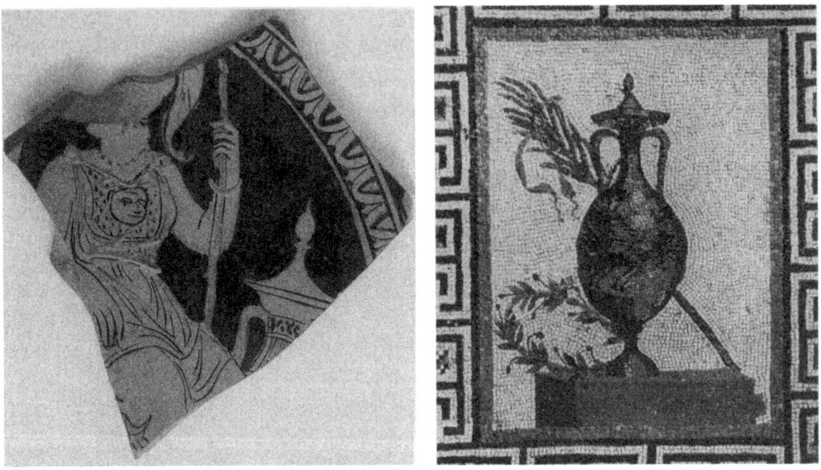

Source: Frielinghaus 2001, pl. 41, 1; Valavanis 2001, pl. 45, 7

In addition to technical aspects, there is another element that has not yet been taken into account, one which is closely linked to the content of the images and the production occasion: the shape of the vase. The iconography of the prize amphorae was not adopted in all the abovementioned media; for example, only its form was reproduced on coins, architectural ornamentation, and, in part, on vase paintings. The form was also adopted in Lucanian/Apulian Southern Italy, where it became the main form of the red-figure amphorae around the middle of the 5th century B.C. (Schierup 2017: 198), and where it underwent its own development (Schierup 2017: 206). It was mainly the heroic-athletic values of the aristocracy that were served by this reception, and the vessel was used particularly frequently in funerary contexts. Only with further development, due to local productions, did the shape of the vessel acquire a more general meaning for a larger clientele, so that its original meaning as a prize of the Athens polis was superimposed and entered into the socio-political culture of the Greeks and ancient Italians (Schierup 2017: 206-207). While the form was being adapted, the decoration and technology were now clearly different from the original models.

Figures 5 and 6: Lucanian Amphora, San Simeon 529.9.614; Lucanian Amphora, Paris, Louvre CA 308

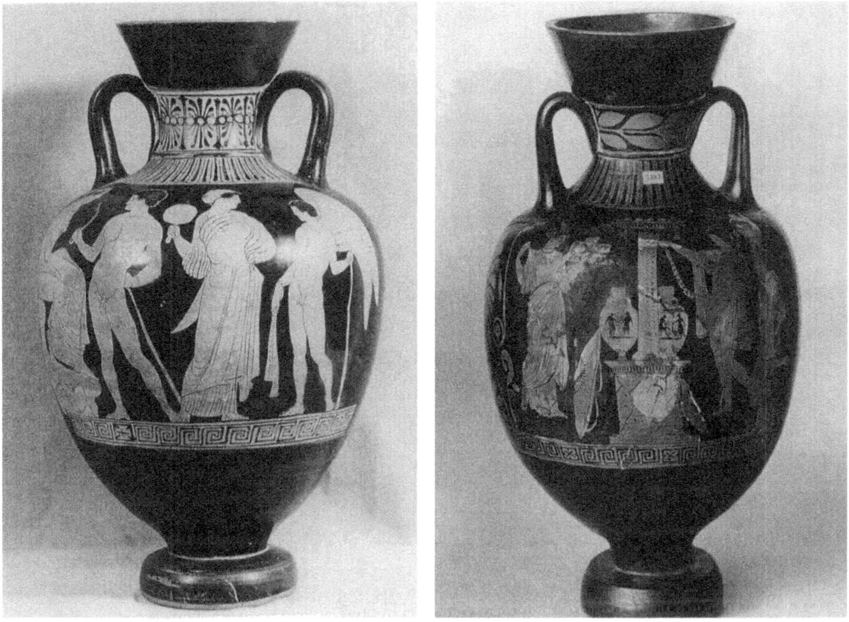

Source: Neils 2001, pl. 36, 1; Neils 2001, pl. 36, 2

The iconography and intended use of the vases, produced for the local market, were modified in such a way that they stand in clear contrast to their original use in Greece (see *Figs. 5* and *6*).[14] However, the form as a reference to the heroic bravery and *arete* of the deceased shows that Greek athletics influenced the burial customs of the Italic community (Schierup 2012). How complex such transfer processes could also be in the Greek motherland is demonstrated by the adaptations of Athenian elements to local vessel forms in neighboring Boeotia, which were used there despite socio-political conflicts (Sabetai 2012). There is thus a connection between visual and content-related aspects (Winkler-Horaček 2003: 227) and therefore also a close connection between vessel (form), image (theme), and practice (use) (Heinemann 2009).[15]

14 The vessels in Southern Italy were no longer produced with the black-figure technique, but rather with the red-figure technique. They were not used as prize vessels, but were most commonly found in funerary contexts. The themes depicted were freely chosen by the painter and client and were not restricted to representations of competitions and Athena. The ornamental decor was also modified.

15 For a brief explanation of the history of research on the connection between the shape of the vessel and the paintings on it, see van de Put (2017).

This connection extends so far that it has been suggested that the shape of the vessel may be interpreted as a symbol of Athens, as the coins of the New Style from the 2nd century B.C. demonstrate (Thompson 1961). On these coins, the vase is represented together with another central symbol of Athens: the owl.

CONCLUSION

The amphorae produced in Athens for the Great Panathenaia were generally accessible to a broad audience due to their occurrence in sacral, profane, and sepulchral contexts in the Mediterranean region. Their pictorial program takes up two of the most popular elements of Greek life: sports and myth. The images were depicted in an easily understandable visual language; and thanks to their open-ended semantic quality, the recipients were able to adapt and (re-)interpret them according to their individual living environments, an aspect that also determined the popularity of the medium. In addition to the iconography of the pictures, both the medium, i.e., the vase shape, and the technique (black-figure vase painting) were closely linked to the contents, so that they could sometimes also be used additively. In this way, the vase shape became so popular that its content was clearly linked to the agonal core of the Panathenaic festival and the polis and could, therefore, be understood without the iconography. In addition to these medium-inherent aspects, it should be noted that viewer's respective context and cultural imprint were central to the process of reception and were closely connected with iconography and meaning (van de Put 2017: 73-74; and Schmidt 2003).

The prize amphorae show that "images on vases cannot be approached as documents constructed with the aim of conveying straightforward information" (Sabetai 2009: 103). Nevertheless, they transmit a wealth of information on and associations with central aspects of Greek culture, which they conveyed to a large audience through their occurrence in almost all areas of life in the ancient world. "Die Gefäße und Gefäßbilder als Medien sind es jedoch, mit denen Wissen und Deutungen in Bewegung gesetzt werden konnten." ("It is with the vessels and the images on those vessels, considered as media, that knowledge and interpretations could be set in motion." Schweizer 2012: 22).

REFERENCES

Bentz, Martin (1998): Panathenäische Preisamphoren. Eine athenische Vasengattung und ihre Funktion vom 6.-4. Jahrhundert v. Chr., Basel: Vereinig. d. Freunde antiker Kunst.

Bol, Renate (2008): "Die Statue des Apollon Termintheus aus Myus und ihre Neuaufstellung in Milet: Ein Beispiel von Kultuübertragung oder Zeugnis politischer Annexion?" In: Renate Bol/Ursula Höckmann/Patrick Schollmeyer (eds.),

Kult(ur)kontakte: Apollon in Milet/Didyma, Histria, Myus, Naukratis und auf Zypern; Akten der Table Ronde in Mainz vom 11.-12. März 2004. Internationale Archäologie Arbeitsgemeinschaft, Tagung, Symposium, Kongress 11 Rahden, Westf.: Leidorf, pp. 145-153.

Bumke, Helga (2008): "Zeugnisse für Kulturtransfer aus dem archaischen Apollonheiligtum von Didyma." In: Renate Bol/Ursula Höckmann/Patrick Schollmeyer (eds.), Internationale Archäologie Arbeitsgemeinschaft, Tagung, Symposium, Kongress: Vol. 11. Kult(ur)kontakte: Apollon in Milet/Didyma, Histria, Myus, Naukratis und auf Zypern; Akten der Table Ronde in Mainz vom 11.-12. März 2004, Rahden, Westf.: Leidorf, pp. 87-96.

Carpenter, Thomas H. (2009): "The Darius Painter: Text and Context." In: Schmidt/Oakley 2009, pp. 153-159.

Eschbach, Norbert (2017). Panathenäische Preisamphoren aus dem Kerameikos zu Athen. Kerameikos: Vol. 21. Wiesbaden: Reichert Verlag.

Fless, Friederike (2003): "Akzeptanz und Integration attischer Keramik des 4. Jhs. v. Chr. in die materielle Kultur griechischer Poleis und außergriechischer Gesellschaften – Pantikapion als Beispiel." In: Berhald Schmaltz/Magdalene Söldner (eds.), Griechische Keramik im kulturellen Kontext: Akten des Internationalen Vasen-Symposions in Kiel vom 24. –28. 9. 2001, Münster: Scriptorium, pp. 241-243.

Frevel, Christian (2007): "Geschenke für die Götter: Votive als Kommunikationsindikatoren in Heiligtümern der Bronze- und Eisenzeit in Palästina/Israel." In: Christian Frevel/Henner von Hesberg (eds.), Schriften des Lehr- und Forschungszentrums für die Antiken Kulturen des Mittelmeerraumes – Centre for Mediterranean Cultures (ZAKMIRA): Vol. 4. Kult und Kommunikation: Medien in Heiligtümern der Antike, Wiesbaden: Reichert, pp. 183-245.

Frielinghaus, Heide (2001): "Darstellungen panathenäischer Preisamphoren in der attischen Vasenmalerei." In: Martin Bentz/Norbert Eschbach (eds.), Panathenaika. Symposion zu den Panathenäischen Preisamphoren. Rauischholz hausen 25.11.-29.11.1998, Mainz: Philipp v. Zabern, pp. 147-159.

Gehrke, Hans-Joachim (2016): "Von der Materialität zur Identität. Methodologische Überlegungen zu einem zentralen Problemfeld der archäologisch-historischen Wissenschaften." In: Holger Baitinger (ed.), RGZM-Tagungen: Band 27. Materielle Kultur und Identität im Spannungsfeld zwischen mediterraner Welt und Mitteleuropa: Akten der Internationalen Tagung am Römisch-Germanischen Zentralmuseum Mainz. 22.-24. Oktober 2014, Mainz: Verlag des Römisch- Germanischen Zentralmuseums, pp. 1-13.

Giuliani, Luca (2003): Bild und Mythos: Geschichte der Bilderzählung in der griechischen Kunst, Munich: C H Beck.

Heinemann, Alexander (2009): "Bild, Gefäß, Praxis: Überlegungen zu attischen Salbgefäßen." In: Schmidt/Oakley 2009, pp. 161-175.

Kotsidu, Haritini (2001): "Zur Verbreitung der Preisamphoren." In: Martin Bentz/Norbert Eschbach (eds.), Panathenaika. Symposion zu den Panathenäischen Preisamphoren. Rauischholzhausen 25.11.-29.11.1998, Mainz: Philipp v. Zabern, pp. 55-61.

Krausse, Dirk L. (2003): "Griechische Keramik nördlich der Alpen. Überlegungen zur Funktion der attischen Trinkschalen aus dem Kleinaspergle." In: Bernard Schmaltz/Magdalene Söldner (eds.), Griechische Keramik im kulturellen Kontext: Akten des Internationalen Vasen-Symposions in Kiel vom 24.-28. 9. 2001, Münster: Scriptorium, pp. 208-210.

Langner, Martin (2005): "Barbaren griechischer Sprache? – Die Bildwelt des Bosporanischen Reiches und das Selbstverständnis seiner Bewohner." In: Friederike Fress/Mikhail Treister (eds.), Bilder und Objekte als Träger kultureller Identität und interkultureller Kommunikation im Schwarzmeergebiet. Kolloquium in Zschortau/Sachsen vom 13.2.-15.2.2003, Rahden, Westf.: Leidorf, pp. 53-66.

Lewis, Sian (2009): Athletics on Attic pottery. Export and imagery. In V. Nørskov/L. Hannestad (eds.), Analecta Romana Instituti Danici Supplementum: Vol. 41. The world of Greek vases, Roma: Ed. Quasar, pp. 133-148.

Lissarrague, François (2009): "Reading Images, Looking at Pictures, and After." In: Schmidt/Oakley 2009, pp. 15-22.

Lund, John (2014): "Pots and Politics: Reflections on the Circulation of Pottery in the Ptolemaic and Seleukid Kingdoms." In: Pia Guldager Bilde/Mark L. Lawall (eds.), Black Sea studies: Vol. 16. Pottery, peoples and places: Study and interpretation of Late Hellenistic pottery, Aarhus: Aarhus Univ. Press, pp. 297-305.

Mannack, Thomas (2012): "Liebesverfolgungen in Unteritalien." In: S. Schmidt/A. Stähli (eds.), Beihefte zum Corpus Vasorum Antiquorum Deutschland: Vol. 5. Vasenbilder im Kulturtransfer: Zirkulation und Rezeption griechischer Keramik im Mittelmeerraum, Munich: Beck, pp. 51-58.

Meyer, Marion (2009): Zur Relevanz bildlicher Darstellungen mythischer Figuren. In: Schmidt/Oakley 2009, pp. 23-32.

Musco, Stefano/Piacentini, Mario/Candida Felici, Anna/Fronterotta, Gabriele/Molari, Cristina/Rizzo, Giorgio/Vendittelli, Margherita (2011): "Le anfore pseudo-panatenaiche da Roma e dintorni: Un fenomeno di atticismo nell'artigianato tardo-ellenistico locale?" In: Sabrina Gualtieri (ed.), Scienze della terra: Vol. 44. La ceramica e il mare: Il contributo dell'archeometria allo studio della circolazione dei prodotti nel mediterraneo; atti della XII Giornata di Archeometria della Ceramica, Genova, 10-11 aprile 2008, Roma: Aracne, pp. 69-85.

Paleothodoros, Dimitris (2003): "Dionysiac Imagery on Attic Red-Figured Vases Found in Italy." In: B. Schmaltz/M. Söldner (eds.), Griechische Keramik im kulturellen Kontext: Akten des Internationalen Vasen-Symposions in Kiel vom 24.-28. 9. 2001, Münster: Scriptorium, pp. 220-222.

Petermandl, Werner (2005): "Geht ihr aber ins Stadion… Ein althistorischer Blick auf das Sportpublikum wie es war, wie es ist und wie es immer sein wird." In: Mathias Marschik (ed.), Das Stadion: Geschichte, Architektur, Politik, Ökonomie, Vienna: Turia + Kant, pp. 127-153.

Pirson, Felix (2008): "Austausch und Inspiration – Zusammenfassende Gedanken zur Bedeutung von Kulturkontakten als Auslöser architektonischer Innovationen." In: Pirson/Wulf-Rheidt 2008, pp. 312-319.

Pirson, Felix/Wulf-Rheidt, Ulrike (eds.) (2008): Diskussionen zur archäologischen Bauforschung: Vol. 9. Austausch und Inspiration: Kulturkontakt als Impuls architektonischer Innovation; Kolloquium vom 28.- 0.4.2006 in Berlin anlässlich des 65. Geburtstages von Adolf Hoffmann, Mainz: von Zabern.

Puritani, Laura (2012): "Amazonen in Etrurien. Zur Rezeption attischer Vasenbilder am Beispiel einer Hydria des Polygnotos." In: Schmidt/Stähli 2012a, pp. 103-112.

Rohn, Corinna (2008): "Olympisches in Aizanoi – Griechisches in den anatolischen Bergen." In: Pirson/Wulf-Rheidt 2008, pp. 174-183.

Sabetai, Victoria (2009): "The Poetics of Maidenhood: Visual Constructs of Womanhood in Vase-Painting." In: Schmidt/Oakley 2009, pp. 103-114.

Sabetai, Victoria (2012): "Looking at Athenian Vases through the Eyes of the Boeotians: Copies, Adaptions and Local Creations in the Social and Aesthetic Culture of an Attic Neighbour." In: Schmidt/Stähli 2012a, pp. 121-137.

Schierup, Stine (2012): "A Heroic Emblem: The Cultural Transformation of the Panathenaic Amphora in Southern Italy." In: Stine Schierup/Rasmussenm Bodil Bundgaard (eds.), Gösta Enbom Monographs: Vol. 2. Red-Figure Pottery in its Ancient Setting: Acts of the International Colloquium Held at the National Museum of Denmark in Copenhagen, November 5-6, 2009, Aarhus: Aarhus University Press, pp. 118-132.

Schierup, Stine (2017): "The Reception of an Attic Prize Vessel: On Import and local Production of Amphorae of Panathenaic Shape in Southern Italy." In: D. Rodríguez Pérez (ed.), Greek art in context: Archaeological and art historical perspectives, London and New York: Routledge, pp. 198-210.

Schmidt, Stefan (2003): "Zur Funktion der Bilder auf weißgrundigen Lekythen." In: B. Schmaltz/M. Söldner (eds.), Griechische Keramik im kulturellen Kontext: Akten des Internationalen Vasen-Symposions in Kiel vom 24.-28. 9. 2001, Münster: Scriptorium, pp. 179-181.

Schmidt, Stefan (2009): "Hermeneutik der Bilder." In: Schmidt/Oakley 2009, pp. 9-14.

Schmidt, Stefan/Oakley, John H. (eds.) (2009): Beihefte zum Corpus Vasorum Antiquorum Deutschland: Vol. 4. Hermeneutik der Bilder: Beiträge zur Ikonographie und Interpretation griechischer Vasenmalerei, Munich: Beck.

Schmidt, Stefan/Stähli, Adrian (eds.) (2012a): Beihefte zum Corpus Vasorum Antiquorum Deutschland: Vol. 5. Vasenbilder im Kulturtransfer: Zirkulation und Rezeption griechischer Keramik im Mittelmeerraum, Munich: Beck.

Schmidt, Stefan/Stähli, Adrian (2012b): "Griechische Vasenbilder als Medium des Kulturtransfers." In: Schmidt/Stähli 2012a, pp. 9-13.

Schollmeyer, Patrick (2008): "Apollon in der zyprischen Kleinplastik: Ein Pradigma interkulturellen Religionstransfers?" In: Renate Bol/Ursula Höckmann/Patrick Schollmeyer (eds.), Internationale Archäologie Arbeitsgemeinschaft, Tagung, Symposium, Kongress: Vol. 11. Kult(ur)kontakte: Apollon in Milet/Didyma, Histria, Myus, Naukratis und auf Zypern; Akten der Table Ronde in Mainz vom 11.-12. März 2004, Rahden, Westf.: Leidorf, pp. 207-214.

Schweizer, Beat (2012): "Bilder griechischer Tongefäße in Mittelitalien und nördlich der Alpen: Medien der Hellenisierung oder Mediterranisierung, der Akkulturation oder der kulturellen Interaktion, der interkulturellen Kommunikation oder der Konstruktion kultureller Identität?" In: Schmidt/Stähli 2012a, pp. 15-25.

Seeher, Jürgen (2008): "Innovation im Bauwesen als Indikator für Kulturkontakt – Hethiter und Mykener als Fallbeispiel." In: Pirson/Wulf-Rheidt 2008, pp. 1-15.

Steingräber, Stephan (ed.). (1985): Etruskische Wandmalerei, Stuttgart et al.: Belser.

Thompson, Margaret (1961): The new style silver coinage of Athens, New York: American Numismatic Society.

Valavanis, Panos (2001): "Panathenäische Amphoren auf Monumenten spätklassischer, hellenistischer und römischer Zeit." In: M. Bentz/N. Eschbach (eds.), Panathenaika. Symposion zu den Panathenäischen Preisamphoren. Rauischholzhausen 25.11.-29.11.1998, Mainz: Philipp v. Zabern, pp. 161-173.

Van de Put, Winfred (2017): "Lost in Translation? Theoretical Implications of Considering Iconography in Context." In: Diana Rodríguez Pérez (ed.), Greek art in context: Archaeological and art historical perspectives, London and New York: Routledge, pp. 73-80.

Wehgartner, Irma (2012): "Die Sammlung Feoli: Attische und etruskische Vasen von der «Tentua di Campomorto» bei Vulci." In: Schmidt/Stähli 2012a, pp. 59-68.

Winkler-Horaček, Lorenz (2003): "Der geflügelte Menschenlöwe (Sphinx): Ein Bildmotiv in der frühgriechischen Vasenmalerei und sein Verhältnis zu den östlichen Vorbildern." In: Bernhard Schmaltz/Magdalene Söldner (eds.), Griechische Keramik im kulturellen Kontext: Akten des Internationalen Vasen-Symposions in Kiel vom 24.-28. 9. 2001, Münster: Scriptorium, pp. 225-228.

Popular Knowledge and its Rhetorical Use in Aristotle

María J. Martín-Velasco

ABSTRACT

This article describes and evaluates Aristotle's rhetorical use of popular scientific knowledge with a special focus on doxography and dialectic argumentation. To that end, we first review the definitions of 'direct investigation' (ἱστορία [*historía*], i.e., knowledge acquired by investigation) and 'indirect investigation' (ἔνδοξα [*éndoxa*], i.e., knowledge acquired from reputable popular opinions, which we know because this is what they tell us). After highlighting the role of Aristotle as a doxographer, we review those passages in which he indicates how indirect investigation, based on probabilities, should be used in scholarly argumentation. We also discuss how this idea of progress in the development of thought contrasts with his vision, in other works, of discontinuity and the destruction of societies, and how he considers wisdom (σοφία [*sophía*]) to be the engine of *palingenesis*, that is, of the rebuilding and evolution of thought and society. Finally, we refer to the different forms of treatment that direct and indirect investigation receive in different disciplines, be they academic, informative, or general in nature.

INTRODUCTION

In debates on the genealogy of popular science, Aristotle's contribution plays a fundamental role. He was the first scientist who considered that different types of sciences required different forms of investigation, argumentation, and dissemination. For some of them, direct observation of the facts (680a2-4) and a taxonomic register were considered to be sufficient. In other cases, induction and deduction would be necessary in order to establish general statements that make it possible for science to advance (491a12-28). In some sciences, he considered that the starting point for valid research should be the *éndoxa* (ἔνδοξα), the opinions of his wise predecessors. To this end, he undertook the great task of gathering knowledge prior to his own and using it as the basis of subsequent research. This repertoire has been used as an essential reference in all fields of popular science. Finally, we must emphasize the importance he attaches to dialectical argumentation, to the form of generalization that is the basis of popular science.

Aristotle's treatises, both those concerning the study of nature and those dealing with the human sciences, start, in accordance with his realist philosophy, from a point that seeks to delimit the subject area, either through an exposition of direct investigation or a compilation of what has been said by his predecessors. They then follow an argumentative or dialectical line which is considered to be most appropriate to that starting point and the nature of the issue under discussion. What we propose here is to describe and evaluate this whole process. First, we will consider the starting point, defining the notions of 'direct investigation' (ἱστορία) and 'indirect investigation' (ἔνδοξα), the scientific postulates defended by previous thinkers, following a line of reasoning based on premises whose truthfulness is only probable, not certain. The very nature of this method makes it possible to approach postulates that cannot be reasoned on the basis of the undisputed truth. We will then turn to those passages in which Aristotle indicates how indirect investigation should be used in scholarly argumentation. Closely linked to this, we will also address the issue of the continuity and discontinuity of thought and society. Finally, we will consider how ἱστορία and ἔνδοξα, the observation and verification of data and the examination of experts' popular opinions, receive different treatments depending on the discipline, be they of an academic, informative, or general character. Here, we will look in particular at the collection of historical data and its use in the human sciences.

THE ETYMOLOGICAL MEANING OF THE TERM ἹΣΤΟΡίΑ (HISTORÍA)

Let us begin by analyzing the initial starting point of Aristotle's treatises: data taken directly from observation and its subsequent classification. This preliminary work is referred to as ἱστορία (historía, Rey 2001: 314), a term with Indo-European origins (Muller 1926). The root *wid-/*weid-, 'see,' is the same root that we find in lat. 'uideo,' in gr. '-ιδ-' (aorist of the verb ὁράω, 'to see'), and also in gr. οἶδα, 'to know,' which comes from the past perfect of the same root (Louis 1955: 39). Thus, it conveys the sense of 'to know' as direct knowledge of 'having seen' (Moradiellos 2001: 36). The same root can be found in ἵστωρ, 'he who has seen,' 'the direct witness.' Hence the word also means 'observation,' 'inquiry,' 'investigation,' and, in some sense, could also imply 'verification' due to the link that the term establishes between knowledge and direct testimony. In the Aristotelian corpus, the word ἱστορία acquires different traits depending on the context (Weil 1960: 89), and is also found frequently with other terms.

In *Parts of Animals* (*PA* 680a2-4), Aristotle affirms that "observation" (τὰ δὲ πρὸς τὴν ὄψιν) and 'verbal description' (τὰ μὲν γὰρ τῷ λόγῳ) are an adequate and sufficient method for certain disciplines. We find the word *historía* (ἱστορία) used with this double methodological meaning in *History of Animals* (*HA* 491a12-28) and in *On the Soul* (*De An.* 402a1-4). In other texts such as *Prior Analytics* (*APr.* 46a24-27), the term, with the sense of historical survey, refers to the next process of investigation, that is, of 'verification' (ἀπόδειξις) of the data, which follows observation: This is a similar process to what in the *First Analytics* (*APr.* 46a17-22) is called ἐμπειρία, 'experience,' described as the initial stage in which the basic principles pertaining to each subject are given, which is the necessary preliminary stage for achieving any artistic or scholarly knowledge (περὶ ἄλλην [...] τέχνην τε καὶ ἐπιστήμην). Finally, in the treatises on human sciences, as in *Rhetoric* (*Rh.* 1360a37.) and *Poetics* (*Po.* 1451a36-b1; *Po.*1459a17-32), it acquires the meaning of 'historical work.' This derivation brings some clarity to the present study, as we will discuss subsequently.

The linking point of these three meanings (Rey 2001: 317) brings us closer to the first phase of the dialectical treatment that Aristotle applies to his data. In the Aristotelian doctrine of the different and diverse stages of knowledge, as explained in *Metaphysics (Metaph.) A* and *Posterior Analytics* (*APo*) *2* (cf. Le Blond 1973), the first stage of knowledge is sense-perception (αἴσθησις). From this, an image (φαντασία) is created, and when a copy of this image is itself made and stored in the mind, it becomes what we call *memory* (μνήμη). The evocation of the memory of a specific thing is *experience* (ἐμπειρία, *APo.* 100a5 y *Metaph.* 980b27), described as 'the universal now stabilized in its entirety within the soul' (παντὸς ἠρεμήσαντος τοῦ καθόλου ἐν τῇ ψυχῇ, *APo.* 100a6), which remains whole and unique to itself amidst all the transitory feelings of each particular being (τοῦ ἑνὸς παρὰ τὰ πολλά, ὃ ἂν ἐν ἅπασιν ἓν ἐνῇ ἐκείνοις τὸ αὐτό, τέχνης ἀρχὴ καὶ ἐπιστήμης, ἐὰν μὲν περὶ γένεσιν, τέχνης, ἐὰν δὲ περὶ τὸ ὄν, ἐπιστήμης *APo.* 100a8). Experience is fixed and whole, in that it unifies and stabilizes sensorial and mnemonic evocations when much time has elapsed (πλῆθος χρόνου in *Nicomachean Ethics* (*EN*) 1142a15-16).

It is this reference to the passing of time (*Metaph.* 981A7-9), along with the non-generalized objective fixing of ideas, that these three meanings of the term ἱστορία have in common, since they describe in each context a type of knowledge of specific, past facts, and as a consequence are neither universal nor necessary, but rather a type of preliminary knowledge of the scientific or philosophical, as well as the knowledge on which these are themselves based.

At the end of his article, Pierre Louis (1955: 44) wonders whether ἱστορία can ever mean ἐπιστήμη, and concludes: "Absolutely not [...]. The same difference separates ἱστορία, knowledge, from ἐπιστήμη, science: the latter can only relate to the general, while the other merely notes the particular facts with all possible accuracy." The same idea is found in Sainte Croix (1992: 24).

THE KNOWLEDGE OF EXPERTS, THE ἘΝΔΟΞΑ (*ÉNDOXA*) OR OPINION

Indirect experience is constituted by the opinions of experts, the so-called ἔνδοξα (Chichi 1997), and, as will be described below, is used on the one hand as a starting point (Le Blond 1973: 251), and on the other as a basis for any type of investigation, thus acquiring the same value as the knowledge of experience (ἐμπειρία) that we have already described, and serving as a source of argumentation analogous to that based on the factual and verified data provided by direct investigation (Harper 1998). In *Nicomachean Ethics* (*EN* 1143b11), Aristotle explains the reasons for this approach:

> "[...] the unproven assertions and opinions of experienced and elderly people, or prudent men, are as much deserving of attention as those which they support by proof; for experience has given them an eye for things, and so they see correctly." [1]

The inclusion of other opinions as a starting point is one of Aristotle's most enduring contributions to the history of thought. It implies that a specific linear historical process exists (Bueno 1980: 98), a development in thought over time and, conse-

1 Translation by H. Rackham. ὥστε δεῖ προσέχειν τῶν ἐμπείρων καὶ πρεσβυτέρων ἢ φρονίμων ταῖς ἀναποδείκτοις φάσεσι καὶ δόξαις οὐχ ἧττον τῶν ἀποδείξεων· διὰ γὰρ τὸ ἔχειν ἐκ τῆς ἐμπειρίας ὄμμα ὁρῶσιν ὀρθῶς. Occasionally, the English translations in this paper use outdated English conventions. In such cases, or in cases where punctuation is likely missing, the English was edited by Kareem James Abu-Zeid and revised by the author to make the text clearer.

quently, a new type of philosophical conscience which is both more responsible and more complex (Jaeger 1993: 11).

We now turn to the question of how argumentation can be conducted from the starting point of indirect investigation, from knowledge transmitted and supported by premises that do not assert something true or certain, but which are based on gathering the opinions of expert and reputable thinkers of the past. This is what is known as 'dialectic argumentation' (Vega Renon: 1998), differing from 'syllogistic argumentation' in that the latter is based on truth whereas the former is based on 'verisimilitude.' Its nature would be similar to that of rhetorical enthymeme, whose premises generally come from the description of habits and the more-or-less shared opinions of the listeners which the speaker knows and which are used in support of persuasive argumentation. Hence, this constitutes the field of the plausible, of the probable, of that which happens most of the time (ὡς ἐπὶ τὸ πολύ).

Aristotle always defines this by comparing it with logical argumentation, as explained in the *Topics* (*Top.* 100a 30-b 24), where it is stated:

> "The syllogism is 'dialectical' if it reasons from opinions that are generally accepted. Things are 'true' and 'primary' which are believed on the strength not of anything else but of themselves: for in regard to the first principles of science it is improper to ask any further than the why and wherefore of them; each of the first principles should command belief in and by itself. On the other hand, those opinions are 'generally accepted' which are accepted by everyone or by the majority or by the philosophers—i.e., by all, or by the majority, or by the most notable and illustrious of them."[2]

Alexander of Aphrodisias, in his commentary on the *Topics*, made the meaning of this notion even more precise, and revealed how to carry out an investigation that starts with these premises (Wallies 19 1891: 22-27).

Let us now look at some of the passages in which Aristotle recommends the use of this dialectical method in the different stages of the investigation. In *Top-*

---

2 Translation by W. A. Pickard—Cambridge. Διαλεκτικὸς δὲ συλλογισμὸς ὁ ἐξ ἐνδόξων συλλογιζόμενος. Ἔστι δὲ ἀληθῆ μὲν καὶ πρῶτα τὰ μὴ δι´ ἑτέρων ἀλλὰ δι´ αὑτῶν ἔχοντα τὴν πίστιν (οὐ δεῖ γὰρ ἐν ταῖς ἐπιστημονικαῖς ἀρχαῖς ἐπιζητεῖσθαι τὸ διὰ τί, ἀλλ´ ἑκάστην τῶν ἀρχῶν αὐτὴν καθ´ ἑαυτὴν εἶναι πιστήν), ἔνδοξα δὲ τὰ δοκοῦντα πᾶσιν ἢ τοῖς πλείστοις ἢ τοῖς σοφοῖς, καὶ τούτοις ἢ πᾶσιν ἢ τοῖς πλείστοις ἢ τοῖς μάλιστα γνωρίμοις καὶ ἐνδόξοις.

ics (*Top.* 101a 25), he explains the three fields in which dialectical reasoning can be used: "intellectual training, casual encounters, and the philosophical sciences."[3]

So, firstly, it can be used in a general manner (Di Camilo 2004: 205) for intellectual debates and as a first step toward philosophy (i.e., toward "intellectual training" [πρὸς γυμνασίαν] or "casual encounters" [ἐντεύξεις]) with the clear function of "guiding a plan of inquiry" (μέθοδον γὰρ ἔχοντες ῥᾷον) to argue on the basis of opinions that have already been expressed (cf. *Top.* 101a 27 and *Top.* 101a 30).

We now turn to its usefulness in scientific knowledge and to the way of putting it into practice, following different fragments of Aristotle's work. More specifically, in discussing philosophic investigations, Aristotle states in *Metaphysics* (*Metaph.* 995a24-30) that this dialectical method should be used at the beginning of the argumentation, both to state and to specify the issue accurately and also to know if somebody before us had already considered it and had found the solution (Evans 1977: 13; *Metaph.* 995a24-30). In a second phase of the argumentation, he encourages the use of ἔνδοξα as indicators that the investigation is proceeding along the right path:

"However, let us avail ourselves of the evidence of those who had approached before us the investigation of reality and philosophized about Truth. For they too recognize certain principles and causes, and so it will be of some assistance to our present inquiry if we study their teaching; because we shall either discover some other kind of cause or have more confidence in those which we have just described."[4] (*Metaph.* 983b 1-7)

In the *Topics* (*Top.* 101a 34) he stresses the same idea. He refers to the usefulness of ἔνδοξα as a means of recognizing the truth or falsity of an *aporía* and to know where that truth or falsity lies. Due to its probable nature, dialectical reasoning does not force us to focus on a subject from only the true point of view (Sim 1999: 22-23), but rather allows us to focus on the problem from two opposing points of view. If we seek difficulties on both sides of a subject, we will be able to detect more easily "the truth and error" (τἀληθές τε καὶ τὸ ψεῦδος) re-

3 Translation by W. A. Pickard—Cambridge. Ἔστι δὴ πρὸς τρία, πρὸς γυμνασίαν, πρὸς τὰς ἐντεύξεις, πρὸς τὰς κατὰ φιλοσοφίαν ἐπιστήμας.

4 Translation by Hugh Tredennick—Cambridge. ὅμως δὲ παραλάβωμεν καὶ τοὺς πρότερον ἡμῶν εἰς ἐπίσκεψιν τῶν ὄντων ἐλθόντας καὶ φιλοσοφήσαντας περὶ τῆς ἀληθείας. δῆλον γὰρ ὅτι κἀκεῖνοι λέγουσιν ἀρχάς τινας καὶ αἰτίας: ἐπελθοῦσιν οὖν ἔσται τι προὔργου τῇ μεθόδῳ τῇ νῦν· ἢ γὰρ ἕτερόν τι γένος εὑρήσομεν αἰτίας ἢ ταῖς νῦν λεγομέναις μᾶλλον πιστεύσομεν.

garding the various issues that arise (*Top.* 101a 34). In another sense, recourse to the opinions of experts allows us to establish first principles (the ultimate bases of principles used in various sciences) that, through being non-demonstrable, remain outside the field of what can be reasoned and that which has been established from the truth (Berti 1996: 107-108; Top. 101a35-101b). The indications on how to approach the principles of science based on the plausible premises are found in *Topics VIII*. In the following excerpt (*Top.* 163b1), he explains the need to resort to the dialectical method in order to be able to address a thesis correctly. To reach the truth, it is necessary first to defend a position and its opposite and to respond to the objections of one and the other, whether we are dealing with real speakers or thinking about and replying to questions ourselves. Irwin (1995: 174-176) calls this the "ordinary" dialectic:

"Always, in dealing with any proposition, be on the look-out for a line of argument with both pro and con: and on discovering it, immediately set about looking for the solution of it: for in this way you will soon find that you have trained yourself at the same time in both asking questions and answering them. If we cannot find anyone else to argue with, we should argue with ourselves." (*Top.* 163b1)[5]

We then have to deduce the arguments of both hypotheses and compare them (*Top.* 163b4). This as a means of being able, finally, to choose the right statement and reject the false one (Irwin 1995: 174-176), as we read in *Top.* 163b11:

"It only remains to make the right choice of one of them. For a task of this kind, a particular natural ability is required: in fact, the real natural ability is the power to choose the true and shun the false. Men of natural ability can do this; for by a right liking or disliking for whatever is proposed to them, they rightly select what is best."[6]

---

5 Translation by W. A. Pickard—Cambridge. Πρὸς ἅπασάν τε θέσιν, καὶ ὅτι οὕτως καὶ ὅτι οὐχ οὕτως, τὸ ἐπιχείρημα σκεπτέον, καὶ εὑρόντα τὴν λύσιν εὐθὺς ζητητέον· οὕτω γὰρ ἅμα συμβήσεται πρός τε τὸ ἐρωτᾶν καὶ πρὸς τὸ ἀποκρίνεσθαι γεγυμνάσθαι, κἂν πρὸς μηδένα ἄλλον ἔχωμεν, πρὸς αὑτούς.

6 Translation by W. A. Pickard—Cambridge. λοιπὸν γὰρ τούτων ὀρθῶς ἑλέσθαι θάτερον. Δεῖ δὲ πρὸς τὸ τοιοῦτον ὑπάρχειν εὐφυᾶ, καὶ τοῦτ´ ἔστιν ἡ κατ´ ἀλήθειαν εὐφυΐα, τὸ δύνασθαι καλῶς ἑλέσθαι τἀληθὲς καὶ φυγεῖν τὸ ψεῦδος· ὅπερ οἱ πεφυκότες εὖ δύνανται ποιεῖν· εὖ γὰρ φιλοῦντες καὶ μισοῦντες τὸ προσφερόμενον εὖ κρίνουσι τὸ βέλτιστον.

Once a principle has been established, it must be revised again with regards to the experts' opinions and, when faced with these different points of view, the arbiter must decide who is right. In *On the Heavens* (Cael. 279b 2-12), he uses the metaphor of litigants in a trial whose arguments should be impartially listened to by an arbitrator (διαιτητής) and not by an opponent (ἀντίδικος), in order for a judge to be in a position to give an impartial verdict (Cleary 1995: 199-205). Because if these opinions are not taken into account, it would be like making accusations without letting the accused defend themselves (*On Breathing* [*Iuu.*] 470b 12):

"Hence these points must first catch our attention, in order that we may not be thought to make unsubstantiated charges against authors no longer alive."[7]

In the final stage of a line of argumentation, the use of ἔνδοξα is also necessary to reinforce the position that we have reached after having studied the case, as stated in *Nicomachean Ethics* (*EN* 1145b 1-7):

"Our proper course with this subject will be the same as we have followed in other occasions. First we will present the various views and resolve the difficulties they involve. Then we will establish if possible all or, if not all, the greater part and the most important of the opinions generally held with respect to these states of mind; since if the discrepancies can be solved, and a residuum of current opinion left standing, the true view will have been sufficiently established."[8]

Now is not the moment to offer a judgment on the grade of veracity with which Aristotle transmits the thoughts of his predecessors. Cherniss (1957: 104-114 and 1991: 10-13) claims that Aristotle only mentions them in order to present those that confirm his theory to be correct and those that do not to be wrong. Guthrie (1970: 240), on the other hand, defends Aristotle's intellectual honesty, stating that the philosopher carefully distinguishes between his own theories and the opinions of others and that he does not manipulate them in order to defend his own beliefs but in fact uses them in support of his investigation, his sole purpose being to clarify the truth. Our interest here is to highlight his role as a dox-

7 Translation by G. R. Thompson Ross. ὥστ' ἀναγκαῖον περὶ τούτων πρῶτον ἐπελθεῖν, ὅπως μὴ δοκῶμεν ἀπόντων κενὴν κατηγορεῖν.

8 Translation by H. Rackham. Δεῖ δ', ὥσπερ ἐπὶ τῶν ἄλλων, τιθέντας τὰ φαινόμενα καὶ πρῶτον διαπορήσαντας οὕτω δεικνύναι μάλιστα μὲν πάντα τὰ ἔνδοξα περὶ ταῦτα τὰ πάθη, εἰ δὲ μή, τὰ πλεῖστα καὶ κυριώτατα: ἐὰν γὰρ λύηταί τε τὰ δυσχερῆ καὶ καταλείπηται τὰ ἔνδοξα, δεδειγμένον ἂν εἴη ἱκανῶς.

ographer and, above all, his contribution to the history of thought and to the idea that continuity and development of thought exist, and that the opinions of the experts constitute the basis for and the touchstone of all correct argumentations.

THE CONCEPT OF ΣΟΦίΑ (*SOPHÍA*) AS THE ENGINE OF THE EVOLUTION OF THOUGHT AND SOCIETY

The idea of progress in the development of thought contrasts with the vision of discontinuity that Aristotle speaks of in other passages. In *On Meteorology* (*Mete.*) he presupposes, as if it were a stated fact that civilizations periodically disappear for various reasons and in different ways. For this reason, there is a rupture in tradition which prevents us from maintaining what we might call historical memory. He explains this clearly in an exciting passage (*Mete.* 351b9-13):

"However, the whole vital process of the earth takes place so gradually and in periods of time which are so immense compared with the length of our life that these changes are not observed, and before their course can be recorded from beginning to end whole nations perish and are destroyed. Of such destructions, the most utter and sudden are due to wars, but pestilence or famine cause them too. Famines, again, are either sudden and severe or else gradual. In the latter case, the disappearance of a nation is not noticed because some leave the country while others remain; and this goes on until the land is unable to maintain any inhabitants at all. So a long period of time is likely to elapse from the first departure to the last, and no one remembers and the lapse of time destroys all record even before the last inhabitants have disappeared. In the same way, a nation must be supposed to lose account of the time when it first settled in a land that was changing from a marshy and watery state and becoming dry. Here, too, the change is gradual and lasts a long time and men do not remember who came first, or when, or what the land was like when they came."[9]

9 Translation by E. W. Webster. ἀλλὰ διὰ τὸ γίγνεσθαι πᾶσαν τὴν φυσικὴν περὶ τὴν γῆν γένεσιν ἐκ προσαγωγῆς καὶ ἐν χρόνοις παμμήκεσι πρὸς τὴν ἡμετέραν ζωήν, λανθάνει ταῦτα γιγνόμενα, καὶ πρότερον ὅλων τῶν ἐθνῶν ἀπώλειαι γίγνονται καὶ φθοραὶ πρὶν μνημονευθῆναι τὴν τούτων μεταβολὴν ἐξ ἀρχῆς εἰς τέλος. μέγισται μὲν οὖν φθοραὶ γίγνονται καὶ τάχισται ἐν τοῖς πολέμοις, ἄλλαι δὲ νόσοις, αἱ δὲ ἀφορίαις, καὶ ταύταις αἱ μὲν μεγάλαι αἱ δὲ κατὰ μικρόν, ὥστε λανθάνουσι τῶν γε τοιούτων ἐθνῶν καὶ αἱ μεταναστάσεις διὰ τὸ τοὺς μὲν λείπειν τὰς χώρας, τοὺς δὲ ὑπομένειν μέχρι τούτου μέχριπερ ἂν μηκέτι δύνηται τρέφειν ἡ χώρα πλῆθος μηδέν. ἀπὸ τῆς πρώτης οὖν ἀπολείψεως εἰς τὴν ὑστέραν εἰκὸς γίγνεσθαι μακροὺς χρόνους, ὥστε μηδένα μνημονεύειν, ἀλλὰ σῳζομένων ἔτι τῶν ὑπομενόντων ἐπιλελῆσθαι διὰ χρόνου πλῆθος.

The idea that nature is subject to periodic destruction caused by cataclysms or gradual natural phenomena is already found in Plato *(Ti.* 23a, *Plt.* 269b9 y *Lg.* 676), and Aristotle develops it in one of his first dialogues (Megino 2016: 59), in a fragment that has been attributed to him only with difficulty (Vallejo 2005: 273), and that Rose (1886) in fact does not include in his work (Ross, Fr. 8b *On Philosophy*):

> "It must be kept in mind that men perish in various ways. They die, in effect, because of epidemics, famines, earthquakes, wars, diseases of various classes and for other causes, but especially by cataclysms even more devastating, like that which is said to have taken place in the times of Deucalion, which was great, although it did not spread its effects on all things."[10]

According to this view, it would be impossible to establish historical continuity, given that the total rupture that exists between one civilization and the following one prevents the new generations from knowing or remembering what constitutes their cultural heritage (Berti 1997). Thus, everything that represents the identity of the people is lost: technical knowledge, political configurations, and the collective historical beliefs and myths about their own past and the existence of deities and their nature (Longo 1984: 49).

More important than the theory of the destruction of civilizations is that of the subsequent consequences—here, *palingenesis*, the reproduction of the historical course. It is, once again, in the dialogue *On Philosophy* (Megino 2016: 59; *On Philosophy* Ross, Fr. 8b) where Aristotle describes the process that goes from the destruction of a civilization to the formation of a new society (Vallejo 2005: 275-276), in different stages of development (Weil 1960: 328) that correspond to different ideas that can be understood as "wisdom" (σοφία). The same idea, outlined in far greater detail, would later be developed in *Politics* (*Pol.*) 1252b, where Aristotle describes, with precise terminology, the different stages of development from the state of the initial cataclysm to the full formation of the polis. A natural community, the house, arises first (οἰκία πρώτη), to satisfy eve-

---

τὸν αὐτὸν δὲ τρόπον χρὴ νομίζειν καὶ τοὺς κατοικισμοὺς λανθάνειν πότε πρῶτον ἐγένοντο τοῖς ἔθνεσιν ἑκάστοις εἰς τὰ μεταβάλλοντα καὶ γιγνόμενα ξηρὰ ἐξ ἑλωδῶν καὶ ἐνύδρων· καὶ γὰρ ἐνταῦθα κατὰ μικρὸν ἐν πολλῷ γίγνεται χρόνῳ ἡ ἐπίδοσις, ὥστε μὴ μνημονεύειν τίνες πρῶτοι καὶ πότε καὶ πῶς ἐχόντων ἦλθον τῶν τόπων.

10 Translation by W.D. Roos. Χρὴ γὰρ εἰδέναι, ὅτι φθείρονται μὲν ἄνθρωποι διαφόρως· καὶ γὰρ ὑπὸ λοιμῶν καὶ λιμῶν καὶ σεισμῶν καὶ πολέμων καὶ νόσων ποικίλων καὶ ὑφ' ἑτέρων αἰτιῶν, μάλιστα δὲ ὑπὸ κατακλυσμῶν ἀθροώτερον· οἷος εἶναι λέγεται ὁ ἐπὶ Δευκαλίωνος, μέγας μὲν, οὐ πάντων δὲ κατακρατήσας·

ryday needs. When dangers must be faced that put survival at risk, a broader community, the village (κώμη), arises, which is the nucleus from which the polis will derive, whose purpose is not only survival but also well-being (οὖσα δὲ τοῦ εὖ ζῆν). The development of the social community is therefore seen as a natural and necessary process, and hence is timeless, atopical, and repeated in different historical moments (*Pol.* 1329b25-31). The same perception can be seen in relation to beliefs and opinions that have come to be believed and transmitted countless times. He develops this idea in *Meteorologica*, 339b 27-33, and in *On the Heavens*, *Cael.* 270b19-21, where he says: "The same ideas, one must believe, recur in men's minds not once or twice but again and again."[11]

In *Metaphysics* 1074b3 he also proves the existence of the gods, in that this is a belief that rests on the dual authority of tradition and *palingenesis*. He refers to proverbs in *On Philosophy* (Megino 2016: 59), stating that these are the relics of the ancient philosophy that withstood the great destructions of humanity and that were saved "because of their concision and perspicacity" (διὰ συντομίαν καὶ δεξιότητα, On Philosophy Ross, Fr. 8a; Vallejo 2005: 275-276). The structure of human history is thus linked to the cosmos of the sublunary world (Chátelet 1962: 182), which would be the only thing affected by this cyclic destruction and regeneration, in that it is not governed by need but by contingency, and, in this specific sense, by fate (Bueno 1980: 96; on the circular movement of the divine body, cf. *Cael.* 286a8).

THE DIFFERENT NATURE OF EACH SCIENCE

Aristotle states in *Topics* 105b 20 that there are three divisions of propositions and problems (εστι δ΄ ὡς τύπῳ περιλαβεῖν τῶν προτάσεων καὶ τῶν προβλημάτων μέρη τρία). Some are ethical, others come from natural philosophy, and others are logical (αἱ μὲν γὰρ ἠθικαὶ προτάσεις εἰσίν, αἱ δὲ φυσικαί, αἱ δὲ λογικαί). We have to recognize each of them by means of the familiarity attained through induction (τῇ δὲ διὰ τῆς ἐπαγωγῆς συνηθείᾳ πειρατέον γνωρίζειν ἑκάστην αὐτῶν), and the method is to exam them in light of the illustrations given above (κατὰ τὰ προειρημένα παραδείγματα ἐπισκοποῦντα).

We now turn again to texts (7) and (8) mentioned in the first section of this essay to explain why the different nature of each science is what determines the way information initially gathered is to be used (Weil 1960: 163). In these passages, Aristotle states that historians should limit themselves to narrating past

11 οὐ γὰρ ἅπαξ οὐδὲ δὶς ἀλλ' ἀπειράκις δεῖ νομίζειν τὰς αὐτὰς ἀφικνεῖσθαι δόξας εἰς ἡμᾶς.

occurrences (τὰ γενόμενα λέγειν) as particular events (τὸ δὲ καθ' ἕκαστον). It is not their task to draw generalizations from them about "what might happen according to likelihood or need" (οἷα ἂν γένοιτο καὶ τὰ δυνατὰ κατὰ τὸ εἰκὸς ἢ τὸ ἀναγκαῖον). Neither is it their mission to seek a causal relationship (ὧν ἕκαστον ὡς ἔτυχεν ἔχει πρὸς ἄλληλα) between what else occurred during the same period of time (ἑνὸς χρόνου). When comparing poetry and history (Sainte Croix, 1992, p. 24), he considers the former (Wartelle 1985) to be more philosophical (φιλοσοφώτερον) and more important (σπουδαιότερον) than the latter (Wartelle, 1982), in that it admits a certain degree of generalization and deduction, dealing as it does with "universal things" (τὰ καθόλου), with "what could happen," with "what should be done or said to a certain type of person in accordance with the plausible or necessary" (τῷ ποίῳ τὰ ποῖα ἄττα συμβαίνει λέγειν ἢ πράττειν κατὰ τὸ εἰκὸς ἢ τὸ ἀναγκαῖον).

Aristotle's valuation here is surprising, to say the least, in that he was the first person to study documents in which facts were recorded directly, and indeed he carried out a significant amount of work compiling (Jaeger 1993: 377 and Weil 1960: 95), classifying, and describing the material under investigation (Cassani 1971: 33).

Yet he values this task of collecting data, which we have called ἱστορία, only as a preliminary process, similar to what might be done in the description of a natural phenomenon or with information relating to the dissection of an animal. The human sciences were built upon such data (Weil 1960: 97), especially those sciences that study the definable conditions for the attainment of happiness in ethics and of the higher state in politics. In the natural sciences a generalizing principle is drawn from data provided through causal inference. Then, a necessary and universal law is applied to particular cases. In the human sciences, however, the incorporation of this individual material must be carried out following a different method. The reason for this is that human actions, by their nature, are not likely to come into causal relationship with one another and thus cannot be submitted to need; instead, they are contingent (Bueno 1980: 95); for they operate in the field of practical reasoning, of φρόνησις, in which necessity does not have a place. Each human action is always a product of free will and is therefore unique.

The fact that it is not possible to establish a necessary principle that determines the events expected in human acts does not mean that the human sciences must always be studied as a set of unique and specific information. Pointing out a certain degree of generalization is necessary, and this indeed is the mission of historical knowledge, as Aristotle explains in *Metaphysics* (*Metaph.* 981a1-24-30):

"Experience seems very similar to science and art, but actually, it is through experience that men acquire science and art; for as Polus rightly says, 'experience produces art, but inexperience chance'. Art is produced when from many notions of experience a single universal judgment is formed concerning similar objects. To have a judgment that when Callias was suffering from this or that disease this or that benefited him, and similarly with Socrates and various other individuals, is a matter of experience; but to judge that it benefits all persons of a specific type, considered as a class, who suffer from this or that disease (e.g., the phlegmatic or bilious when suffering from burning fever) is a matter of art."[12]

Historical knowledge, therefore, is knowledge of experience, of generalization on similar cases that, when applied to the human sciences, allows us to theorize about human behavior and to make an object of science which would otherwise only be susceptible to observation as a phenomenon undergoing random influences. Historical research would only be a necessary instrument for making descriptions and analyses of specific situations in terms of universal components or factors. The natural sciences determine universal taxonomic types. These types are then applied as models to analyze and to render comprehensible the empirical and contingent behavior of organisms. In the same way, human history determines universal taxonomic types. They can be related to forms of political organizations: oligarchies, tyrannies, democracies, or to types of behavior that determine the principles of ethics.

12 Translation by Hugh Tredennick—Cambridge. καὶ δοκεῖ σχεδὸν ἐπιστήμῃ καὶ τέχνῃ ὅμοιον εἶναι καὶ ἐμπειρία, ἀποβαίνει δ' ἐπιστήμη καὶ τέχνη διὰ τῆς ἐμπειρίας τοῖς ἀνθρώποις· ἡ μὲν γὰρ ἐμπειρία τέχνην ἐποίησεν, ὡς φησὶ Πῶλος, ἡ δ' ἀπειρία τύχην. γίγνεται δὲ τέχνη ὅταν ἐκ πολλῶν τῆς ἐμπειρίας ἐννοημάτων μία καθόλου γένηται περὶ τῶν ὁμοίων ὑπόληψις. τὸ μὲν γὰρ ἔχειν ὑπόληψιν ὅτι Καλλίᾳ κάμνοντι τηνδὶ τὴν νόσον τοδὶ συνήνεγκε καὶ Σωκράτει καὶ καθ' ἕκαστον οὕτω πολλοῖς, ἐμπειρίας ἐστίν· τὸ δ' ὅτι πᾶσι τοῖς τοιοῖσδε κατ' εἶδος ἓν ἀφορισθεῖσι, κάμνουσι τηνδὶ τὴν νόσον, συνήνεγκεν, οἷον τοῖς φλεγματώδεσιν ἢ χολώδεσι ἢ πυρέττουσι καύσῳ, τέχνης. πρὸς μὲν οὖν τὸ πράττειν ἐμπειρία τέχνης οὐδὲν δοκεῖ διαφέρειν, ἀλλὰ καὶ μᾶλλον ἐπιτυγχάνουσιν οἱ ἔμπειροι τῶν ἄνευ τῆς ἐμπειρίας λόγον ἐχόντων (αἴτιον δ' ὅτι ἡ μὲν ἐμπειρία τῶν καθ' ἕκαστόν ἐστι γνῶσις ἡ δὲ τέχνη τῶν καθόλου, αἱ δὲ πράξεις καὶ αἱ γενέσεις πᾶσαι περὶ τὸ καθ' ἕκαστόν εἰσιν· οὐ γὰρ ἄνθρωπον ὑγιάζει ὁ ἰατρεύων ἀλλ' ἢ κατὰ συμβεβηκός, ἀλλὰ Καλλίαν ἢ Σωκράτην ἢ τῶν ἄλλων τινὰ τῶν οὕτω λεγομένων ᾧ συμβέβηκεν ἀνθρώπῳ εἶναι· ἐὰν οὖν ἄνευ τῆς ἐμπειρίας ἔχῃ τις τὸν λόγον, καὶ τὸ καθόλου μὲν γνωρίζῃ τὸ δ' ἐν τούτῳ καθ' ἕκαστον ἀγνοῇ, πολλάκις διαμαρτήσεται τῆς θεραπείας· θεραπευτὸν γὰρ τὸ καθ' ἕκαστον).

Data arising from historical investigation would also be one of the sources for deliberation, one of the standards for the formation of prudent judgments about what is best, both in the field of personal reflection and in collective discussion, in terms of correctly guiding the decisions of a government (Bueno 1980: 101). A simple illustration of this is how it is used in the construction of syllogisms, as can be seen in *Apo.* 94a36:

"'What cause originated the waging of war against the Athenians?' and the answer is, 'Because they raided Sardis with the Eretrians', since this originated the war. Let A be war, B unprovoked raiding, C the Athenians. Then B, unprovoked raiding, is true of C, the Athenians, and A is true of B, since men make war on the unjust aggressor. So A, having war waged upon them, is true of B, the initial aggressors, and B is true of C, the Athenians, who were the aggressors. Hence here too the cause – in this case, the efficient cause – is the middle term."[13]

Aristotle builds a syllogism in which the central premise questions sociological, anthropological, or moral necessity; the minor premise is a mere description of historical fact and is, therefore, individual and contingent; the conclusion, consequently, must be equally contingent. This minor conclusion is contingent not because it is individual, as a universal cause can indeed function within the individual, but because it is dependent on a political decision of a prudent nature: 'to invade Medes,' which cannot be included in a general plan.

Related to this is the use made in *Rhetoric* of historical examples in the deliberative discourse. This discourse is always oriented toward the future, and the orator alludes to past events by using them as a persuasive resource that has an impact on the audience as a criterion of verisimilitude (κατὰ τὸ εἰκός): something similar can occur within the scope of 'what happens most of the time.' This is a type of certainty on which knowledge is based, and that refers to what is characteristic of human nature (HA 727b29-30 μάλιστα κατὰ φύσιν) that by its own essence cannot be necessary, but from which scientific and general knowledge can be derived. Because if a criterion of verisimilitude does exist in

13 Translation by W.D. Ross. Τὸ δὲ διὰ τί ὁ Μηδικὸς πόλεμος ἐγένετο Ἀθηναίοις; τίς αἰτία τοῦ πολεμεῖσθαι Ἀθηναίους; ὅτι εἰς Σάρδεις μετ' Ἐρετριέων ἐνέβαλον· τοῦτο γὰρ ἐκίνησε πρῶτον. πόλεμος ἐφ' οὗ Α, προτέρους εἰσβαλεῖν Β, Ἀθηναῖοι τὸ Γ. ὑπάρχει δὴ τὸ Β τῷ Γ, τὸ προτέροις ἐμβαλεῖν τοῖς Ἀθηναίοις, τὸ δὲ Α τῷ Β· πολεμοῦσι γὰρ τοῖς πρότερον ἀδικήσασιν. ὑπάρχει ἄρα τῷ μὲν Β τὸ Α, τὸ πολεμεῖσθαι τοῖς προτέροις ἄρξασι· τοῦτο δὲ τὸ Β τοῖς Ἀθηναίοις· πρότεροι γὰρ ἦρξαν. μέσον ἄρα καὶ ἐνταῦθα τὸ αἴτιον, τὸ πρῶτον κινῆσαν.

reference to historical facts, this would be due to the fact that a certain regularity in human behavior exists which, if not determined by need, is at least determined by what is possible (ὡς ἐπὶ τὸ πολύ), due to what happens most of the time.

CONCLUSIONS

Following Berti's statements (Berti 1996: 129), we can conclude that while Aristotle's conception of dialectic is complicated, it doesn't change, neither in his earlier or later works; in matters of ethics and physics, this method is fully capable of ensuring knowledge, because these sciences concern objects that are not susceptible to a more exact demonstration. In matters of metaphysics, of first philosophy, the philosopher uses dialectics to develop the significant points of contention, to distinguish between the different meanings of being. In all these sciences the dialectical method achieves rigorous results.

It is difficult to evaluate the extent to which Aristotle's conception of investigation and his use of dialectic argumentation influenced the opinions of other authors from the distant past. Many of them did transmit and adapt his views and methods, extending them beyond exclusively philosophical areas, and making a kind of popular episteme of his valuable doxography.

What we want to note here is how Aristotle's methods came into widespread use, and that his not always scientific views were popularized by writers such as didactic poets, and later by rhetoricians. His conception of scientific knowledge, as we have discussed, is complex. Moreover, it changes when applied to different sciences. For this reason, his methods were in some ways amenable to the aims of popular science. This is particularly the case with ethics and physics, where dialectic argumentation is considered to be a wholly sufficient method to ensure knowledge, in that these sciences deal with objects that are not susceptible to more exact demonstration. Having evaluated the whole process of direct and indirect research and argumentation, we can claim that to some extent Aristotle did indeed establish a line of thinking which allowed for scientific knowledge to develop into popular science.

There is undoubtedly a connexion between doxography and popular science. Aristotle can be considered the most important doxographer of antiquity. Thanks to him we know what previous authors contributed to the history of thought in different subjects. Implied here is that the quotations he used were inevitably mentioned in studies of a diverse nature, most of the time with an uncritical spirit. His commentaries were used gratuitously, decontextualized most of the time, and without mentioning the rhetorical function they played in the elaboration of his thought. Furthermore, Aristotle's connection with popular science becomes clear in many contributions from the present volume. One theme linked to the scientific method of study and disclosure is the way of presenting that method, either in prose or in verse. In a well-known passage from *Poetics* (1447b), Aristotle refers to it in these terms: "For if people publish medical or scientific treatises in meter the custom is to call them poets. But Homer and Empedocles have nothing in common except the meter, so that it would be proper to call the one a poet and the other not a poet but a scientist." This is precisely what Rossetti's paper is about—Arato's scientific exposition in his poem and how it was disseminated by his Latin translators.

In this volume, Dorit Engster refers to the importance of Aristotle's study of nature, which had a major influence on scientific studies for a long time. Despite its methodological deficiencies, it was influential throughout all of antiquity. After Aristotle, biological studies were pursued further and partially extended. The scientific focus on nature was thus pushed into the background. Scientific topics were published in 'popular-science scripts.' The systematic method of the Peripatetic school was abandoned and scholars tried to present the contents of marine life from a great variety of perspectives.

Parrusel insists on the social phenomenon of popular science that led to the birth of the Second Sophistic. On this ocassion, culture was identified with rhetoric, and in this sense the movement also owes a debt to Aristotle, who established the foundations of persuasive speech.

Aristotelian doxography seeks for the statements of the experts as a way of searching for origins. In different cultural contexts, there seems to be an assumption that those who can explain the origin of a phenomenon have fully understood, and possess complete knowledge of, the phenomenon in question.

REFERENCES

Aristotle. Editions and Translations

History of Animals (HA)

Cresswell, Richard (ed.) (repr. 2005): Aristotle. History of Animals (HA), Adamant Media Corporation

Metereologica (Mete.)

Webster, Erwin (1923) (ed.): Aristotle. Meteorologica, Oxford: Clarendon Press.

Metaphysics (Metaph.)
Ross, William David (ed.) (1924): Aristotle's Metaphysics, Oxford: Clarendon Press.

Nicomachean Ethics (NE)
Bywater, Ingram (ed.) (1894): Aristotle's Ethica Nicomachea, Oxford: Clarendon Press.
Rackham, Harris (ed.) (1934): Aristotle in 23 Volumes, Vol. 19, Cambridge and London: Harvard University Press and William Heinemann Ltd.

On Breathing (Iuu.)
Ross, Robert Thompson (ed.) (2004): On Youth and Old Age, on Life and Death, on Breathing, Belle Fourche: NuVision Publications.

On the Heaven (Cael.)
Allan, Donald James (ed.) (repr. 2005): Aristotle. De Caelo, Oxford. Clarendon Press.

On the Soul (De Anima)
Smith, John Alexander (ed.) (1931): The Works of Aristotle: De Anima, Oxford: Clarendon Press.

Parts of Animals (PA)
Ogle, William (ed.) (1882): Aristotle. On the Parts of Animals, London: K. Paul, French & co.
Ross, William David (ed.) (repr 1973): Aristotle. On the Parts of Animals, Oxford: Clarendon Press.

Poetics (Po)
Fyfe, William Hamilton (ed.) (1932): Aristotle in 23 Volumes, Vol. 23, London: William Heinemann Ltd.
Kassel, Rudolf (ed.) (1966): Aristotle's Ars Poetica, Oxford: Clarendon Press.

Politics (Pol.)
Ross, William David (ed.) (1957): Aristotle's Politica, Oxford: Clarendon Press.

Posterior Analytics
Barthélemy Saint-Hilaire, Jules (ed.) (1842): Logique d'Aristote. Derniers Analytiques, Paris: Librairie Philosophique de Ladrange.

Prior Analytics (APr.)
Barthélemy Saint-Hilaire, Jules (1839): Logique d'Aristote. Premiers Analytiques, Paris: Librairie Philosophique de Ladrange.
Jenkinson, Arthur J. (eds.) (2006): The Works of Aristotle Translated into English, Vol. 1 [only the translation of *Prior Analytics* is by Jenkinson]. Oxford: Clarendon Press.

Rhetoric (Rh)
Ross, William David (ed.) (1959): Ars Rhetorica. Aristotle, Oxford: Clarendon Press.

Topics (Top.)
Ross William David (ed.) (1958): Aristotelis: Topica et Sophistici Elenchi. Oxford Classical Texts.

Books and Articles

Berti, Enrico (1962): La filosofia del primo Aristotele, Padova: CEDAM.
Berti, Enrico (1972): La dialettica in Aristotele. In: L'attualitá della problematica aristotelica. Atti del Convegno franco-italiano su Aristotele, Padova: Antenore, pp. 33-80.
Berti, Enrico (1977): Aristotele: dalla dialettica alla filosofia prima, Padova: CEDAM.
Berti, Enrico (1986): "Sul carattere 'dialettico' della storiografia filosofica di Aristotele." In: Giuseppe Cambiano (ed.), Storiografia e dossografia nella filosofia antica, Torino: Tirrenia, pp. 101-125.
Berti, Enrico (1996): "Does Aristotle's Dialectic Develop?" In: William Wians (ed.), Aristotle's Philosophical Development, London: Rowman & Littlefield Publishers, pp.105-130.
Bonitz, Hermann (1955): Index Aristotelicus, Graz: Akademishe Druck- und Verlagsanstalt. (2nd ed.)
Bueno Martínez, Gustavo (1980): El Individuo en la Historia. Comentario a un texto de Aristóteles. Poética 1451b. Discurso inaugural del curso 1980-81, Oviedo: Universidad de Oviedo.
Cassani, Jorge Luis/Pérez Amuchástegui, Antonio J. (1968): Del epos a la historia científica, Buenos Aires: Editorial Nova.
Châtelet, François (1962): La Naissance de l'Histoire, Paris: Ed. de Minuit.

Cherniss, Harold (1957): The History of Ideas and Ancient Greek Philosophy. In: Estudios de Historia de la Filosofía 1, Tucuman: Universidad Nacional de Tucumán, pp. 104-114.

Cherniss, Harold (1991): La crítica aristotélica a la filosofía presocrática, Mexico: UNAM.

Chichi, Graciela. (1997): "El concepto aristotélico de éndoxon según la técnica de discusión de los Tópicos." In: Actas del Congreso Aristóteles, Instituto de Filosofía, Mendoza, pp. 180-187.

Cleary, John J. (1995): Aristotle & Mathematics. Aporetic Method in Cosmology and Metaphysics, Leiden: Brill.

Di Camillo, Silvana G. (2004): "El carácter dialéctico de la historiografía aristotélica. Estrategias argumentativas en Metafísica I, 9." In: María Isabel Santa Cruz (ed.), Diálogo con los griegos, Buenos Aires: Ediciones Colihue, pp. 201-216.

Evans, John David Gemmill (1977): Aristotle's Concept of Dialectic, Cambridge, UK: Cambridge University Press.

Guthrie, William Keith Chambers (1970): Studies in Pre-Socratic Philosophy, Vol. I, Cambridge, UK: Cambridge University Press, pp. 239-254.

Harper, Edward (1998): "Poetry, History, and Dialectic." In: Paideia Ancient Philosophy, Twentieth World Congress of Philosophy, Boston.

Irwin, Terence (1995): Aristotle's First Principles, Oxford: Oxford University Press.

Jaeger, Werner (1993): Aristóteles. Bases para la historia de su desarrollo intelectual, Mexico: Fondo de Cultura Económica.

Le Blond, Jean-Marie (1973): Logique et méthode chez Aristote. Paris: J. Vrim.

Longo, Oddone (1984): Memoria culturale: discontinuitá temporali nell l'antica Grecia. In: Bollettino del centro internazionale di storia dello spazio e tempo, pp. 45-51.

Louis, Pierre (1955): "Le mot ἱστορία chez Aristote." In: Revue de Philologie 29, pp. 39-44.

Megino Rodríguez, Carlos (2016): Edición crítica, traducción y comentario del diálogo 'Sobre la Filosofía' de Aristóteles. Universidad Complutense de Madrid. [Doctoral Thesis]

Moradiellos García, Enrique (2001): Las Caras de Clío: una introducción a la historia, Madrid: Siglo XXI.

Muller, F. (1926). De 'Historiae' vocabulo et notione. Mnemosyne, 54, pp. 234-257.

Rey Puente, Fernando (2001): Os sentidos do tempo em Aristóteles, Saô Paulo: Loyola.

Ross, William David (1979 [1955]): Aristotelis Fragmenta Selecta, Oxford: Oxford University Press.

Rose, Valentin (1967 [1886]): Aristotelis Qui Ferebantur Librorum Fragmenta, Leipzig: Teubner.

Sainte Croix, G.E.M. de (1992): "Aristotle on History and Poetry (Poetics, 9. 1451a36-b11)." In: A. Oksemberg, (ed.), Essays on Aristotle's Poetics, Princeton: Princeton University Press, pp. 23-32.

Sim, May (ed.) (1999): From Puzzles to Principles? Essays on Aristotle Dialectics, Maryland: Lexinton Books.

Solmsen, Friedrich (1960): Aristotle's System of the Physical World: A Comparison with His Predecessors, Ithaca: Cornell University Press

Vallejo Campos, Álvaro (2005): Aristóteles. Fragmentos, Madrid: Gredos.

Vega Renon, Luis (1998): "Aristotle's endoxa and plausible argumentation." In: Argumentation 12, pp. 95-113.

Wallies, Maximilian (1891) (ed.): "Alexandri Aphrodisiensis in Aristotelis *Topicorum* commentaria." In: Commentaria in Aristotelem Graeca 2/2. Berlin, pp. 17-22.

Wartelle, André (1985): Lexique de la "Poétique" d'Aristote, Paris: Les Belles Lettres.

Weil, Raymond (1960): Aristote et l'histoire. Essai sur la Politique, Paris: Klinksieck.

Ékphrasis as a Device for Knowledge Dissemination in Euripides

Sara Matías Pérez

ABSTRACT

The importance of Euripidean *ékphrasis* in rhetorical terms lies not only in its effect on the dramatic action or the audience; this resource also works as a means of information. The analysis of five ekphrastic passages (from *Electra*, *The Phoenician Women*, *Ion*, and *The Suppliants*) reveals the usefulness of *ékphrasis* both for capturing the attention of the public through visual vividness and as a device for disseminating historical, mythological, and cultural knowledge to the public at that time.

EURIPIDES AND THE ECPHRASTIC TRADITION

Ékphrasis, or description (lat. *descriptio*), is a literary and rhetorical resource that has its own idiosyncrasy within the tragedies of Euripides and differs from previous literary models, e.g., Homer and Hesiod. The concept of *ékphrasis* evolved throughout the Classic and Hellenistic periods. Even though it was not yet theoretically defined, one element prevails as a universal and invariable factor to any period: visual vividness through specific language. While Plato and Aristotle shed some light on the idea of vividness through language from a stylistic and philosophical point of view, *ékphrasis* does not appear named as such in their works.

The theorization of *ékphrasis* was particularly prominent during the Roman) Iron Age, when rhetoric again became crucial within the cultural and literary period known as the Second Sophistic. *Ékphrasis* was defined as a technical-rhetorical mechanism by the rhetoricians (*rhetores*), one that also systematized the exercises of school rhetoric (*progymnásmata*) to which *ékphrasis* belongs. Theon, Aphthonius, and Hermogenes, who revised the existing rhetorical procedures and took up the concept of *ékphrasis*, defined its characteristics and typology. Theon (1st century A.D.) was the first, as far as we know, to define the concept and establish a classification of the elements described: person, fact, place, time, object, events, and also a mixed type which combines two of the former categories. He underlines the beauty, utility, and gentleness that *ékphrasis* should produce (cfr. Kennedy 2003: 45-47).

On the other hand, Hermogenes (2nd to 3rd centuries A.D.) adds the *ékphrasis* of event or occasion (*kairós*) and argues that *ékphrasis* should be characterized by the unimaginable or original element (*parádoxos*; ibid 2003: 86). Lastly, Aphthonius (4th century A.D.) includes the *ékphrasis* of animals and plants, and he details how a description should be structured, e.g., by describing from head to toe (ibid 2003: 117). Later, Nicholas of Myra (5th century A.D.) adds statues and paintings to the *ékphrasis* types. From this point on, the concept will be limited to the description of crafted objects, which is similar to the sense utilized today as the description of works of art.

The systematization of *ékphrasis* in the *progymnásmata* and specific references to them make us think that Euripides could have served as a literary model in the way that Homer and Hesiod did. This would reflect the importance and permanence of Euripidean tragedy from the rhetorical point of view. Nevertheless, a matter that remains open to discussion is the dating of these very progymnasmatic exercises. Although these school rhetoric exercises were fully incorporated into the education of the Roman Iron Age, this does not necessarily mean that they were created at that time. Indeed, these school exercises probably date back to earlier times. The discovery of new papyri in recent years has shed light on this matter, since we now have evidence for progymnasmatic exercises from the Hellenistic period, such as the maxim (*gnṓmē*, e.g., in P. Berol inv. 12318) or the encomium (*enkṓmion*, e.g., in P. Mil Vogl. III 123), both from the third century B.C.[1] This has been highlighted by Professor Fernández Delgado in his contribution to the current volume. There is also evidence of school exercises in the Greek papyri. Therefore, the chronological limits of the *progymnásmata* have been expanded, and their first appearance now dates back to at least the Hellenistic Period.

The purpose of this article, on the one hand, is to emphasize the importance of Euripidean *ékphrasis* as a technical-rhetorical resource whose permanence is found even in school rhetorical exercises; on the other hand, it seeks to highlight the use of *ékphrasis* as a key element for knowledge dissemination among con-

1 Cf. Fernández Delgado/Pordomingo 2007: 227-238; 2008: 167-92.

temporary recipients. The mythological, historical, or cultural information that was transmitted through ekphrastic discourse constituted a way of informal learning. Finally, *ékphrasis* —for us modern recipients— allows us to approach modes of thought and perception from antiquity.

I have chosen five passages from four Euripidean works. Our analysis will focus exclusively on the description of crafted objects, with the aim of answering the central question of this article: Which ekphrastic passages can be considered mediums of historical, scientific, or mythological knowledge for the audience of that time?

ECPHRASTIC PASSAGES BETWEEN DESCRIPTION AND SYMBOLISM

The Euripidean Version of Achilles' Weapons

I will begin my analysis with an *ékphrasis* from *Electra* (451-477; cf. translation by Cropp 2013). Before the *anagnórisis* between Orestes and Electra, the chorus describes Achilles' weapons. This description of a crafted object seems detached from the main subject. Nevertheless, its background—the Trojan War—is connected to the dramatic action. First, the chorus relates the images of the shield, which constitutes the most extensive description. Then it details the hero's cuirass and spear. Despite the frequent evocation of Homer (*The Iliad* [*Il.*] 18.478-612) and even Hesiod (*The Shield of Heracles*), the excellence of Euripides resides in the capacity to create an extremely vivid *ékphrasis* with very few details.

The frame of the ekphrastic passage starts on line 442 (translation by Cropp 2013), where the Nereids bring Achilles the golden shield forged by Hephaestus. This passage differs from the Homeric one, where Thetis hands the shield to her son (*Il.* 19.12). Hereafter, the description starts (452 ff.) with the clarification of what is going to be related.[2] The chorus itself repeats something that it has heard but not seen. However, this *ékphrasis* does not lack detail. In this description, as well as in all descriptions of objects in this play, the order of elements remains the same, a phenomenon which the progymnasmatic theory would later express. For this reason, the descriptions begin with the figures in the orb, followed by the ones in the center of the shield.

---

2 *Electra* 452 ff.: Ἰλιόθεν δ᾽ ἔκλυόν τινος; "from a man of Troy sojourning [...] I heard." (trans. Cropp 2013)

In the orb of the shield we see Perseus with winged sandals, holding the severed head of the Gorgon, and Hermes next to him. The epithets characterizing the hero (459-60)[3] and the god (460-61)[4] provide an epic tone that evokes Homer and Hesiod. Besides the euphony of these lines, the term σήματα (*sēmata*), 'sign,' is significant, since Euripidean *ékphrasis* always introduces symbolic elements that concern the central action in the tragedy.

Figure 1: Achilles Receiving His Weapons on a Black-Figure Hýdria, ca. 575-550 BC[5]

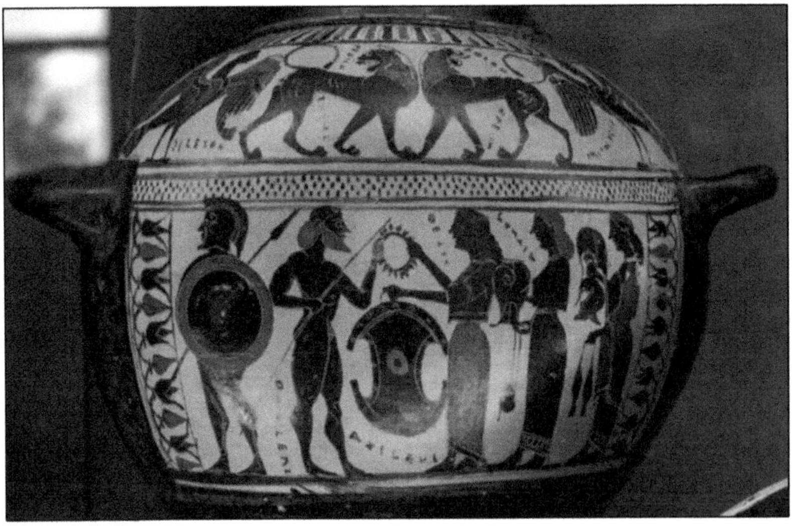

Source: Musée du Louvre; Photographer: Egisto Sani

The chariot of the sun, the Pleiades, and the Hyades appear in the middle of the shield. These lines serve as an example of the continuous use of adjectives and verbs in the imperfect indicative tense in ekphrastic passages (e.g., 464 κατέλαμπε, "shone down"). In this way, adjectives (464-469) specify how the chariot of the sun appears and what stars appear in the center of the shield.[6] Besides, the adjective χρυσός is a characteristic element in each of Achilles' weap-

3 *Electra* 459-60: Περσέα λαιμοτόμαν ὑπὲρ/ ἁλὸς ποτανοῖσι πεδί-/ λοισι; "was Perseus over the sea with flying sandals." (trans. Cropp 2013)

4 *Electra* 461-462: Διὸς ἀγγέλῳ σὺν Ἑρ-/ μᾷ, τῷ Μαί-/ ας ἀγροτῆρι κούρῳ; "Zeus' herald Hermes/ the rustic child of Maia." (trans. Cropp 2013)

5 Cf. Streicher's contribution to the present volume, in which he underlines the connections between Greek ceramics, myth, and the cultural context.

6 *Electra* 464-469: ἐν δὲ μέσῳ κατέλαμπε σάκει φαέθων/ κύκλος ἀελίοιο; "on the buckler's centre radiant shone down/ the circle of the sun." (trans. Cropp 2013)

ons—except the spear—that brings visual vividness to the description. Euripides constantly highlights the gold in the arms of Achilles, and this feature is a significant factor in his originality that distances him from his literary models (Fernández Delgado 2016: 127).

Another adjective that has the same effect is τροπαῖοι (*tropaĩoi*), 'of a turning,' in the description of the Pleiades and Hyades (468-9): "to turn back the eyes of Hector." We could think that the adjective suggests the escape of the enemy, but what τροπαῖος (*tropaĩos*) actually connotes is the blinding the weapon causes in others.

In the golden helmet (470) other figures are detailed, like the Sphinx (471),[7] which carries a trophy that it won for its singing. Once again, we observe how the adjectives concentrated in one line provide all the necessary accuracy to visualize the object of description. Thus, through the term χρυσότυπος (*chrysótypos*), 'beaten gold,' two characteristics of the helmet are indicated: it is made of gold, and it is forged with relief figures. On the sides of the cuirass (472-3) we find the figure of a lioness breathing fire and escaping from Pegasus.[8] This doubtlessly alludes to the mythic episode of the death of the Chimera at the hands of Bellerophon, who rode on Pegasus. The Chimera, as well as other creatures in Euripidean *ekphráseis*, implies a very concrete negative meaning that interrelates to the central action of the tragedy. The ekphrastic passages on Euripides contain words with powerful symbolism that allude to the main subject, as in the case of the Chimera or the Gorgon, which in this passage are prefigurations of Clytemnestra.

Lastly, the brief description of the spear (476-7 ἄορι δ' ἐν φονίῳ, "on the murderous sword") highlights its offensive quality. Furthermore, the action of the horses jumping and kicking up dust in their race is very precisely defined, despite the brevity of the description (477).[9] The interpretation of this line has provoked debate among modern scholars. Some editors and commentators have drawn attention to the word δορί (*dorí*), 'spear,' contained in manuscript (L). Some scholars emphasize the enormous difficulty that would be involved in forging such detailed figures on a spear, and opt instead for the term ἄορι,

7 *Electra* 470-71: ἐπὶ δὲ χρυσοτύπῳ κράνει/ Σφίγγες ὄνυξιν ἀοίδιμον ἄγραν; "on the helm of beaten gold/ were Sphinxes bearing in talons their song-trapped prey." (trans. Cropp 2013)

8 *Electra* 472-3: περιπλεύ-/ ρῳ δὲ κύτει πύρπνοος ἔσπευ-/ δε δρόμῳ λέαινα χαλαῖς; "and on the hollow corslet, breathing fire,/ sped at, run the lioness on clawed feet." (trans. Cropp 2013)

9 *Electra* 477: κελαινὰ δ' ἀμφὶ νῶθ' ἵετο κόνις; "and dust was billowing dark about their blacks." (trans. Cropp 2013)

'sword' (cf. Headlam 1901: 98-108; Denniston 1987: 108; Cropp 2013: 133), since images forged on a sword are much more likely.

Shields with Symbolic Images in *The Phoenician Women*

The second ekphrastic passage I will examine belongs to *The Phoenician Women* (1106-1140; cf. translation by Craik 1988). Some scholars have considered these lines as an interpolation since they do not seem to connect with the rest of the *rhesis*. However, our task is not to enter this discussion (cf. Mastronarde 1994: 456). The description is expressed by the messenger, who tells Jocasta of the mortal fight that is going to take place between her two sons, Eteocles and Polynices. Line 1106 gives way to the catalog of Argive heroes at each of the seven gates of the Theban wall.

This passage is exciting for two reasons. On the one hand, the *ékphrasis* of the shields of the Argive warriors has evident reminiscences of Aeschylus (*Seven against Thebes* 369-676). Although Euripides does not follow the style of Aeschylus, he too achieves a very detailed description with fewer descriptive elements. On the other hand, it seems that two types of *ekphráseis* are combined in these lines: a description of *prāgma* (i.e., of "action," the most common; cf. Garzya 1997: 41; Webb 1999: 12), detailing a war event with a catalog of the warriors; and within this we find another *ékphrasis*, on this occasion one of objects, which specifies what type of weapons each warrior carries and what emblems each one has.

There are seven warriors described with their respective weapons. The first of them is Parthenopaeus, who carries his family emblem on his shield, that is, Atalanta (1108) "with far-shooting shafts overwhelming," killing the Calydonian Boar (translation by Craik 1988). The second is Amphiaraus (1111 ὁ μάντις Ἀμφιάραος, "divine Amphiaraus"), detailed as a man who acts prudently by not carrying arms with emblems (1112). The third warrior is the sovereign Hippomedon (ἄναξ, "sovereign"), who had painted in the middle of his shield the giant Argos with a hundred eyes (1114-5).[10] Through a correlation, the description specifies how the giant's eyes never close completely (1116-7).[11] Fourth is the description of the exterior

10 *The Phoenician Women* 1114-5: ἐν μέσῳ σάκει/ στικτοῖς Πανόπτην ὄμμασιν δεδορκότα; "Lord Hippomedon/ went with an emblem on the middle of his shield:/ Panoptes seeing with faceted eyes." (trans. Craik 1988)

11 *The Phoenician Women* 1116-7: τὰ μὲν σὺν ἄστρων ἐπιτολαῖσιν ὄμματα/ βλέποντα, τὰ δὲ κρύπτοντα δυνόντων μέτα; "one set of eyes looking with the rise of the stars/ one set closing with their setting." (trans. Craik 1988)

and interior parts of the shield of Tydeus. Atop the shield, he carries a lion's skin with a curly mane (1120-21), and in the middle of the shield the figure of the Titan Prometheus holding a torch.[12]

The fifth warrior is Polynices, the son of Jocasta, described as enraged (1124). He carries the Potnian mares as the emblem on his shield, whose description is especially vivid (1125-1127). From a subjective viewpoint (1126) the messenger interprets the image, relating the sound that the mares seem to transmit running (1127 ὥστε μαίνεσθαι δοκεῖν, "so as to seem crazed"). The sixth warrior that appears is Capaneus. In the description of his weapon it is specified that the figures are forged for the first time (1130).[13] On the shield of this warrior, a giant appears holding an entire city on his arms. Lastly, there is Adrastus, whose shield is not forged but painted (1135) with the hundred serpents of the Hydra. Only in Adrastrus can we observe in which hand the warrior carries the shield (1136). Furthermore, this shield is unique because the image refers to an action against the Thebans: the Hydra snatching the city's children (1137-38).[14]

From the literary point of view, this double *ékphrasis pragma*-object has a meta-textual sense; in painting, it would be called meta-painting, that is, a picture represented within another. Each emblem of each of the shields tells a story in itself that influences the central action. To what degree? The images of the shields are elements that the messenger himself interprets and uses to compose two prolepses. The first one is enunciated in lines 1125-6, when he describes the shield of Tideus: Prometheus seems to be setting fire to the city of Thebes. The second prolepsis appears in lines 1130-3 when the shield of Capaneus is described.[15] In this passage, we can verify that Euripidean *ékphrasis* is a resource that, besides detailing how a subject appears (whether a place, event, person, or object), sometimes serves as a σῆμα or symbol for the central action. In relation to this, Paul Friedländer (1912) has contributed some interesting comments about *ékphrasis* in Aeschylus' *Seven against Thebes*, arguing that the images of the *ékphrasis* characterize warriors.[16]

---

12 *The Phoenician Women* 1120-21: λέοντος δέρος ἔχων ἐπ ἀσπίδι/ χαίτῃ πεφρικός; "with on his shield a lionskin/ bristling with hairs." (trans. Craik 1988)

13 *The Phoenician Women* 1130: σιδηρονώτοις δ' ἀσπίδος τύποις, "iron-backed outline of his shield." (trans. Craik 1988)

14 *The Phoenician Women* 1137-38: ἐκ δὲ τειχέων μέσων/ δράκοντες ἔφερον τέκνα Καδμείων γνάθοις; "from the middle of the Walls/ the snakes carried of the children of Kadmean in their jaws." (trans. Craik 1988)

15 *The Phoenician Women* 1133: ὑπόνοιαν ἡμῖν οἷα πείσεται πόλις, "a hint to us of the kind of thing a city will endure." (trans. Craik 1988)

16 "Aeschylus läßt in den ‚Sieben gegen Theben' seine Boten Nachricht von den angreifenden Helden erstatten, und bei jedem wird das Schildzeichen lebhaft beschrieben [...]; die Bilder

The purpose of this ekphrastic passage is clear: to arouse in the audience the desire and furor of the warriors that try to defeat the city of Thebes. From the rhetorical point of view, it is possible to move the audience by bringing it near to the interpretation of the reality of the person making the description, in this case the messenger. In addition to this, the character of each warrior is outlined through the description of the weapons, and their lineage is detailed through the description of the emblems.

The Dialogic Form of Object *Ékphrasis*

In *Ion* (184-218; cf. translation by Lee 1997), the first description of an object is related by the chorus, composed of servants of Creusa who contemplate the Temple of Delphi as if it had never been there before (cf. Zeitlin 1994:159). For this reason, we continuously observe the admiration of the chorus when they describe the seven images with familiar mythical scenes, although we do not know the medium in which they are executed (in relief or painted on the walls).

This *ékphrasis* has a dialogic form. The chorus asks about and responds to each image it sees inside the temple. Here, the visual vividness becomes present again through imperatives and adjectives (193),[17] affirmation sentences (205-206),[18] or questions (211-13).[19] Euripides is able to create a familiar image (the Temple of Delphi) with which the audience identifies its cultural reality.[20] In this way, the public is transported to that place and made participants in what the chorus is watching.

---

dienen nicht dem Schmuck des Berichtes, sondern charakterisieren ihre Träger, die ja nur von außen und aus einer gewissen Entfernung geschildert werden und doch eindrucksvoll geschildert werden sollen." (Friedländer 1912: 23)

17 *Ion* 193: φίλα, πρόσιδ᾽ὄσσοις,/ ὁρῶ; "my friend, take a close look." (trans. Lee 1997)

18 *Ion* 205-206: πάντᾳ τοι βλέφαρον διώ-/ κω; "as you see, I cast my eye in every direction." (trans. Lee 1997)

19 *Ion* 211-13: τί γάρ; κεραυνὸν ἀμφίπυρον/ ὄβριμον ἐν Διὸς/ ἑκηβόλοισι χερσίν;/ ὁρῶ; "What about this? The mighty thunderbolt flaming at both ends in Zeus' far-shooting hands?" (trans. Lee 1997)

20 Stieber 2011: 314: "in the tradition of most ekphrases, Euripides' stakes the genre's effectiveness upon a certain degree of plausibility in the object/s described. The intentional extraordinariness of the literary artifact is only made apparent if an audience can successfully summon to mind visual comparanda in the form of real achievements of real craftsmen, if not necessarily an actual match."

The first figure described is the Hydra that the son of Zeus defeats with his golden sickle (191-192).[21] The second image that the chorus sees is the lancer Iolaus, as well as Hercules himself, with his usual attributes, killing the Hydra of Lerna (201-204).[22] The image of the battle against the Hydra probably was not in the temple but would have been added by Euripides. Also part of the descriptive gaze are the battle of the Giants (206) and Athena with the head of Medusa on her shield (209-210).[23] Zeus, as the supreme god, is characterized by several adjectives that specify how he carries the thunderbolt in his "far-shooting hands" (212-214).[24] In response, the other part of the chorus provides information about the action represented, i.e., Zeus killing the giant Mimas (215-216).[25] Another god who kills a giant is Bacchus, who also appears in the Temple of Delphi; this god is characterized by his 'unwarlike' ivy staff (217).

This passage has a particular moral meaning since by describing several battles between monsters and gods it highlights the division between divine will and superiority. The public, through these mythological stories, realizes the difference between gods and mortals. This, in turn, interacts with the central action of the tragedy, especially in the life of the protagonist Ion, who is the son of the god Apollo.

In the second passage of *Ion* (1122-1165), a tapestry is described. On this occasion, the servant relates how Ion prepares a banquet in honor of the gods. Creusa was not in agreement with the adoption of Ion and in that celebration—where she still does not know that Ion is her own son—she will try to kill the boy.

In the servant's *rhēsis*, two elements are described: first (1132-1140), the measures that Ion draws to assemble the tent where the celebration will take place; second (1141 ff.), the preparations for the banquet, especially the decorations, i.e., the sacred tapestries. One of the tapestries is hung on the ceiling of the tent as if it were wings (1143); this tapestry belonged to Heracles, who in turn took it from the Amazons. For this reason, it is perhaps the most important ob-

21 *Ion* 191-192: Λερναῖον ὕδραν ἐναίρει/ χρυσέαις ἅρπαις ὁ Διὸς παῖς; "The son of Zeus slays the hydra of Lerna with his golden sickle." (trans. Lee 1997)

22 *Ion* 201-204: τὰν πῦρ πνέουσαν ἐναίρει/ τρισώματον ἀλκάν; "the mighty three-bodied creature/ that breathes fire." (trans. Lee 1997)

23 *Ion* 209-210: λεύσσεις οὖν ἐπ᾽ Ἐγκελάδῳ/ γοργωπὸν πάλλουσαν ἴτυν; "Do you see, then, the woman brandishing the gorgon-faced shield in front of Enkelados…?" (trans. Lee 1997)

24 *Ion* 212-214: κεραυνὸν ἀμφίπυρον/ ὄβριμον ἐν Διὸς/ ἑκηβόλοισι χερσίν; "The mighty thunderbolt flaming at both ends in Zeus' far-shooting hands?" (trans. Lee 1997)

25 *Ion* 215-216: τὸν δάϊον/ Μίμαντα πυρὶ καταιθαλοῖ; "The dread Mimas he burns to ashes with fire." (trans. Lee 1997)

ject of this ekphrastic passage; and indeed, its description occupies 17 lines, as opposed to the other tapestries, which remain very much in the background.

The figures that appear in the first tapestry are the most significant stars and constellations, along with the personifications of Heaven (Οὐρανός), the sun (Ἥλιος), the moon (Σέληνη), or the dawn (Ἕως), and the servant lists them in the following order (cf. Mirto 2009: 231-233):

a) Uranus gathering the stars in the circle of Ether (1147).[26]
b) Helios driving his horses (1149).[27]
c) One of the Pleiades and the lancer Orion through the Ether. Above them, the Bear twisting its golden tail (1152).[28]
d) The full circle of Selene throwing its rays upwards (1155).[29]
e) The Hyades, the clearest signal for navigators (1156-7),[30] and Eos, bearer of light, pursuing the stars (1157-8).[31]

This *ékphrasis* stops the central action (i.e., the preparation of the banquet) to narrate the course of a night from sunset to sunrise. The description is vivid, not only because of the recurrent use of adjectives in every enumerated element, but also because of the combination of verbs in both personal and participle forms. In these lines, we observe four participles that combine simultaneous actions, of which line 1149 constitutes an example, where Helios drives his chariot through the Ether and at the same time leaves behind the light of Hesperus. The other example is found on line 1154. Here, the Pleiades and Orion are simultaneously described as walking while the Bear twists its tail. However, the participles that appear at the beginning and end of the *ékphrasis* may play a different role.

26 *Ion* 1147: Οὐρανὸς ἀθροίζων ἄστρ᾽ ἐν αἰθέρος κύκλῳ; "Heaven was assembling the stars in the circle of the sky." (trans. Lee 1997)

27 *Ion* 1149: Ἥλιος, ἐφέλκων λαμπρὸν Ἑσπέρου φάος; "Helios was driving his horses to their flaming goal." (trans. Lee 1997)

28 *Ion* 1152: Πλειὰς μὲν ᾔει μεσοπόρου δι᾽ αἰθέρος/ ὅ τε ξιφήρης Ὠρίων, ὕπερθε δὲ/ Ἄρκτος στρέφουσ᾽ οὐραῖα χρυσήρη πόλῳ; "The Pleiades moved on a path through the middle of the sky,/ as did Orion with his sword, and above them was the Bear turning at the pole its golden tail." (trans. Lee 1997)

29 *Ion* 1155: κύκλος δὲ πανσέληνος ἠκόντιζ᾽ ἄνω/ μηνὸς διχήρης; "The full circle of the moon darted her beams upwards/ as at mid-month." (trans. Lee 1997)

30 *Ion* 1156-7: Ὑάδες τε, ναυτίλοις/ σαφέστατον σημεῖον; "Hyades, the clearest/ sign to sailors." (trans. Lee 1997)

31 *Ion* 1157-8: ἥ τε φωσφόρος/ Ἕως διώκουσ᾽ἄστρα; "Dawn put the stars to flight." (trans. Lee 1997)

On the one hand, on line 1144 Uranus is gathering the stars in the Ether, and from here the stars are located and their particular action is detailed. At the end of the description, the same structure from the beginning is recalled, but this time it is Eos who is chasing the stars. Therefore, these participles are elements that help frame the descriptive passage and provide a beginning and an end for it.

The Shortest *Ékphrasis* in *The Suppliants*

The last passage belongs to *The Suppliants* (1197-1204; cf. translation by Morwood 2007), whose *ékphrasis* of an object is the shortest of our analysis and the only one pronounced by a goddess: Athena. The goddess instructs Theseus on how to perform an oath and details the ritual objects he must use, and especially a tripod. First she details the material of the object (1197),[32] and later clarifies its origin. Then Athena explains to the hero what kind of oath he must perform and the part of the tripod on which he must engrave that oath (1201),[33] advising him to give the tripod to the god Apollo in the Temple of Delphi.

We should consider only two lines (1197-1199) to be properly ekphrastic, while the rest of the discourse would be considered an enumeration of guidelines that Theseus should follow to perform the oath. Despite this, the compound adjective χαλκόπους (*chalkópous*), 'with feet of bronze,' gives extremely detailed information in a very short *ekphrasis*. For this reason, I have included this *ékphrasis* in our analysis. We should not overlook the short *ekphráseis*, although studying this type of description is more arduous.

CONCLUSION: *ÉKPHRASIS* AS A KNOWLEDGE SOURCE FOR A TARGET GROUP

The analysis of these ekphrastic passages has served to verify the usefulness of *ékphrasis* as a source of knowledge for the audience at that time. I am aware that this study should be expanded, but in a first approach, we can determine that the Euripidean *ékphrasis* presents the main characteristics that are later collected in the *progymnásmata*, putting the object before the eyes of the audience. At the same time, this *ékphrasis* probably served to give information to the public about

32 *Ion* 1197: ἔστιν τρίπους σοι χαλκόπους, "a bronze-legged tripod." (trans. Lee 1997)

33 *Ion* 1201: ἔγγραψον ὅρκους τρίποδος ἐν κοίλῳ κύτει; "inscribe the oath on the curving hollow of the tripod." (trans. Lee 1997)

an object it did not know, such as the shield of Achilles, the tapestries in the Temple of Delphi, or the shields of the warriors on the Theban walls.

We have also observed in these Euripidean passages how the use of *ékphrasis* responds to rhetorical purposes. The descriptions of Euripides occasionally show some variations in the story (e.g., the shield of Achilles or the description of images in the Temple of Delphi), which is always replete with symbolic words. The audience is aware of attending a new representation of the mythical story, related to the critique of values that Euripides intended to convey.

Finally, I believe that *ékphrasis* in the Euripidean tragedy has achieved its rhetorical purposes, and furthermore that this resource has managed to transmit mythological, historical, and cultural knowledge by means of presenting objects or events. This applies not only to the audience of Euripides' time but also to us, modern recipients that fail to perceive the cultural references that the Euripidean *ékphrasis* entirely conceals.

REFERENCES

Craik, Elizabeth M. (1988): Phoenician Women, Warminster: Aris & Phillips.

Cropp, Martin (2013): Euripides. Electra, Oxford: Oxbow Books.

Denniston, John D. (1987): Euripides. Electra, Oxford: Clarendon Press.

Diggle, James (1981): Euripides Fabulae, Vol. II, Oxford: Oxford University Press.

Fernández Delgado, José Antonio (2016): "Inversión moral y cósmica, premonición y ékphrasis en el segundo estásimo de la Electra de Eurípides." In: José Guillermo Montes Cala/Rafael J. Gallé Cejudo/Manuel Sánchez Ortiz de Landaluce/Tomás Silva Sánchez (eds.), Fronteras entre el verso y la prosa en la literatura helenística. Homenaje al Prof. José Guillermo Montes Cala, Bari: Levante Editori, pp. 379-89.

Fernández Delgado, José Antonio/Pordomingo, Francisca (2007): "Topics and Models of School Exercises on Papyri and Ostraca from the Hellenistic Period: P.Berol. inv. 12318." In: Traianos Gagos (ed.), Proceedings of the Twenty-Fifth International Congress of Papyrology, Ann Harbor.

Fernández Delgado, José Antonio/Pordomingo, Francisca (2008): "PMilVogl 1, 20: bocetos de progymnásmata." In: Zeitschrift für Papyrologie und Epigraphik 67, pp. 167-92.

Friedländer, Paul (1912): Johannes von Gaza und Paulus Silentiarius: Kunstbeschreibungen Justinianischer Zeit, Leipzig: Teubner.

Garzya, Antonio (1997): "L'ekphrasis nella tragedia greca." In: Antonio Garzya (ed.), La parola e la scena. Studi sul teatro antico da Eschilo a Plauto, Naples: Bibliopolis, pp. 47-58.

Headlam, Walter (1901): "Notes on Euripides, II." In: The Classical Review 15, pp. 98-108.

Kennedy, George A. (2003): Progymnasmata. Greek Textbooks of Prose Composition and Rhetoric, Translation with Introduction and Notes, Leiden/Boston: Brill.

Lee, Kevin H. (ed.) (1997): Euripides. Ion, Warminster: Aris & Phillips.

Mastronarde, Donald J. (ed.) (1994): Euripides. *Phoenissae*, with Introduction and Commentary, Cambridge, UK: Cambridge University Press.

Mirto, Maria Serena (2009): Euripide. Ione, introduzione, traduzione e commento, Milan: BUR.

Morwood, James (2007): Euripides. Suppliant Women, Oxford: Aris & Phillips.

Stieber, Mary C. (2011): Euripides and the Language of Craft, Leiden/Boston: Brill.

Webb, Ruth (1999): "Ekphrasis Ancient and Modern: The Invention of a Genre." In: Word and Image 15, pp. 7-18.

Zeitlin, Froma (1994): "The Artful Eye: Vision, Ekphrasis and Spectacle in Euripidean Theatre." In: Simon Goldhill; Robin Osborne (eds.), Art and Text in Ancient Greek Culture, Cambridge, UK: Cambridge University Press, pp. 138-96, 295-304.

Argument Schemes Related to Popular Science in the Second Sophistic

Maurice Parussel

ABSTRACT

The genealogy of popular science can be traced back to antiquity. The rhetorical nature of the phenomenon of popular science is especially noticeable during the Second Sophistic, a literary-historical movement that had its roots in the 1st century and lasted until late antiquity. Among the several speeches of the 2nd and 3rd centuries AD in which the representatives of the Second Sophistic gained much influence, the *Apologia*, a defense speech by the famous Sophist Apuleius of Madaurus, comes to the fore. The special conditions of the origin of this speech, in which Apuleius had to defend himself against the accusation of having used magic, meant that he was addressing a heterogeneous audience. Therefore, in his speech, Apuleius combines complex scientific contents with elements that nowadays are regarded as part of popular science: highly simplified scientific explanations, shortened quotations from famous authorities of ancient natural science and philosophy, and demotic ideas.

HISTORICAL BACKGROUND: THE SECOND SOPHISTIC

Nearly every examination of the genealogy of popular science which deals with the exploration of the roots of this phenomenon inevitably has to hark back to the historical and cultural developments of antiquity. Nevertheless, the occurrence of popular science in this age has to be seen as a minor phenomenon, one that only appears in connection with a small number of intellectual currents. According to this, the transmission of knowledge to the less educated social classes, whose presentation and contents were adapted to this target audience, can be defined as popular science in a first approach, but cannot be seen as a characteristic feature of ancient science on a more global level. Instead, despite all the cultural and political differences between the various ancient societies, science, in general, has been the holding of the particular elite of every ancient society. Taking this into account, the small number of developments in antiquity in which a specific form of science occurs that does not require extensive knowledge gain eminent significance for the reconstruction of the social contexts in which the genesis of popular science seems to be possible.[1]

In our search for those specific developments in the rhetoric of antiquity which exhibit an adaptation of the conveyed contents and the mode of presentation to less educated people, the intellectual current of the so-called Second Sophistic inevitably comes to the fore. The Second Sophistic, which never occurred as an actual historical unity, and which was first called by this name in the 1970s to distinguish it from the first occurrence of Sophists in antiquity during the fifth and fourth centuries BC, can approximately be dated to the second and third centuries AD. Related to this, the occurrence of the so-called Second Sophistic in this period of antiquity, especially in contrast to the first appearance of Sophists, allows some important inferences about the social conditions enabling the genesis of elements of popular science within ancient scholarship. An important reason for this development can be found in the growing sphere of influence of the so-called Second Sophistic. While the Sophists of the fifth and fourth centuries BC had mainly operated in Greece and in the Greek cities on the western coast of Asia Minor, the representatives of the Second Sophistic gained influence, fame, and prominence within all social classes in nearly all the provinces of the Roman Empire (Bowie 2004: 72). Furthermore, the representatives of the Second Sophistic extended the field of interest from rhetoric, which the Sophists of the fifth and fourth centuries BC had primarily dealt with, to other fields of science, such as religion, history, and biology. This growing influence of the Sophists and their well-explored connections to one another are closely connected to the increase of cultural exchange between the various provinces of the Roman Empire, which also reached its climax during the second century AD. Thus, the increase of exchange of all kind of goods on the one hand, and of ideas, knowledge, and literature on the other hand, can be seen as the two major circumstances which guaranteed the power of the Sophists and the knowledge presented by them in the second and third centuries AD (Baumbach/Möllendorff 2017: 62).

Considering the noteable presence which had been attained by many Sophists by acting as speakers in public or even as writers of various text forms, it becomes possible to draw conclusions about the cultural backgrounds which

1 Another example of the adaptation and simplification of scientific contents to the target audience in antiquity can be found in Martín-Velasco's essay on popular scientific knowledge in Aristotle in the present volume.

obliged the sophistic scholars to adapt the contents and the modes of presentation of their speeches and texts to gain such enormous influence. In this context, many ancient reports on the courses of public speeches of different Sophists show that the audience of these public appearances were heterogeneous in nature (Korenjak 2000: 50-52), because many people from the lower classes participated (Schmitz 1997: 160-161). Nevertheless, the heterogeneous composition of the audience of the Sophists' public speeches cannot be regarded as a common reason for a general adaptation of the contents and the mode of presentation dealing with scientific topics in the speeches to a less educated audience. Instead, in observing the style and the contents of the extant speeches of the Second Sophistic, it becomes obvious that only a few speeches contained difficult scientific contents which would need to be made comprehensible to a less educated audience. Therefore, all three primary genres of speeches from the Second Sophistic—the ἐπιδεικτικοὶ λόγοι (speeches for display), the διαλέξεις (lectures), of which there are only a few extant examples, and the *declamationes* (speech exercises)—demonstrate an obvious lack of scientific contents. Especially considering the speeches which can be categorized as ἐπιδεικτικοὶ λόγοι (speeches for display), or those which can be categorized as *declamationes*, a focus on the rhetorical presentation becomes obvious, while scientific contents apparently take a back seat. Thus, the ἐπιδεικτικοὶ λόγοι (speeches for display) are generally focused on the concrete occasions of the speeches, such as the glorification of a deceased person, while the *declamationes* are mostly confined to an abstract question which has been previously provided, such as the moral assessment of the actions of a person from mythology (Whitmarsh 2004: 70-73).

APULEIUS' *APOLOGIA*

Taking all this into account, the defense speech of the famous rhetorician Apuleius of Madaurus (123-170 AD, cf. *Fig. 1*), which has been called the *Apologia* or even *Oratio pro se de magia*, gains an important position among all sophistic speeches which have been handed down to the modern age. While the other speeches given by representatives of the Second Sophistic, in general, arose in the context of voluntary public performances of rhetorical skills, mostly arranged by a client, the *Apologia* appears as the result of a real court process in which the Sophist Apuleius had to defend himself against the accusation of the forbidden use of magic. Moreover, the extraordinary position of the *Apologia* among all other speeches of the Second Sophistic is reinforced by the fact that it is the only court speech from the time of the Roman Empire which has survived in a complete form all the way to the present day.

Figure 1: Contorniate Medallion with a Portrait of Apuleius from the 4th Century Showing his Enormous Popularity

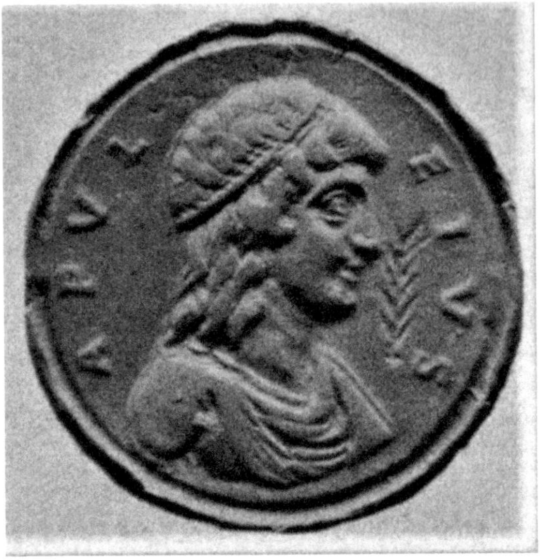

Source: Bernoulli 1882: Plate 5,117; Photo: Public Domain

Despite this, the degree of similarity between the written version of the speech which was published by Apuleius after the court process, and which seems to be the version that we have access to today, and the version which was provided in the court process, cannot be precisely determined anymore. Moreover, there is no evidence, apart from the written version of the speech itself, that the court process, which seems to have been the impetus for the speech, actually took place (Hammerstaedt 2002: 10-16). Finally, in contrast to many other written versions of ancient speeches, such as those of Cicero, there is some evidence for a high degree of similarity between both versions. First of all, the circumstances of the court case offer an impossibility for the defendant Apuleius to alter long passages of the speech, which has been held in comparison to the later published version. Thus, during the Second Sophistic, in contrast to the first spreading of the Sophistic during the fifth century BC, the importance of oral speeches in comparison to written speeches increased enormously (Anderson 1993: 16-18). The importance of a Sophist's speech for the city concerned (Baumbach/Möllendorff 2017: 62) suggests that Apuleius' appearance in a court case in the presence of a large audience that included all social classes of the society[2] was regarded as an occasion for the famous Sophist to prove his eloquence (Butler/Owen 1914: XVI). Taking this into account, it can be assumed that Apuleius, in all proba-

2 Possibly even women and slaves (Korenjak 2000: 41-50).

bility, thoroughly prepared the version of the speech he was going to deliver. Moreover, it can be assumed that he refrained from major changes after the oral presentation of the speech, which the audience had at that point already heard him deliver in court. Furthermore, even the numerous interactions between Apuleius and the judge, the accusers, and the bailiffs (Constantini 2019: 10-11), which sometimes complicate the perspicuity of the written version, suggest a close accordance between the oral version and the written version of the speech (Abt 1908: 80-83).

Taking into account that these indications for a close accordance between both versions are based on considerations about the plausibility of different possible theories on the genesis of the written version of the speech, the obvious existence of elements of popular science in the speech provides an essential proof for the reliability of the written version of the speech. Therefore, the adaptation of the contents and the presentation of the scientific elements of the speech to a less educated audience clearly reveal Apuleius' goal to reach a large number of recipients by using his position as a defendant in a public court case. Through this means, the speech becomes a significant example of the adaptation of the mediation of knowledge to heterogeneous groups of recipients during the Second Sophistic.

STRUCTURE OF THE SPEECH

The division of the speech in two different parts reflects this adaptation of the presentation. While the first part of the speech (Chapters 1-65) mainly contains discussions on general topics which were mentioned by the accusers, such as magic, marine biology, and even the significance of epilepsy, in the second part (Chapters 66-103) Apuleius refers directly to the arraignment and tries to reject it by discussing it point by point.[3] The style of the argumentation in both parts of the speech differs accordingly (Constantini 2019: 16-17). Thus, the first part of the speech mainly contains long explanations of the various topics that were mentioned in the arraignment, and often refer to more general topics of ancient philosophy or literature, with Apuleius using the authority of older sources to support his argumentation. In contrast to this, Apuleius changes over to a hermeneutic argumentation in the second part of the speech and begins to refute the accuser's argumentation point by point (Helm 1955: 98-99). Although the mode of argumentation in the second part of the speech is generally regarded as an important example for eloquence among the representatives of the Second Sophistic,[4] the speech mainly gains its extraordinary position among the other speeches

3 For an overview of the individual charges leveled against Apuleius, cf. Gindhart 1996: 17.

4 Hunink (1997: 20-22) provides a detailed description of the speech's partition, including an extensive overview of references.

from the Second Sophistic through the insertion of elements of popular science into the first part of the speech.

The numerous explanations given by Apuleius which refer to the topics mentioned in the arraignment take on a double function, namely that of both guiding and pursuading the audience. First of all, Apuleius, when dealing with topics mentioned in the arraignment, initially rejects the individual allegations, before later entering into a discussion of the individual accusations (Schenk 2002: 26-30). Despite this, the long explanations at the beginning of the speech mainly belong to an important strategy of argumentation which can be regarded as the main peculiarity of the speech. Apuleius uses the explanations at the beginning of the speech to simplify the complex contents, such as magic or epilepsy, that he is going to deal with in the subsequent argumentation. In doing so, he is mainly trying to provide the less educated people in the audience access to his argumentation, since they would otherwise not be able to follow his complicated reasoning, which is based on scientific and philosophical knowledge.

This characteristic feature of the *Apologia* becomes obvious when we consider three passages from the first part of the speech, in which Apuleius adapts his explanations of the nature of magic (Apul. *apol.* 25.4-27.2), his reasons for research on marine animals (Apul. *apol.* 29-41), and his explanations of the 'falling sickness' (most likely epilepsy, Apul. *apol.* 42-52) to the everyday reality of the audience. By using the same structure in all three passages, even though they differ greatly in length according to the topic explained, Apuleius strengthens the perspicuity of the explanations. Thus, Apuleius starts all three explanations with an exact determination of the subject he is going to deal with, always referring to a phrase or an opinion of a well-known author to base his argumentation on an authority who would be known even by less educated people.

Apuleius thereafter distinguishes between the explanation of the topic given by the accusers and the ideas of the authority who he has cited, in order to show the falsehood of the arraignment (Gindhart 1996: 17-18). Finally, he concretizes his idea of the subject by closely referring to the authority he has cited, so that his extensive knowledge of ancient philosophy becomes obvious for the audience. Although this structure of argumentation is used for explaining all three topics, the insertion of elements of popular science becomes most obvious in the explanations dealing with the falling sickness.[5]

5 In describing the illness the slave suffers from, Apuleius later uses the Latin denomination *morbus comitialis* (i.e., "illness of the convention," Apul *apol.* 50.1) and the Greek denomination ἱερὸς νόσος ("divine illness," Apul. *apol.* 50.5). Furthermore, considering the symptoms of the illness described by Apuleius, it seems quite probable that the illness was in fact epilepsy. Because this illness had many different connotations in antiquity, it is not, however, possible to equate the illness described by Apu-

EXPLANATION OF THE FALLING SICKNESS

The explanations of the falling sickness find their origin in an allegation against Apuleius provided in the arraignment. There, Apuleius is accused of having made a young slave faint in public by using some form of magic. To underline the absurdity of this allegation, Apuleius notes that the accusers should have reported that the young slave made many prophecies in front of a large audience as a consequence of this supposed use of magic (*enim fabula ut impleretur, addendum etiam illud fuit, puerum eundem multa praesagio praedixisse* / To complete the story, they even should have added that this boy had made many prophecies, Apul. *apol.* 42.3).[6]

Referring to this accusation, Apuleius initially deals with the possibility of the existence of accurate prophecies as a result of the appearance of divinity in the earthly world (Apul. *apol.* 42.3-5). He then mentions the ideas of Plato on this topic, and concludes that this kind of presence of divinity on earth is generally possible (Apul. *apol.* 43.1-2). Especially considering the mode of argumentation used by Apuleius in this passage, his endeavor to simplify the required background information becomes clear. Knowing that the question of the presence of the so-called *δαιμόνιον* ("divinity") within all people on earth is a very difficult topic, and one that plays an important part in Platonic philosophy and was dealt with by Apuleius in his book *De deo Socratis*, Apuleius tries to reduce his explanations of this topic to a few catchy pieces of information. Thus, Apuleius restricts his explanations of this topic exclusively to the statement that Plato has already confirmed the presence of divinity among the people on earth (*quamquam Platoni credam inter deos atque homines natura et loco medias quasdam divorum potestates intersitas, easque divinationes cunctas et magorum miracula gubernare* / Although I believe Plato when he says that there are certain divine powers in a position between humans and gods which have a split character and control all divinations and miracles, Apul. *apol.* 43,1), leaving out any further argumentation or quotations from other authors. Apart from the intelligibility of this passage, which is ensured by simplifications, the reference to Plato serves two further functions. First of all, the mention of Plato as a guarantor for the existence of divinity on earth ensures the reliability of any further argumentation on this topic, because it can be generally assumed that Plato is a well-known figure to the whole audience and that the whole audience accepts Plato as an authority for philosophical and theological topics. However, the little

---

leius with the illness nowadays known as epilepsy with absolute certainty (Temkin 1971: 15-21).

6 All quotations from the Latin text of the *Apologia* are given in the form of Helm's critical text edition (1959). All English translations from the Latin are my own.

knowledge of Plato and his philosophy that the less educated people in the audience presumably have is not to be equated with a profound understanding of Platonic philosophy.[7] Furthermore, it becomes obvious that Apuleius, by mentioning Plato, is trying to target the less educated addressees in particular. For this purpose, he uses the audience's sparse knowledge of Platonic philosophy by complementing this knowledge with some short and simple explanations, to make them feel that they have understood a basic element of Platonic philosophy. Through this, the conveyed knowledge of Platonic philosophy, on the one hand, becomes a door opener for the understanding of any further argumentation and, on the other hand, makes even the less educated people among the audience feel confident enough to understand the more elaborate parts of the argumentation.

After this preliminary explanation, Apuleius transitions to the main question of the passage, which deals with the relationship between the existence of divinity among the people on earth and the falling sickness. Thus, Apuleius explains that, based on the postulation that divinity generally can appear on earth, the epiphany of this divine power, which is perfect per se, inevitably has to appear within a perfect human being. Here, Apuleius is specifically thinking of a knave who is perfect in his looks, intellect, and speech (*debet ille nescio qui puer providus, quantum ego audio, et corpore decorus atque integer deligi et animo sollers et ore facundus* / To be chosen, this prophetic boy—as far as I have heard—must have a beautiful and impeccable body, and has to be shrewd and eloquent, Apul. *apol.* 43.3). In contrast to this, the young slave Thallus, who is mentioned in the arraignment as the victim of Apuleius' witchery, is deformed, crippled, and heavily wounded as a result of his permanent seizures (*est enim miser morbo comitiali ita confectus, ut ter an quater die saepe numero sine ullis cantaminibus corruat omniaque membra conflictationibus debilitet, facie ulcerosus, fronte et occipitio conquassatus, oculis hebes, naribus hiulcus, pedibus caducus / The poor boy is so afflicted with the falling sickness that he falls three or four times a day without magic influence and hurts his body heavily by his convulsions.* His face is covered with ulcers, his head is swollen front and back, his eyes are dull, his nostrils are wide open, his legs made for stumbling, Apul. *apol.* 43.5)—seeing him, people always vomit and avoid eating from the same bowl or drinking from the same cup as him (Apul. *apol.* 44.1).

In light of this last part of the explanation in particular, Apuleius' aim of making his argumentation intelligible for the whole audience becomes obvious. Apuleius no longer argues on the basis of Platonic philosophy, but harks back to demotic beliefs, such as the belief in the magical power of a virgin knave (Abt 1908: 184-185), the trust in the protecting power of spittle (Helm 1977: 160),

7 Novotný (1977: 272-275) has argued convincingly for this estimation of the prominence of Plato and his philosophy in the Roman Empire.

and the public's imagination of the transmission of mental diseases through close contact with the afflicted. Moreover, the demotic character of this passage is intensified by the fact that there are no references to scientific literature here, even though these are abundantly used in other passages of the speech. Apart from the intended simplification, Apuleius' lack of scientific references here means that he is basing his argumentation exclusively on demotic beliefs, omitting all scientific knowledge of the falling sickness (Gruppe 1906: 886-894).[8] It should be taken into account that the falling sickness had been examined in several well-known scientific books which the well-educated Sophist Apuleius was certainly familiar with. Therefore, it is obvious that he intentionally omits all this scientific knowledge. Moreover, Apuleius shows another strategy of argumentation in this passage. While the first passage of his treatment of the accusation of having bewitched a knave to faint mainly operates by using simplifications, this passage mainly operates with referenes to the everyday lives of the audience. Mentioning these demotic ideas, which were well known to the whole audience, Apuleius appeals particularly to less educated people, whose demotic beliefs play an important role in their lives, so that for them the following explanations are based on a reliable and intelligible foundation.

Apuleius continues this mode of argumentation in the following passage of his speech as well. In this passage, he reacts to the accusation that, in addition to the hex on the young slave Thallus, he also put the same hex on a free-born woman while pretending to examine her illness (Apul *apol.* 48.1). Apuleius, when dealing with this accusation, at first polemically rejects the charge of having put a hex on the woman by pointing out that the effect of putting the hex of the falling sickness on a woman who already suffers from the falling sickness would have been quite little (Apul *apol.* 48.2-5). Then he moves on to a discussion of the reasons for and the indications of the falling sickness (Apul. apol. 48.6-52).

Again, Apuleius does not refer to the various—and in his time frequently quoted—medical texts which deal with the origins of the falling sickness, such as the *Corpus Hippocraticum* or the treatises of Aretaeus (Flashar 2016: 230-232). Instead, he simplifies the whole topic and reduces his explanation of the origin of the sickness to a quotation from Plato's *Timaeus*. Thus, Apuleius quotes from the *Timaeus* that sicknesses within the human body can have three potential reasons, among which the falling sickness results from an accumulation of slime in the body, which moves into the head and deactivates the functions of

8 Although the ideas pertaining to the origin and significance of the falling sickness which are presented by Apuleius are often mentioned in literature, they are always marked as demotic beliefs and clearly distinguished from the scientific knowledge of this disease. Cf. Petron. 63; Plaut. Capt. 535; Plin. *nat.* 10.69; Plin. *nat.* 28.35.

the brain during the seizures (Apul. apol. 50.2-5). Considering this origin of the disease, Apuleius further explains his treatment of the woman. At first, he located where the slime in the woman's body was temporarily located, and then he examined on which side of the body the symptoms had appeared for the first time. He did this because Aristotle had already pointed out that a disease starting at the right side of the body was much heavier (Apul. *apol.* 51.1-3).

Observing the way Apuleius explains his treatment of the woman, it again becomes obvious that in this passage of his speech he is mainly addressing the less educated people in the audience. Therefore, the most important feature of his argumentation can be found in his consequent omission of any reference to the medical literature dealing with the falling sickness. This is explained by his statement that philosophy itself can generally answer any question.[9] Besides this glorification of philosophy, which can be regarded as a central motive of the whole speech (Hunink 1997: 142), this statement also reveals the two basic elements of the rhetoric strategy used by Apuleius in the first part of his speech (Chapters 1-65). On the one hand, Apuleius again bases his persuasive strategies on simplicity and brevity, which first of all include the omission of any references to the varied medical literature on the topic. The references to Plato and Aristotle also ensure the comprehensibility of his argumentation. Therefore, Apuleius strictly abridges the references and picks out catchy passages from the long explanations which can be understood by everyone without any knowledge of Plato's and Aristotle's theories. On the other hand, Apuleius again refers to demotic ideas which are widely known among his audience. Thus, he connects the philosophical theory of the origin of diseases developed by Plato in the *Timaeus* with observations of people suffering from the falling sickness.[10] In doing this, he succeeds in giving a plausible explanation of the origin of this disease which corresponds with the impressions of the falling sickness among people

---

9 Apul. *apol.* 51.5: haec idcirco commemoravi, nobilium philosophorum disputata simul et libros sedulo nominavi nec ullum ex medicis aut poetis volui attingere, ut isti desinant mirari, si philosophi suapte doctrina causas morborum et remedia noverunt / "I have mentioned these things, I have intentionally talked about the discussions of the philosophers and their books, and I have avoided mentioning physicians or poets, so that these men stop wondering that philosophers have acquired knowledge about the reasons for diseases and their remedies."

10 Apul. *apol.* 50.5: eum nostri non modo maiorem et comitialem, verum etiam divinum morbum, ita ut Graeci ἱερὰν νόσον, vere nuncuparunt, videlicet quod animi partem rationalem, quae longe sanctissimast, eam violet / "People from our nation have not only called it the 'great sickness' and the 'sickness of the assemblies,' but also rightly, like the Greeks, the 'divine sickness,' because it attacks the part of the soul which is the holiest one."

who are not conversant with medical literature (Abt 1908: 272-273). By connecting these different types of knowledge, Apuleius again creates a basis for his further argumentation, which consists of a combination of science and demotic ideas. Because of this, he enables even the less educated people in the audience to understand his argumentation.

ELEMENTS OF HIGHER EDUCATION

Observing this passage of the first part of the speech makes the changes in the references for the addressees obvious. Thus, the *Apologia* can be regarded as a complex weaving of various elements of the Sophists' rhetoric, which further is combined with typical elements of a defense speech (Schenk 2002: 46-47). While these typical elements of a defense speech can mainly be found in the second part of the speech (Chapters 66-103), the first part of the speech (Chapters 1-65) primarily shows, apart from various elements of the Sophists' rhetoric, noticeable changes in the references for the addressees, which make the particular passages interesting and comprehensible for different target audiences. Thus, the biggest part of the speech appears much more complicated than the three passages mentioned before, in which elements of popular science can be found. Instead, most passages from the first part of the speech are full of quotations from different types of literature, mostly from philosophical books (Gindhart 1996: 17-18), which becomes especially obvious considering Apuleius' justification for writing love poems (Apul. *apol.* 9-13,2). In rejecting the criticism of having used his eloquence for writing lewd poems (Apul. *apol.* 9,1-2), and despite the brevity of his explanation, Apuleius cites nine passages from six different authors (Solon, Diogenes, Plato, Catullus, Hadrian, and Lucius Afranius) who are known as writers of love poems. The choice of the authors cited, the choice of the particular passages, and the order of the quotations, which has gained much attention even among modern scholars, was also discussed—not without some controversy—among ancient scholars. As a consequence, this striking accumulation of quotations can be regarded as an expression of extraordinary knowledge which Apuleius inserted into the speech to guarantee its reception by the well-educated elite of the Roman Empire (Helm 1977: 8-9). In contrast to this, the less educated people among the audience certainly neither knew the passages cited nor the particular meaning of each quotation, so that it would have been nearly impossible for them to understand this step of the argumentation.

Taking this into account, it becomes obvious that in spite of the various uses of simplifications and references to demotic ideas, the *Apologia* cannot be exclusively regarded as an example for the involvement of less educated people in scientific affairs. Instead, the *Apologia* has to be primarily regarded as an expression of its author's desire for glory among the sophistic scholars in the Roman Empire. Apart from this, Apuleius uses his unique status as a defendant to prove his eloquence and his knowledge in the presence of a heterogeneous audience. Moreover, Apuleius is forced into changing specific parts of the structure of argumentation usually used in Sophists' speeches (Schenk 2002: 46-47). Therefore, Apuleius had to supplement his speech with elements that guaranteed the comprehensibility of the reasoning and the structure of argumentation for an audience which mainly consisted of less educated people. These changes made by Apuleius reveal similarities to the ideas behind modern popular science. Thus, the structure of speech and its contents show simplifications as well as adaptations to the everyday reality of the target audience. Moreover, the famous authors and philosophers cited by Apuleius take on an important position and serve to strengthen the argumentation through their generally accepted authority, although their books and philosophical theories certainly were not known by the main part of the audience.

Taking all this into account, the *Apologia* of Apuleius of Madaurus marks an important development within the genealogy of popular science. Indeed, the *Apologia* has to be regarded as a result of the presence of science in the society of the Roman Empire during the second century AD. This social development made it necessary to adapt the presentation of scientific contents to an audience which did not have any previous knowledge. Despite the numerous expressions of sophistic eloquence and a large number of intertextual markers which were mainly addressed to an extensively educated target audience, the *Apologia* mainly has to be regarded as the result of a real court case, which guarantees its extraordinary position among all other speeches of the Second Sophistic. Therefore, the example of the *Apologia* makes it obvious that popular science had a real presence in antiquity and that the occurrence of this specific form of popular science generally accords with modern forms of this scientific phenomenon.

REFERENCES

Abt, Adam (1908): Die Apologie des Apuleius von Madaura und die antike Zauberei. Beiträge zur Schrift de magia, Gießen: Verlag Alfred Töpelmann.

Anderson, Graham (1993): The Second Sophistic. A Cultural Phenomenon in the Roman Empire, London and New York: Routledge.

Baumbach, Manuel/Möllendorff, Peter von (2017): Ein literarischer Prometheus. Lukian aus Samosata und die Zweite Sophistik, Heidelberg: Winter Verlag.

Bernoulli, Johann Jacob (1882): Römische Ikonographie, Vol. 1. Die Bildnisse berühmter Römer mit Ausschluss der Kaiser und ihrer Angehörigen, Stuttgart: W. Spemann.

Bowie, Ewen (2004): “The geography of the Second Sophistic. Cultural variations.” In: Barbara Borg (ed.), Paideia. The World of the Second Sophistic, Berlin and New York: Walter de Gruyter, pp. 65-75.

Butler, Harold Edgeworth/Owen, Arthur Syng (1914): Apulei Apologia sive pro se de magia liber. With introduction and commentary, Oxford: Clarendon Press.

Constantini, Leonardo (2019): Magic in Apuleius’ *Apologia*. Understanding the Charges and the Forensic Strategies in Apuleius’ Speech, Berlin and Boston: Walter de Gruyter.

Flashar, Hellmut (2016): Hippokrates. Meister der Heilkunst. Leben und Werk, Munich: C. H. Beck.

Gindhart, Marion (1996): “Von Erntezauber und Meerestiermagie. Das crimen magiae in der römischen Republik und Kaiserzeit und der Prozeß gegen Apuleius von Madauros.” In: Augsburger Volkskundliche Nachrichten 2/2, pp. 7-34.

Gruppe, Otto (1906): Griechische Mythologie und Religionsgeschichte, Munich: C. H. Beck.

Hammerstaedt, Jürgen (2002): “Der Autor. Apuleius. Leben und Werk.” In: Peter Habermehl/Jürgen Hammerstaedt/Francesca Lamberti/Adolf Ritter/Peter Schenk (eds.), Apuleius. Über die Magie, Darmstadt: Wissenschaftliche Buchgesellschaft, pp. 9-22.

Helm, Rudolf Wilhelm (1955): “Apuleius’ Apologie. Ein Meisterwerk der zweiten Sophistik.” In: Das Altertum 1, pp. 86-108.

Helm, Rudolf Wilhelm (ed.). (1959): Apulei Platonici Madaurensis opera quae supersunt. Vol. II, Fasc. 1. Pro se de magia liber (Apologia), Leipzig: B. G. Teubner.

Helm, Rudolf Wilhelm (1977): Apuleius. Verteidigungsrede, Blütenlese, Berlin: Akademie-Verlag.

Hunink, Vincent (1997): Apuleius of Madauros. Pro se de magia (Apologia). Volume II. Commentary, Amsterdam: J. C. Gieben.

Korenjak, Martin (2000): Publikum und Redner. Ihre Interaktion in der sophistischen Rhetorik der Kaiserzeit, Munich: C. H. Beck.

Novotný, František (1977): The Posthumous Life of Plato (J. Fábryová, Trans.), The Hague: Martinus Nijhoff.

Schenk, Peter (2002): “Die Schrift. Einleitung.” In: Peter Habermehl/Jürgen Hammerstaedt/Francesca Lamberti/Adolf Ritter/Peter Schenk (eds.), Apuleius. Über die Magie, Darmstadt: Wissenschaftliche Buchgesellschaft, pp. 23-57.

Schmitz, Thomas (1997): Bildung und Macht. Zur sozialen und politischen Funktion der zweiten Sophistik in der griechischen Welt der Kaiserzeit, Munich: C. H. Beck.

Temkin, Owsei (1971): The Falling Sickness. A History of Epilepsy from the Greeks to the Beginnings of Modern Neurology, Baltimore and London: The Johns Hopkins Press. (2nd ed.)

Whitmarsh, Tim (2004): Greek Literature and the Roman Empire. The Politics of Imitation, Oxford: Oxford University Press.

Knowledge about the Sea and its Creatures in the Roman Empire

Dorit Engster

ABSTRACT

The foundation of modern biology was laid in the 4th century BC. Aristotle's study of nature was a central work for a long time. After him, biological studies were pursued further and partially extended. Scientific topics were published in 'popular science scripts,' wherein the sea environment was of great interest, in particular the habits and behaviors of fish. The authors of these texts were accused of pseudo-scientific and methodological mistakes. In fact, myths, anecdotes, and curiosities were the focus of these depictions. The systematic method of the peripatetic school was abandoned and one tried to present the contents of marine life from a perspective that was as varied as possible. These popular science scripts addressed the academic upper class, not a small cadre of scientists. Thus, the focus was on amusing and curious stories because the knowledge gained was used in the context of banquets and soirées.

Ancient mosaic depictions form interesting parallels to the descriptions of fish in poetry and prose. Here, the sea environment was a popular topic as well. A guest was able to demonstrate his knowledge by explaining the depictions and establishing the link between the fish on the mosaic and the served meal. This article aims to show the relevance of popular-scientific scripts and depictions for the Roman upper class and their debates.

INTRODUCTION

> At Baculo in the Baiae district, the pleader Hortensius had a fishpond containing a moray which he fell so deeply in love with that he is believed to have wept when it expired. At the same country house, Drusus's wife Antonia adorned her favorite moray with earrings, and its reputation made some people eager to visit Baculo.
> *(Plin. NH IX,172)*[1]

This passage from the *Naturalis historia* by Pliny the Elder illustrates the significance and importance of fish and pisciculture for the Roman elite. It is also typical of Pliny's method and focus. In his monumental overview, comprising 37 volumes, Pliny treats natural phenomena in detail—from the shape of the universe to the tiniest animal. The passage quoted above belongs to the context of his description of the different species of fish and their characteristics. It is exemplary for his way of presenting animal life in his natural history, which is certainly one of the most influential scientific works of antiquity.

Pliny devotes the 9th book of his *Naturalis historia* to the discussion of fish, concentrating less on biological details and instead attaching great importance to myths and historical anecdotes. A systematic presentation / classification of marine fauna is missing (cf. Bodsen 1986: 100-101). In this respect, Pliny's natural study resembles the works of other authors writing during the Principate, who engaged in the description of animals, including their shape, organs, and habits.

Like Pliny, the Roman author Aelian collected in his work *De natura animalium* stories of the remarkable behavior and conduct of animals. Characteristic of his description is that animals' actions and their manner of living are judged according to human standards, and that animals are, to some extent, even presented as being morally superior to men.[2] The poet Oppian, in turn, writing in the 2nd century AD, stands in the tradition of Greek didactic poetry. In his

1 Plin. *HN* 9.172: "apud Baulos in parte Baiana piscinam habuit Hortensius orator, in qua murenam adeo dilexit, ut exanimatam flesse credatur. in eadem illa Antonia Drusi murenae, quam diligebat, inaures addidit; cuius propter famam nonnulli Baulos videre concupiverunt." Unless otherwise noted, all translations are from the Loeb Classical Library.

2 Hübner (1985: 173) speaks of a "religiös-sentimentalen Aufwertung der Tierwelt" [religious-sentimental valorization of the animal kingdom] and an "Abwertung der Menschen zugunsten einer idealisierten Tierwelt" [devaluation of humans in favor of an idealized animal kingdom] respectively. Cf. also Müller-Reineke (2010: 118-119).

Halieutika, he describes in detail the manner of living, sexual habits, and hunting behavior of fish, and also discusses the conflicts between different species.[3]

All of these authors have in common the fact that curiosities play an important role in their descriptions of the natural world. Scientific learning is combined with anecdotal reports and moralizing discourses. For this reason, the works have often been considered dubious and unreliable.

In general, the scientific literature of the Roman Imperial period is frequently regarded as unscientific. Their points of reference are usually the works of Aristotle, which are considered methodically and conceptually superior. In the following, the specific topics, prioritization, and methodology of the authors mentioned above will be examined in greater depth. The starting point of the analysis will be several sea creatures presented in particular detail in their works, and which several authors pay special attention to. The approach, method, presentation, and focus of the Imperial author will be compared to that of Aristotle, and the particularities and characteristics of the later works will be identified.

ARISTOTLE AND THE HISTORY OF ANCIENT BIOLOGY

Aristotle is considered the founder of ancient biological research. In the 4th century, he laid the theoretical foundations for the establishment of biology as a science. With his *Historia animalium*, in which he described the animal world, he created a work that set new standards and remained significant for centuries. Despite some methodological weaknesses, it shaped scientific language and literature throughout antiquity. Aristotle formulated the first systematic approach to the world of nature—the distinction between genera and species—which is still valid today. Ancient natural scientists subsequently used his work, frequently referring to his methodology and concepts.

The authors of the Roman Imperial era also based their descriptions on Aristotelian observations, but modified them, adding new elements, for example, details concerning the behavior of the animals and references to mythical and historical events. This different approach is particularly evident concerning ordinary fish—these include especially fish that were used as food, i.e., those species which were familiar to the authors and their readers. In the following section, some select examples will be analyzed in some depth.

3 For the works of Oppian, cf. also Drummond 1818: 3-459; Martinez/Silva 2003: 219-230.

The Mullet

First of all, descriptions of the mullet and its behavior patterns will be analyzed more closely. The mullet is a very common type of fish which can be found mainly near coasts. Even today, it is still a very popular source of food.

In his *Historia animalium*, Aristotle extensively comments on this fish, stating that the mullet has two fins on the gills (Arist. *HA* 504b) and providing further information concerning the shape of the fish and its internal organs.[4] Moreover, the mullet's habitat and way of life of are discussed.[5] Aristotle is also especially interested in its spawning behavior[6] and food intake.[7] In his writings on biology, he repeatedly refers to the mullet, although his remarks are sometimes rather short. However, Aristotle goes into some detail and expands on specific characteristics of the mullet, describing how it can be caught, for example. According to Aristotle, one possible way of catching it is to use a trapped fish to attract other specimens. Following a description of the mullet's mating behavior, Aristotle describes a form of fishing practiced in Phoenicia that takes advantage of sexual attraction: A mullet is put in a trap and entices members of the opposite gender (Arist. *HA* 541a.).

Furthermore, Aristotle presents additional details about the mullet, which is characterized as "the most voracious and insatiable" fish. Thus, a specific defense mechanism is mentioned: When the mullet feels threatened, it hides its head in the mud and thereby feels safe and secure (Arist. *HA* 591b). Another topic is the relationship of the mullet with other fish: According to Aristotle, the seabass is the natural enemy of the mullet, but both species may temporarily join together to form a single shoal (Arist. *HA* 610b).

---

4 Arist. *HA* 508b (concerning the shape of the intestine), 434a (concerning the faculty of hearing), 537a (referring to the deep sleep of the mullet which provides the opportunity to kill it with a trident).

5 Arist. *HA* 598a (the mullet lives near land); 601b (it also swims in rivers); 602a (rain can hurt and blind it); 607b (the mullet becomes unedible when pregnant); 620b (mullets are the swiftest species of fish).

6 Arist. *HA* 543a and b; 567a; 569a—also mentioning the generation of mullets from mud; 570b; 621b.

7 Arist. *HA* 591a—stressing that the mullet, in contrast to other fish, refrains from eating fellow members of the species. It despises meat, and instead feeds on seaweed, sand, and mud, but can be baited by bread.

The explanations in Pliny are based on the Aristotelian descriptions, but reveal a different set of priorities.[8] He also refers to the mullet's habit of hiding its head in the sand when in danger, but adds a critical commentary and underscores the ridiculousness of this behavior.[9] Similarly judgmental is his account of the fishing practices in Phoenicia. Pliny describes the behavior of the mullet in neither a neutral nor distanced tone, but speaks of the unwary lasciviousness (*incauta salacitas*) which induces the fish to follow a female member of the same species and causes them to be caught in the trap (Plin. *HN* 9.59).

The story of the enmity between the mullet and the seabass is also taken up by Pliny and dramatized to a greater extent than in Aristotle.[10] The account of Pliny also includes additional information about the mullet not mentioned by Aristotle.[11] In particular—and this applies not only to the mullet—he discusses the medicines and remedies that can be made from the fish or parts of it.[12]

Similar reevaluations and embellishments can be found in Aelian's descriptions. He bases his account on that of Aristotle but adds several new elements.[13]

---

8 Cf. Beagon 1992: 129-131. For the relationship of Pliny to his Greek sources, see Darab 2012: 153-155. She emphasizes the negative assessment of Greek authors by Pliny and his intention to show the superiority of Roman science. The works of Aristotle, however, are appreciated by him; see Fögen 2009a: 231-234. Cf. Wallace-Hadrill 1990: 92-95, for the objective to combine scientific research with Roman ways of thinking and Roman values. Overall, Wallace-Hadrill regards Pliny's attitude towards Greek science as ambivalent. Cf. Healy 1999: 109-111; Beagon 1992: 18-20.

9 Plin. *HN* 9.59: mugilum natura ridetur in metu capite abscondito totos se occultari credentium. ("The mullet when frightened hides its head and thinks it is entirely concealed.")

10 Plin *HN* 9.185: mugil et lupus mutuo odio flagrant, conger et murena, caudam inter se praerodentes. […] praerodere caudam mugili lupum eosdemque statis mensibus concordes esse; omnes autem vivere, quibus caudae sic amputentur. ("Violent animosity rages between the mullet and the sea-bass, and between the conger and the moray, which gnaw each other's tails." […] "the sea-bass gnaws at the tail of the mullet, although they are friendly together in certain months, but that all the mullets with their tails amputated in this way continue to live.")

11 Plin. *HN* 9.29 (swarms); 31 (agility, ways of catching them); 54 (swiftness, leaping over ships); 144 (swiftness).

12 Plin. *HN* 32.104 (ash made from the head of mullets combined with honey heals sore points).

13 Aelian frequently bases his accounts on the works of Aristotle. He also uses the writings of other authors, some of whom have not come down to us. Relevant sources for his passages on fish and the ocean world in general are, among others, Leonidas of Byzantium and Oppian. Cf. Keydell 1937; Müller-Reineke 2010: 118.

In his elaboration on the account of fishing techniques in Phoenicia, the essential elements are identical, but the fish's behavior is now judged entirely from the human point of view. Human motivation and emotions are attributed to the animal world—and are combined with moral judgments. The focus in Aelian is on sexual attraction, which the fishermen can exploit.[14] Aelian compares this to a young couple. The male mullet is compared to a "licentious youth" who follows a beautiful girl, "mad with sexual desire." Aelian furthermore speaks of the "urgent lust of the fish" and notices that the female mullet must be attractive to arouse the male's love and the "desire for sexual intercourse."

A similar approach can be found in the didactic poem of Oppian. He also mentions—if only briefly—the tricks that can be used to catch mullets and the "ἔρως" (i.e., Eros — "sexual attraction") that causes harm to the fish.[15] The mullets are compared to young men following a beautiful girl "ὑπὸ ῥιπῆς Ἀφροδίτης." (i.e., by the force of Aphrodite).

The Barbel

Another fish that is frequently mentioned in the ancient scientific writings is the barbel. The basic details are given by Aristotle, who describes the fish's spawning[16] and feeding behavior.[17] He also addresses the position of the internal organs— for example, the intestine—and the fish's outer appearance.[18] The barbel is, however, one of the fish that Aristotle discusses rather briefly; in the Roman Imperial period, this fish is covered quite differently.

In his *Naturalis historia*, Pliny provides the basic facts about the fish, but also some information that does not belong to the field of biology in the strictest sense. He comments on the record prices that were paid for this fish in Rome. The former consul Asinius Celer is said to have paid 8,000 sesterces for a single barbel. In this context, Pliny critically discusses the increasing tendency to spend enormous sums on luxury products, and the competition among members of the elite to outdo one another with unusual and exquisite dishes (Plin. *HN* 9.67).[19]

---

14 Ael. *NA* 1.12. Cf. 13.12 for a different way of catching a mullet.

15 Opp. *H.* 4.127-135. Cf. also III,108-114; 501-506; 512-518.

16 Arist. *HA* 543a, 570b—spawning behavior.

17 Arist. *HA* 591a (diet, carnivore); 591b (wallowing in the mud); 610b (gregarious animal); 621b (cycle of life).

18 Arist. *HA* 508b (shape of the intestine).

19 Pliny connects the reference to this episode with a critique of the competition among the members of the elite regarding luxury goods (Plin. *HN* 9.67). Cf. Wallace-Hadrill

Deliberations of this nature are typical of Pliny, and should be considered in the context of his criticism of luxury and decadence.[20] The modest ancestral Roman way of life—a very common cultural attitude in Roman literature which Pliny is adopting here—had been abandoned after Rome's rise to hegemony. Individuals had aquired great power and wealth, which they used not for the public good but for their personal pleasure and enjoyment. The barbel and its use as a status symbol is given as an example for this moral decline. But Pliny also provides other details about this specific fish—for example, information regarding its medical uses, which he also discusses with regard to the mullet.[21]

The opinion of Oppian, whose assessment of the barbel is almost polemic, is radically different. He characterizes barbels as greedy and gluttonous, feeding on everything, even on human corpses. Because of their voracity and their habit of living in the mud of the seafloor, Oppian compares them to pigs (Opp. *H.* 3.432-442). The barbel is not at all characterized as an object of desire or an epitome of luxury. On the whole, his description is also much more judgmental than Pliny's—the barbel's conduct is not assessed as natural animal behavior, but measured by human standards.

However, both descriptions have in common the fact that they focus on the sensational and exceptional. On the one hand, the barbel is a symbol for luxury and even decadence, i.e., for transgressions regarding the appropriate expenses for banquets. On the other hand, in what is almost a reversal of Pliny's description, Oppian characterizes the barbel as not being choosy concerning food and as transgressing social norms because it does not refrain from eating corpses.

1990: 91; Hölkeskamp 2005: 190-196. Pliny also reports that a barbel weighing 80 pounds was caught in the Red Sea.

20 Cf. his remarks concerning garum, a kind of sauce or liquor made from fermenting fish, Plin. *HN* 31.95. The manufacturing had become more sophisticated and garum had become a luxury product—more and more sorts of garum were fabricated (transiit deinde in luxuriam, creveruntque genera ad infinitum) ("then *allex* [a kind of sediment produced when making garum] became a luxury, and its various kinds have come to be innumerable") and exquisite starting products—as for example barbel—were used.

21 Plin. *HN* 28.82 and 32.44 (ash made from the head of a barbel is an antidote to poisons); 32.25 (the barbel can be used to heal people hurt by the stingray) 32.91 (crushed barbel can be taken as an emetic); 32.127 (ash made from the flesh of barbel heals fistulae).

The Bream

Another fish which is frequently mentioned by the ancient authors is the bream. As with other fish, Aristotle primarily decribes the external and internal features typical of the species. Overall, however, he confines himself to few and rather brief remarks.[22] Pliny partly takes over the descriptions of Aristotle, for example with regard to reproduction and spawning.[23] He also—as in the case of the barbel—discusses the production of medicine from this fish.[24]

In this case, Oppian's depictions are considerably more detailed and extensive. He differentiates between several types of breams and classifies them according to physical characteristics, way of life, and character traits.[25] Unlike Aristotle, he elaborates at length on the different types of breams, which are each distiguished by special features and specific adjectives.

In another passage, he addresses the ways and means to attract the fish and then catch them. Since they usually appear in groups, fishermen have to entice them one by one. The behavior of the breams is compared with that of soldiers, since they swim in formation. They are also compared to children: Just as children become excited about their toys, the fish are happy about the bait and ignorant of the imminent danger (Opp. *H.* 3.610-619). In general, how to capture the fish is an important topic for Oppian.[26] Men have to outwit the fish, to which

22 Arist. *HA* 543b, 570b (spawning behavior); 508b (shape of the intestine); 489b (possesses four fins); 537a (deep sleep which provides the opportunity to kill it with a trident); 591b (carnivore); 5598a (lives near land and in lakes near the sea); 602a (sensitive to cold).

23 Plin. *HN* 9.56: According to Pliny, no male seabreams exist, all specimens that are caught are egg-bearing. Cf. 9.57 (vulnerable to very cold winters—may go blind: multi caeci capiuntur. itaque his mensibus iacent speluncis conditi ["many are caught blind, consequently in the winter months they lie hidden in caves."]); 9.58 (hiding in summer because sensitive to heat); 9.162 (spawning behavior—twice a year, in spring and autumn); 9.166 (possesses a uterus); 9.182 (behavior when caught—freeing itself).

24 Plin. *HN* 32.43 (antidote to poisonous honey); 32.101 (antidiarrheal medicine).

25 Opp. *H.* 1.93-118. The passage covers the habitat and the dietary habits of various types of fish. Characteristic is the attempt to describe the marine world in every detail and the effort to demonstrate learning und stylistic virtuosity.

26 Cf. Opp. *H.* 3.338-352, describing the ways to trick and catch a bream using a specific bait.

Oppian attributes a certain cleverness. Eventually, their voracity and lack of self-control prove to be their demise (Opp. *H.* 3.347-370; 3.443-481).

Oppian is also preoccupied with the fish's sexual orientation, especially if this deviates from normal standards. He thus writes about the erotic attraction that bream allegedly have toward goats. When shepherds bathe goats in the sea, the breams—attracted by the smell of the goats—approach and mix with them. The behavior of the fish is compared to that of human lovers, and the physical contact between the fish and the goats is described in detail (Opp. *H.* 4.308-403).[27]

Aelian also focuses on the love life of the fish, comparing their mating behavior to that of men fighting for women. He even draws a parallel to the fight over Helen between Paris and Menelaus (Ael. *NA* 1.36).

All of the abovementioned fish were common sources of food and were often served for dinner, especially at the banquets of the elite. In the following, three different types of sea creatures which were of special interest to the Roman authors will be introduced.

The Moray

As stated in the introduction, the moray plays a special role among the fish—at least for the Romans. This prominence is not to be found in the descriptions of Aristotle, in whose zoology the moray is but one among many fish. As in the case of the other fish, Aristotle's focus is on its outer shape and details like the fins and gills, or the location of the gallbladder in the body.[28] The fish's colorful outward appearance is commented on, as is its mode of reproduction[29] and its mode of life in general.[30] Aristotle also—in the context of a description regarding the enmity between certain species—refers to the moray and its archenemies, the conger eel and also the perch.[31]

In the work of Pliny, the moray figures much more prominently, as was mentioned above. He provides basic biological information concerning the shape of

27 Oppian stresses that the spectacle is exceptional and unusual.

28 Arist. *HA* 489b, 504b (no fins, shape of gills); 505a (four gills); 506b (position of the gall); 517b (shape of eggs).

29 Arist. *HA* 543a (spawning behavior, offspring).

30 Arist. *HA* 599b (hibernation); 540b (entanglement of the animals); 591a (carnivore); 598a (lives off the coast).

31 Arist. *HA* 610b (eel and murray are enemies; the murray bites off the tail of the eel).

the fish, its specific way of moving in the water, its mode of reproduction, etc.[32] He takes up information about the fight between the moray and other eels, but combines this description with an extensive narrative about fights between other creatures—for example, the struggle between the crawfish and the octopus. The enmity between the creatures is not, according to Pliny, motivated solely by their feeding behavior. He speaks of the "burning hatred" that drives the fish. Human emotions like fear and anger are attributed to them, and normal creaturely behavior is thus dramatized (Plin. *NH* 9.185).[33] This also applies to other areas of life. For example, the moray allegedly has a special preference for snakes, and is said to mate with them.[34] Additional information in the work of Pliny relates—as in the case of the other fish—to its uses as a remedy or medicine.[35]

The main focus of Pliny's study is on the role of the moray as a status symbol in Roman society. He describes in detail the developement of fish farming in Rome, which led to ever-increasing competition among members of the elite. Pliny reports extensively on how the senator Sergius Orata (a cognomen meaning "gilt-headed seabream") introduced professional oyster farming.[36] Following this example, we learn that Licinius Murena (a cognomen referring to the moray) started to develop the first fishponds—to ensure that fresh fish could be served at any time. This led, as Pliny further reports, to real competition among the members of the senatorial order, and record prices were paid for the fish.[37] A special type of pond for morays was introduced by Gaius Hirrus, who allegedly was able to provide Caesar with 6,000 morays for his banquets.[38] Again, the statements of

---

32 Plin. *HN* 9.40 (soft skin); 9.73 (no fins, no gills, moving like snakes); 9.194 (position of the gall); 9.158 (quick developement of the eggs); 20.261 (ferula is toxic for them).

33 For the central importance of fighting and conflict in the depiction of nature by Pliny, cf. Ash, 2011, p. 15. Typical is the naming of the animals in combination with their natural enemies, i.e., the forming of antagonistic pairs.

34 Plin. *NH* 9.76-77. After explaining the spawning behavior of the murray, Plinius refers to the popular belief that the murray comes to shore and then engages in a sexual relationship with the snake.

35 Plin. *NH* 9.58 (the ash of a murray is a remedy against bites of the same fish).

36 Plin. *HN* 9.168-169. Pliny observes that the introduction of oyster cultivation was motivated not by connoisseurship but by greed (nec gulae causa, sed avaritiae). For the structure of the narrative, cf. Clemence Schultze 2011: 180-181; Bannon 2014: 173.

37 Plin. *HN* 9.170: Pliny mentions in this context Philippus, Hortensius, and Lucullus, who built a facility that was worth 40.000 sesterces.

38 Plin. *HN* 9.171-172; Pliny reports moreover that the fishponds increased the value of his moderate estate in such a way that it could be sold for 4.000.000 sesterces. Pliny

Pliny have to be considered in light of his criticism of excessive luxury, although he adds several touching anecdotes about the close relationship of some senators to their fish—the abovementioned passage about the orator Hortensius being one example. The moray seems to have been a sort of ancient koi. In this respect, Gaius Hirrus started a new trend.[39] Pliny also mentions examples of the drastic misuse of the fish. The Roman eques Vedius Pollio is said to have punished escaped or criminal slaves by feeding them to the morays.[40]

The importance of the moray is also reflected in the descriptions of other authors from the Imperial period. They mention the relationship between the fish and their owners, which could be so close that the morays would come if called by name. Even prominent politiciens are said to have been infected by the passion for morays. Aelian tells the story of how the orator Lucius Licinius Crassus cried because of the death of his favorite moray.[41] This anecdote ends, however, with a punchline. When Crassus was criticized by another senator because of his emotional reaction, he is said to have answered jokingly that his colleague had already buried three wives without shedding a tear.

Aelian and Oppian also discuss the reputed sexual attraction between the snake and the moray which Pliny had mentioned. Aelian and Oppian expand on the affair between the two creatures and elaborate on the details, describing how the moray goes ashore and begins a relationship with the snake. The account is similiar to that of a human relationship and includes a moral evaluation. Aelian speaks of the moray's "longing for love," which causes it to leave the water. Similarly, the snake comes to the shore to call the fish. Moray and snake "hug each other" and have sexual intercourse on the bed of the snake—Aelian comments that this union brings together creatures from different habitats, which are drawn together by nature or a natural desire (Ael. *NA* 1.50).

---

concludes: invasit dein singulorum piscium amor ("subsequently affection for individual fishes came into fashion"). See also Varro, Rust. 3.17,2; cf. Bannon 2014: 173; Stein-Hölkeskamp 2005: 166-168, regarding the piscinarii; cf. Schmölcke, Nikulina 2008: 48 and 50-51.

39 Pliny *HN* 9.172. Cf. for example Cic. *Att.* 2.1,7; Mart. 5.30, 2-8; Sen. *de ira* 3.40; Cass. Dio 54.23, 1-2; Plut. *De soll. an.* 976a; Macrob. *Sat.* 3.15, 3; Ael. *NA* 8.4 for the close relationship between fish and their owners and the tameness of the animals.

40 Plin. *HN* 9.77. The passage serves to criticize decadence, cruelty, and degeneration.

41 Ael. *NA* 8.4; cf. Plut. *De soll. an.* 976a; see also the analysis of Keydell 1937: 412.

Figure 1: Roman Mosaic from House VIII.2.16, Pompeii

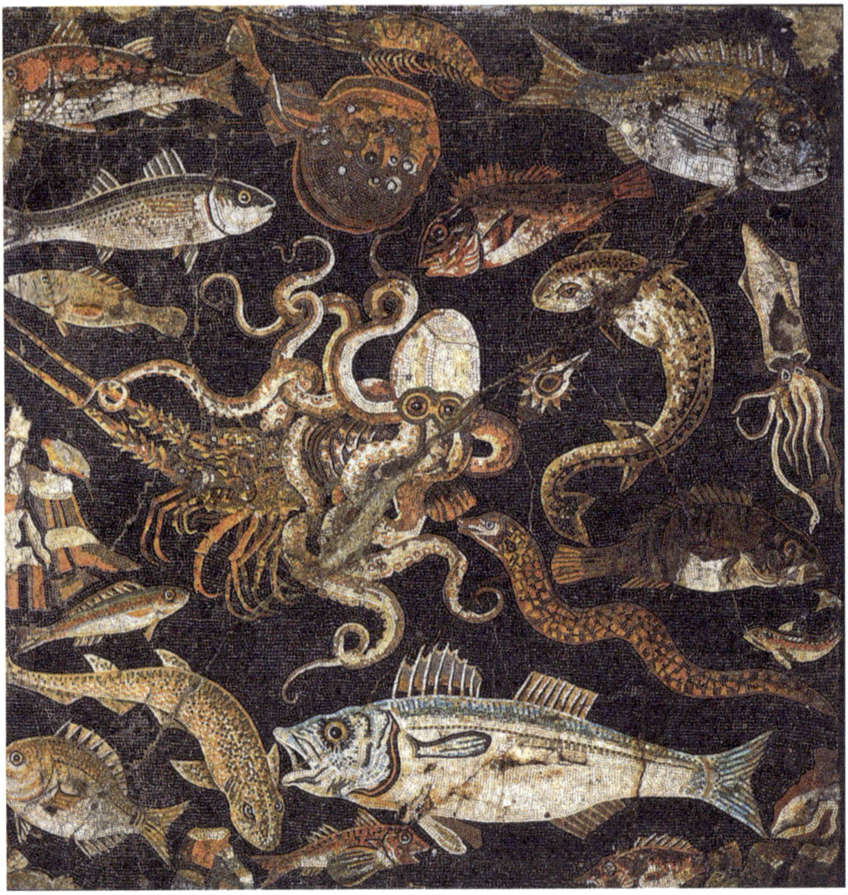

Source: Museo Archeologico Nazionale, Naples, inv. no. 120177; Photo: Wolfgang Rieger, Public Domain

Oppian depicts the love affair between the snake and the moray in even more detail. Their desire and passion are described extensively (Opp. *H.* 1.554-562). Oppian discusses the sexual act in poetic terms, in which the moray assumes the role of the "bride." He also describes the consequences of the relationship. The moray returns to the sea, but the snake, which spit out its venom before the physical union, might die because it is now defenseless.

Oppian also discusses one of his favorite topics—fighting between members of different species. One of the moray's archenemies is the octopus. In the course of his listing of the enmities between specific fish, he mentions the "repugnat moray," which feeds on the flesh of the octopus.[42] The assessement of the Greek author thus differs from Pliny's characterization. While from the Roman point of view the moray is a friendly, tame, and very popular fish, Oppian's main focus is on the lack of self-control—strange sexual desires and gluttony with regard to food.

Squid and Octopus

In addition to fish in the proper sense, some other types of sea creatures that particularly attracted the attention of the ancient authors shall now be taken into consideration, namely: the squid and the octopus. Aristotle shows considerable interest in these creatures. In many passages, he discusses the characteristics and special features of these molluscs—their physical form, as well as their manner of movement and reproduction.[43] Regarding squid and octopi, Artistotle deviates, to a certain extent, from his usual, strictly factual mode of presentation, adding information concerning their hunting behavior and their struggles against other sea dwellers. According to Aristotle, the octopus' greatest adversary is the crawfish (*Fig. 1*).[44] The struggle between the animals is vividly described and part of a lengthier discussion of the feeding behavior of sea creatures.

The octopus and the squid (*polypus*) also figure promently in the work of Pliny. He provides information concerning their sense of smell (Plin. *NH*

42 Opp. *H.* 3.183-189. The main topic of this passage is how gluttony leads to disaster. According to Oppian, fish are especially prone to this weakness and indulgence—they can be allured by bait, each type of fish by its favorite prey. The murray, for example, longs for the flesh of the octopus. Cf. also 2.289-294: Oppian describes in detail how the murray brutally lacerates and devours the octopus.

43 Arist. *HA* 489a (bloodless); 489b (swimming behavior); 490a (way of movement); 490b (molluscs); 523b (bloodless); Arist. *HA* 529a (comparison with squids); 534b (squids and octopus are caught with bait); 541b (contact, i.e., interlacing of members of the same species); 524a-525, 527a (shape and behavior of molluscs); 544a; 550a; 567b (mating habits, insemination, breeding behavior); 610b (forming of herds); 621b (slyness of the squid, uses its ink to hide and not only when frightened—like the octopus and calamari).

44 Arist. *HA* 590b-591a: He describes a food chain: the crayfish is overpowered by the octopus and is afraid of it, but the crayfish also for its part subdues the eel. The eel in turn feeds on the octopus. Aristotle subsequently describes other aspects of the hunting behavior, stating that the squids are able to overpower even large fish and that the octopus feeds on seashells which it collects and brings to its lair.

10.194), their internal organs,[45] and their modes of reproduction, and stresses the particularly high fertility rates of these creatures.[46] Their search for food and feeding behaviors[47] are addressed, as well as the manufacture of medicine using the octopus or parts of it.[48] However, Pliny also includes "horror stories," in which the conflict between humans and octopi or the intrusion of octopi into the human sphere are depicted. He dramatically describes the fate of castaways who fall victim to the octopus. Pliny emphasizes the creature's cruelty: It grips humans with its tentacles and suction cups and viciously kills them.[49]

Even more spectacular is his report about a giant squid that came ashore at night and even climbed over fences and trees to rob fisherman of their catch. Pliny describes how the surprised people fought against the "monster," and were ultimately vicorious. The enormous head and body of the beast were afterwards publicly displayed.[50]

Similiar stories about predatory octopi and squid seem to have been widespread and popular. Aelian also reports that octopi go ashore and climb on trees.[51] Oppian even speaks of an octopus that leaves the water to pillage the fruit and olive trees (Opp. *H.* 1.305-312; cf. Lytle (2011: 334-335). Oppian attributes human emotions

---

45 Plin. *NH* 9.83-84 (mollusc, shape of the body, tentacles); 84 (behavior of the male and the female octopus); 9.85 (different types of polyps and their behavior); 87 (colour change, adaption to the environment); 158 (mating behavior); 11.8 (ink of the squid); 11.199 (position of the intestine).

46 Plin. *NH* 9.163 (oviparous, fertility); 164 (way of spawning and breeding).

47 Plin. *NH* 9.86 acknowledges that the animals act cleverly and effectively. They collect mussels, bring them to their den and eat them. With the remaining shells they attract fish and catch them.

48 Plin. *NH* 32.71 (bone of the octopus heals inflammation of the eyelids), 85 (ash of the octopus heals freckles and other skin conditions); 88 (crushed bones heal ear infections); 103 (eggs of the octopus are diuretic); 125 (ash of the octopus helpful in case of arrow wounds).

49 Plin. *NH* 9.91. Cf. also Healy 1999: 67-68.

50 Plin. *NH* 9.92-93 (reliquiae adservante miraculo pependere pondo DCC ["its remains, kept as a curiosity, were found to weigh 700 lbs"]). Pliny adds that other animals of similiar size are said to have washed up on that coast.

51 Ael. *NA* 9.45, describing the break-in at a warehouse and the subsequent devastation.

and behavior to the octopus: It embraces the trees with its tentacles like a boy hugging his wet nurse (Opp. *H.* 4.279; Cf. Bartley 2005: 274-275).[52]

In accordance with his interests and prioritizations, Oppian also describes the sexual behaviour of the octopus and its tragic consequences. The male octopus dies after the sexual act due to exhaustion, and is then eaten by other sea creatures (Opp. *H.* 1.536ff). The female octopus in turn dies as a result of anatomical peculiarities during childbirth.

A typical part of Oppian's repertoire is the depiction of the fight between animals. He eloborates on the struggle between the octopus and the crawfish already reported by Aristotle, and extensively describes the brutal fight. The octopus surprises the crawfish and grips it with its tentacles. The crawfish desperately tries to defend itself, but is strangeled by the octopus in the end (Opp. H. 2.384ff). Oppian dramatically describes how the octopus eats its dead victim: It sucks in the flesh like a child drinking milk from the breast of the wet nurse. Finally, the brutal action of the octopus is compared to that of a robber who ambushes and robs a drunk at night (cf. Bartley 2005: 219-220).

The Dolphin

While the octopus is viewed rather negatively by the ancient authors, the dolphin is one of the most popular animals in antiquity. Accordingly, Aristotle gives a detailed description of this animal. Stating that dolphins are mammals, he describes how they give birth to their offspring (Arist. *HA* 489b, 504b, 566b) and breastfeed them (Arist. *HA* 566b). Furthermore, he emphasizes the special care they give to their young. Aristotle, however, also discusses the problem of clearly categorizing the dolphin. Due to its peculiar way of breathing, it is considered a special case among sea creatures (Arist. *HA* 589b). It has no gills but breathes through a blowhole (Arist. *HA* 697a). According to Aristotle, the dolphin possesses a voice but lacks a tongue and lips to actually talk (Arist. *HA* 536a). Aristotle also reports on some curiosities, for example that the dolphin snores when sleeping.

Aristotle repeatedly draws comparisons with human character traits, and later authors increasingly focus on this aspect. In general, the dolphin is not regarded as a wild animal, but is attributed human preferences and characteristics.[53] A

52 For Oppian's tendency to anthropomorphize animals in general, see Wilkins 2008: 315-328; cf. also Kneebone 2008: 34, for the comparison between fish and different types of people.

53 The love life and mating behavior of dolphins are also compared to those of humans. Cf. Oppian *hal.* 1.580-585.

number of writers stress the dolphin's love of music.[54] Euripides already mentions this, and the description is adopted by the authors of the Imperial era. Thus, Plutarch, for example, mentions that dolphins like to listen to music and follow people who sing or make music (Plut. *Conv. sept. sap.* 18).

Aelian also stresses the dolphin's love of music.[55] In similiar fashion, Pliny declares that the dolphin is a friend of man and loves music. It is attracted by concerts (especially those of the water organ).[56] Another specifically human trait of the dolphin mentioned by the ancient authors is its habit of dancing and playing in the sea.[57] Oppian describes a herd of dolphins that performs a dance in the sea. As in the case of the other sea creatures, he draws parallels to human behavior—the dolphins resemble a procession of dancers.[58]

Another topic is the emotional bond between individuals and dolphins. The so-called dolphin riders, i.e., mythical or historical figures that formed friendships with dolphins, play an important role here (cf. Opp. *H.* 5.448-518; cf. Hindermann 2011a: 347-348; 2011b; Bartley 2005: 87-94). These relationships could even resemble love affairs.[59] A popular story that a number of authors refer to took place in Iasos. A boy named Hermias is said to have formed a close relationship with a dolphin. The friendship, however, ended tragically. The boy,

---

54 Pind. *Pyth.* 2.51; Eur. *Hel.* 1454-1456; Ael. *NA* 2.6; 11.12; 12.6 and 12.45; Plut. *De soll. an.* 984b; Luk. D. Mar. 8 (ἐπαινῶ σε τῆς φιλομουσίας ["your love of music does you great credit"]); Solin. 12.6-7; Mart. Cap. 9.927.

55 Ael. *NA* 11.12. For the general characterization of the dolphin by Aelian, cf. Williams 2013: 208-210; Hübner 1985: 163. Cf. Ael. *NA.* 12.6 und 45. See also Plut. *de soll an..* 19, 26, 29, 32, 36; Luk. D. Mar 8. Keydell (1937: 430-422) assumes that Leonidas of Byzantion was the source for the accounts in Aelian and Oppian.

56 Cf. Plin. *NH* 9.24. Pliny also stresses that dolphins accompany ships and are not afraid of men. For the dolphin's love of music, cf. Stebbins 1929: 91-92. Beagon (1992: 137ff.) considers the descripton of Pliny less moralizing than that of Aelian. Though the animals were credited with human qualities, Pliny would not have regarded them as equal to humans.

57 For a similiar description of dolphins dancing in the calm sea, see Plut. *Septem* 18; cf. also Plut. *Septem.* 19. For the philanthropy of the dolphin, cf. also Plut. *de soll. an.* 984c; symp. 704-705. See also Hübner 1985: 166. See also the popular story of the poet Arion who was saved by a dolphin: Hdt. 1.24; Plut. *Septem.* 18.

58 Opp. *H.* 1.670-685. He describes the swarms of dolphins: The young males are compared with unmarried men, the young dolphins with dancers, and the older dolphins with shepherds looking after little lambs. Cf. Bartley 2005: 62-67.

59 For the role of Eros and the idealization of these relationships, cf. Smith 2013: 74-76; 80-84.

who often rode on the creature's back, died in a storm while on one of these rides. The dolphin, feeling guilty because of the accident, committed suicide by beaching itself.[60] Similar stories are mentioned by several ancient authors, the main focus being on the emotional bond between boy and dolphin. The popularity of this motif is also apparent in the work of Pliny, who reports an incident that took place in his own time. In the city of Hippo Diarrhytus, there appeared a tame dolphin that allowed people to ride on its back. This attracted the attention of large crowds and even of the Roman governor, who wanted to see the spectacle for himself. Pliny reports that the magistrate performed a kind of ritual with the dolphin—with the consequence that the frigthened creature avoided human company for some time. When it finally returned and again attracted large numbers of visitors, this seems to have overstretched the capacities of the city. The inhabitants therefore decided, according to Pliny, to secretely kill the dolphin (Plin. *NH* 9.26; cf. Ael. *NA* 6.15).

General Characteristics of Roman Imperial Zoology

Based upon the analysis of the selected passages, the following observations can be made: Typical for the authors of Imperial period is the tendency to humanize sea creatures and their habits. To explain these creatures' patterns of behavior, Pliny, Aelian, and Oppian frequently draw parallels to human manners. This applies to aggressive behavior towards members of the same species and towards prey, but also to mating habits and specific preferences.

The greater the similarity to human behavior, the more favorable and more interesting a species is rated. Science's prosaic forms of presentation and prosaic language are abandoned in favor of dramatizations, pointed emphasis, and value judgements. These narrative strategies can be observed in Pliny and—even more

---

60 Plin. *HN* 9.27; Aelian, *NA* 6.15, reports that the citizens of Iasos built a grave and a memorial for the pair. Cf. also Ael. *NA* 2.6—the story of a dolphin rider near Poroselene. As Smith (2013: 80-84 and 87) has pointed out that it is the strong emphasis on the physical aspects of the relationship which leads to tragedy. Relationships that were based on mutual affection and friendship, on the other hand, would be considered positive by Aelian. For the dolphins' habit of swimming together with children, see Plut. *Conv. sept. sap* 19.

distinctly—in Aelian and Oppian.[61] The animals are credited with human character traits and virtues. Thus, Aelian attributes virtuous behavior to fish and other sea creatures and sometimes even characterizes them as morally superior to men.[62] The conduct of the creatures is judged by moral standards and implicitly contrasted with human deficiencies.[63] Not only are sea creatures compared to humans, their conduct is also illustrated by the insertion of anecdotes and digressions.[64] These are often accompanied by moral judgments and aims—for example, the criticism of luxury and decadence.[65]

The moralizing digressions and comparisons with human behavior are remarkable in view of the valuation of animals that was prevalent at the time. According to the prevailing Stoic view, animals (including fish and other sea creatures) were devoid of reason.[66] Sea creatures in particular were regarded as inferior and lower-ranking when compared with creatures living on the land.[67] Though in general Pliny and Aelian incline towards Stoic ideas, they acknowledge that animals at least possess some level of reason and intelligence.[68] With the inclusion of anecdotes about clever and humanlike animals, they implicitly argue against Stoic views and conceptions (cf. Hübner 1985: 155-

61 For the humanizing tendencies and the attribution of human qualities to animals in Pliny, cf. Fögen 2007: 184-198; Beagon 1992: 133-142. Korhonen (2012: 70) contrasts this with the Stoic convictions.

62 Cf. Hübner (1985: 160-162) for an analysis of relevant passages (cf. also Smith, 2013: 73-74). The superiority of the animals is based on their lifestyle, i.e., their existing in accord with nature (cf. also Kindstrand 1998: 2964-2965).

63 For the anthropomorphization of animals in Aelian, see Hübner (1985: 159-160), who distiguishes between different types of human roles and behavioral patterns. Cf. also Hindermann (2016: 83-84); Gilhus (2006); For intelligent and virtuous behavior of animals, see Hübner (1985: 164-165). Fögen (2009a: 188) stresses the ethical element and the tendency to anthropomorphize animals, especially in the work of Aelian. Even more frequently than in Pliny, animals are regarded from a moralizing perspective and corresponding mirabilia are included. Cf. also Fögen 2009b: 50.

64 For the moralizing anecdotes and digressions, see Darab (2014: 207).

65 Cf. Darab 2012, concerning Pliny's remarks on bronze statues from Corinth.

66 For the Stoic attitude towards animals, see Dodds (1933: 104).

67 Hübner (1985: 168) argues that Aelian, for this very reason, focusses on fish—because their humanoid behavior is all the more astonishing in view of their lack of *logos*.

68 For the influence of Stoic concepts, especially the ideas of Poseidonios, on the descriptions of Pliny, cf. Paparazzo 2008: 52-53; Beagon 1992: 125-133; 153-154; Paparrazzo 2011: 89-111.

156; Gill 1969: 404-409).[69] Even more apparent is the tendency to humanize animals in Oppian, who frequently draws parallels between human customs and animal behavior.[70] The purpose of these digressions and mirabilia was, apart from instruction, to entertain and amuse the reader.[71] Anecdotes about intelligent creatures and reports of curious incidents are therefore combined with scientific observations. The sensational and spectacular are adduced with explicit references to alleged eyewitnesses. Human-animal relationships are also a prominent topic, including those of a quasi-sexual character.[72] For Pliny in particular, the relation of the stories to events from his own lifetime is of great importance. For this reason, stories like those of the dolphin riders are included.

The works of the Imperial authors have been regarded as inferior by modern-day researchers because they did not conform to the standards set by Aristotle and his immediate successors. The following are normally regarded as general criteria for scholarliness and scientificity: a systematic approach, observation, empiricism, objectivity, and replicability of results.[73] Against this background, authors like Pliny and Aelian in particular have been treated contemptuously, their works being regarded as undertheorized and lacking a consistent scientific approach, with the main problem being the insertion of mirabilia.

Their handling of the anecdotes and fantastic stories has been a matter of debate, however. It has been noted, for example, that Pliny sometimes comments quite critically and even sceptically on at least some of the reports.[74] Thus, Stei-

69 Kindstrand (1998: 2966-2968), however, states that Aelian's attitude in this respect is inconsistent.

70 For the sources for the passages about the cleverness of fish and animal fights in Oppian and Plutarch, cf. Keydell 1937: 419-422; Kindstrand 1998: 2974-2975.

71 Cf. also the characterization of the *naturalis historia* by Pliny the Younger, 3.5; 4.16 and 20; Darab 2012: 15-16.

72 Korhonen (2012: 70) assumes that peripatetic works in particular were the sources for the stories about zoophilia. For the kinds of human-animal relationsships, cf. ibid: 65-77. See also Hübner 1985: 164.

73 For the distinction between a scientific, i.e., empirical, and a philosophical approach, see Paparazzo (2008: 41, 48).

74 Beagon (1992: 11) notes a certain scepticism on the part of Pliny concerning the mirabilia. Healy (1999: 63-65) also emphasizes that Pliny, on the one hand, considers nothing impossible in the realm of nature, but on the other hand expresses doubts concerning marvelous stories reported by Greek authors. Fögen (2009a: 224-230), observes a certain ambivalence in Pliny regarding the mirabilia. He identifies his sources—partly to shift

ner (1955) points to some limiting, relativizing, and critical remarks (Steiner (1955: 137-143). In her opinion, Pliny expresses doubts about mirabilia and portents (see also Naas 2002: 66). In some instances, however, Pliny closely follows his sources, accepting even completely incredible and implausible reports.[75] Steiner explains this through Pliny's dependence on different sources, prevailing opinions, and traditions, as well as on his own personal experience. Accordingly, Paparazzo sees the work of Pliny as "a representative example of what one could call a postscientific culture," characterized by a sharp distinction between philosophy and natural science—with a preference for the philosophical approach.[76] Pliny's specific approach could, however, also be explained by his overarching goal. His stated aim is to cover nature in all its facets, and the mirabilia serve the purpose of illustrating the variety of natural phenomena.[77] At the same time, they are meant to arouse the interest of the reader.[78]

In the introduction to the *Naturalis historia*, Pliny addresses the requirements and the limits that are specific to the genre.[79] He emphasizes, however, the innovative nature of his work ("the path is not a beaten track, nor one which the mind is much disposed to travel over").[80] Pliny's novel approach has been thoroughly discussed in recent years.[81] In this context, the *Naturalis historia* is frequently

---

responsibility, partly to critically discuss them. Fögen speaks in this regard of a "Selbstinszenierung als Sachautor" (self-presentation as a factual author).

75 Healy (1999, 109-111) points out that Pliny frequently accepts incorrect information, because he respects the statements of recognized authorities. Steiner (1955: 142) observes, depending on the sources, a fluctuation between credulity and scepticism. Cf. also Nutton (1986: 30-58), who qualifies the extent of the criticism to be found in Pliny.

76 Paparazzo (2008: 53-54). He notices, however, pp. 49-50, a combination of philosophical and empirical approaches. Similiarly, cf. Bodson (1986: 104) for the combination of a theoretical and a practical approach.

77 Cf. Naas (2002: 60); Beagon (1992: 62). For the concept of nature as benefactress of mankind, see Wallace-Hadrill (1990: 83-85).

78 See Beagon (1992: 76-84). For the association of natural history with human cultural history, see Wallace-Hadrill (1990: 81).

79 Cf. Darab (2002: 150-151), who emphasizes that Pliny in his introduction rejects exactly those stylistic devices that he employs later in his work.

80 Plin. *NH praef.* 13-15; cf. Fögen 2009a: 208, 217-220.

81 Regarding the objective of Pliny, the concept of the *enkýklios paideia* and the question of his educational ideal have been brought into focus. Cf. Doody (2009: 3); as Doody (2009: 14) stresses, the term in Pliny denotes those things that are essential to know. But in her opinion it would be anachronistic to assume that Pliny intended to write an

addressed as the first ancient encyclopedia.[82] But this label has also been questioned, mainly because the description of natural phenomena is frequently interrupted or expanded by the insertion of anecdotes and mythical stories.[83] Thus, Healy (1999: 40) considers the work not so much a handbook but more a "series of extended 'essays' on topics within the major field of applied science." A particularly conspicuous aspect of the work is the explicit reference to other authors. The naming of his sources is an integral part of the descriptions. It probably serves to impart a feeling of reliability as well as competence and to demonstrate erudition (cf. Doody 2009: 113-114).

As for Aelian, his lack of structure has frequently been addressed. In his rather short introduction to *De natura animalium*, he provides no guidance for the reader regarding the structure or the concept of his work.[84] Instead, he states that his purpose is to prove that animals possess remarkable qualities.[85] To achieve this, he collected many different kinds of stories about animals. Hindermann (2016) thus emphasizes that, in comparison to Pliny, the work of Aelian is characterized by frequent changes of themes, as well as an emphasis on colorfulness

encyclopedia. Instead Pliny, according to Doody (2009: 16), imparts the knowledge of those things that one has to learn before studying nature. Cf. Beagon (1992: 12), stating that Pliny is here referring to Aristotle and his concept of the basic knowledge that could be required of a citizen and that similiar ideas were brought forward by Quintilian. Cf. Murphy (2004: 33); Taub (2017: 73-74).

82 Beagon (1992: 12-13) regards the concept of a comprehensive encyclopedia as a Roman phenomenon. Cf. König (Woolf: 23-27); Schultze (167-186). See also Darab (2012: 149); Fagan (2006: 190). Cf. also Naas (2002). Though Doody (2009: 2) states that Pliny is interested in the encyclopedic approach, he regards the labelling of "encyclopedia" for his work as problematic. Cf. also Healy (1999: 36-37); Murphy (2004: 11-15) for the difference between the work of Pliny and a modern encyclopedia.

83 Concerning digressions, Darab (2002: 149-150) differentiates between mythical or anecdotal aetiologies, interesting details connected to the materials, and anecdotes with a moral purpose. Regarding the encyclopedic approach, see also Carey 2003; König/Woolf 2013: 23-63; Healy 1999: 73-74.

84 Fögen (2009b: 50) detects no "coherent organising principle" in Aelian's work, but discerns instead an impressionistic and eclectic character as well as a striving for variation.

85 Ael. *NA* praef.

and variety.[86] Aelian inserts cross-references, but not consistently—his objective being to portray the complexity of nature by dispensing with structure—, and instead employs the stylistic device of *poikilia* ('variegation').[87] Accordingly, Müller-Reineke (2010) regards the work as an "interesting mixture of scientific facts and miraculous stories" whose objective was "to educate and at the same time entertain his readership" (Müller-Reineke 2010: 118).[88]

In his prologue to the *Halieutika*, Oppian states that his objective is to describe marine life in all its aspects. In doing so, he intends to touch upon the life of ordinary fish but also upon monsters and the secrets of the ocean depths (Opp. *H.* 1.1-92). He stands in the tradition of Greek didactic poetry with regards to both structure and style. Accordingly, his account is combined with detailed instructions concerning fishing.[89] Contrary to the depictions by Pliny and Aelian, here the sea is characterized as a realm of strange and friendly, but also dangerous and aggressive, creatures (cf. Opp. *H.* 1.9). To convey this message to the reader, Oppian concentrates on sensational and spectacular accounts.

THE TARGET AUDIENCE

We have yet to consider whether the works of Pliny, Aelian, and Oppian should be viewed as unscientific and deficient, or if the inclusion of and anecdotes serves a specific purpose. The target audience has to be taken into consideration in order to effectively evaluate their approach.

86 Hindermann (2016: 76-83). Thus she speaks, pp. 88-95, of an "Eindruck des Willkürlichen und Zufälligen" (an impression of the arbitrary and accidental).

87 Fögen (2007: 188) observes that Aelian does not aim at a logically coherent presentation of his subject matter, but instead provides a great deal of variation in order to avoid monotony. See also Kindstrand 1998: 2962-2963.

88 Fögen (2007: 187) states that Aelian "did not aim for a scientific zoology or a precise factual account. His work is a collection of stories that combine the transmission of less sophisticated and less detailed information with entertainment." Concerning the characteristics of Aelian's work, Korhonen (2012: 69-70) states that he rather stands in the tradition of paradoxography than in that of natural science.

89 Opp. *H.* 1.7-9: ἁλίης τε πολύτροπα δήνεα τέχνης / κερδαλέης, ὅσα φῶτες ἐπ᾽ ἰχθύσι μητίσαντο / ἀφράστοις. ("And the crafty devices of the cunning fisher's art—even all that men have devised against the baffling fishes.")

In his introduction to the *Naturalis historia*, Pliny stresses that he is writing not for the elite but for artisans and peasants.[90] This statement has, however, been questioned.[91] Borst (1993) thus emphasizes that Pliny conceptualized his work with regard to the needs of the general public, and therefore especially considered technical and practical aspects, modes of implementation and practicality.[92] On the other hand, Murphy (2004) stresses that Pliny imparts no practical knowledge and no skills that could be useful for one's career or trade (Murphy 2004: 211). Pliny is also conspicuously interested in the prices and qualities of goods and products, which could be an indication of the target audience (cf. Murphy 2004: 212).[93] It can be assumed that these considerations were of interest for members of the elite rather than for ordinary citizens.

The language and style of the work also allow us to make conclusions with regard to the target audience.[94] For Pliny, the contradiction between his rejection of digressions in the introduction and his use of this stylistic device has frequently been stressed.[95] Darab (2014) ascribes the function of exempla to the anec-

90 Plin. *NH praef.* 6: humili vulgo scripta sunt, agricolarum, opificum turbae, denique studiorum otiosis. ("It was written for the common herd, the mob of farmers and of artisans, and after them for students who have nothing else to occupy their time.") Wallace-Hadrill (1990: 91) links this with the criticism of luxury that is typical of Pliny. For Pliny's introduction, cf. Köves-Zulauf 1973: 134-184; Doody 2009: 113-114.

91 Beagon (1992: 13-14) considers the work of Pliny as a combination of theory and practice addressed to an educated reader, who is "neither an intellectual expert nor a specialist." Similiarly, Healy 1999: 71; Nikitinski 1998: 346-349. Fögen (2007: 190) considers the target audience of Pliny to be not a small circle of specialists but a larger group of "interested laymen." Similarly, Paparazzo (2008: 53) regards the work as "intended not for a learned audience, but mostly for the average, not well-read subject of the Roman Empire."

92 Taub (2017: 79-81) also stresses Pliny's interest in practical things like agriculture, astronomy, and meteorology. Cf. Bost 1993: 34.

93 Beagon (1992: 75-76) notes Pliny's appreciation of long-distance trade, which might be motivated by considerations of his readership; Pliny also focusses on the date on which certain goods were imported for the first time; cf. Murphy 2004: 49-51; 96-113.

94 Cf. Pinkster (2005: 240-244) for the abscence of complex sentence structure, the nominalization of adjectives, and the effort to achieve clarity and brevity. For the language of Pliny, cf. also Hine 2011: 624-654.

95 Healy (1999: 38-40) sees the avoidance of boredom as the primary objective. Fögen (2009a: 191-192), however, stresses that the emphasis on content instead of form can be regarded as topical. Cf. also Murphy 2004: 37-38; Fögen 2009a: 208-209.

dotes (cf. Darab 2014: 293-295). In her opinion, Pliny made use of them—instead of technical terms—because his work followed the guidelines of ancient rhetoric. The inclusion of mirabilia and wondrous stories therefore has to be seen as a rhetorical strategy which would satisfy even demanding and sophisticated readers.

Pliny's frequent mention of his sources is also conspicuous. These references were meant to enhance the prestige of the author (cf. Murphy 2004: 53-61). The readers would also be able to refer, in their conversations, not only to the work of Pliny but also to the theories of other authors. Thus, Scarborough (1986: 61) states, regarding the readership that "Pliny is writing for an educated public."[96]

Aelian's target audience was, according to the author himself, a wider readership, not limited to a public only interested in scientific facts or scientists (Müller-Reineke 2010: 118). In his introduction to *De natura animalium*, he explicitly states that he intends to write comprehensibly and to use "untechnical language."[97] In his epilogue, he states that he intends to provide not only information but also entertainment.[98] The members of the elite can, however, also be considered part of Aelian's potential target audience (Kindstrand 1998: 2963). In addition to emphasizing the easy readability of his work, he refers to its high scientific approach and the wide range of its research.[99]

As for Oppian, it has already been shown that he sees himself in the tradition of didactic poetry. Even though he provides practical advice regarding fishing, especially concerning bait and traps, his target audience is not the ordinary citizen. As allusions to Homer and other prominent authors prove, he, like Pliny and Aelian, addressed his work to educated members of the elite who were familiar with philosophy and history.[100]

---

96 Doody (2009: 113-114) stresses that the reader that Pliny addresses is on the one hand an expert who consults his work with a specific question in mind, and on the other hand an educated amateur who reads the complete work for pleasure. Naas (2002: 67) comes to the conclusion that the work of Pliny represents a "different kind of knowledge: one more attractive, which appeals to a larger audience and which is accessible to it."

97 Ael. *NA* praef. For his use of the Greek language, see Rodríguez-Noriega Guilliés 2011: 94-95.

98 Fögen (2009b: 59-61) assumes that Aelian discussed questions of terminology in order to emphasize the complexity and artistic claim of his work. For the statements in the epilogue, see also Hübner 1985: 157.

99 Ael. *NA* praef.

100 For the language and style of Oppian, see Gow 1968: 60-68; cf. also Bartley 2005 for the rhetoric and stylistic elements in Oppian and the orientation toward the contemporary taste of the intellectual elite.

Thus, though the works show some features of popular science, they were probably not addressed to the whole population. The insertion of mirabilia and anecdotes may have instead been motivated by the preferences and tastes of the elite.[101]

THE ROMAN BANQUET

We must still examine why the Roman elite would show such a keen interest in fish and their way of living, or why specialized knowledge of this sort could be profitable. A context in which such expertise could be relevant would be the symposion or banquet. The prestigious, exclusive, and expensive dinner parties played an important—social and political—role for the self-presentation of members of the elite. For them, the ostentation of wealth served the purpose of distinguishing themselves and displaying their own status. The effort to serve expensive and rare fish, which were considered the epitome of luxury, belongs in this context.[102]

Apart from the food, an integral part of the banquet was cultivated conversation. The reading of scientific works could prepare the elite for this type of discussion. The participants of the banquets were expected to talk casually, but also to demonstrate their education and expertise in the fields of philosophy, literature, and science. In this way, it was possible to express social distinction and demonstate possession of status.[103] As Varro explains, the conversation should be pleasant, entertaining, and useful.[104] Similiar considerations apply to the Greek symposion which, according to Plutarch, was likewise regarded as a pos-

101 Beagon (1992: 14-15) sees the work of Pliny as typical for the time and the contemporary educational ideal of the elite. Darab (2014: 291-292) also emphasizes that the anecdotes could be read on a concrete as well as on an abstract level. Thus they could satisfy the demands of an educated audience.

102 For the association of the criticism of luxury with fish and fishponds in particular, see Bannon 2014: 166 and 172.

103 The importance of sophisticated and intellectual conversation can be demonstrated with a counter-example: The nouveau riche Trimalchio, described by Petronius in his satyricon, states at the beginning of his banquet that he is a learned man and able to conduct a sophisticated conversation, but then fails spectacularly, demonstrating his lack of education and thereby his low social status.

104 Varro *Sat. Men.* 337; cf. 336; 338. cf. Stein-Hölkeskamp 2005: 220-221.

sible occasion for polite conversation and education (cf. Meeusen 2016: 195-196). Plutarch recommends scientific topics as suitable subjects of conversation.[105]

Figure 2: Roman Mosaic with Fishes and Ducks from the House of the Grand Duke of Tuscany, IX, 2, 27, Pompeii

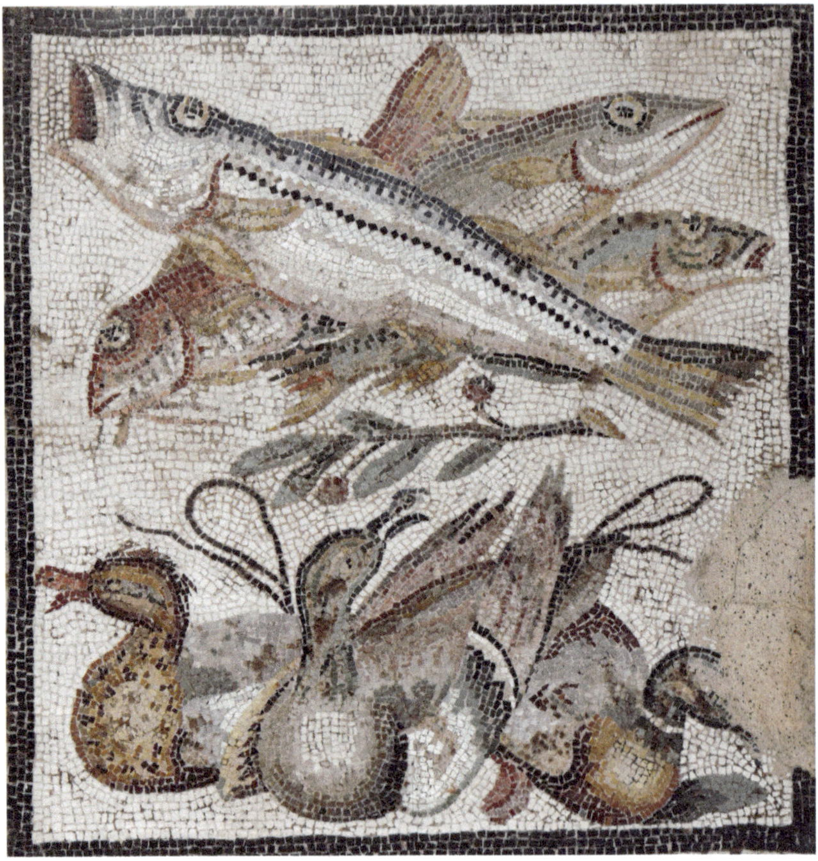

Source: Museo Archeologico Nazionale, Naples, inv. no. 109371; Photo: Marie-Lan Nguyen, Public Domain

The objective of works like those of Pliny, Aelian, and Oppian was to enable the reader to make explanatory and intellectually stimulating comments on the fish

105 See for example his dialogue *de tuenda sanitate praecepta*, esp. 133d and f, where scientific questions are characterized as suitable topics. Cf. the analysis of Meeusen 2016: 196-199, esp. pp. 197-198.

that were served. They were intended to offer starting points for subsequent discussion. The mention of anecdotes or myths about fish could offer the opportunity for further discussions—about human and animal behavior, ethical and moral questions, the divine order of the world, etc.

The main objective of the treated works was not to impart dry theoretical scientific knowledge but to enable the reader to hold an intellectually satisfying conversation, to respond to the statements of others, and to comment on the dishes being served.[106] Against this background, it becomes understandable why authors like Pliny mention recipes or discuss possibilities for preparing fish. Furthermore, fish were frequently used as an aphrodisiac or as medicine, which offered other opportunities for debate.[107] In general, food, illness, and love—topics that are frequently addressed by Pliny, Aelian, and Oppian—were ideally suited for table talk.

A passage in Horace can be regarded as evidence that such issues were discussed during the banquets. In one of his satires, the banquet of a certain Nasidienus is described. When a particularly exquisite fish, a moray, is served, he comments as follows: "'This,' said he, 'was caught before spawning; if taken later, its flesh would have been poorer.'"[108]

Nasidienus knows precisely when the flesh of the lamprey is best enjoyed. Subsequently, he even explains what goes into the sauce, demonstrating that he can not only afford to hold expensive banquets but also has a sense of good taste and possesses knowledge about food.[109]

This episode shows that the members of the senatorial elite were expected to distinguish themselves as connoisseurs. The focus was not primarily on an in-depth analysis of scientific problems. Subtle and pedantic discussions regarding the differences between crustaceans and molluscs would have been regarded as inappropriate and boring. Instructive in this respect is a comment by Cicero on a banquet that he attended (Cic. *Att.* 14.12,3): "I have put this together on the 22nd

106 Naas (2002: 66) speaks of a "compilation at the expense of research" and an "inventory of marvels."

107 Scarborough (1986, 61) compares the pharmaceutical passages in Pliny with a modern handbook about herbology, the focus being primarily on the remedies and not on the disease. Cf. also Locher (1986: 20-29); Fögen (2009a: 190-191 and 203). For the criticism of Greek doctors and their overpriced drugs, cf. Beagon (1992: 202-224); Fögen (2009a: 238-240).

108 "haec gravida" inquit/ "capta est, deterior post partum carne futura." (Hor. *sat.* 2.8,43f)

109 See McInerney (2014: 260-261). For the relevance of culture and education for the identity of the Roman elite, also and especially in connection with the banquet, see Meeusen (2016: 204-206).

at dinner with Vestorius, a practised mathematician however ignorant of dialectics."[110] The host had obviously bored his guests by presenting numbers and calculations. What those guests desired instead was a sophisticated and fruitful conversation.

There are further indications that zoological writings were highly relevant for the participants of the banquets and that biological questions were taken into account when planning an event. Thus, apart from scientists, cooks are also quoted as sources of biological information.[111] Furthermore, in order to provide entertainment, the courses were served in the most spectacular way, including as biological "experiments."[112] The striving for the extraordinary and exceptional—regarding both the dishes and the topics of the conversation—is also reflected in the special interest in exotic creatures in the works of Pliny, Aelian, and Oppian.[113]

The focus on gastronomical questions and culinary pleasure could, however, also be viewed as problematic. The trade in fish, for example, did not meet the expectations which were associated with the senatorial rank.[114] The interest in exotic and expensive creatures could be regarded as a sign of decadence and luxuria. A recurring motif and element of the narratives are the consequences of Roman rule and superiority.[115] Thus, Pliny, for example, views the impact of

110 haec conscripsi x Kal. accubans apud Vestorium, hominem remotum a dialecticis, in arithmeticis satis exercitatum.

111 Ael. *NA* 10.7.

112 Cf. Pliny's description of the killing of a mullet at the table and the analysis of the change of color during the process (9.66); this procedure is, however, criticized as decadent. See also Seneca *nat. quest.* 3.18. Cf. Richardson-Hay (2009: 85), who stresses the event character of the banquets.

113 For Pliny's interest in exotic animals and mirabilia in particular, see Beagon 1992: 128-129. For Aelian's preference for exotic animals, cf. Müller-Reineke 2010.

114 As Bannon (2014: 167-181), also referring to passages in Cicero, Varro, Valerius Maximus, Columella, and Macrobius, has shown, men like Orata challenged the traditional values and the foundation of elite identity. According to Bannon, he "symbolizes an unresolved conflict about the proper relationship between wealth and status in elite identity." By displaying his wealth, he had chosen an alternative way to demonstrate his status. For the fish and fishponds as symbols of luxury and as status symbols, see Geoffrey Kron (2014: 192-202). Cf. also McInerney 2014: 259-261; Schmölcke/Nikulina 2008: 46-48 and 50.

115 This partially finds expression in the description of the natural world, as Darab (2012) has shown in her analysis of Pliny's account of amber and bronze. Pliny on the one

Rome's expansion of power in a critical light.[116] Though he acknowledges the benefits of peace and stability, he notices a lack of freedom and interest in scientific research and complains about the excessive desire for luxury.[117] This type of criticism was also a popular topic of conversation.[118] Such moral reflections, which play an important role in the works of the Imperial authors, could provide ideas and arguments in this respect as well.

In addition to the descriptions of the fish in literature, archaeological evidence can also be used to examine knowledge about fish in antiquity. Special attention should be paid to the mosaic floors which decorated the dining rooms of the Roman villas. These frequently show complex depictions of the marine world. A wide range of fish and other sea creatures are depicted, and these are easily identifiable due to the realistic representation. Conspicuous and probably not accidental is the fact that mosaics frequently depict those species that figure prominently in the zoological literature. In some instances, it is possible to detect direct connections between the literary descriptions and the pictorial representations—thus, the deadly fight between the octopus" and its archenemies, the moray and the crawfish, is frequently depicted in the center. An example of this type of composition is the mosaic from the House of the Faun in Pompeii.[119]

It was not only the dishes served that could stimulate sophisticated conversation at the table; the floor decorations could offer inspiration as well. The educated guest could refer to the fish depicted on the floor and distinguish himself

hand criticizes the erroneous explanations of the Greek authors and on the other hand focusses on Roman rule over Greece.

116 Cf. Plin. *NH* 2.117-118. Beagon (1992) in particular argues that the selection of information by Pliny and the structure of his work in general have to be seen as a response to Roman rule. Jones-Lewis (2012: 51-74) points out that Pliny illustrates the expansion of Roman power by mentioning things as different as dangerous poisons and works of art from foreign countries. Cf. Darab 2014: 208-209; see also Murphy 2004: 49-51; for the comparison of the account of Pliny with a triumphal procession, see pp. 154-163. Beagon (1992: 9-11) also sees the *mirabilia* in the context of the expanding Roman Empire. For Pliny's "imperialistic perspective," see Naas 2002: 57-70. Cf. also Fear 2011: 21-34; Taub 2017: 76.

117 Cf. Naas 2002: 68-69. Cf. Murphy 2004: 69-72; Fögen 2009a: 217-220; Healy 1999: 76-77.

118 For the criticism of luxury, cf. Wallace-Hadrill 1990: 85-92; regarding luxury and the criticism of it as a means of expressing social distinction, cf. Lao 2011: 35-56.

119 Museo Archeologico Nazionale, Naples, No. 120177. Cf. Andreae 2003: 148-160; Mielsch 2005: 126-127.

by showing his expertise.[120] At the same time, the host could demonstrate that he was well acquainted with the biodiversity of the sea. The works of Pliny and Aelian could offer suggestions for the embelishment and decoration of Roman villas. The mosaics could be seen as a kind of illustration of their works. The mosaic depictions may also explain why the authors omit detailed descriptions of the external features of the fish. The animals were omnipresent—on the table as dishes and on the floor as part of the decorations.

To summarize, it can be said that the works of the authors of the Imperial era may not comply with the standards of Aristotle. Nevertheless, they should not be devaluated or regarded as inferior. Their content and form were in line with the expectations of their specific readership, i.e., educated members of the elite. In their conversations, the knowledge of curiosities and mirabilia was of great importance. The display of erudition could serve to demonstrate status and excellence.

The works of Pliny, Aelian, and Oppian were therefore widely read in antiquity, and were even relevant for the elite in the 20th century. Thus, for example, the Austrian poet Hugo von Hoffmannsthal reflects, in a fictional letter, on the importance of the moray in Rome and the emotional attachment of Crassus to his pet—and expects his readers to know about the background of this episode.

REFERENCES

Primary Sources

Aelian (1958): Characteristics of Animals, Vol. 1, trans. A.F. Schofield, Cambridge, MA: Harvard University Press.

Aelian (1959): Characteristics of Animals, Vol. 2, trans. A.F. Schofield, Cambridge, MA: Harvard University Press.

Cicero (1999): Letters to Atticus, Vol. 1, trans. D.R. Shackleton Bailey, Cambridge, MA: Harvard University Press.

120 For tableware or rather its adornments (freqently depictions of fish) as useful points of departure for interesting conversations, cf. Stein-Hölkeskamp 2005: 144-146 and 152-154. Apart from the banquet, other semi-public spaces in which such discussions could take place are conceivable, for example the public baths where representations of fish are also very common. As Fagen (2006: 190-207) has shown, Pliny frequently discusses bathing and its healthy effects, especially in connection with the description of other remedies.

Horace (1929): Satires, Epistles, Ars Poetica, trans. H. Rushton Fairclough, Cambridge, MA: Harvard University Press.

Lucian (1961): Vol. 7, trans. M.D. Macleod, Cambridge, MA: Harvard University Press (LCL).

Oppian, Colluthus, Tryphiodorus (1928), trans. A.W. Mair, Cambridge, MA: Harvard University Press.

Pliny (1949), Natural History, Vol. 1, trans. H. Rackham, Cambridge, MA: Harvard University Press.

Pliny (1963): Natural History, Vol. 8, trans. W.H.S. Jones, Cambridge, MA: Harvard University Press.

Pliny (1989 [2nd ed.]): Natural History, Vol. 3, trans. H. Rackham, Cambridge, MA: Harvard University Press.

Plutarch (1928): Moralia, Vol. 2, trans. F. Cole Babbitt, Cambridge, MA: Harvard University Press.

Secondary Sources

Andreae, Bernard (2003): Antike Bildmosaiken, Mainz: Philipp von Zabern.

Ash, Rhiannon (2011): "Pliny the Elder's Attitude to Warfare." In: Roy Gibson/Ruth Morello (eds.), Pliny the Elder: Themes and Contexts, Leiden and Boston: Brill, pp. 1-20.

Bannon, Cynthia J. (2014): "C. Sergius Orata and the Rhethoric of Fishponds." In: The Classical Quarterly 64, pp. 166-182.

Bartley, Adam (2003): "Stories from the Mountains, Stories from the Sea: The Digressions and Similes of Oppian's *Halieutica* and the *Cynegetica*." In: Hypomnemata 150, Göttingen: Vandenhoeck und Ruprecht.

Bartley, Adam (2005): "'What's Fishing Like?' The Rhetoric of Similes in Oppian's *Halieutica*." In: Classics Ireland 12, pp. 1-12.

Beagon, Mary (1992): Roman Nature. The Thought of Pliny the Elder, Oxford: Claredon Press.

Bodson, Liliane (1986): "Aspects of Pliny's Zoology." In: Roger French/Frank Greenaway (eds.), Science in the Early Roman Empire: Pliny the Elder, His Sources and Influence, London and Sydney, Croom Helm, pp. 98-110.

Borst, Arno (1993): Das Buch der Naturgeschichte. Plinius und seine Leser im Zeitalter des Pergaments, Heidelberg: Winter.

Carey, Sorcha (2003): Pliny's Catalogue of Culture. Art and Empire in the Natural History, Oxford and New York: Oxford University Press.

Darab, Ágnes (2012): "Corinthium Aes versus Electrum: The Anecdote as an Expression of Roman Identity in Pliny the Elder's *Naturalis Historia*." In: Hermes 140/2, pp. 149-159.

Darab, Ágnes (2014a): "Natura, ars, historia: Anecdotic History of Art in Pliny the Elder's *Naturalis Historia*. Part I." In: Hermes 142/2, pp. 206-224.

Darab, Ágnes (2014b): "Natura, ars, historia: Anecdotic History of Art in Pliny the Elder's *Naturalis Historia*. Part II." In: Hermes 142/3, pp. 279-297.

Dodds, Eric Robertson (1933): "The Portrait of a Greek Gentleman." In: Greece & Rome 2, pp. 97-107.

Doody, Aude (2009): "Pliny's *Natural History: Enkuklios Paideia* and the Ancient Encyclopedia." In: Journal of the History of Ideas 70/1, pp. 1-21.

Doody, Aude (2011): "The Science and Aesthetics of Names in the Natural History." In: R. Gibson/K. Roy /R. Morello (eds.), Pliny the Elder: Themes and Contexts, Leiden and Boston (Mass.): Brill, pp. 113-129.

Drummond, William Hamilton (1818): "An Essay on the Life and Writings of Oppian." In: The Transactions of the Royal Irish Academy 13, pp. 3-45.

Fagan, Garrett G. (2006): "Bathing for Health with Celsus and Pliny the Elder." In: The Classical Quarterly 56/1, pp. 190-207.

Fear, Andrew (2011): "The Roman's Burden." In: Roy Gibson/Ruth Morello (eds.), Pliny the Elder: Themes and Contexts, Leiden and Boston: Brill, pp. 21-34.

Fögen, Thorsten (2007): "Pliny the Elder's Animals. Some Remarks on the Narrative Structure of *Nat. Hist.* 8-11." In: Hermes 135/2, pp. 184-198.

Fögen, Thorsten (2009a): "Wissen, Kommunikation und Selbstdarstellung. Zur Struktur und Charakteristik römischer Fachtexte der frühen Kaiserzeit." In: Zetemata. Monographien zur klassischen Altertumswissenschaft 134, Munich: Beck.

Fögen, Thorsten (2009b): "The Implications of Animal Nomenclature in Aelian's *De natura animalium*." In: Rheinisches Museum für Philologie 152/1, pp. 49-62.

Gigon, Olof (1966): "Plinius und der Zerfall der antiken Naturwissenschaft." In: Arctos 4, pp. 23-45.

Gilhus, Ingvild Sælid (2006): Animals, Gods and Humans. Changing Attitudes to Animals in Greek, Roman and Early Christian Ideas, London and New York: Routledge.

Gill, James E. (1969): "Theriophily in Antiquity: A Supplementary Account." In: Journal of the History of Ideas 30/3, pp. 401-412.

Gow, A. S. F. (1968): "On the Halieutica of Oppian." The Classical Quarterly 18/1, pp. 60-68.

Healy, John F. (1999): Pliny the Elder on Science and Technology, Oxford: Oxford University Press.

Hindermann, Judith (2011a): "Verliebte Delphine, schwimmende Inseln und versiegende Quellen beim älteren und jüngeren Plinius." In: Gymnasium 118, pp. 345-354.

Hindermann, Judith (2011b): "Zoophilie in Zoologie und Roman: Sex und Liebe zwischen Mensch und Tier bei Plutarch, Plinius dem Älteren, Aelian und Apuleius." In: Dictynna 8, pp. 1-27.

Hindermann, Judith (2016): "Aelian und die ΠΟΙΚΙΛΙΑ. Ordnung und Unordnung in *De natura animalium*." In: Rheinishces Museum für Philologie 159/1, pp. 71-98.

Hine, Harry Morrison (2011): "'Discite … Agricola': Modes of Instruction in Latin Prose Agricultural Writing from Cato to Pliny the Elder." In: The Classical Quarterly 61/2, pp. 624-654.

Hübner, Wolfgang (1984): "Der Mensch in Aelians Tiergeschichten." In: Antike und Abendland 30, pp. 154-176 (= 1985, Trierer Beiträge 15, pp. 69-76).

Jones-Lewis, Molly Ayn (2012): "Poison: Nature's Argument for the Roman Empire in Pliny the Elder's *Naturalis Historia*." In: The Classical World 106/1, pp. 51-74.

Keydell, Rudolf (1937): "Oppians Gedicht von der Fischerei und Aelians Tiergeschichte." In: Hermes 72/4, pp. 411-434.

Kindstrand, Jan Fredrik (1998): "Claudius Aelianus und sein Werk." In: Aufstieg und Niedergang der römischen Welt (ANRW) II.34.4, pp. 2954-2996.

Kneebone, Emily (2008): "The Poetics of Knowledge in Oppian's *Halieutica*." In: Ramus 37/1-2, pp. 32-59.

König, Jason/Woolf, Greg (2013): "Encyclopaedism in the Roman Empire." In: Jason König/Greg Woolf (eds.), Encyclopaedism from Antiquity to the Renaissance, Cambridge/New York: Cambridge University Press, pp. 23-63.

Korhonen, Tua (2012): "On Human-Animal Sexual Relationships in Aelian's De Natura Animalium." In: Arctos 46, pp. 65-77.

Köves-Zulauf, Thomas (1973): "Die Vorrede der plinianischen 'Naturgeschichte.'" In: Wiener Studien 86, pp. 134-184.

Kron, Geoffrey (2014): "Ancient Fishing and Fish Farming." In: Gordon Lindsay Campbell (ed.), The Oxford Handbook of Animals in Classical Thought and Life, Oxford: Oxford University Press, pp. 192-202.

Lao, Eugenia (2011): "Luxury and the Creation of a Good Consumer." In: Roy Gibson/Ruth Morello (eds.), Pliny the Elder: Themes and Contexts, Leiden and Boston (Mass.): Brill, pp. 35-56.

Locher, A. (1986): "The Structure of Pliny the Elder's Natural History." In: Roger French/Frank Greenaway (eds.), Science in the Early Roman Empire. Pliny the Elder, His Sources and Influence, London: Croom Helm, pp. 20-29.

Lytle, Ephraim (2011): "The Strange Love of the Fish and the Goat: Regional Contexts and Rough Cilician Religion in Oppian's *Halieutica* 4.308-73." In: Transactions of the American Philological Association 141, pp. 333-386.

Martínez, Sebastián/Silva, Tomás (2003): "Opiano, un poeta o dos?" In: L'antiquité classique 72, pp. 219-230.

McInerney, Jeremy (2014): "Civilization, Gastronomy, and Meat-Eating." In: G. Campbell (ed.), The Oxford Handbook of Animals in Classical Thought and Life, Oxford: Oxford University Press, pp. 248-268.

Meeusen, Michiel (2016): Plutarch's Science of Natural Problems, Leuven: Leuven University Press.

Mielsch, Harald (2005): Griechische Tiergeschichten in der antiken Kunst, Mainz: Philipp von Zabern.

Müller-Reineke, Hendrik (2010): "Oriental Animals as Moral Examples in Aelian's *De natura animalium*." In: Graeco Latina Brunensia 15/2, pp. 117-126.

Murphy, Trevor (2004): Pliny the Elder's *Natural History:* The Empire in the Encyclopedia, Oxford and New York: Oxford University Press.

Naas, Valérie (2011): "Imperialism, 'Mirabilia' and Knowledge: Some Paradoxes in the *Naturalis Historia*." In: Roy Gibson/Ruth Morello (eds.), Pliny the Elder: Themes and Contexts, Leiden and Boston: Brill, pp. 57-70.

Nikitinski, Oleg (1998): "Plinius der Ältere: Seine Enzyklopädie und ihre Leser." In: Wolfgang Kullmann/Jochen Althoff/Markus Asper (eds.), Gattungen wissenschaftlicher Literatur in der Antike, Tübingen: Narr, pp. 341-359.

Nutton, Vivian (1986): "The Perils of Patriotism: Pliny and Roman Medicine." In: R. French/F. Greenaway (eds.), Science in the Early Roman Empire: Pliny the Elder, His Sources and Influence, London and Sydney: Croom Helm, pp. 30-58.

Paparazzo, Ernesto (2008): "Pliny the Elder on Metals: Philosophical and Scientific Issues." In: Classical Philology 103/1, pp. 40-54.

Paparazzo, Ernesto (2011): "Philosophy and Science in the Elder Pliny's *Naturalis Historia*." In: Roy Gibson/Ruth Morello (eds.), Pliny the Elder: Themes and Contexts, Leiden and Boston: Brill, pp. 89-112.

Pinkster, Harm (2005): "The Language of Pliny the Elder." In: Tobias Reinhardt/John Norman Adams/Michael Lapidge (eds.), Aspects of the Language of Latin Prose, Oxford: Oxford University Press, pp. 239-256.

Richardson-Hay, Christine (2009): "Dinner at Seneca's Table. The Philosophy of Food." In: Greece & Rome 56/1, pp. 71-96.

Rodríguez-Noriega Guillén, Lucía (2011): "Greek and Latin in the Late Second and Early Third Centuries CE: Athaeneus of Naucratis and Claudius Aelian." In: José Bernardino Torres Guerra (ed.), Vtroque sermone nostro. Bilingüismo social y literario en el imperio de Roma / Social and Literary

Bilingualism in the Roman Empire, Collección Mundo Antiguo 14, Pamplona: Ediciones Universidad de Navarra, pp. 83-98.

Scarborough, John (1986): "Pharmacy in Pliny's *Natural History*: Some Observations on Substance and Sources." In: Roger French/Frank Greenaway (eds.), Science in the Early Roman Empire: Pliny the Elder, His Sources and Influence, London and Sydney: Croom Helm, pp. 59-85.

Schmölcke, Ulrich/Nikulina, Elina A. (2008): "Fischhaltung im antiken Rom und ihr Ansehenswandel im Licht der politischen Situation." In: Schriften des Naturwissenschaftlichen Vereins für Schleswig-Holstein. 70, pp. 36-55.

Schultze Clemence (2011): "Encyclopaedic Exemplarity in Pliny the Elder." In: Roy Gibson/Ruth Morello (eds.), Pliny the Elder: Themes and Contexts, Leiden and Boston: Brill, pp. 167-186.

Smith, Steven D. (2013): "Monstrous Love? Erotic Reciprocity in Aelian's *De natura animalium*." In: Ed Sanders/Chiara Thumiger/Christopher Carey/Nick Lowe (eds.), Eros in Ancient Greece, Oxford: Oxford University Press, pp. 73-80.

Stebbins, Eunice Burr (1929): The Dolphin in the Literature and Art of Greece and Rome, Menasha: The George Banta Publishing Company.

Stein-Hölkeskamp, Elke (2005): Das römische Gastmahl. Eine Kulturgeschichte, Munich: Beck.

Steiner, Grundy (1955): "The Scepticism of the Elder Pliny" In: The Classical Weekly 48, pp. 137-143.

Taub, Liba (2017): Science Writing in Greco-Roman Antiquity, Cambridge, UK: Cambridge University Press.

Wallace-Hadrill, Andrew (1990): "Pliny the Elder and Man's Unnatural History." In: Greece & Rome 37, pp. 80-96.

Wilkins, John (2009): "Animals in the Romano-Greek Culture of the Second Century A.D." In: Annetta Alexandridis/Markus Wild/Lorenz Winkler-Hora ček (eds.), Mensch und Tier in der Antike. Grenzziehung und Grenzüberschreitung, Wiesbaden: Reichert, pp. 315-328.

Williams, Craig A. (2013), "When a Dolphin Loves a Boy: Some Greco-Roman and Native American Love Stories." In: Classical Antiquity 32, pp. 200-242.

The Celestial Axis in Manilius' *Astronomica*: Making the Invisible Visible

Matteo Rossetti

ABSTRACT

This article deals with the description of the celestial axis in Manilius' *Astronomica* (1.274-293), an astronomical/astrological Latin didactic poem of the 1st century AD. The celestial axis guarantees the regular and uniform motion of the Cosmos and in Manilius' explanation represents a sort of 'agent of providence' which controls the Universe. By tracking down Manilius' poetic models and considering the stylistic features of the passage, this essay analyzes the ways that Manilius exposes and popularizes a complex subject. The poet seems to emphasize the immobility and immateriality of the axis, as opposed to the mobility and materiality of the Universe, through the use of the device of paradox. The sense of paradox is accentuated by a description of the Universe in its sublime greatness. By studying the presence of the sublime in the exposure of the axis, this article also takes the cultural models of Manilius (in particular Lucretius) into account.

THE CELESTIAL AXIS FROM ARATUS TO MANILIUS

Manilius,[1] in the first book of *Astronomica*[2] after the description of the zodiac (vv. 255-274), begins the treatment of the celestial axis (1.275-293), i.e., the im-

1 Marcus Manilius wrote, possibly at the end of the Augustan age, a poem in five books entitled *Astronomica*, on astronomy and astrology. According to Manilius, astrology is the most suitable intellectual tool to know not only the future fate but also the inner structure of

aginary line, orthogonal to the celestial equator, which intersects the two poles of the sphere. Manilius' exposition is divided into three parts: 1. A frame in which the poet focuses his attention on the poles, which are identified through the two Bears (i.e., *Ursa Major* and *Ursa Minor*, 1.275-280). 2. The treatment of the incorporeality and immobility of the axis, opposed to the whirling of the celestial sphere's mass (1.281-291). 3. Two etiological verses at the end of the discussion (1.292-293) that summarize the entire didactic exposition and propose an explanatory etymology of the celestial axis.

The concept of axis of the Cosmos is present in Plato (*Tim.* 40C see also *Res.* 616D-617B) and in Aristotle (*De cael.* 293b, 296a), although both authors, as Kidd (1997: 178) has noted, adopted the term πόλος ("pivot," "pole").[3] Aratus, instead, used ἄξων / ἄξις ("axis") in the sense of 'celestial axis' for the first time (*Phaen.* 22). In Aratus' *Phaenomena*, and also in Latin translations, the treatment of the axis, placed immediately after the Proem, precedes the description of

---

the Cosmos. The central idea of the poem is that man, endowed with all rational capacities, is a sort of small image (or microcosm) of the Universe. We have no definite biographical information about Manilius; the name, with some uncertainty, is restored by a few manuscripts, and many doubts remain about the dating of the poem. Scholars are still undecided whether to date the text to the Augustan age or to the Tiberian age. The most cautious thesis remains that of Housman, who dates the poem between the two emperors. While there are no ancient authors who explicitly mention Manilius, he certainly influenced the debates on astrology and astronomy throughout the imperial age: there are traces of the *Astronomica* in Seneca, Lucan, Petronius, Calpurnius Siculus, Juvenal, and above all in Firmicus Maternus' *Mathesis*, a work of 'Christian astrology.' The *Astronomica* was discovered in 1417 by the Florentine humanist Poggio Bracciolini, and from that moment on the poem was copied in many manuscripts and aroused the interest of many philologists and scientists (the first edition was made by the astronomer and mathematician Regiomonatus in 1473, and the first commentary by the astrologer Lorenzo Bonincontri in 1484). Between the 16th and 18th centuries, some of the most important philologists studied and made editions of the *Astronomica*, including J. J. Scaliger (1540-1609) and R. Bentley (1662-1742), key figures in the culture of their time. For a general introduction to Manilius, cf. Hübner 1984 and Volk 2009; 2011a: 11-13.

2 The first book of the poem, the subject of the present contribution, consists in a general astronomical introduction to the astrological matter presented in Books 2-5. At the beginning of the first book, after the proem, Manilius takes up the origin and the shape of the Cosmos (1.118-245), then describes the constellations and circles of the celestial sphere (1.255-804). In the final part of the book, the planets (1.805-808, the collocation of these verses is uncertain) and comets (1.809-926) are remembered.

3 The term was used for the first time in the meaning of 'pole' by Anaxagoras. On the history of the concept, cf. Kauffman *RE*.

the constellations.[4] The poet insists on the immobility of the axis, in opposition to the movement of the stars on the sphere, which is constant over time. In fact, at v. 20 Aratus does not hesitate to accumulate terms indicating temporal continuity in expressions like πάντ' ἤματα συνεχὲς αἰεί. Through this exposition, the poet provides the reader a simple cosmographic introduction and establishes two fundamental principles in its astronomy: the whirling of the celestial sphere and its stable equilibrium. Aratus' exposition represents the starting point for the axis' description in Manilius. The Latin poet, however, amplifies his model in a less synthetic treatment. Manilius, instead, shapes his didactic exposition through the insistent repetition of two concepts: the static nature and the immateriality[5] of the axis opposed to the whirling nature and materiality of the Cosmos.

At qua fulgentis caelum consurgit ad Arctos,
Omnia quae summo despectant sidera mundo
Nec norunt obitus unoque in uertice mutant
In diuersa situm caelumque et sidera torquent,
Aera per gelidum *tenuis* deducitur axis
Libratumque regit diuerso cardine mundum,
Sidereus circa medium quem uoluitur orbis
Aetheriosque rotat cursus, immotus at ille
In binas Arctos magni per inania mundi
Perque ipsum terrae derectus constitit orbem.
Nec uero e solido stat robore corporis axis

4 *Phaen.* 19-26: οἱ μὲν ὁμῶς πολέες τε καὶ ἄλλυδις ἄλλοι ἐόντες / οὐρανῷ ἕλκονται πάντ' ἤματα συνεχὲς αἰεί· / αὐτὰρ ὅγ' οὐδ' ὀλίγον μετανίσσεται, ἀλλὰ μάλ' αὔτως / ἄξων αἰὲν ἄρηρεν, ἔχει δ' ἀτάλαντον ἁπάντη / μεσσηγὺς γαῖαν, περὶ δ' οὐρανὸν αὐτὸς ἀγινεῖ. / Καί μιν πειραίνουσι δύω πόλοι ἀμφοτέρωθεν· / ἀλλ' ὁ μὲν οὐκ ἐπίοπτος, ὁ δ' ἀντίος ἐκ βορέαο ὑψόθεν ὠκεανοῖο. ("The numerous stars, scattered in different *directions*, sweep all alike across the sky every day continuously for ever. The axis, however, does not move even slightly from its place, but just stays for ever fixed, holds the earth in the center evenly balanced, and rotates the sky itself. Two poles terminate it at the two ends; but one is not visible, while the opposite one in the north is high above the horizon") (trans. Kidd 1997: 73-74).

5 Aratus does not mention the materiality of the axis (and neither do his Latin translators), but we know from the *scholia to Phaenomena* that the idea of its immateriality is implicit in the poem. Cf. *Schol. Arat.* pp. 69-70 Martin: (ἄξων) γραμμὴ δέ ἐστιν εὐθεῖα [ἡ] ἀπὸ σημείου ἀρξαμένη <καὶ εἰς σημεῖον λήγουσα> [...] ἡ δὲ εὐθεῖα γραμμὴ τοῦ ἄξονος τοὺς δύο πόλους ἔχει [τοὺς] ἀπὸ σημείου εἰς σημεῖον. ("[The axis] is a straight line beginning from the sign <and terminating to the same sign> [...]. The axis' straight line holds the two poles, from one sign to another").

Nec graue pondus habet, quod onus ferat aetheris alti,
Sed cum aer omnis semper uoluatur in orbem
Quoque semel coepit totus uolet undique in ipsum,
Quodcumque in medio est, circa quod cuncta mouentur,
Vsque adeo tenue ut uerti non possit in ipsum
Nec iam inclinari nec se conuertere in orbem,
Hoc dixere axem, quia motum non habet ullum,
Ipse uidet circa uolitantia cuncta moueri.[6]

PARADOX: THE WEIGHTLESS IMMOBILE AXIS HOLDS THE UNIVERSE

At first glance, the Manilian exposition does not seem very incisive and gives the impression of extreme repetitiveness. We can see the iteration of terms that indicate the whirling motion of the sphere: *torquere* ("to twist," 1.278), *uoluere* ("to orbit," 1.281 and 287), *uolare* ("to fly," 1.288 and 293), *mouere* ("to move"); or the use of the same expressions: *tenuis* (1.279, 290), *cuncta mouentur – cuncta moueri* (1.288, 293), at the end of the verse. Incidentally, it should also be pointed out that the tendency toward repetition is characteristic of the style of the Manilius' explanations, and that it can be seen in another cosmological piece: the demonstration of terrestrial sphericity (1.173-246; cf. Abry 2005; Henderson 2009).

For the objectives of this paper, it is good to proceed with an analysis of the passage on the axis, in order to reveal some of the text's expressive features. On line 279, after the preamble on the poles, Aratus mentions the subject of discussion; it seems that the poet, putting the term "axis" at the end of the verse, is trying

6 My emphasis. Manil. 1.275-293: "Now where heaven reaches its culmination in the shining Bears, which from the summit of the sky look down on all the stars and know no setting and, shifting their opposed stations about the same high point, set sky and stars in rotation, from there an insubstantial axis runs down through the wintry air and controls the universe, keeping it pivoted at opposite poles: it forms the middle about which the starry sphere revolves and wheels its heavenly flight, but is itself without motion and, drawn straight through the empty space of the great sky to the two Bears and through the very globe of the Earth, stands fixed. Yet the axis is not solid with the hardness of the matter, nor does it possess massive weight such as to bear the burden of the lofty firmament; but since the entire atmosphere ever revolves in a circle, and every part of the whole rotates to the place from which is in the middle, about which all moves, so insubstantial that it cannot turn round upon itself or even submit to motion or spin in circular fashion, this men have called the axis, since motionless itself, it yet sees everything spinning about it." (trans. Goold 1977: 27)

to create a sense of expectation in the reader. The first quality of the axis pointed out by Manilius through the adjective *tenuis* ("subtile") and the verb *deducere* ("to draw") is its incorporeal nature. It is interesting that the verb *deducere* refers to the vocabulary of writing and drawing, and, in the vocabulary of geometry, can indicate the tracing of a line (cf. *TLL* [Thesaurus Linguae Latinae] 5.1.279, 50-57). These lexical observations show that Manilius could have had in mind the image of the axis as a geometric line (this image, as observed in nt. n. 5 in this article, is also present in aratean exegesis).[7] Moreover, the use of a term like *deducere* shows how Manilius attempts to trace, through his verses, the "picture" of a line passing from one pole to another through the celestial sphere.[8] We can also note marks of a vocabulary of geometry at line 284 ([axis] *derectus constitit orbis*).[9] As *deduco*, as well as *derigo*, can mean "to draw a line" in Latin, the verb is widely used, in this sense, in mathematical and geometrical texts.[10] The participle *derectus* ("straight, vertical, standing at right angles") has clear spatial value and conveys the geometric idea of perpendicularity.[11] The axis—according to Manilius—pivots at the two ends of the sphere, intersects the globe of the earth in the middle, and forms a right angle with the polar points (*binas Arctos* 1.283). The sense of *derectus*, moreover, is completed by a verb of state (*constitit*),[12] which well represents the motionless nature of the axis, and anticipates the subject of the discussion at 1.290-294. In addition, it is important to note that poet, at 1.284, combines technical language with precise literary

7 On ancient debates about the axis' materiality, see Achilles' *Introduction to Aratus* 28.

8 Axis is also represented as a line in a much later text, Isidore of Seville's *Ethymologiae* 3,36 = 13.5, 3-4 *axis est septentrionis linea recta, quae per mediam pilam sphaerae tendit* ("The northern axis is a straight line that stretches through the center of the ball of the sphere," trans. Barney, Lewis, Beach, Berghoff 2006: 100).

9 Postgate (1897: 21) conjectured *derectus* ("straight"), which was adopted by Housman and Goold, whereas the manuscripts have *directus* ("to give a particular direction").

10 Cf. Boeth. *Arit.* 2, 22 *si a tetragona basi proficiscantur et ad unum uerticem eius lineae dirigantur* ("if they depart from a tetragon basis and draw lines to one point on the vertex"); Chalc. *Comm.* 78 *littera Θ, per quam lineae duae dirigantur* ("the point *theta* through which two lines are drawn"); Ps. Boeth. *Geom.* p. 387 (= Grom. p. 388) *si in circulo per centrum linea quaedam recta dirigatur* ("if a straight line is drawn in the center of a circle").

11 See also *TLL* 5.1.1237,27, which proposes the comparison with Lucr. 2.197-198 *nam quo magis ursimus altum / derecta et magna ui multi pressimus aegre* ("for the deeper we have thrust them and pushed them right down laboriously with full force and many together," trans. Rouse-Ferguson Smith 1975: 111). On Lucretius, see Fowler 2002: 284-285.

12 *Constitit* is an emendation made by Joseph J. Scaliger; the manuscript tradition presents *conspicit*.

echoes. The hexameter ending in *constitit orbem* recalls Ixion's wheel in *Georgics* 4.484 (*atque Ixionii uento rota constitit orbis*).[13] The Virgilian context is naturally different from the Manilian one, but we can presume that the reader would recognize this important quotation reused in a didactic exposition.

The axis keeps the Cosmos in balance on the two opposite poles (v. 280, *libratumque regit diuerso cardine mundum*). Manilius in this verse is very close to Germanicus' *Phaen.* 20-21 ([*axis* scil.] *libratasque tenet terras et cardine firmo/ orbem agit*),[14] which translates Aratus *Phaen.* 22 (ἔχει δ' ἀτάλαντον ἁπάντη).[15] *Librare* ("to hold suspended") indicates the equilibrium of the celestial sphere and above all of the earth, and has done so ever since Cicero;[16] it is a part of a Latin cosmological vocabulary, used even in poetry, in a didactic and scientific context (not only in Germanicus and Manilius). The stability of the earth in the middle of the Universe is an important basis of ancient cosmology and at the same time is evidence of the presence of an immanent order, as opposed to the original chaos in which the elements were devoid of order, as in Ov. *Met.* 1.12-13 *nec circumfuso pendebat in aere tellus / ponderibus librata suis.*[17] Behind this representation of the Cosmos, there is a very concrete metaphor, taken from an architectural lexicon. The Universe, instead, is conceived as a scale (*libratus* represents a very close calque of the Greek ἀ-τάλαντος[18] "equal in weight") in which the cosmic mass lies in constant equilibrium. In this way, it is not surprising that expressions widely used in cosmological context, such as *pendere* or *librare ponderibus*,[19] occur in the vocabulary of weights and measures. Therefore, it would not seem out of place to hypothesize that Manilius, even in the treatment of the celestial axis, could refer to a metaphorical horizon that conceives the Cosmos as a large building constructed and regulated by a rational Intelligence

13 "And Ixion's wheel was stayed by the still wind." (trans. Fairclough 1956: 253)

14 "[The axis] holds the earth in equilibrium and makes the sky rotate about its steady pivot." (trans. Gain 1976: 53, translation modified)

15 Ἀτάλαντος is an adjective used also by Empedocles (cf. fr. 17 DK v. 19), where it indicates the balance of Love and Strife in the Sphere. Cf. Kidd 1997: 178 and Martin 1998: 153-155; see also Hardie 1983.

16 Cf. for example *Tusc.* 5, 69 (= Arist. *De Phil.* fr. 13a) *unde terra et quibus librata ponderibus* ("Hence sprang the investigation into the beginnings... to find out... what the origin of the earth, what weights preserve its equilibrium," trans. King 1960: 497).

17 "Nor was the Earth hanging in the surrounding air, balanced by its own weight," trans. Hill.

18 In Greek, τάλαντον means "balance," and ἀ- is a copulative prefix.

19 We could also quote Manil. 1, 173, *quod ni librato penderet pondere tellus* ("and did not the Earth's weight hang poised," trans. Goold), and Sen. *Phaedr.* 973-974 *uasti pondera mundi / librata suos ducunt orbes* ("the masses of the mighty heavens trace in balance their proper orbits," trans. Fitch 2008: 499).

(see, for example, 1.247-254; cf. Volk 2009: 218-223).[20] The cosmic axis seems to be a sort of 'agent of Providence' (cf. Hardie 1983: 24), as can be inferred from the expression *mundum regere* ('to hold the world'), which is borrowed from the language of philosophy and theology. Instead, it is the instrument through which nature realizes its rational order, which is a tangible image and teleological proof of the presence of divinity in nature.

In Cicero,[21] for example, or in Quintilian,[22] in a philosophical context, the rational intelligence of the providential God controls the Universe, like the axis in the Manilian exposition. In the following verse (1.281 *sidereus circa medium quem uoluitur orbis*), the poet's attention moves to the subject of cosmic motion. Through a hyperbaton, Manilius places particular emphasis on the celestial sphere and its motion, the object of the argumentation of these verses. In fact, at 1.282, the expression *uoluitur orbis* is recalled by *rotat cursus*:[23] The whirling movement of the sphere of the sky keeps the Universe and the earthly globe in balance, like in Ovid *Fasti* 6.271 (*ipsa uolubilitas libratum sustinet orbem*, "its own power of rotating keeps its orb balanced," trans. Frazer 1931: 339). The rhetorical construction at 1.282 is particularly remarkable: We could note the anastrophes *immotus at*,[24] which puts the adjective in contact, in antinomy, with *rotat cursus*. The pause of the *hepthemiral caesura* (i.e., the pause after the seventh half-foot) accentuates the contrast between the two terms. And at 1.285, the second movement of the argumentation begins. Manilius returns to subjects he had previously explained, repeating the leitmotif of the immateriality and immobility of the axis. The style of these verses, compared with the previous ones, seems to proceed with a more didactic and argumentative course. At 1.287, the adversative conjunction *sed* follows the anaphoric repetition of *nec* at 1.285-286, where Manilius notes the axis' incorporeity. In these verses, the poet throws together a series of terms with marked material and bodily meanings, which indicate the

20 For the metaphor of the Universe as body cf. Habinek 2007.

21 Cf. *De leg*. 2.32, *Nat*. 1.4; 2.74, 77, 85; 3.6, *Rep*. 6.13.

22 Cf. *Inst*. 5.10, 14 *sic enim fiet argumentum: 'cum prouidentia mundus regatur, administranda res publica <est: sequitur ut administranda res publica> sit, si liquebit mundum prouidentia regi'* ("Thus for instance it may be argued that since the world is governed by providence, the state should similarly be governed by some controlling power, once it is clear that the world is governed by providence," trans. Butler 1921: 209).

23 Cf. Cic. *Arat*. 232-233 *uerum haec, quae semper certo uoluuntur in orbe / fixa* ("truly what revolves in a circle always remains fixed") and Varr. *Menip*. fr. 201 Cèbe *ut sidera caeli / diuum, circum terram atque axem / quae uoluntur motu orbito* ("as stars of the sky of the gods that rotate around the earth and the axis in a circular motion").

24 The nexus recalls Germ. *Phaen*. 19, *axis at immotus semper uestigia seruat* ("the axis, however, does not move but always stands in the one place," trans. Gain 1976: 53).

qualities that the axis does not possess. Over the course of two lines, the adjectives *solidus*, *grauis* ("solid, heavy") and the substantives *uis*, *corpus*, *pondus* ("force, body, weight") recur, words that are often used to represent the Universe and its mass. *Pondus* is a widely-used term from the astronomical lexicon that indicates the cosmic mass. *Corpus*[25] and *onus* (Manil. 2.896) ("mass") are also used in this sense. Therefore, the axis, which is defined by the negation of different characteristics of the Cosmos, is able to sustain what is heavier in nature, that is, the Universe.

At 1.287, Manilius returns to the argument of the axis' immobility: Similarly to the previous verses, we can here see a *coaceruatio* of terms indicating, this time, the whirling movement of the cosmic sphere as *uoluo in orbem* (cf. 1.281), *uolo*, *moueo*. Verse 1.289, therefore, summarizes what has been expressed by the previous hexameters and serves as a link with the subsequent exposure, focused, instead, on the axis' immateriality. Indeed at 1.290, Manilius repeats the adjective *tenuis* as a keyword that identifies the axis. The sense of the adjective is underlined by the consecutive clause *adeo ut* that closes, in the *Ringkomposition* (i.e., the chiastic structure of the text), the second part of the didactic exposition at 1.290-291.[26] The consecutive clause, which plays an important role in the economy of the passage, shows the consequences of the axis' tenuity: the object is so slight that it cannot turn around itself. At 1.292-293, an etiological etymology of the axis closes the complete exposition: The imaginary line—Manilius says—is called an axis because it remains immobile and sees the celestial vault whirling.[27] The static nature of the axis, which in this verse seems to be personified, is rendered through a clever play of allusions. The characterization of the movement is accentuated by the pleonastic presence of *uolito* and *moueo*, which is contrasted by a verb of sight, which seems to make the axis an immobile spectator placed at the center of the great spectacle of the Cosmos. In the passage, the visual dimension decreases because the incorporeal axis has no image; however, the author does not give up trying to describe an invisible object. The *tenuitas* ("thinness") is, in fact, a fundamental qualitative characteristic of this object: The axis, devoid of matter, is also devoid of image. The description works, however, on the level of the effects that the pivot of the sky has on the whole Cosmos: balance and movement. A double order of polarities, therefore, constructs

---

25 See, for example, Cic. *Tim.* 14 *Universi corpus* ("the world's body").

26 1.290 was deleted by Bentley, who considers it spurious because of its repetitiveness. Housman (1903), on the other hand, defends the text with good arguments.

27 A similar etymology for the axis is also present in Achilles' *Introduction to Aratus* 28, 4 ὠνόμασται δ' ἄξων διὰ τὸ περὶ αὐτὸν ἄγεσθαι καὶ περιδινεῖσθαι τὸν οὐρανόν ("is called 'axis' since the universe revolves around it and is set in motion"). Cf. Malchin 1893: 18-20 and Goold 1977: 26-27.

the entire didactic exposition: the first between the immaterial lightness of what holds the Universe and the heaviness of the matter that forms the Cosmos, the second between the immobility of the axis and the movement of the sky. The strong opposition constructed by the poet through a repetitive style, in which a few clear concepts are marked, suggests a sense of paradox to the reader. The equilibrium of the Cosmos and its movement are guaranteed by an object that has no form or weight, but that is nevertheless capable of supporting an enormous mass of matter. The paradoxical effect is underlined by the manner in which Manilius presents and describes his qualities: the axis is described through the negation of its qualities. It does not move, but sees the Universe moving (*motum non habet ullum / ipsa uidet circa uolitantia cuncta moueri*); it has no matter (*nec uero e solido stat corpore ... / nec graue pondus habet*), but holds the world (*regit mundum*).

SUBLIME: THE AXIS IN THE GREAT VOID OF THE UNIVERSE

The cosmological picture that Manilius depicts in the passage dedicated to the axis seems to suggest the idea of a grandiose world, a Universe that stands out to the reader in its enormous dimensions. In this exhibition, therefore, in order to accentuate the idea of the greatness of the Cosmos, the poet refers to a sublime imaginary, as can be seen from some textual examples. We have to consider, in this way, the expression at 1.283 *magni per inania mundi*, "through the empty space of the great sky" (as translated by Goold). The syntagma also recurs in other points of the first book of the poem, for example, in the discussion of the Universe's origins.[28] The expression, in fact, as at 1.200, indicates the empty sidereal space in which the celestial bodies are suspended and orbit around the earth. The void is a particularly important point of Stoic reflection concerning incorporeal beings. First of all, when analyzing this problem, a distinction must be made between the void inside the Cosmos and the void outside it (Todd 1982; Algra 2003; de Harven 2015). If it was necessary for the Stoa to conceptualize the void outside the Cosmos (cf., e.g., Achil. *Isag.* 8 = Crys. SVF II 610), this wasn't at all admissible inside the Universe, because an internal void would undermine the cosmic unity (cf. Cleomed. *Cael.* 1.1 = Crys. SVF II 546). J. Porter, who studied the manifestations of the sublime in ancient poetics, analyzes the

28 Cf. 1.153-154 *proximus in tenuis descendit spiritus auras / aeraque extendit medium per inania mundi* ("air next sank down to become the tenuous breezes and spread out the atmosphere midway through the empty spaces of the sky," trans. Goold 1977: 17), a passage strictly connected to Virgil's sixth *Eclogue* (see vv. 31-32).

text of Manilius, without, however, paying close attention to the passage under examination. For Porter, the treatment of the void by Manilius would be a significant sign of the presence of a "material sublime" of Lucretian origin, which is achieved through precise linguistic echoes and allusions to concepts that are, at least in appearance, eccentric to the Stoic cosmology that informs the *Astronomica*. The idea of a sidereal void, over the course of the poem, is reinforced by the image of a chasm,[29] which seems to reconcile poorly with the two-sphere cosmological system on which the *Astronomica* is based, and refers instead to an atomistic-Epicurean scenario. We can, therefore, assent with Porter (2016: 494)[30] that the poet's intent is to represent "the finite Universe in all its vast infinity." This Manilian assertion constitutes a partial inconsistency[31] with a supposed Stoic orthodoxy, which can be explained not only as a literary but also as a scientific necessity. The attention to the sublime and the connection to concepts that refer to an atomistic scenario are to be seen in the light of the allusion to a poetic dialogue with Lucretius, the Latin "progenitor" of didactic poetry. The expression *magnum per inane* ("through the big void") is a clear Lucretian verbal echo, as Rösch (1911: 77) and Porter (2016: 493-494, 504) point out. In this regard, consider *DRN* 1.1018 = 1103;[32] *magnum per inane soluta* ("dissolved through the great void," see also 2.65, 105, trans. Rouse and Ferguson Smith 1975: 85) and 1.1108 *per inane profundum*[33] ("through the empty profound," trans. Rouse and Ferguson Smith 1975: 83). Manilius uses the syntagma in the same metrical position as Lucretius, and this choice reveals the desire to refer directly to the *auctoritas* of the model. Moreover, the Lucretian quote is changed through the transfer of the adjective *magnus* to the noun *mundus*: the Manilius' *mundus* is *magnus* because contains the void. As Porter has pointed out, the intertextual reference just noted reveals not a willingness to refute or compete with Lucretius, but rather the intention to use the same expressive and theoretical tools. We should ask, at this point, if the mention of the void is a tribute to Lucretius, or if it could have a deeper meaning in terms of poetic construction and, therefore, does not constitute a simple filler in a context characterized by a marked expressive redundancy. The presence in the passage of a stylistically refined language

29 At the end of the fourth book, the poet says that knowledge of the Cosmos can be implemented only through a descent into the most intimate recesses of nature: cf. 4.877-878 *inque ipsos penitus mundi descendere census / seminibusque suis tantam componere molem* ("and penetrate to the innermost treasures of the sky; to construct the mighty universe from its component seeds," trans. Goold 1977: 75).

30 On the idea of infinity in Lucretius, cf. Morenval 2017: 155-180.

31 For the problem of inconsistency in Manilius, see Volk 2011b.

32 The expression recalls the Democritean μέγα κενόν ("big void;" cf. Bailey 1947: 777).

33 For the vocabulary of the void in Lucretius, see Schulz 1958.

(*sidereus, aetherius*—"starry, ethereal"), combined with terms from the cosmological lexicon (*libratum*), as well as the polished rhetorical structures, are all signs that should arouse our attention. The linguistic remarks just noted demonstrate the great attention Manilius paid to the topic. In order to compose an incisive didactic exposition, the author thus intends to represent the axis in its grandeur through the instruments of the Lucretian sublime. The notion of the axis has without doubt particular relevance in the description of the Cosmos, and it is for this reason that Manilius uses a language that alludes to the great model of sublime didactic poetry. The image of the void, therefore, is not out of place, but rather amplifies the prodigious ability of the axis, which is immaterial but capable of supporting the mass of the *magnus mundus*.

Support for this reading of the axis as a "sublime" object can come from the "third" translator of Aratus: Avienus.[34] The late antique author[35] translated vv. 25-27 of Aratus' *Phenomena*, rendering, in a very literal way, the ὑψόθεν ("from on high") of Aratus with an evocative *sublimis* ("sublime, high"). The term is a sign of this type of poetic language, which is coupled with other terms ascribable to a sublime vocabulary, such as *uertigo*, or *barathrum*.[36] *Barathrum* in particular is a poetic term, a calque of the Greek world βάραθρον ("abyss"), which is well suited to the idea of the void and refers to infernal and chaotic scenarios.[37] These rapid comparisons with Avienus, which have yet to be further explored, reveal, at the level of diction, how the axis was, for didactic poets, a suitable subject for a "sublime exposition." The recourse to the sublime, indeed, can be considered as proof of Manilius' attempt to compose an inspired poem, in which aesthetic purposes and theoretical instances are deeply related.[38] In the Manilian passage, we can see the presence of an "Empedo-

---

34 The passage has many linguistic correspondences with Manilius that have not yet been investigated.

35 *Arat.* 93-95 *Oceano pars / sublime erigitur, subit altera, mersa sub undas, / pars Erebum...* "one high (pole) raises on the Ocean, the other one, dipped under the waves, goes down to Erebus..."

36 *Arat.* 86-87 *sed non axis item curui uertigine fertur / aetheris* ("but the axis is not subject to the whirling motion of the ether"); 97 *ac teres in gemina stridit uertigine cardo* ("and the polished pivot creaks in the double rotation"); 98-99 *latet alter et alto / deprimitur barathro* ("one is hidden and is pressed down into the abyss").

37 The passage by Avienus, from an analysis conducted on Greek and Latin lexical repertoires (*TLL, DGE, LSJ*), seems to be the unique example of the use of *barathron* / βάραθρον in a celestial context.

38 The unity of poetic form and scientific purpose is stated in the beginning of the first book of the poem: cf. 1.20-22 *bina mihi positis lucent altaria flammis, / ad duo templa precor duplici circumdatus aestu / carminis et rerum* ("two altars with flame kindled upon them

clean-Ennian-Lucretian sublime," noted by T. Hardie (2009: 136-152) in the didactic speech of Pythagoras in Ovid's *Metamorphoses*. Manilius' *Astronomica*, ultimately, could be included in a specific literary tradition that, in the Latin world, refers to Ennius and Lucretius, in which the poetic word is the exclusive instrument for communicating the most important scientific contents. The category of the sublime is a device through which the inspired didactic poet[39] can achieve his gnoseological task, which is realized by a text that is configured as the only medium for the knowledge of the Cosmos.[40]

CONCLUSION

To describe the invisible, Manilius doesn't use an analogic process, as Lucretius does (cf. Schiesaro 1990); nor does he use a mythological process that identifies the axis with the figure of Atlas, as Virgil does (*Aen.* 4.481-482 = 6.796-797 *caelifer Atlas / axem umero torquet stellis ardentibus aptum* "heaven-bearing Atlas spins upon his shoulder the heavens' axis," trans. Horsfall 2013: 55; cf. Hardie 1983). First of all, the poet notes the immateriality and immobility of the axis and negatively defines these qualities (see the repetition in anaphora of *nec*). In the passage, the author stimulates the imaginative skills of the reader, who must strive to trace, in his mind, the form of what is incorporeal. In a second moment, the passage concentrates on the effects, qualities, and potential of the axis, through an evocative and sublime cosmological language. For this reason, the poet makes use of an imaginary seemingly foreign to his own philosophical background. The order of the Cosmos, which is realized through the axis, comes to be inserted into the context of an infinite and void Universe, different from the reassuring design of the traditional two-spheres cosmology. However, these effects of order and balance are emphasized in a repetitive didactic exposition, and by a rhetoric of paradox that operates both on the level of content and on that of form.

REFERENCES

Abry, Josèphe-Henriette (2005): "La sphéricité de la terre: un poète aux prises avec la démonstration (Manilius, Astronomiques I, 173-235)." In: Pallas 69, pp. 247-260.

shine before me; at two shrines I make my prayer, beset with a twofold passion, for my song and for its theme," trans. Goold 1977: 7). Cf. Volk 2009: 187-188, 210-212.

39 At 1.23, Manilius defines himself as *uates*.

40 Torre (2007: 58-61) has also identified this characteristic of the didactic poetry of the Augustan age in the philosophical prose of Seneca.

Algra, Keimpe (2003): "Zeno of Citium and Stoic cosmology: some notes and two case studies." In: Elenchos, Rivista di Studi sul Pensiero Antico 24/1, pp. 9-32.

Bailey, Cyril (ed.) (1947): Lucretius, De rerum natura libri sex, I; II; III edited with translation and commentary, Oxford: Clarendon Press.

Barney, Stephen A./Lewis, W J./Beach, J. A./Bergoff Olivier (eds.) (2006): The Etymologies of Isidore of Seville, Translation with Introduction and Notes, Cambridge, UK: Cambridge University Press.

Butler, Harold Edgeworth (ed.) (1921): The Institutio Oratoria of Quintilian, Vol. II, with an English Traslation, Cambridge, MA: Harvard University Press.

de Harven, Vanessa (2015): "How nothing can be something: The Stoic theory of void." In: Ancient Philosophy 35/2, pp. 405-429.

Fairclough, H. Rushton (ed.) (1956): Virgil, Eclogues-Georgics-Aeneid Books I-VI, Cambridge, MA: Harvard University Press.

Fitch, John G. (ed.) (2008): Seneca Tragedies, Volume I: Hercules. Trojan Women. Phoenician Women. Medea. Phaedra, Cambridge, MA: Harvard University Press.

Fowler, Don (2002): Lucretius on Atomic Motion: A Commentary on Lucretius: *De Rerum Natura* Book Two, Lines 1-332, Oxford and New York: Oxford University Press.

Frazer, James (ed.) (1931): Ovid, Fasti, Cambridge, UK: Cambridge University Press.

Gain, David Bruce (ed.) (1976): The Aratus ascribed to Germanicus Caesar. Edited with an Introduction, Translation, and Commentary, London: Athlone Press.

Goold, George Patrick (ed.) (1977): Manilius: *Astronomica*, Cambridge, MA: Harvard University Press.

Habinek, Thomas (2007): "Probing the entrails of the universe: astrology as bodily knowledge in Manilius' Astronomica." In: Jason König/Tim Whitmarsh (eds.), Ordering the Knowledge in the Roman Empire (229-240), Cambridge, UK: Cambridge University Press.

Hardie, Philip (1983): "Atlas and axis." In: Classical Quarterly 33, pp. 220-228.

Hardie, Philip (2009): Lucretian receptions: history, the sublime, knowledge, Cambridge and New York: Cambridge University Press

Henderson, John (2011): "Watch this space (getting round 1.215-46)." In: Steven J. Green/Katharina Volk (eds.), Forgotten Stars: Rediscovering Manilius' *Astronomica*, Oxford and New York: Oxford University Press, pp. 59-84.

Horsfall, Nicholas (2013): Virgil, Aeneid 6, a Commentary, Berlin and New York: De Gruyter.

Hübner, Wolfgang (1984): "Manilius als Astrologe und Dichter." In: Hildegard Temporini/Wolfgang Haase (eds.), Aufstieg und Niedergang der römischen Welt, II 32.1 (126-320), Berlin: De Guyter.

Kidd, Douglas (ed.) (1997): Aratus: Phaenomena, edited with introduction, translation and commentary, Cambridge and New York: Cambridge University Press.

King, James Edward (ed.) (1950): Cicero's Tusculan Disputations, Cambridge, MA: Harvard University Press.

Malchin, Franz (1893): De auctoribus quibusdam qui Posidonii libros meteorologicos adhibuerunt, Rostock: Typis Caroli Boldtii.

Martin, Jean (ed.) (1998): Aratos, Phénomènes, texte établi, traduit et commenté, Paris: Les Belles Lettres.

Morenval, Alexandra (2017): Le tout et l'Infini dans le *De rerum natura* de Lucrèce, Amsterdam: Adolf M. Hakkert.

Porter, James I. (2016): The sublime in Antiquity, Cambridge, UK: Cambridge University Press.

Postgate, John Percival (1897): Silva Maniliana, Cantabrigiae: I. & C. F. Clay.

Rösch, Hans Lüdwig (1911): Manilius und Lukrez, Diss. Kiel.

Rouse, William Harry Denham/Ferguson Smith, Martin (eds.) (1975): Lucretius, *De Rerum Natura*, Cambridge, MA: Harvard University Press.

Schiesaro, Alessandro (1990): Simulacrum et imago. Gli argomenti analogici nel *De rerum natura*, Pisa: Giardini.

Schulz, Peter-Rudolf (1958): "Das Verständnis des Raumes bei Lukrez." In: Tijdschrift voor Philosophie 20/1, pp. 17-56.

Todd, Robert B. (1982): "Cleomedes and the Stoic concept of the void." In: Apeiron 16/2, pp. 129-136.

Torre, Chiara (2006): "Tra Ovidio e Seneca. La traccia dell'Epos di Pitagora nel programma filosofico delle Naturales quaestiones." In: Alessandro Costazza (ed.), La poesia filosofica (46-61), Milano: Cisalpino.

Volk, Katharina (2009): Manilius and his intellectual background, Oxford and New York: Oxford University Press.

Volk, Katharina (2011a): "A century of Manilian scholarship." In: Steven J. Green/Katharina Volk (eds.), Forgotten Stars: Rediscovering Manilius' *Astronomica*, Oxford and New York: Oxford University Press, pp. 1-10.

Volk, Katharina (2011b): "Manilian self-contradiction." In: Steven J. Green/Katharina Volk (eds.), Forgotten Stars: Rediscovering Manilius' *Astronomica*, Oxford and New York: Oxford University Press, pp. 104-119.

Between Pre-Modernity and the Age of Enlightenment

Popular Mechanics: Hero of Alexandria from Antiquity to the Renaissance

Courtney Ann Roby

ABSTRACT

Hero of Alexandria wrote on an astonishing variety of technical topics, ranging from theoretical mechanics and geometry to applications like surveying, pneumatics, and catapult design. Throughout his corpus he asserts that his contribution will be to select the best and most reliable solutions made by his predecessors, add some novelties of his own, and assemble the whole into a form that will be maximally accessible and useful for his reader. This paper will analyze the techniques Hero uses to make his varied subject matter come alive for a broad readership, especially his use of diagram and analogy to explain the mechanics of his artifacts through reference to more familiar objects. Secondly, the paper discusses Hero's reception in the 16th and 17th centuries through humanist translations of his *Pneumatica* and *Automata* that used similar techniques to Hero's to make their material attractive and accessible.

INTRODUCTION

Hero of Alexandria, who probably lived during the first century CE, wrote treatises in Greek on an astonishing range of technical topics, from Euclidean geometry to the design and construction of theatrical automata.[1] Mechanics, as seen

1 Hero's dates have long been the subject of a controversy in which he is variously identified as a contemporary of Ctesibius, a figure from the first century CE, a successor to Ptolemy, and a figure from as late as the third century CE. The most convincing evidence seems to me to place him in the first (perhaps the second) century CE. Giovanna Giardina provides an excellent recent review of the question at

through his eyes, is a wide-ranging, intricately interwoven tapestry of subdisciplines with the power to serve practical needs as well as answer philosophical questions. His own contributions to the discipline took the form of books on a dazzling variety of subjects: surveying, catapult design, techniques for measuring objects of all different shapes, pneumatics, theoretical mechanics, and more. Throughout his works he emphasizes the importance of opening up all these subject areas to anyone who wishes to study them. He acknowledges his predecessors, but criticizes them where they left explanations obscure, or omitted introductory knowledge that a layman might need. In other words, Hero was a staunch advocate of 'popular science' *avant la lettre*. The explanatory techniques he uses to bring his technical subjects to a broad audience include the use of diagrams, employing examples drawn from everyday life, and creating opportunities for the reader to observe mechanical and physical principles at work in familiar objects.

Hero's accessible, wide-ranging texts on mechanics indeed enjoyed renewed popularity in the 16th and 17th centuries.[2] As humanist editors and translators discovered his works on pneumatic devices and theatrical automata, they found a way to link their own expanding courtly culture of technical spectacle with the classical past they sought to revive. After Federico Commandino's 1575 Latin translation of Hero's *Pneumatica*, a thriving industry of translations into Italian and other languages followed suit. These translations were complemented by the incorporation of Hero's pneumatic theories into contemporary debates over matter and void, and of his designs into new treatises on mechanical devices both practical and spectacular.

Hero's works contributed as well to the ongoing discussion about how to reconcile the "craft" aspects of mechanics with its intellectual or philosophical side. His classical pedigree allowed his texts to flourish even as the 'popular science' aspects of mechanics sometimes attracted criticism for being too closely linked to manual labor, corrupting distractions for the common people, and even witchcraft. Bernardino Baldi, a 16th-century humanist with a strong interest in

(Hero 2003: 8-30). Paul Keyser and Bernard Vitrac examine the evidence supporting the first-century date (Keyser 1988; Vitrac 2010: 1), while Nathan Sidoli takes a more cautious approach to assessing Hero's date (Sidoli 2011).

2 In fact, that popularity was never really lost in the intervening centuries, as attested by a robust tradition of Greek works based on Hero's in late antiquity, and Latin, Arabic, and Persian translations and adaptations of his works in the medieval period. However, the scope of this paper does not allow for a detailed review of those fascinating periods in Hero's own "genealogy." Marie Boas (1949) discusses the transmission of his *Pneumatica* through the medieval period, while a recent translation of the medieval Arabic and Persian versions of his *Mechanica* (Hero 2016) describes those traditions in detail.

scientific and mathematical texts from the ancient world, reports in the introduction to his translation of Hero's *Automata* that whereas mechanics in the ancient world was a noble practice that offered insight into natural philosophy, in his own time the discipline has been tainted with the 'brutishness' of profiteers who produce puppet-shows for ignorant crowds (Baldi, *De gli automati* 11r). By making Hero's work available in vernacular translation, Baldi hopes to salvage the practice of mechanics by restoring its classical nobility. Hero's broad appeal had already been experienced by Italian readers of the several translations of his *Pneumatica* that had already been produced, and Baldi's *Automata* ensured the enduring popularity and influence of this other part of his work as well. Baldi would also go on to translate Hero's *Belopoeica*, a text on constructing catapults. Hero's gift for aligning technical material with the prior experience of a broad readership made him one of the principal names in the humanist revival of ancient scientific and technical texts.

MAKING MECHANICS ACCESSIBLE

Hero does not typically represent himself as a technical innovator first and foremost. Instead, the achievement he celebrates most openly in many of his texts is opening up the subdisciplines of mechanics to a broad audience that includes beginners as well as experts. For example, he opens his *Belopoeica* by invoking the previous texts on the subject and explaining how he will improve on them:

> "Those who wrote the many treatises on belopoeica before me created measurements and designs, but not a single one of them laid out either the designs of the machines or their uses in a reasonable way, but as though they created their treatise for all those who know. So I thought it would be good to extract from them and make clarifications concerning building artillery devices, as though from scratch, so that the tradition would be easy for everyone to follow." (*Belopoeica* 2.1-14, building artillery devices)

Even though the catapults he will describe in this text are likely to have been far from cutting-edge technology in Hero's time, he sees his text as making an important contribution by making the designs of catapults understandable even to beginners. In order to do this, he starts at the beginning ("as though from scratch") and provides a kind of evolutionary history of how different types of catapults were designed to address the mounting challenges of the battlefield. These ranged from the relatively simple "belly-bow" (*gastraphétēs*, γαστραφέτης) that scaled up the hand-bow, to more complex torsion catapults that used coiled skeins of sinew cord to fire missiles further.

For each design, Hero promises to speak about the whole machine as well as its subassemblies, about how to construct and deploy it, about its measurements, and about how it was developed. In this process he makes frequent stops to delve into his devices' frames, springs, firing mechanisms, and other components. For example, he sets the scene for the development from early torsion engines to later models equipped with washers as follows:

"The result was that in the aforementioned design the twisting and stretching of the sinews did not really happen, because the cross-beams ΑΔ and ΒΓ were not able to take the spring-cord. So they superimposed connecting-pins on the bore-holes and then did the the same as those described. And again twisting the connecting-pins had a negative result, because the connecting-pin set over the cross-beam did not turn, and touched it at every point. Hence they were driven to add what are called the "washers," about which we will speak in order." (*Belopoeica* 9.1-34)

Because Hero's narrative follows mechanical developments spurred by technical demands like shooting further or raising payload size rather than giving details about the battles where these demands arose or the inventors and patrons who met them, he is free to devote close attention to the structural changes made to the machine.

Even as Hero takes the reader through the technical specifics of the catapults he describes, he remains true to his goal of easy accessibility for a broad readership. The very wealth of detail grants even a reader unfamiliar with catapult technology a comprehensive view of its most important components, taking no prior experience for granted. Perhaps more importantly, letter labels such as appear in the passage above correspond to points in a detailed diagram, similar to their use in Greek mathematical texts. The reader can thus cross-reference the text with the diagram to explore the catapult's form visually as well as verbally. While the surviving manuscripts of the *Belopoeica* date to the tenth century CE at the earliest, and are thus rather remote from Hero himself, his generous allocation of letter labels results in high levels of similarity from one manuscript to another, which may reflect high stability in the tradition of diagrams over time. While we cannot know what the ancient diagrams might have had in common with the surviving Byzantine manuscripts in terms of use of color, conventions for representing three-dimensional objects, and other graphical technologies, at least the letter labels provide a check on the components represented and their relationships to one another.[3] Hero uses letter-labeled diagrams throughout his

3 I discuss the representation of mechanical artifacts in manuscript traditions, the use of letter labels, and the relationship between mechanical and mathematical diagrams elsewhere (Roby 2016: 152-91). Wolfgang Lefèvre (2002) uses case studies of illustrations which survive in multiple forms in the manuscripts to make cautious assertions about the features which may have been present in the original illustrations. Rainer Leng (2004) analyzes the cultural pressures which might have shaped medieval

corpus, adjusting their use where appropriate to suit the needs of a more mathematical text like the *Metrica*, or texts on entertaining artifacts that invite creative interventions, like the *Pneumatica* or *Automata*.

Whereas Hero's engagement with his predecessors in the *Belopoeica* focused almost entirely on the question of the text's ease of use, in other works he gives a stronger sense that his textual interventions are closely linked to correcting the technical subject matter itself. The *Dioptra* instructs the reader about how to carry out surveying projects using the eponymous instrument, a sophisticated piece of machinery that allowed for quite precise sighting of targets and accurate measurements, as long as it was used correctly. However, Hero hints that some of his predecessors gave their readers erroneous advice about the instrument's use, and clouded even the correct information with unnecessary difficulties.

"Since the treatment of dioptrics offers many necessary uses and since much has been said about it, I think it necessary to value the things which, though passed over by those before me, offer utility; to shift what was hard to handle into ready ease; and to bring the things said falsely to correction. I do not consider it necessary to trot out the things put forth in error and difficulty by my predecessors, or those that were complete failures. For it is possible for whoever happens to wish to do so to judge the difference." (*Dioptra* 1.1-11)

The work of correction and enhanced accessibility he promises to apply to his predecessor's work is important, Hero suggests, precisely because the dioptra is such a powerful instrument that it finds applications in many different aspects of human life. It can be used to measure landscapes for constructing buildings or fording rivers; it can sight the downward course of an aqueduct in the making with high precision; it can measure the heights of enemy walls from a safe distance to guide the construction of equipment for scaling them; it is a powerful asset for scientific projects such as geographers' mapping out new lands or astronomers' tracking the locations and progress of celestial bodies. The dioptra's very universality creates the need for a reliable guide to its construction and use, so that even those whose expertise lies in the fields it serves rather than in surveying *per se* can obtain trustworthy measurements with it.

Hero's work on the design and construction of theatrical automata arguably placed the greatest strain on his ability to describe technology in terms that would be accessible to a wide range of readers. Indeed, he stipulates that automaton-making is the culmination of the study of mechanics, since it requires an understanding of so many of its subdisciplines (*Automata* 1.1.1-2). Here he

developments in representing mechanical technologies. Joyce van Leeuwen (2014), Ken Saito (2012), Nathan Sidoli (2012), and Christián Carman (2018) all discuss the vagaries of diagrams in manuscript transmission.

explains how to build two types of theatrical automaton, both driven by a slowly falling counterweight that unwinds a cord that has been complexly wound up and attached to the automaton's various moving features. One type travels along a preprogrammed path before stopping, performing a series of actions (Hero recommends a figurine of Dionysus who strikes the ground with his thyrsus, causing milk and wine to well up), and then returning to its starting point. The second type remains stationary, taking the form of a miniature theater whose doors open and close to reveal a play in multiple scenes with changing backdrops and moving figures (Hero recommends the story of Nauplius). These were complex machines, dependent on skillful craftsmanship as well as clever design for their cascade of actions to play out properly. Even in this work, however, Hero strives to make his material as easy as possible to understand. Where the Hellenistic Philo of Byzantium and Hero's other predecessors had gone wrong, he says, is by making devices that were too complex to run reliably and explaining them in unclear terms (*Automata* 20.1-4). Once again, Hero's role will be to update these devices to remove the needless difficulties for his reader (*Automata* 20.5). Even up to the very culmination of mechanics, the complicated technology of theatrical automata, Hero's priority is to transform knowledge about these devices (and the scientific principles behind their operation) from the privileged possession of a few into a widely accessible and readable form.

MECHANICS OF THE EVERYDAY

Hero's program of popularizing mechanics relies, in part, on strategies of explaining mechanical properties and artifacts to readers using analogies to more familiar objects. By exploiting readers' past experience of the structure and function of natural as well as artificial objects, he can give a relatively concise yet richly evocative introduction to less familiar processes. For example, he tasks himself in his *Pneumatica* with demonstrating to his reader that all matter is infused with tiny voids that separate particles from one another—a daunting task, particularly as he charges philosophers with having worked themselves into an endless loop of discourse on the problem without ever solving it convincingly (*Pneumatica* 1.pr.190-193). He needs to give his reader a way to visualize the interactions of individual particles of matter with their tiny neighboring voids, most unfamiliar territory indeed. He therefore begins his reader's journey into this microscopic world in a familiar place: the beach.

"The particles of the air press against one another; however, they do not coincide in every part, but have some void spaces in between, just like sand at the beach. Here the grains of sand must be understood to be analogous to the particles of the air, while the air in between the grains of sand is analogous to the voids of the air." (*Pneumatica* 1.pr.65-71)

Hero walks his reader through this introductory visualization carefully, making sure they note just how the analogy works, particularly since air is present on both sides of the analogy, a possible pitfall to comprehension. Once the reader has had a chance to imagine compressing a handful of grains of sand into one another, Hero invites them to think down to the microscopic level, gaining a more detailed understanding of the principles of matter that apply there:

"Therefore when any force is applied, the air is subject to compression and collapse into the spaces of the voids, against the nature of the particles squeezed against one another." (*Pneumatica* 1.pr.72-74)

Here he introduces a new wrinkle into the analogy: the particles of matter by "nature" have a stronger drive to regain their original positions than the grains of sand, and the compression acts contrary to that nature. So what happens next to the particles is a bit different than what happens to the sand, but fortunately the process does resemble what happens to another familiar substance:

"Upon relaxation, the air is restored to its original arrangement by the tension of its particles, just as happens to shavings of horn and dried sponges: when compressed and released, they are restored to the same space and return to their same bulk. Similarly, when any force is applied, the bodies of air stand apart from one another and the void space becomes greater than natural; then they run back toward one another." (*Pneumatica* 1.pr.75-82)

I dwell at length on this analogy because Hero does: he carefully walks his reader from the familiar to the unfamiliar, and then explains a discrepancy between the two parts of the analogy through a new analogy. The sand is a clearer example to explain the structure and compressibility of particles and void; the horn shavings are a clearer example to explain the reformative force that brings compressed particles back to their natural density.

Hero then continues to an argument that the "microvoids" between particles are not the only type of void, that an extended void may not be found in nature but can be produced artificially. Once again, his argument takes a familiar experience as its starting point, one that is probably familiar enough even to modern readers:

"Then if someone takes a vessel, as light as possible and with a narrow neck, and holding it by the mouth he sucks out the air and gets rid of it, he hangs the vessel from his lips, with the void attracting the flesh to refill the emptied-out space. Thus it becomes clear from this that a continuous void is present in the vessel." (*Pneumatica* 1.pr.86-91)

The common experience of sucking the air out of a light container to let it hang from the lip here combines a memory of a bodily engagement with a reminder of the physical principles that make it possible to convey the argument that an extended void must be present within the vessel. He then reinforces that argument with a closely related example, this one freighted with the authority of medical science:

"And it is clear in another way too. Take the medical "eggs" which are made of glass and have narrow necks: whenever they want them to be full of a humor, they empty the air from them them by mouth and, holding the neck with the finger, they point them toward the humor. When the finger has been removed, the liquid is drawn into the emptied-out space as an upward impulse develops for the liquid against its nature. And what happens in the case of the gourd-shaped vessel is no different in sense than what has been described above." (*Pneumatica* 1.pr.91-100)

Cupping glasses were widely used by physicians in the ancient world. Since at least the time of Hippocrates, Greco-Roman medicine was based on a humoral theory which called for excess humors to be purged. When the humor in question was blood, the physician would use a heated cupping glass to draw excess blood to the surface of the skin for release through venesection (*Fig. 1*). Although by the first century CE the medical scene was dominated by the 'Methodist' school of physicians, who advocated for the use of gentler treatments, cupping glasses in their various forms would have been familiar to Hero's reader, and he turns them into an opportunity to unfold the physical mechanisms behind a familiar phenomenon, as well as arguing that those mechanisms are largely independent of the form of the vessel.

Now that the reader has been introduced to the physics of compressible matter in a general way, they are ready to hear about how the four elements influence one another. Hero frames the interaction of the elements as driven by the presence or absence of fire, "as other bodies are destroyed by fire and change into smaller entities (I mean water and air and earth)" (*Pneumatica* 1.pr.106-108). When fire is introduced (ἐμβληθὲν, 104) into closed vessels, it breaks down the matter within, allowing its lighter (λεπτότερα) components to separate and flow up and away from the heavier ones. So, when coals burn into embers, they keep their original size if undisturbed, but are hollowed out to a much lower density as their rarer components have escaped in smoke (*Pneumatica* 1.pr.108-119).

Figure 1: Venesection and Cupping Glasses on the Peytel Aryballos, ca. 480-470 BCE

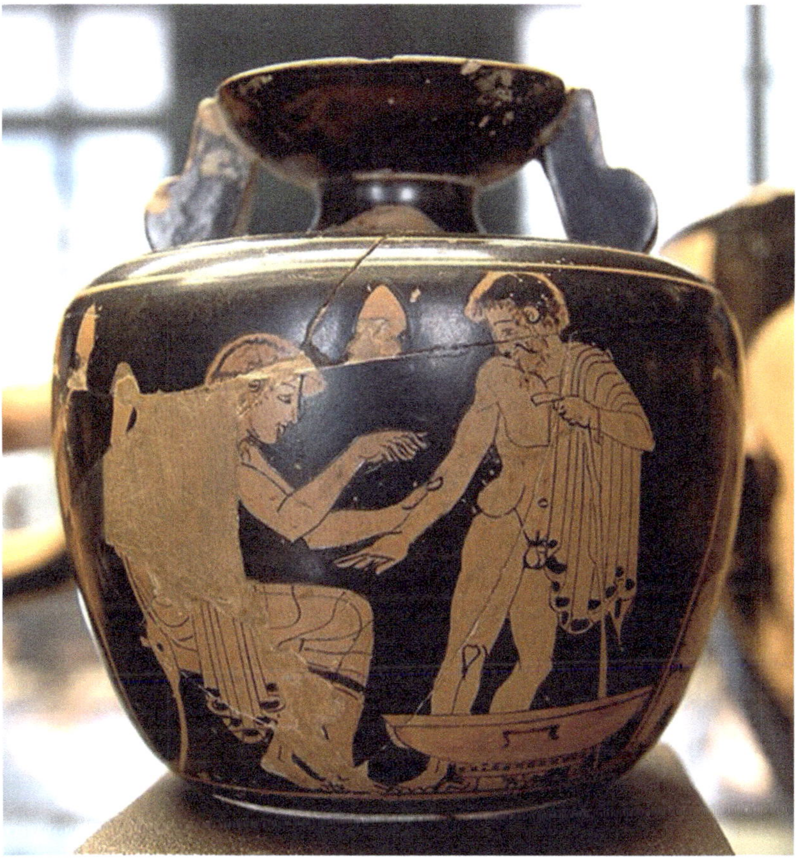

Source: Musée du Louvre, CA 1989-CA 2183; Photo: Marie-Lan Nguyen (CC by 3.0)

Here again, Hero uses familiar examples to induct the reader into the foreign territory of particles in transformation. The application of fire can break down water so its least dense components become air; this can be comprehended, he says, simply by observing how the water in a boiling kettle escapes as steam, "nothing other than refinements of the water into air" (*Pneumatica* 1.pr.121-123). Once the reader has gotten used to picturing the separation of matter into its denser and rarer parts through the application of fire, Hero expands his explanation even further, using those principles to explain that dew is a product of the "exhalation of the more refined part of the water in the earth," and that hot springs are produced by the heat of the sun seeping through the earth and separating off the "sulfurous and bituminous" matter that is particularly susceptible to rising up when heated (*Pneumatica* 1.pr.127-133).

As his discussion expands, Hero is certainly not averse to mining new layers of explanation from previous examples. The "gourd-shaped vessel" from the early part of the preface in fact reappears later on, once the reader has been equipped with an understanding of fire's effects on matter:

"The same thing happens when the air in the "gourd-shaped vessel" has been destroyed and refined by the fire and seeps out through the interstices of the wall: the interior space, having been emptied out, draws back the remaining matter, whatever it happens to be. And when the gourd-shaped vessel admits it, air flows back to the emptied space, and then none of the matter will any longer be drawn back." (*Pneumatica* 1.pr. 183-190)

Hero's preface thus teaches his reader to see even commonplace objects in a new way, illuminating the foreign territory of microscopic elemental activities using familiar substances and objects, from the sand at the beach to the boiling kettle at home to the cupping glass of the doctor's office. Later sections of the preface refer back to objects already introduced, using them as a familiar backdrop against which to unfold new physical principles.

The more complex devices described in the main body of the *Pneumatica* recall the physical principles introduced in the preface in various ways. Hero's account of the "hydrological cycle" in which air and water are transferred around the earth through heating and cooling is conjured up by two devices in the second book. The first is a small ball which floats at the center of a glass sphere half-filled with water, intended to represent the earth's position at the center of the cosmos (*Pneumatica* 2.7), which Aage Drachmann (1948: 79) described as "unintelligible." Indeed, the design is rather unusual for the *Pneumatica*, being static rather than set in motion by compression or heating, but in fact the 'model cosmos' is a neat little demonstration of the principles of buoyancy and of the relative density of water and air already introduced in the preface. The "cosmos" is immediately followed by the "spring," a globe partly filled with water, mounted on a pedestal to which it is connected by a tube, and having an outlet siphon running back into the pedestal through a funnel. When the globe is set in the sunlight, the air inside heats up and pushes out the water through the siphon into the pedestal. When it is put in the shade, as the air condenses the tube draws the water back up again. Hero emphasizes that the cycle will repeat as often as the globe is moved from sun to shade and back again, making the device an elegant mirror for the repeated cycle of heating and cooling, rarefaction and condensation of liquid that the sun exerts upon the earth itself.

The cupping glasses from the preface recur in the second book, where Hero proposes a new model that does not require the application of heat. Instead, this device includes a compartment which can be sucked free of air and then sealed on the outside, partitioned off from a second compartment which is placed against the patient's skin. When the partition is opened up, the air from the second compartment partially vacates the space, allowing the cupping glass to perform its function without heat. This novel cupping glass is followed up with another device that conjures up the physician's office, a syringe that can be used either to draw off pus or other unwanted liquids or to perform injections. Hero introduces the syringe by noting that it shares not only its medical context of use but its operating principle with the heatless cupping glass (*Pneumatica* 2.18.1).

Likewise, the preface's description of oil lamps whose flames gutter and burn out when they run out of fuel (*Pneumatica* 1.pr.168-176) prepares the reader for new and improved lamps that use the principles of pneumatics to maintain a steady oil supply. The first is a lamp that automatically extends its wick as it burns, which works by means of a small rack and pinion geared device attached both to the lamp and to a dish that floats upon the oil in the lamp (*Pneumatica* 1.34). As the oil supply falls, the rack falls with it, drawing the pinion around to unwind more wick as needed. Interestingly, while Hero does not specify the form of the lamp in the text, in the earliest surviving manuscript of the *Pneumatica* it is portrayed as a decorative lamp in the shape of a bird. The playful, decorative aesthetic of the wonders of the *Pneumatica* here makes itself visible even in practical devices like this.

The second book of the *Pneumatica* features a series of three lamps (*Pneumatica* 2.22-24). The first is a lamp mounted on a tall stand, hollowed out to accommodate a vessel of oil and a connecting system of pipes, so that as the oil supply burns down, new oil can be drawn in from the vessel to replace it by means of a valve. The second again includes a stand that can hold oil, but this time the stand includes a small pipe through which air can be blown by mouth to replenish the lamp's oil supply. As in the case of the syringe, Hero emphasizes that the operating principle of the second lamp is already familiar to the reader from the first one, and indeed that the second type is easier to construct (*Pneumatica* 2.23.1-2). The final lamp in the series allows the user to pour in water to raise the internal oil level as needed. These three lamps are all designed to make the physical principles expressed in the preface visible in practical devices enlivened by their decorative form, delightful spectacle, and interactive display of scientific principles.

The artifacts described in the *Pneumatica* are thus important sites for observing physics in action. Yet the "serious" didactic work they do does not obscure the fact that they are also beautiful, playful spectacles: indeed, the compulsion a viewer feels to gaze on these inventive devices becomes itself a spur toward study. Practical devices like siphons appear in the *Pneumatica* alongside automata that play trumpets, sing like birds, and drink water (in one case even after being decapitated). Many of the devices are trick vessels for combining wine and water in various playful ways, suggesting a sympotic context for their exhibition. Some even staged small dramas: birds crowded on a branch sing merrily until an owl turns to look at them, and then resume when the owl turns away; temple doors open up when a fire is lit on their altar; and Hercules fires an arrow at a serpent guarding an apple tree, an allusion to his labor of collecting the golden apples of the Hesperides.

Pneumatic wonders like these would have been rare and expensive in antiquity, but Astrid Schürmann (2002) has persuasively argued that the garden dining rooms of some Roman villas appear to be plumbed so as to support installing them. These elite dining rooms were often decorated with frescoes painted with naturalistic images of plants and birds, creating an elegant play with ideas of 'natural' and 'artificial' that would have been greatly enhanced by an automatic tableau like the singing birds and owl. Such a device would have been quite at home amid the other rarefied displays of playful erudition, such as poetry readings and dramatic performances, presented to elite Roman diners.

HERO AND THE HUMANISTS

Centuries later, Hero was 'rediscovered' by the Italian humanists, whose revival of classical texts had begun in the 14th century and continued to flourish in Italy and beyond well into the 17th century. The Italian humanist Ronald Witt argues that Italy was a natural wellspring for the movement, given the preponderance of secular urban communities politically organized into city-states, who naturally saw something of themselves in the classical world (Witt 1988: 52-53). While most of the early humanists' efforts focused on Latin texts, increasing contact with the Byzantine world (particularly the westward migration of Byzantine scholars following the Turkish conquest of Constantinople) created a market for Greek texts as well, to be translated into Latin and vernacular languages (Kristeller 1988: 10-14). The rise of print culture drastically increased the potential readership for new editions and translations of classical texts, which shaped the development of new library collections throughout Europe, including works both old and new, in manuscript and print, spanning disciplines from literature to science.

Mechanics became a particular focus in the ducal city-state of Urbino, and Hero a central representative of its practical side (whereas Archimedes and Pappus represented its theoretical and mathematical aspects).[4] The 15th-century dukes Federico III da Montefeltro and Guidobaldo da Montefeltro had made their mark as staunch supporters of scientific, technical, and artistic work. Federico had been particularly supportive of mathematical work, both on its own merits and as the foundation of ar-

4 Domenico Bertoloni Meli (1992) describes the diversity of projects carried out by authors in the so-called "Urbino school;" Jessica Wolfe (2004: 29-55) discusses the role played by automata and other mechanical creations at the Urbino court; and Alexander Marr (2011) focalizes his study of the balance between mathematics and its practical applications through the 17th-century Urbino polymath Mutio Oddi.

chitecture (Marr 2011: 29-32). Their successors (plus or minus some interference from the Borgias and de' Medici), Francesco Maria della Rovere, his son Guidobaldo II della Rovere, and his grandson Francesco Maria II della Rovere, carried over these principles to the very end of the Urbino dynasty in 1625, when it became part of the Papal States. By the 16th century, mechanics, and military engineering in particular, had become a special focus at the Urbino court. Bernardino Baldi, a polymathic scholarly fixture at the ducal court, translated Hero's *Belopoeica* as well as his *Automata*, reflecting the court's appetite for knowledge about artillery and fortifications (outdated though Hero's catapults are, as Baldi acknowledges) as well as elegant entertainments.[5]

Baldi and Hero's other humanist translators discovered that his didactic accessibility and the elegant devices his writings describe suited their project of reviving classical texts for an elite readership. His explicit concern for the needs of his readership, and in particular his focus on making his work useful, made him an adaptable figure for later translators seeking to position his technical work (and by extension theirs) in new intellectual, philosophical, and cultural terrain. Federico Commandino, who served as the court mathematician to Francesco Maria della Rovere, produced the first Latin translation of Hero's *Pneumatica* in 1575, Giovan Battista Aleotti used Commandino's translation as the basis for an Italian translation printed in Ferrara in 1589, and Alessandro Giorgi of Urbino published his own Italian translation in 1592.[6] Commandino's student Bernardino Baldi, after his elevation to the position of abbot of Guastalla, translated Hero's *Automata* in 1589 and his *Belopoeica* in 1616. Beyond translations of his own works, Hero (and his pneumatic work in particular) appeared as a supplementary figure in works from Giorgio Valla's 1501 *De expetendis et fugiendis rebus opus* to Salomon de Caus's 1615 *Les Raisons des Forces Mouvantes*.

The translation and propagation of Hero's works in this period involved a whole new transformation of the technologies he describes into 'popular science.' Both the ancient technologies and the texts that described them needed to be revived for a new reading audience. The first step of this revival was the translation itself, from Greek into more widely accessible languages like Latin and various vernaculars. Though Commandino was a pioneer in translating Hero, he does not tell us anything about his process, as his translation only ap-

5 Alfredo Serrai (2002) presents a quite comprehensive guide to Baldi's life and work, including his own literary projects as well as his humanist translations, while Enrico Gamba (2005) focuses in particular on his engagements with ancient and contemporary science.

6 Additionally, Matteo Valleriani (2015: 46-48) discusses the 1582 Italian translation in manuscript by Oreste Vannoccio Biringuccio, available at http://echo.mpiwg-berlin.mpg.de/content/pratolino/sources/Heron_Alexandrinus.

peared soon after his death. The book was dedicated to the cardinal Giulio della Rovere, brother of Commandino's patron Guidobaldo della Rovere, and a brief preface addressed to the patron by Commandino's brother-in-law Valerio Spaccioli explains that any flaws remaining in the text are due to Commandino's premature death, before he had time to edit the text carefully.

Later translators, however, included more explicit notes about the struggles they encountered in their work. Alessandro Giorgi, whose Italian translation of the *Pneumatica* was beaten to the press by Aleotti's by three years, advertised his access to a better selection of Greek manuscripts than Commandino (and therefore Aleotti) had had. Hence he is particularly solicitous about explaining the choices he made as a translator in "annotations" that follow Hero's descriptions of each device. The water organ, a device that used water and compressed air to evoke sounds from pipes of different lengths, was complex enough to require considerable explanation. For the most part Giorgi concentrates his commentary on the terminology of his translation, taking his reader on an elaborate linguistic journey. For example, the pipes of the organ terminated in small individual compartments for the compressed air that produces the organ's sound. Hero calls these compartments *glōssókoma* (γλωσσόκομα), a word that primarily refers to a small case for holding the reeds or "tongues" of woodwind instruments. Aleotti explains the term as follows:

"'Glossocoma' is the Greek word 'glōssokomon,' and can be entirely expressed with another Latin word, as well as a Tuscan one, in an old Latin translation which came my way without an author's name. In all these places where the Greek has 'glossocoma' it is translated 'lingusa,' but properly that means the case where the little tongues are held. It is likewise found used by authors in other meanings quite different from this, applying it sometimes for a vessel, sometimes for a bag, sometimes for a basket." (Giorgi *Spiritali* 78r)

Bernardino Baldi's translation of the *Automata* includes a note on this word as well (*De gli automati*, 44r). Baldi furnishes some additional details: the cases are commonly used by tailors to hold chalk, thimbles, and other small necessities, while the late-ancient medical encyclopedist Oribasius uses the same term to describe a bone-setting machine in a case. Baldi further reports that he himself always thought the name came from its shape (which like the human tongue is wider on one end than the other), but upon further etymological research discovered, like Aleotti, that the "tongues" in question belonged to musical instruments. Baldi thus opens the text up to his reader not only through a comparison to familiar items in the contemporary world, but also through his own story of philological discovery.

Giorgi's next note explains the shape of another component of the water organ, a little tile of horn used to cover up the apertures of inactive pipes. This time, rather than taking the reader on a linguistic excursion, he likens the unfamiliar component to an object that readers may have encountered themselves:

"'Spatula:' the Greek says 'spathion,' which signifies a little knife or scalpel, and also sometimes an instrument of surgery; we have translated it 'spatula,' which is a piece of wood whittled to the shape of a spade (somewhat larger in the middle, and sharp on the sides) such as ladies use to adjust flaxen thread, and we call it by this name 'spatula.'" (Giorgi *Spiritali* 78r)

Particularly given that Giorgi (like Commandino and Aleotti before him) follows the manuscript tradition of illustrating each device with a single image, rather than deviating from tradition to provide close-up illustrations of individual components, invoking a familiar object like this is a most useful strategy for giving a non-technical reader insight into the various components of these devices.

Giorgi's final "annotation" on the water organ acknowledges that perhaps his translation does not quite make the design of the device crystal clear; this, he says, is owing not just to the corruption of the manuscript text, but also the remoteness of the technology:

"Beyond the corruption of the texts, from which it is not possible to extract good sense, the difficulty comes from the material itself: because these devices are not in use in our time, their artifice cannot be completely understood. If Hero had been forced to write as plainly about its parts as the extravagance of the machine requires, those who have experience of the modern organs would be able to understand many things more than others (notwithstanding the ancient and disused vocabulary) from the similarity of their parts." (Giorgi *Spiritali* 78v)

This appears to be a rare failure of Hero's mission to make the devices he describes broadly accessible; not only his remoteness from his humanist translators, but something opaque in his own writing, leaves a gap in comprehension that Giorgi says is better filled by practical experience of the machinery itself than by close reading of the text. Ultimately, Giorgi refers readers with additional curiosity about the water organ to Daniele Barbaro's translation of Vitruvius' *De architectura* (1556, revised in 1567; Vitruvius 1567). Even though Vitruvius' account is simplified compared to Hero's, Giorgi says, Barbaro's commentary fills in some of the blanks.

In Baldi's translation of the *Automata*, he likewise encounters some terminology likely to be unfamiliar to his readers not only in the original Greek, but in any vernacular equivalent he could devise. The Greek *hýsplēx* (ὕσπληξ) or *hysplēngion* (ὑσπλήγγιον), for example, generally means a trap or snare, but in this context refers to a strand of material that stores energy by being twisted up, like the sinew spring-cord that powers a catapult. Small versions of this are used to snap elements of the automaton quickly from one place to another, like a lightning bolt Athena fires down in the stationary automaton's play of Nauplius. Baldi's explanation of the term, which he simply transliterates as *hysplengio*, puts his own knowledge on display even as he builds a bridge to his reader's level:

> "'The hysplengio': this word has many and various meanings among the Greeks. It means 'whip' (that is, a scourge), or the halter stretched in front of those who run a race, which the Latins called 'carceri,' or it means the pricking which goads oxen (though according to the etymology of the word, it is rather shown that it is a whip for pigs). But nevertheless in this place it has none of these senses, but denotes a little wooden rod similar to the metal one that locks a door, which we call 'saltarello' and in Tuscany they call 'saliscende.'" (*De gli automati*, 43r)

Baldi observes that Hero's explanatory note that the automaton's *hysplēngion* works like that of a catapult is unlikely to be helpful to his audience, thanks to the intervening centuries of technology change. Now that the understanding of catapults is "only a little less than lost" (Baldi's translation of the *Belopoeica* would not appear for another quarter century), Baldi merely points to the fact that the reader does not actually need to understand the details on catapult springs Hero gives in order to understand the automaton, and abandons further discussion. It might be noted that Baldi's explanatory note is here somewhat murkier than it might be: there is no use of the Greek *hysplēngion* that matches up to the *saltarello* door-latch, and his thinking here appears to be based on a similarity in their springiness rather than their material.

As Hero's works were updated by his many translators for an early modern readership, so were the explanatory techniques they used to make his technologies, and the physics behind them, comprehensible for a broad audience including laymen. Aleotti follows his translation of Hero's preface with some additional discourse of his own on the impossibility of an extended void existing in nature and on the incompressibility of elemental air. Here he adapts Hero's strategy of using familiar objects to make the principles of pneumatics come alive for his own readership, who of course occupy a somewhat different world of everyday objects than existed in Hero's time. Some of the objects Hero mentions are no longer familiar, while new technologies have appeared that illustrate pneumatic physics in new ways. First Aleotti makes some observations on the effects of compression and evacuation on a musket. If a ramrod that exactly fits the dimensions of the barrel is pushed in and the vent (*fogone*) is covered up, a vacuum will be created that makes it difficult to remove the ramrod; if the vent is first covered and then the ramrod inserted, it will be difficult because the air inside had to be compressed.

He then abruptly shifts from artillery to maternity, picking up the physical principles suggested by Hero's cupping glasses with an updated technology. Though cupping glasses were no longer in vogue by Aleotti's time, a sort of breast pump had been invented whereby women who had recently given birth could siphon off milk in excess of that consumed by their infants. Aleotti describes two variants of this device, both made of glass (Aleotti, *Artifitiosi et Curiosi Moti Spiritali*, 8). The first is made with a body designed to fit over the nipple, and a neck long enough for the woman to reach it with her mouth, "from which they suck the air which is in the vessel, and the milk follows suddenly in its place, escaping out of the breast." The second is a small vessel (*ampolla*) of glass which is heated before being placed over the nipple. In Aleotti's account, the heated, rarefied vapor within the vessel escapes upwards through the tiny pores in the glass; as the pores are too small to admit air from outside, this leaves behind a void which is filled by drawing milk from the breast. The functional principle of this second type, says Aleotti, is the same as in kilns, where the fire creates a vacuum toward which the fire and the heated vapor it produces are both drawn. If the fire is situated at the mouth of the furnace, he argues, the heavier, colder outside air cannot enter the furnace by moving over the lighter, warmer vapor within, so the fire continues to create vapor until it has fully evaporated, leaving a vacuum so that the fire is then drawn up and inside the furnace.

The changing technological environment reflected in Aleotti's updated mechanical analogies in turn provided opportunities for new pneumatic designs, and indeed Aleotti follows up his translation of Hero's *Pneumatica* with four new "spectacles (*theoremi*)" of his own. The first of these is in fact a reworking of Hero's piece where Hercules fires arrows at the dragon guarding the golden apples. In Aleotti's version, Hercules instead strikes the hissing dragon with a club (which it must be said is much closer to Hercules' usual style), at which point the dragon retaliates by spitting water at him. The mechanism makes use of a cone valve, a device introduced into the pneumatic tradition established by Philo and Hero only after its transmission to the medieval Islamic world; these valves characterize the innovations of the Banū Mūsa, three brothers who composed their *Book of Ingenious Devices* in Baghdad in the ninth century.[7] Even though Hero's humanist translators privilege the Greek tradition and ignore the medieval Arabic and Latin traditions that came between, the novel technologies developed within those traditions nevertheless make themselves felt in new devices like Aleotti's updated Hercules tableau.[8] Aleotti's other new *theoremi* blaze a new trail: a figure of Triton rises from the depths of a basin of water to blow a trumpet and then sink down again (in a device that shares some mechanical

7 Donald Hill translated this Arabic work into English along with extensive explanatory notes and background information (Banu Musa Bin Shakir 1978), and elsewhere discusses the distinctive association of the Banū Mūsa with the cone valve (Jazarī 1974: 9).

8 Elly Truitt (2015) discusses the medieval traditions of automaton-making in the Western and Islamicate worlds.

components with Hero's owl-and-birds tableau); four blacksmiths gather around an anvil to hammer a piece of metal as water spritzes up to imitate the flying sparks of a smithy; and finally, a room is constructed with pipes running below its floors, driving compressed air so that those in the room may be refreshed by breezes flowing from figures of the winds mounted around the walls.

Figure 2: Frontispiece of Salomon de Caus' Les Raisons des Forces Mouvantes, *1615*

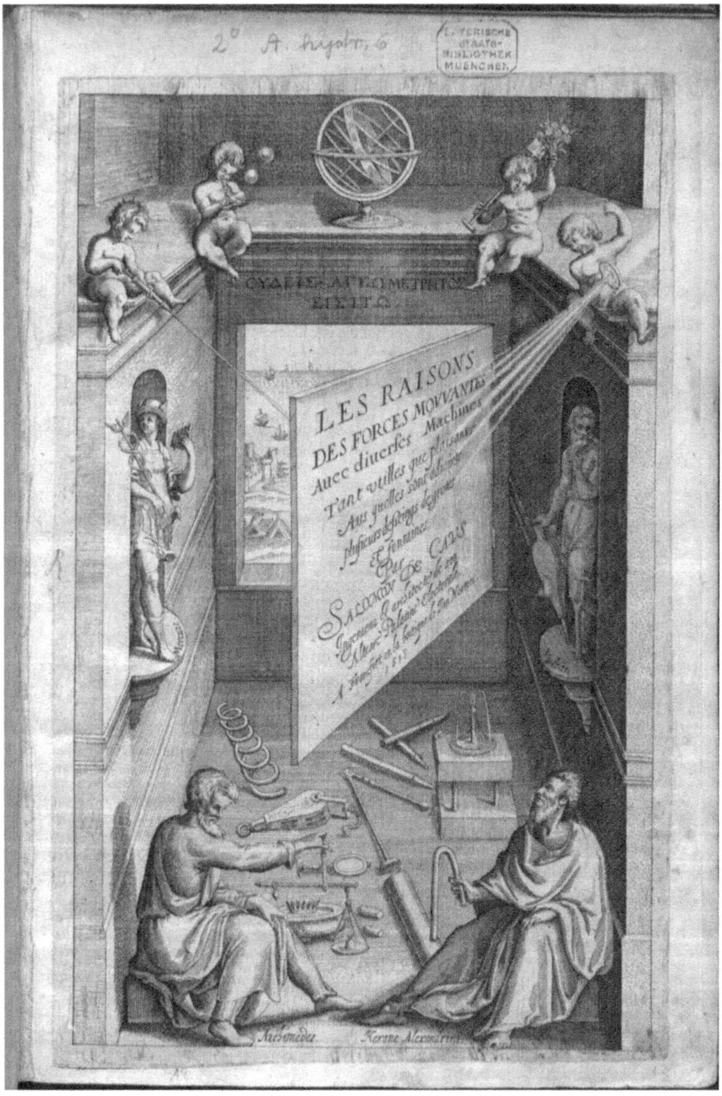

Source: Bayerische Staatsbibliothek München, Res/2 A.hydr. 6-1/3#1, frontispiece, urn:nbn:de:bvb:12-bsb00103174-8

Aleotti's refreshing air-cooled room represents a way for pneumatic devices to delight not only the eyes and ears of elites, but their whole bodies. Over the course of the 16th and 17th centuries, the application of pneumatic technology not just to miniature devices and automata, but to immersive environments, became increasingly popular in the gardens and grottoes of wealthy patrons in Italy, France, and elsewhere. The French landscape architect Salomon de Caus was the most prominent representative of the vogue for developing pneumatic wonders as points of interest for gardens and grottoes.[9] His *Raisons des Forces Mouvantes* (1615) builds up a pneumatic toolbox, starting from Hero and working toward a large array of pneumatic devices both useful and entertaining. Its frontispiece (*Fig. 2*) features a portrait of Hero working alongside Archimedes, Archimedes with a balance and Hero with a siphon, surrounded by the products of their labor and ingenuity. Archimedes has his water screw, the multiple-pulley system called the *polýspaston* (πολύσπαστον), a burning mirror, and the crown whose alloy composition he is meant to have discovered in the bath.[10] Hero, on his side of the room, is flanked by a pneumatic piston, a device from the *Pneumatica* that makes a little ball dance on a jet of water, and the pipes of a water organ.

De Caus' garden designs were only part of a larger culture of elaborately engineered landscapes in this period, and (like Aleotti, and Hero himself) he was clearly inspired by earlier efforts. Notable among these was the sprawling garden of Pratolino, just north of Florence, which de Caus strove to surpass with his "Parnassus" at Somerset House in London (Morgan 2007: 115-23). Originally commissioned by the Grand Duke Francesco I, whose interests in engineering, metallurgy, and alchemy appeared perennially more attractive to him than the work of governing the duchy. Much of the design of the garden was delegated to the talented Bernardo Buontalenti, whose similarities to Hero have themselves been remarked on by Matteo Valleriani (2014: 138).

The garden once featured a wide variety of pneumatic devices to enliven the experience of visitors, from relatively staid elements like fountains and automata to features that would surprise guests by spraying them playfully with water. While the remains of these devices do not survive today, Valleriani has reconstructed much of the content of the garden from documentary evidence (2014: 142-47). Most important here, he shows that indeed Buontalenti used Hero's *Pneumatica* as a model for the design of the garden, drawing not only on the basic technologies Hero describes but on their principles of aesthetics and composition. Several of his designs closely mirror devices in Hero's text: the drink-

9 Luke Morgan (2007) offers a detailed study of de Caus' garden designs and his debt to classical culture.

10 Mary Jaeger (2008) examines all these stories and their afterlives in Roman culture.

ing animals, the water organ, and even the owl-and-birds tableau were all constructed for Pratolino (Valleriani 2014: 147-54). Moreover, some of the adaptations of Hero that Aleotti had conceived of a few years earlier seem to have made it into the garden as well; Valleriani cites the resemblance between the "Forge of Vulcan" and Aleotti's design of the smithy (Valleriani 2014: 163-64), and Giovanni Guerra's designs for the garden include a grotto featuring a Hercules tableau which strongly resembles Aleotti's first design (*Fig. 3*).[11]

Figure 3: Aleotti's "Hercules" Theorema, Gli Artifitiosi Et Cvriosi Moti Spiritali Di Herrone *(1589), Detail at Page 89*

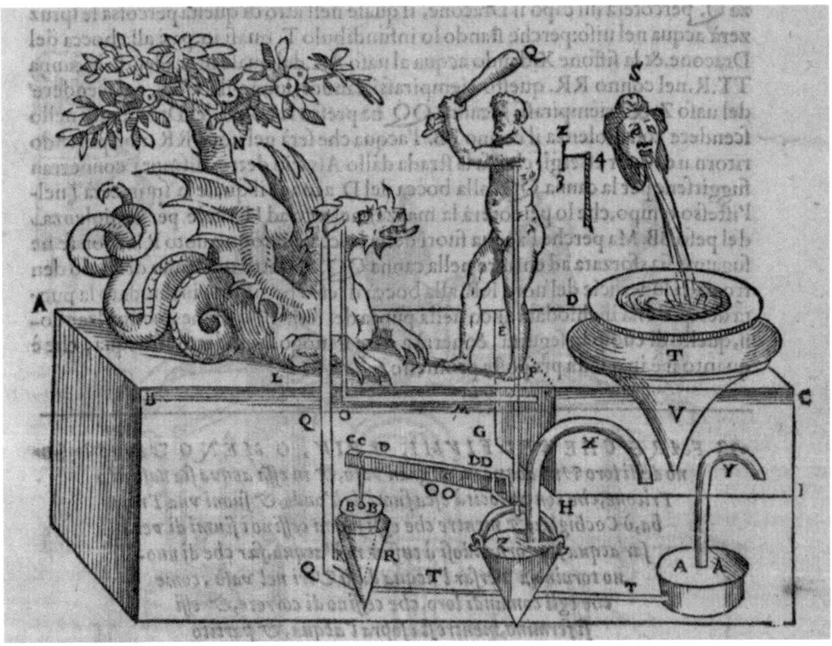

Source: Bayerische Staatsbibliothek München, 4 A.gr.b. 698, p. 89, urn:nbn:de:bvb:12-bsb10150779-4

The humanist revival of Hero's works thus led to concrete effects in the world quite beyond most other texts from antiquity. The principles of pneumatics that Hero strove to make accessible to a wide audience fueled the exploration of physics in the Renaissance and the early modern period. At the same time, Hero's analogical explanations, which emphasize that physical principles can be observed at work both in everyday objects and finely crafted automata, drove the creation in

11 Guerra's grotto design can be viewed at http://echo.mpiwg-berlin.mpg.de/ECHOdocuView?url=/permanent/echo/pratolino/Guerra_Albertina/index.meta&pn=9.

the 16th and 17th centuries of pneumatic technologies both old and new, designed to make those physical principles live again before the eyes of a new audience.

CONCLUSION

The journey taken by Hero's pneumatic devices, and the texts that carried them, from Greco-Roman antiquity to the early modern period are testament to the enduring success of his mission to make science and technology accessible to a broad audience. His frequent use of extensively letter-labeled diagrams allowed readers to explore his machinery visually at the same time as they studied his verbal descriptions. Even while those images changed to reflect the shifting graphical paradigms between the 'flat' drawings of the medieval manuscripts to the perspective drawings of the early modern printed books (and had probably changed already during the ancient transition from papyrus rolls to codex form), their close ties to the text made them a perennially useful didactic aid.

The reader's ability to visualize even the exotic realm of elemental physics was likewise aided by Hero's deft comparisons of the activities of microscopic particles to familiar experiences from the real world: picking up a fistful of sand at the beach, feeling the shift from the sun's warmth to the dew-producing overnight cool, or being treated with a heated cupping glass at the doctor's office. These 'hooks' from the unfamiliar to the familiar, the abstract to the practical, in turn provided models for Hero's translators to do the same. Even though many features of their technological landscape had changed—catapults had given way to cannons, cupping glasses to breast pumps, and so on—translators like Aleotti and Baldi were able to find ways to relate those new technologies to the invariant physical principles that drove them.

Hero's richly interdisciplinary approach to mechanics built bridges between some very different contexts of technical practice, spanning the utility of the dioptra for surveying problems, the competitive struggle of catapult design, and the refined elegance of the pneumatic devices (and the automata that built upon their principles). In creating a brand of mechanics that would be intelligible for practitioners of these disciplines as well as interested lay readers, he simultaneously bridged the gap between the esoteric intricacies of individual disciplines and the everyday experiences shared by all readers, turning analogical connections between the two into didactic opportunities. Hero embodies a kind of "popular science" that aims to teach first and foremost, to entertain where possible, and to be flexible enough to be applicable to a variety of disciplines—and, as it happened, for a variety of historical contexts. His efforts to connect mechanics to mathematics and philosophy fueled his longevity as well, as his style of mechanics won the approval of his humanist translators even in an intellectual environment where 'popular' practical mechanics had fallen into disrepute. Just as Hero had become the voice of his predecessors, incorporating their methods and designs into his highly readable works, so he himself was incorporated into new works that made his technologies shine again for a new audience.

REFERENCES

Banu Musa Bin Shakir (1978): The Book of Ingenious Devices, New York: Springer. (Translation, introduction, and notes by Donald Hill)

Bertoloni Meli, D. (1992): "Guidobaldo dal Monte and the Archimedean revival." In: Nuncius 7, pp. 3-34.

Boas, M. (1949): "Hero's Pneumatica: A Study of Its Transmission and Influence." In: Isis 40, pp. 38-48.

Carman, Christián C. (2018): "Accounting for overspecification and indifference to visual accuracy in manuscript diagrams: A tentative explanation based on transmission. Hist." In: Math. 45, pp. 217-236.

Drachmann, Aage G. (1948): Ktesibios, Philon and Heron; a study in ancient pneumatics, Copenhagen: Munksgaard.

Gamba, Enrico (2005): "Bernardino Baldi e l'ambiente tecnico-scientifico del Ducato di Urbino." In: E. Nenci (ed.), Bernardino Baldi (1553-1617) studioso rinascimentale: poesia, storia, linguistica, meccanica, architettura: atti del convegno di studi di Milano, 19-21 novembre 2003: FrancoAngeli, pp. 339-351.

Hero (2003): Erone di Alessandria: le radici filosofico-matematiche della tecnologia applicata: Definitiones: testo, traduzione e comment, Catania: CUECM.

Hero (2016): The Baroulkos and the Mechanics of Heron, Firenze: L.S. Olschki.

Jaeger, Mary (2008): Archimedes and the Roman imagination, Ann Arbor: University of Michigan Press.

Jazarī, I. ibn al-Razzāz (1974): The book of knowledge of ingenious mechanical devices, Dordrecht: Reidel. (Translation, introduction, and notes by Donald Hill)

Keyser, Paul (1988): "Suetonius Nero 41.2 and the date of Heron Mechanicus of Alexandria." In: Class. Philol. 83, pp. 218-220.

Kristeller, Paul Oskar (1988): "Renaissance humanism and classical antiquity." In: A. Rabil, (ed.), Renaissance Humanism: Foundations, Forms, and Legacy: University of Pennsylvania Press, pp. 5-16.

Leeuwen, Joyce van (2014): "Thinking and Learning from Diagrams in the Aristotelian Mechanics." In: Nuncius 29/1, pp. 53-87.

Lefèvre, Wolfgang (2002): "Drawings in Ancient Treatises on Mechanics." In: Giuseppe Castagnetti/Annamaria Ciarallo/Jürgen Renn (ed.), Homo Faber: Studies on Nature, Technology, and Science at the Time of Pompeii: "L'Erma" di Bretschneider, pp. 109-120.

Leng, Rainer (2004): "Social Character and Pictorial Style, and the Grammar of Technical Illustration in Craftsmen's Manuscripts in the Late Middle Ages." In: W. Lefèvre (ed.), Picturing Machines 1400-1700: MIT Press, pp. 85-111.

Marr, Alexander (2011): Between Raphael and Galileo: Mutio Oddi and the mathematical culture of late Renaissance Italy, Chicago: University of Chicago Press.

Morgan, Luke (2007): Nature as model: Salomon de Caus and Early Seventeenth-Century Landscape Design, Philadelphia: University of Pennsylvania Press.

Roby, Courtney Ann (2016): Technical Ekphrasis in Greek and Roman Science and Literature: The Written Machine between Alexandria and Rome, New York; London: Cambridge University Press.

Saito, Ken (2012): "Traditions of the diagram, tradition of the text: A case study." In: Synthese 186, pp. 7-20.

Schürmann, Astrid (2002): "Pneumatics on Stage in Pompeii: Ancient Automatic Devices and their Social Context." In: G. Castagnetti (ed.), Homo Faber: Studies on Nature, Technology, and Science at the Time of Pompeii: "L'Erma" di Bretschneider, pp. 35-56.

Serrai, Alfredo (2002): Bernardino Baldi: la vita, le opere, la biblioteca, Milano: S. Bonnard.

Sidoli, Nathan (2011): "Heron of Alexandria's Date." In: Centaurus 53, pp. 55-61.

Sidoli, Nathan/Saito, Ken (2012): "Diagrams and arguments in ancient Greek mathematics: Lessons drawn from comparisons of the manuscript diagrams with those in modern critical editions." In: Karine Chemla (ed.), The History of Mathematical Proof in Ancient Traditions: Cambridge University Press, pp. 135-162.

Truitt, Elly R. (2015): Medieval robots: mechanism, magic, nature, and art, Philadelphia: University of Pennsylvania Press.

Valleriani, Matteo (2014): "Ancient pneumatics transformed during the early modern period." In: Nuncius 29, pp. 127-173.

Valleriani, Matteo (2015): "Sixteenth-century hydraulic engineers and the emergence of empiricism." In: T. Demeter, K. Murphy, and C. Zittel (eds.), Conflicting Values of Inquiry: Ideologies of Epistemology in Early Modern Europe: Brill, pp. 41-68.

Vitrac, Bernard (2010). Héron d'Alexandrie et le corpus métrologique: état des lieux. Accessed March 3, 2020. https://hal.archives-ouvertes.fr/hal-00473981/document.

Vitruvius (1567): I Dieci libri dell'Architettura di M. Vitruvio, Venice: appresso Francesco de'Franceschi Senese et Giovanni Chrieger Alemano compagni.

Witt, Ronald G. (1988): "Medieval Italian culture and the origins of humanism as a stylistic ideal." In: A. Rabil (ed.), Renaissance Humanism: Foundations, Forms, and Legacy: University of Pennsylvania Press, pp. 29-70.

Wolfe, Jessica (2004): Humanism, machinery, and Renaissance literature, Cambridge, UK: Cambridge University Press.

Knowledge Order and Knowledge Popularization in Pre-Modern Encyclopaedism*

Mathias Herweg

ABSTRACT

Ever since ancient times, encyclopedias have been organizing cross-disciplinary knowledge and breaking it down for a general public. In doing so, they reflect both the identities and the self-awareness of their respective cultures and eras. This essay seeks to shed light on the evolution of the genre through its cultural and medial changes, from its beginnings in Roman times to the medieval 'golden age' of the genre in the 13th century, with a third focus on the epistemological and medial revolutions at the dawn of modernity. The article examines the development of the organization and presentation of knowledge, which is fanned out by systematic topics and disciplines rather than alphabetically. Circle, tree, treasure, house of knowledge, mirror, image, theatre of the world: metaphors like these, evoked by the titles of early encyclopedias, indicate a close interplay of text and image, which opens a large panorama of the genre that shaped knowledge and education over a period of almost 2000 years. Moreover, the pre-modern origins, practices, and popularity of the encyclopedia still live on in its current, internet-based successors.

* This article was translated by Ian Pepper and the author. Any errors are the sole responsibility of the author.

INTRODUCTION

> "L'univers soit réel soit intelligible a une infinité de points de vue sous lesquels il peut être représenté, et le nombre des systèmes possibles de la connaissance humaine est aussi grand que celui de ces points de vue. Le seul, d'où l'arbitraire serait exclu, c'est, comme nous l'avons dit dans notre Prospectus, le système qui existait de toute éternité dans la volonté de Dieu."[1]

Encyclopedias are among the earliest media that organize and popularize knowledge. At the same time, they reflect the identities and self-perceptions of the respective organizing cultures.[2] Encyclopedias have always shaped knowledge in breadth (i.e., at the synchronic level of 'educated society') and in length (i.e., at the diachronic level of intellectual history). A century ago, one would still have referred to such knowledge as 'conversational' in order to underscore that it essentially pertains to those kinds of cognitive and informational inventories and contents that were required by the *homo eruditus*, the citizen in the emancipatory 19th-century sense of the word, for adequate discursive performance, and for the sake of competent participation in the scientific, cultural, and political processes of the respective epochs. Societal changes have altered the contents and the 'conversational' status of such knowledge, but have never rendered it wholly obsolete. All the way till the present day, encyclopedias, for all intents and purposes, have continued to depict cultural memory and the scientific worldview, while simultaneously shaping both of them.

The order in which things are presented does not predate encyclopedias—it is, instead, originated by them (among other factors). To make unstructured knowledge accessible and permanently fixed in this way requires didactic, technical, and medial strategies which are in part subject to historical changes (we are experiencing this to a particularly large extent in our days), but which have at the same time remained remarkably persistent since their very beginnings. This essay is devoted both to the functional constants as well as to the epistemic

1 *Encyclopédie*, cited: Œuvres de Denis Diderot, Tome Deuxième, Ire partie, Paris 1818, p. 519 ("The universe, either real or intelligible, has an infinite number of points of view from which it can be represented, and the number of possible systems of human knowledge is as large as that of these possible points of view. The only one from which arbitrariness would be excluded is [...] the system that has existed for all eternity in the will of God;" unless otherwise noted all translations are my own).

2 For an introductory discussion, cf. Diers 1977; Wiethölter et al. 2005: 1-51; Ribémont 1997, esp. 53-54; with a focus on the 'omne scibile:' Fowler 1997.

changes experienced by this literally archi-disciplinary and archi-discursive genre, which unites the investigation and transmission of knowledge. Paradigmatic works are singled out for this purpose, while particular attention is paid to the supra-individual strategies and structures that are constituent of the genre as a whole. Furthermore, I investigate the popularization effects of the encyclopedia, which aims to convey specialized knowledge of many disciplines to the educated parts of the population. Through teaching, preaching, and, ever since Gutenberg's times, supported by specific 'mass media' such as pamphlets and leaflets, this knowledge, obviously filtered, could trickle down to illiterate recipients, too.

It should be emphasized that precisely for the sake of the aims sketched above, genre history is understood and described here, to a certain extent, as material and medial history. This is not at all an external factor, but instead an undeniably determinative aspect that strongly demarcates the very possibilities, boundaries, and limits of the genre. The basic medium is the material written or printed on, which predetermines the durability and mobility of the stored material: clay tablet, papyrus, parchment, and paper succeed(ed) one another, before digital media deprived the inscribed material of its monopoly status. In form, the mutation of substance corresponds to the medial 'aggregate conditions.' And this has often underappreciated consequences for content and for user options. Ancient encyclopedias were originally published on papyrus scrolls (*rotuli*), which did not strictly exclude cross-references, but certainly hindered them: It is impossible to leaf through a roll, and widely separated lemmata can be compared with one another only awkwardly. The transition from *rotulus* to codex, achieved in late antiquity, must therefore be regarded as a milestone in media history that precedes the more familiar caesura-like event of the (Western) invention of printing.[3] Nor, however, did the for the most part multivolume codices allow for the full utilization of all encyclopedic options, as can be seen from today's perspective. It is here that the step towards the Internet—which cross- and interlinks text, image, and sound virtually ad infinitum—represents the ultimate keystone (at least for the moment) of encyclopedism.

3 Cf. Chartier/Cavallo 1999: 125-133; Blanck 1992: 75-101, esp. p. 97 ("an event as important and far-reaching as the invention of the printing press with movable types by Johannes Gutenberg and the victory of the printed book over the manuscript").

ENCYCLOPEDIC 'BASIC TECHNIQUES' FOR KNOWLEDGE TRANSFER

The relationship between encyclopedia and reality—with the latter understood as the empirically preexistent totality of the world—is determined diachronically by two basic forms of access: the lemmatization and the ordering of the material (see Michel 2002: 35-37). Through lemmatization, the almost infinite amount of material (characterized by Goethe as the "million-fold Hydra of empiricism"[4]) is bundled together into categories, classes, and ideas, and is consequently conceptualized. In the ordering process, that which has been registered in categories is structured, hierarchized, and contextualized, so that elms, alders, and firs, for example, are grouped together under the lemma 'tree.' Emerging at this level is the knowledge network that is implied by the concept of the encyclopedia, and the extreme relevance of the second step in particular becomes manifest if we imagine a library without catalogues, or a voluminous book without such paratexts as a table of contents or an index.

The two basic processes of lemmatization and ordering simultaneously guide and are guided by knowledge. Nature itself obviously doesn't prescribe any exact categories, so the ordering of knowledge is culturally relative—or, according to Michel Foucault's *The Order of Things*, determined by the discursive structure of the respective epochs. Encyclopedias from diverse periods and/or cultures therefore seek and find divergent ways of ordering the seemingly identical world. This comes immediately, and at the same time paradoxically, to light in the strange taxonomy which Jorge Luis Borges, in his essay *The Analytical Language of John Wilkins*, attributes to "a certain Chinese encyclopedia" (and which Michel Foucault, in his preface to *The Order Of Things*, cites with significant inferences); there, according to Borges, the realm of the animals is 'divided' as follows:

> "[animals] (a) belonging to the Emperor, (b) embalmed, (c) tame, (d) sucking pigs, (e) sirens, (f) fabulous, (g) stray dogs, (h) included in the present classification, (i) frenzied, (j) innumerable, (k) drawn with a very fine camelhair brush, (l) *et cetera*, (m) having just broken the water pitcher, (n) that from a long way off look like flies." (Foucault 1970: xv)

We may be skeptical of Borges' putative source, but when we consider the degree of relativity and cultural dependency that characterizes *any* order of knowledge, the well-known aphorism seems to apply: *Se non è vero, è ben*

4 Letter from Goethe to Schiller dated 16/17 August 1797; ed. Beetz 1990: 393. For the quote, see also Michel 2002: 35.

trovato. And despite all invariants, it is precisely this relativity that also emerges (albeit in ways far less amusing or bewildering than in that "certain Chinese encyclopedia") in the more credible examples drawn from the history of European encyclopedia, which will be surveyed in the following sections.

AN EXCURSION TO THE ANCIENT ORIGINS OF ENCYCLOPEDIA

As a terminus technicus, the word 'encyclopedia' became common only in the 15th to 16th centuries[5]—significantly during that period of proto-modernity that saw the great voyages of commerce and discovery. The origins of the term, however, can already be found in Greek antiquity. The term *enkýklios paideía* (cf. Christes 1997; Fuchs 1962), meaning 'circle of learning,' later translated into Latin as *orbis doctrinae*, referred to the universal education that was not just worthy of the citizen of the polis, but which even defined him. In late antiquity, as in the Middle Ages, this meant first and foremost the *septem artes liberales*, i.e., the *Trivium*, composed of grammar, rhetoric, and dialectics, together with the *Quadrivium*, consisting of astronomy, (theory of) music, arithmetic, and geometry.[6] This disciplinary canon—which was expanded subsequently to include civil (Roman) and canon law, medicine, and theology, and which could perhaps be referred to as the most successful curricular concept in Occidental educational history—shaped higher education, reading and learning requirements, and from the 13th century until the present day the organization of university departments as well.[7]

But both the Greek as well as the Latin terms—*enkýklios* and *orbis*, respectively—referred not yet to the genre itself, i.e., to the textual compilation and presentation of universal knowledge in book form, but instead to the cosmos of knowledge per se, to its subdivision into distinct disciplines and *artes*. The metaphorization of the concept of the encyclopedia from 'universal knowledge' to 'universal book' was first accomplished only in the early modern period. Before this question of the concept is returned to later in this essay, a fundamental *auctoritas* of the genre should be investigated with regard to the concept it inscribed

5 On the conception and genesis of the term, see Dierse 1977: 9-15.

6 On the liberal arts both as a whole and individually, see the good introductory articles in the handbook by Grafton et al. 2010; for the Middle Ages, see also Englisch 1994 and Stolz 2004: here esp. Vol. 1, 6-85.

7 Cf. here part I and II of the pertinent anthology by Schwinges 1999.

onto that genre. Leaving aside the Greeks, whose reception in the 'Latin' medieval West was limited due to a lack of linguistic competence, the determining source was Pliny the Elder (23-79 CE) and his monumental *Naturalis historia* (Natural History), which united the universal knowledge of its time in no less than 37 books, all of which have been preserved until today.[8] Individuals responsible for such a work can evidently not be authors in a strict sense, but emerge as compilers, as 'assemblers' of material that has already been researched and proofed—and Pliny is no exception to that rule: according to Arno Borst, he did not want "to write a scientific reference work that could serve as the basis for progressive knowledge, but to impart educational knowledge, the canon of which was regarded as complete."[9] Pliny himself outlines this view of the compilatory author—which has remained a constant condition of encyclopedic writing to this day—in his preface dedicated to Emperor Titus (39-81 CE), in which he reflects at the same time on the difficulties and limits that face the 'ideal' of every real author (whereby Pliny's practice, too, presents itself as being far more differentiated, subjective, and narrative than the 'encyclographic' ideal formulated here):

"[The present work] does not admit of talent [...], nor does it allow of digressions, nor of speeches or dialogues, nor marvellous accidents or unusual occurrences – matters interesting to relate or entertaining to read. My subject is a barren one – the world of nature, or in other words life [*rerum natura, hoc est vita*]; and that subject in its least elevated department, and employing either rustic terms or foreign, nay barbarian, words [*externis, immo barbaris*] that actually have to be introduced with an apology. Moreover, the path is not a

8 The only complete bilingual edition of *Naturalis historia* is still the one of the Loeb Classical Library with the Latin text and its translation into English. It was started in 1938 by Harris Rackham, who edited six volumes (1-5, 9), and it was completed in the 1960s by different editors: W.H.S. Jones (Vols. 6-8), D.E. Eichholz (Vol. 9), E.A. Warmington (8-9), and Webster (9); The Latin text used is the one by Detlefson (1-2) and Mayhoff (3-10); cf. Pliny 1967.

9 Borst 1994: 12 ("[...] kein wissenschaftliches Fachbuch schreiben, das als Grundlage fortschreitender Erkenntnis diente, sondern Bildungswissen vermitteln, dessen Kanon als abgeschlossen galt"). In this respect, Friedrich Schiller's judgment, long determinative for the reception of Pliny, misses the target: According to Schiller, Pliny was, to be sure, in possession of tremendous learning, "but I fear that beyond this tremendous reading, excerpting, and dictating, he really had no adequate time to engage in free reflection" (letter of Aug. 18, 1802, addressed to Goethe; "[...] aber ich fürchte, er hatte über dem ungeheuren Bücherlesen, Exzerpieren und Diktieren zum freien Nachdenken nicht recht Zeit"). If Pliny had needed time for this purpose, he would surely have found it.

beaten highway of authorship, nor one in which the mind is eager to range. [...] A large part of us seek agreeable fields of study, while topics of immeasurable abstruseness treated by others are drowned in the shadowy darkness of the theme. Deserving of treatment before all things are the subjects included by the Greeks under the name of 'Encyclic Culture' [*enkýklios paideia*]; and nevertheless they are unknown, or have been obscured by subtleties, whereas other subjects have been published so widely that they have become stale. It is a difficult task to give novelty to what is old, authority to what is new, brilliance to the common-place, light to the obscure, attraction to the stale, credibility to the doubtful, but nature to all things and all her properties to nature. Accordingly, even if we have not succeeded, it is honourable and glorious in the fullest measure to have resolved on the attempt." (Pliny 1967: Praefatio, 8-11; transl. by Harris Rackham)

Even with these restraints, Pliny's work deserves our highest respect: According to the author's own testimony in the dedicatory preface, the work is based on the consultation of approximately 2,000 books, "very few of which, owing to the abstruseness of their contents [*propter secretum materiae*], are ever handled by students [*studiosi*, i.e., scholars]" (ibid: 12-13). Written by about 100 different authors, they provide some 20,000 'noteworthy facts' to which Pliny adds further information that was hitherto ignored and recently invented, or even gained through his own experience (*adiectis rebus plurimis quas aut ignoraverant priores aut postea invenerat vita*). The result is overwhelming, but nevertheless the author again emphasizes with modesty: "Nor do we doubt that there are many things that have escaped us also; for we are but human, and beset with duties, and we pursue this sort of interest in our spare moments, that is at night" (ibid).

The topos of humility in this context constitutes an equally serious, timeless, and pragmatic reference to the discrepancy between the task of the encyclopedist and the premise of individual authorship. It demarcates the limits of *any* encyclopedia—limits which, over the next two millennia, authors would continually and repeatedly put to the test. And nothing confirms these limits as strikingly as Wikipedia, that never-ending, perpetually growing postmodern online encyclopedia. In 79 CE, when Pliny died near Misenum during the famous eruption of Mount Vesuvius (a precise description of which we owe to the author's homonymous nephew Pliny the Younger), only 10 of the total 37 volumes had yet appeared; the others followed posthumously.

According to the sources included within it, Pliny's archetypical work—which we can refer to as archi-generical as well as archi-textual (in Gérard Genette's terms)—literally replaces an entire library. And just as a library requires a catalogue and pertinent signature marks, Pliny provides access to the content of his 'book of all books' by numerous paratextual additions. Book 1 as a whole, for example, offers an index that indicates the contents of all subsequent volumes, along with the most important sources upon which they draw. Many of these sources are now lost, which pointedly perpetuates the work's postulated status as a substitute for a whole library. Not just in his work's poly-historical, totalizing impetus, but stylistically as well, Pliny stipulates tendencies for the encyclopedia that remain valid even today: Without pathos or poetic ambition, but consistently underpinned by narrative devices,[10] the genre deals programmatically with "facts." Frequently, the beginnings of articles/lemmata are linked by cross-references, and terms and categories are pointed out in the abundance of entries, within which, wherever possible, systematic and historical approaches are combined. A description of a specific technology, for example, is always followed by a reference to its *primus inventor*, as well as to any subsequent developments it has undergone till Pliny's own time.[11]

The breadth of the universal knowledge recorded by Pliny is suggested by a simple keyword summary for the individual books of the work: Book 1 contains, as mentioned above, a comprehensive table of contents for the entire compendium and a meticulous list of sources; Book 2 deals with the cosmos and the universe; Books 3-6 summarize the geography and ethnography of the various parts of the world, and each bears the same title, namely, "On the situation of the countries, their inhabitants, seas, towns, harbors, mountains, rivers, the distances of places from one another, and on the peoples who are still there or who were once there" (following the respective names of about 150 countries and provinces). Book 7 is devoted to human anthropology, physiology and medicine; Books 8-11 treat zoology (terrestrial and aquatic animals, birds, insects); Books 12-27 deal with botany (including viniculture, orcharding, and agriculture, as well as medical herbalism); Books 28-32 offer veterinary medicine and its remedies; Books 33-37 treat the science of metals and minerals, with their refining also addressed by a brief history of the techniques of the fine arts (35).

Pliny placed his stamp on the history of knowledge and science for one and a half millennia. Countless scholars—cartographers, historians and art historians, technicians, natural historians, lexicographers, linguists, etymologists, and encyclopedists—invoked him, adopting the structures and the tremendous repository of sources contained in his magnum opus alongside its contents and methods. During the Latin Middle Ages, this work served as a model for the most im-

10 This is manifest, among other places, in Book 35, which shapes the biographies of major artists such as Zeuxis and Apelles by means of diverse anecdotes and arabesques. Accordingly, epic poets are among the sources used by Pliny.

11 On the tradition of this concept, cf. Thraede 1962: 158-186.

portant 'world books,' which were progressively supplemented by new fields and sources of knowledge.[12] Isidore of Seville in the 7th century, Hrabanus Maurus in the 9th, and Bartholomaeus Anglicus, Thomas of Cantimpré, and Vincent of Beauvais in the 13th, the 'golden age' of pre-modern encyclopedia: all of these figures either perpetuated the Plinian tradition or extended the sphere of knowledge about nature beyond it.[13] They ensured a remarkable continuity in the history of science between antiquity and the European Middle Ages. In this domain, ultimately, no Renaissance ever took place. Instead, through Pliny and his successors, readers, and revisers, the classical tradition became an integral part of post-antiquity knowledge, which incidentally acknowledged its debt to its antecedents. In the scholastic milieu of Bernard of Chartres, medieval scholars referred to themselves as "dwarves on the shoulders of giants"—which we are justified to interpret as an expression of self-confidence, not self-denial: Looking out over the heads of these giants, the post-antique scholars saw further and more sharply than their great forbearers, even as "dwarves."[14] But even so, the phrase expresses the utmost respect, which is confirmed by the abundance of copies and adaptations of antique texts through which cloisters, monasteries, cathedral schools, and, later, medieval universities exploited, preserved, and obliquely popularized ancient knowledge by appropriating, teaching, preaching, and translating.

ENCYCLOPEDIAS AND ENCYCLOPEDIC LEARNING IN THE MIDDLE AGES: MAIN FEATURES AND CASE STUDIES

In the aftermath of Pliny's pioneering act, the medieval encyclopedia[15] is at once a literary genre, a source of general (i.e., cross-disciplinary) knowledge, and a

---

12 An essential source that pursues the contours of the reception of Pliny through various epochs is Borst 1994.

13 With Meier 1984 (modified by Ernst 2000), it becomes possible to distinguish, with regard to the Middle Ages, between a Plinian type of encyclopedia based on a systematic approach and another type of encyclopedia based on the accumulation of knowledge, with outstanding examples by Thomas of Cantimpré on the one side, Vincent of Beauvais on the other side.

14 On the trajectory of this metaphor through the history of science, cf. Merton 1985.

15 For orientation and further study, from among the abundant research we refer to: Binkley 1997 (in particular the contributions of Fowler, Ribémont, Chr. Meier, and Rivers); Franklin-Brown 2012; Meier 1984, 2002; Michel 2002: 35-83; Ribémont 1996; Ventura 2014. The encyclopedic texts discussed are contextualized by Collison 1964.

holistic concept of the world. Its career began in the early Middle Ages, when the most significant innovations consisted of the emergence of new theological domains and, in some cases, of new exegetic methods, e.g., allegorical interpretation, for which there existed older traditions in patristic biblical exegesis, in the Greek *Physiologus*, etc., which could be linked with adopted encyclopedic practices in a functionally convincing way. Among the authors, Isidore of Seville and Hrabanus Maurus stand out. But the medieval encyclopedia arrived at its apogee only two centuries later, when scholasticism was on the rise. Concentrated, readily accessible knowledge from the most diverse topics and subjects had become an indispensable desideratum for teachers and students at the renowned cathedral schools and universities, and not least for ecclesiastics who were confronted with escalating intellectual expectations. This may explain the wide dissemination of copies of old and new texts by this time, for example, the [*Liber*] *de natura rerum* by Thomas of Cantimpré (more than 180 manuscripts and several versions), the [*Liber*] *de proprietatibus rerum* by Bartholomaeus Anglicus, or the *Speculum maius* in three voluminous parts by Vincent of Beauvais (all of them 13th century). An early vernacular pendant is the Middle High German *Lucidarius* (12th/13th century), which follows the model of the *Elucidarium* of Honorius Augustodunensis. With the approach of the later Middle Ages, the 'encyclopedic drift' towards the vernacular was strengthened through Jacob van Maerlant (see *Fig. 1*: Illuminated description of the monstrous beings of the Far East) and Conrad of Megenberg, accompanied by, or resulting from, changing functional concepts and gradually expanding publics.[16]

Ordering Systems and Designations

The knowledge accumulated by these and other authors was grouped together according to quite diverse systems of arrangement and classification, with mixed forms (e.g., based on alphabetical or subject-based systems) appearing frequently. Significantly, all empiricism retreats behind the written sources, most of them ancient ones: regarding its self-understanding and conception, the encyclopedia was (and ever remained) a predominantly conservative genre, a circumstance that was causally connected to the self-image and practices of the authors—here, Diderot/d'Alembert form a groundbreaking exception and usher in a new epoch of the genre.

16 Regarding the last example, to which we will return, cf. Feistner 2011: 43-109.

Figure 1: Illuminated Description of the Monstrous Beings of the Far East, from Der naturen bloeme *by Jacob van Maerlant*

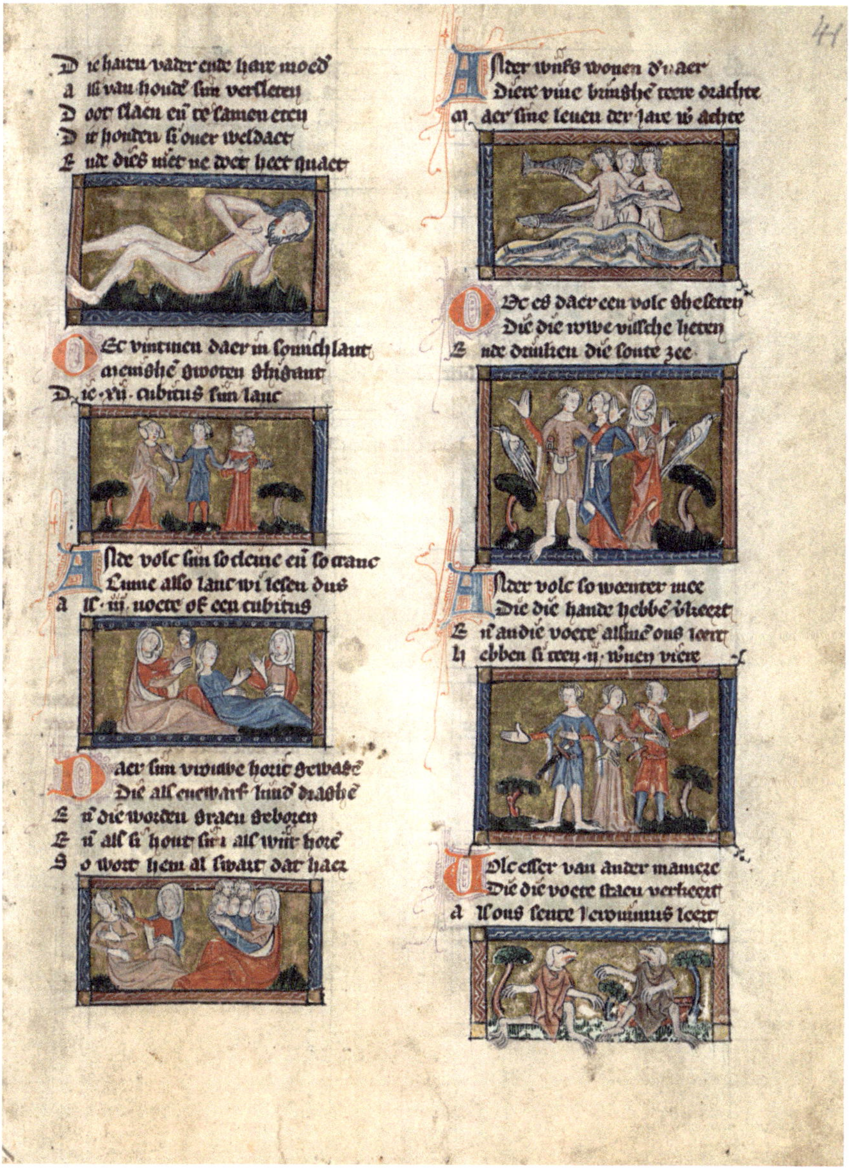

Source: Koninklijke Bibliotheek, KB KA 16 – f. 41r; Public Domain

We shall dwell briefly on this point before investigating the system of ordering and organizing in greater detail. Up until the present day, encyclopedic labor has remained for the most part compilatory in nature, that is to say, a process of sifting and excerpting. An early figure such as Isidore of Seville (ca. 560-636 CE) was not interested in capturing the new, but rather in collecting the secure stocks of knowledge, submitting them to a guiding interpretation, and making them accessible to broader circles, including non-learned ones, by a didactically prepared form for the sake of preaching and instructing.[17] The 'whole world between two book covers,' the Creation in a written form—nothing more and nothing less was the claim, and it acquired its essential justification through the premise that Nature in some sense represented God's 'second book.' The encyclopedia—less by virtue of its programmatic impetus than by its pragmatic intentions—therefore took its place alongside the Bible. Many encyclopedias already bear this strong claim in their title, or in titular qualifications reflecting on their status. Pliny, not without a sardonic undertone, already invoked the felicitous metaphors of the Greeks, who called their encyclopedic works, for example, "honeycomb," "horn of plenty," "meadow," "painting," even "the epitome of all knowledge" (*pandéktai*), "titles," as Pliny adds, "that might tempt a man to forfeit his bail." (Pliny 1967: Praefatio, 14-17; transl. by Harris Rackham) In the Middle Ages and in the early modern period, we encounter designations such as *De universo* (Hrabanus), *Imago mundi* (Honorius Augustodunensis), *Speculum (maius)* (Vincent of Beauvais; see also Jacob van Maerlant: *Spieghel historiael*), *Mare* (Giovanni Colonna), *Thesaurus* (Brunetto Latini), or even, simply, *Summa* (Honorius Augustodunensis)[18]—the latter a term oscillating eloquently between

17 On the pragmatic aims of the medieval encyclopedia, cf. Meier 1984 and Ribémont 1996; according to Meier (ibid: 475), "the encyclopedia is also a 'world book,' a book par excellence, in the sense that it can replace a library, a large number of specialized works" ("...auch in dem Sinn 'Weltbuch', Buch par excellence, daß sie eine Bibliothek, eine große Zahl von Spezialwerken ersetzen kann"), and, concerning its function, Ribémont (ibid: 392) postulates: "The encyclopedia locates itself between the *accessus ad auctoritates* and a condensed vision of the universe of Creation, a mirror that speaks through itself. [...] The encyclopedist addresses those who do not have the time (such as the powerful) or the knowledge to access scholarly books" ("L'encyclopédie se situe donc entre l'accessus ad auctoritates et une vision condensée de l'univers de la Création, miroir qui parle par lui-même... L'encyclopédiste s'adresse à ceux qui n'ont pas le temps, comme les puissants, ou le savoir suffisant pour accéder aux livres savants").

18 Honorius drew up his *Summa totius de omnimoda historia* as a complement to *Imago mundi*; cf. Collison 1964: 53.

'the whole' (i.e., the totality of knowledge) and 'the highest' (i.e., the summit of knowledge).[19] Today, one may think of 'universal lexicon' or 'cosmography' for measuring the connotations of the mentioned titles and genre designations.

Such titles already indicate what encyclopedias actually claim to be and generate: world view, world appropriation, world interpretation. Paradoxically only at a first glance (for brevity is the necessary condition for the desired totality), many titles and prologues simultaneously express programmatic stringency and scarcity. Terms as *Summa brevis*, *Breviarium*, and *Compendium* are widely disseminated.[20] This implies that whoever strives to comprehend the world in its totality has to choose a concise, comprehensible, objective style, as free as possible from autopoetic affectation and self-referentiality. In this sense, Conrad of Megenberg, in the introductory poem to his pioneering German-language work of the genre, the *Buch der Natur* (circa 1350), proclaims that the flow of speech should be clear and even ("diu red schol unverschertet sein, mit clârheit schôn umbslungen"), emphasizing shortly thereafter that truth is opened only by the key of correct speech ("rehter rede slüzzel;" Pfeiffer 1962: 1).

Even the recourse to authorities and sources, which guarantees the scientific authenticity of the genre, is subordinated to this 'imperative of laconism,' as we may call it, and to the guiding concept of popularization: The authorities relied upon are often unnamed or cited imprecisely, making verification difficult—a peculiarity that has been retained in popular encyclopedic compendia up to this day, and something that should be unthinkable in 'serious' scientific contributions. Here as well, Conrad of Megenberg provides us with numerous instances in his German version of Thomas von Cantimpré's encyclopedia. He consistently refers to mere names, even if there are huge oeuvres behind them (as with Aristotle and Augustine), and rarely offers any specific titles. At times, he alludes even more vaguely to "the Jews," "the Egyptians," or simply to "several people" in order to authenticate a given piece of information.

Yet this nonchalance in dealing with authors and authorities is by no means attributable solely to brevity or to contemporary practices of citing sources at second or third hand. It should instead be understood programmatically, and in particular including the encyclopedists themselves: For them, as for their informants, the priority belongs to the objects and to the Creator standing behind them, and in a second step to the encyclopedic 'world book.' Both marginalize the 'random figure' of the author: "L'auteur [...] importe peu, c'est l'ouvrage

19 Ribémont (1996: 391) adumbrates the encyclopedic concept just in these two senses of the summa-term as "totalité et point culminant" ("totality and culmination").

20 Cf. the selection of titles in Hofmann 1987: here esp. 398-399.

lui-même qui compte et qui veut 'éclairer' le lecteur ignorant."[21] (In this regard, a phenomenon such as Wikipedia, which no longer even mentions the authors of an article, stands in an unconscious but close relationship of continuity to pre-modern genre conventions.)

The function—constitutive for this genre—of making specialized knowledge accessible to non-specialists and promoting knowledge transfer between separate disciplines, as well as the poly-historical impetus (which includes, for the Middle Ages, salvation history and theology), create direct transitions to other, to some extent complementary, genres: to the chronicle, whose Creation chapters regularly include cosmo-ethno-geographic sections (e.g., the *Weltchronik* of ca. 1250, by Rudolf von Ems, or the famous printed *Nuremberg Chronicle* of 1493, by Hartmann Schedel); to biblical and ancient epics (e.g., concerning the Trojan War or Alexander the Great); to travel novels of the Hellenistic type (e.g., Apollonius of Tyre); and to spiritual and didactic poetry and prose. In transmedial terms, the encyclopedia is unquestionably closest to cartography, that enormous store of knowledge on medieval maps, which illustrated encyclopedias themselves sometimes integrate.[22]

Practices of Organizing Universal Knowledge

Characteristic of the way in which cross-linked knowledge is organized are impressive metaphors in which the encyclopedists describe their work. Here, incidentally, the analysis of pre-modern encyclopedic practices emerges as productive for the assessment of the 'new' information technologies and media of our time. The sea (*mare*), into which countless large and small rivers flow, has already been mentioned—and still today we are 'surfing,' although the metaphor no longer quite fits, because what works on the sea does not work in a web. The web, too, is a quite common metaphor in the Middle Ages (for etymologically, *textus* [text] means nothing other than that), along with the tree (*arbor*)[23] with its roots, limbs, branches, and leaves. Other metaphors include the house (*domus*; a mnemotechnic topos that goes all the way back to an anecdote by the poet Si-

21 Luff 2003: 57 ("The author [...] doesn't matter, it is the book itself which counts and which aims to enlighten the ignorant reader").

22 On interdependencies between cartography and encyclopedism, cf. von den Brincken 1968; Simek 1992: 55-65.

23 Cf. the illustrations of the *Arbor sapientiae* in Stolz 2004: Vol. 2, Fig. 37-38, pp. 787-788 (with commentary in Vol. 1, 250-252).

monides of Ceos, ca. 557-467 BCE; cf. Mierke/Fasbender 2013: esp. 1-2)[24]—and of course the all-encompassing ring, which inspired the Greek terminus technicus *enkýklios*. Aside from the universality for which they also stand, all of these conceptual images refer to structural features of the genre. The most diverse forms of medial transfer appear metaphorically side by side (e.g., in the image of ring [*kýklios*], mirror [*speculum*], or tree [*arbor*]). Text, map (*mappa*), image, and later graphic add-ons complement each other—and sometimes the connection also creates subtle discrepancies or inconsistencies. Columns, directories, indices, and rubrics create user-friendly structures.[25] The 'user,' by the way, is by no means inactive. He deepens and completes the given tools with his own entries and notes, including the familiar pointing hand, an invention of medieval monks to mark important text passages or key words.[26] Finally, cross-references link different articles as well as text and graphic inserts with each other. Images and pictograms can have a text-complementary, a mnemonic, or simply an illustrative function, with their utility value depending also upon the intellectual conditions and interest horizons of the individual user (see *Fig. 2*: Explaining figures in Isidore's of Sevilla encyclopedia).

The tools and options just mentioned are well-reflected reactions that had already been developed during the pre-Gutenberg age in order to tackle a fundamental problem, namely, that interconnected data in book form can only be presented in a linear sequence. These tools and options are recipes designed to counteract the deficiency of opening a medium that is actually predestined for cover-to-cover reading also to users who consult it selectively—which is the default case for users of an encyclopedia at any time. Ever since late antiquity, the transition from the scroll (*rotulus*) to the codex was accompanied by progress in this respect: It is not possible to leaf through *rotuli*, and their users were only able to pursue cross-linkages, cross-references, and connections between lemmata to a restricted degree and with considerable effort. The breakthrough to the codex, then, converted the looking up, leafing back and forth, or annotating of

24 This anecdote can be found in Quintilian and Cicero (Cicero, *De oratore* 2.353-354; Quintilian, *Institutio Oratoria* 11.2.11-17). On the meaning of the anecdote as a rhetorical, instructive device, known in the *progymnásmata* manuals as *chreía*, see Fernández Delgado's article in this volume, pp. 98-100.

25 In place of individual examples, we invoke the illustrations of encyclopedic and didactical provenance assembled in Stolz 2004: Vol. 2, 765-860.

26 For an example, see in James le Palmer's autograph of the encyclopedia *Omne bonum* (1360/75), reproduced in Meier 2002: 564. On the significance of the hand in premodern and modern digital knowledge cultures, cf. Wenzel 2003.

the text from a challenging possibility to a simple and common method. In addition, extensive texts could now be stored in a single 'volume' (etymologically nothing other than a roll).

Figure 2: Geometric Diagrams Added to Book III of Isidore's Etymologiae

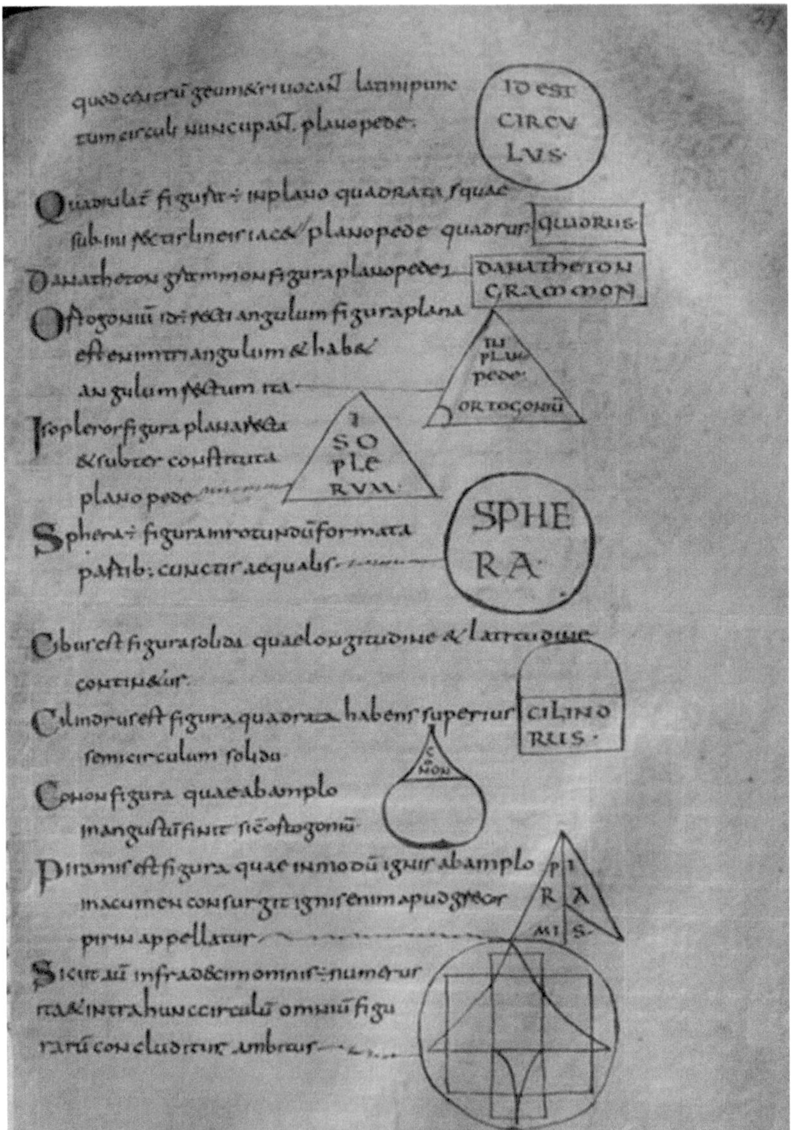

Source: Basel, Universitätsbibliothek, F III 15, f. 27r – Isidorus, *Etymologiae*, lib. II-XIX (https://www.e-codices.ch/en/list/one/ubb/F-III-0015)

Today, Pliny's 37 books fit without difficulty into two 'voluminous' books. Henceforth, increasingly refined access aids, index types, and rubrics would come into use. However, what the book never managed in medial terms, and what probably represents the most blatant progress (and the greatest advantage) of its digital successors, is firstly the option of interconnecting different topics and presentation media by hyperlink, and secondly the possibility of automated (including, of course, misleading or failing) searches. But the abovementioned possibilities of 'pre-modern' mediality already allowed the active use of encyclopedias to become a common phenomenon, and also enabled the user, with his/her motives and strategies, to gain increasing influence on the contents and structures of the works, in short: on the genre's design.

It would be quite simple and obvious here to insert comprehensive studies on the texts of, e.g., Thomas of Cantimpré or Vincent of Beauvais. But in this limited context we prefer a macro-analytical approach; and since a benchmark has already been established with Pliny's ancient prototype, one of the most ambitious encyclopedic projects of the Middle Ages may now serve for a concise comparison, even though it never progressed past its preliminary stages: Hugh of Saint Victor's *Didascalicon de studio legendi* (Introduction to the Study of Reading, ca. 1128).[27] This work, in six books, is also remarkable as a representative of an explicitly systematic scientific exposition of the material (in contrast to the natural and cultural exposition of Pliny or Isidore). "In modern terms [...] it would probably be called a doctrine of science and an introduction to study."[28] The macrostructural plan covers four large fields of knowledge in a quite independent adaptation of the traditional *artes* scheme, which are subdivided into individual disciplines (and further subdivided into topics):

- the *Ars Logica*, with the linguistic-philological "trivium," consisting of grammar, rhetoric, and dialectics;
- the *Practica*, which is subdivided, in a manner dating back to Aristotle and Boethius, into *ethica*, *oeconomica*, and *politica*;
- the *Theorica*, with its mathematical-natural-scientific "quadrivium," consisting of arithmetic, geometry, astronomy, and (theory of) music, along with physics and, remarkably from a modern perspective, theology as well;

27 Offergeld 1997; English transl. (cited) by Taylor 1991.

28 Offergeld 1997: Introduction, 40-41. ("mit modernen Begriffen ... wäre es wohl als eine Wissenschaftslehre und Studieneinführung zu bezeichnen"). Among its predecessors, it stands closer to Cassiodorus' Institutiones (6th c.) in this regard than to Isidore's Etymologies (7th c.).

- the *Artes Mechanicae* (in analogy to the liberal arts), namely weaving, weapon-making, navigation, agriculture, hunting, medicine, and the dramatic arts.[29]

Book 3, Chapter 1 explicates this order, and concludes as follows:

> "In this division only the divisive parts of philosophy are contained; there are still other subdivisions of such parts, but those given may suffice for now. [...] We read that different persons were authors of these sciences. They originated the arts – some by beginning them, others by developing them and others by perfecting them: and thus for the same art a number of authors are frequently cited. Of these I shall now list the names of a few."[30]

Overlooking the virtual matters, nearly everything that encyclopedias still offer today could be fitted into the diachronically authenticated system outlined in this way: botanical-zoological subjects in the specialized sections on agriculture and hunting; anthropology in the sections on medicine; history (as history of religion and salvation) in the sections on theology; history (as dynastic and world history) in the sections on politics; and history (as history of war) in the sections on the art of weaponry, etc.

With regard to interconnecting and cross-linking, it is obvious which opportunities, but also which constraints, are inherent in this structure. No less obvious, however, is the extent to which they would have required intermedial diversification: Pictures and drawings, lists, tables of kings and popes, as well as rubrics and references, would have to run through the hundreds of chapters and allow leaps through the linear progression of the text, for in Hugh of St. Victor's view 'everything is connected to everything' in both the macrocosm and the microcosm. It is well known that medieval manuscript culture was basically capable of such refinements in codicological setup and layout. In practice, however, an encyclopedia of this claim as the work of *one individual person* was not feasible. It would have required a complex workshop, a human 'network' of chroniclers, theologians, philosophers, and naturalists, of excerptors, compilers, miniaturists, proofreaders, and editors, who the author 'only' had to preside over as a *spiritus rector*. It's not surprising that the most ambitious plans did not come to fruition, neither in Hugh of St. Victor's case nor in any later cases. This fate is shared by many of the large-scale world chronicles and cosmographies of the time.

29 Cf. the content overview in Offergeld's edition (1997: 51-58); for an overview of structure and contents, also see Collison 1964: 47-49.

30 Cited translation: Taylor 1991: 83; Latin text in Offergeld 1997: 216.

Modern encyclopedias since Diderot/d'Alembert, in contrast, have been mercantile enterprises on a large-scale, with countless authors and contributors,[31] and postmodernism is testing the possibilities of this venerable genre to its very limits by virtually allowing every user and every reader to participate in the circle of knowledge by correcting or supplementing texts, or by registering as an author. Not only in terms of access, but also in terms of the concept of authorship, the encyclopedia in the World Wide Web itself has become encyclopedic. But looking back again from this (provisional) *télos* of genre development to the individual achievements of a Pliny, an Isidore, or a Vincent of Beauvais, there is good reason for humility. The sentence quoted above, formulated in the scholastic milieu of Chartres, still applies to our present day: We too are "dwarves on the shoulders of giants," seeing further and more deeply than ever before, yet by no means standing on our own two feet.

THE EARLY MODERN PERIOD: A CHANGING GENRE IN A NEW MEDIUM

It would be a fallacy to claim, in the sense of the sometimes quite sensible concept of a 'long Middle Ages,' that the medial changes and the significant breaks in science, religion, society, and government during the 16th/17th centuries had no decisive impact on the history of the encyclopedia. Nevertheless, there are continuities, too, in terms of content, planning methods, and presentation, which have been shaping the past and future image of the genre. For even under the influence of the printing press, the expansion of knowledge, and far-reaching discoveries, encyclopedias generally remained works by individual personalities who deliberately receded behind their universal material. Of course, new topics broadened the traditional framework, but this had for a long time a gradual rather than a fundamental impact on the adopted classification system. One of the most successful encyclopedias of the Incunabula period, the *Margarita Philosophica* by Gregor Reisch (c. 1495), a Carthusian from Freiburg who was close to notable humanists such as Thomas Murner, Beatus Rhenanus, and Jakob Wimpheling, held on to the schema of the *artes liberales*, and extended it pragmatically rather than programmatically to include the *artes mechanicae* and new disciplines such as the study of nature and the soul (Reisch 1973; on the author's vita, cf. ibid: vi-vii).

31 For exemplary work on the 'corporate character' of the modern encyclopedia, cf. Darnton 1979: 1-37.

Figure 3: Woodcut entitled Typus logic[a]e, *from Gregor Reisch,* Margarita philosophica *(Basel 1517), Book II tractatus I, page 67*

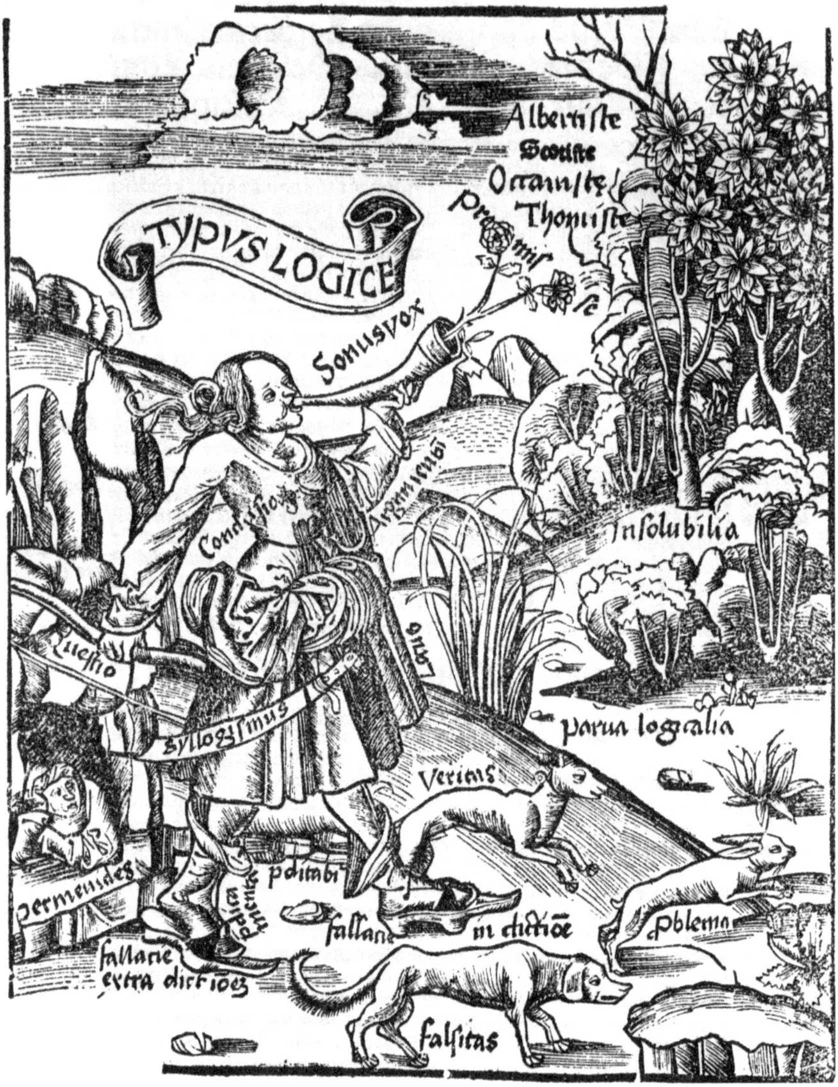

Source: Reisch 1973: 67; Photo: Public Domain

Even the abundant and lavish inclusion of illustrations, graphics, and maps, along with the transmedial layout resulting from this, and the indexing by running titles, are all still reminiscent of the *artes*- and technical writings of the manuscript era.[32] The introductory woodcuts on individual disciplines such as astronomy, arithmetic, rhetoric, geometry, music, and logic are bound to traditional iconography and allegoresis, as shown strikingly by the *TYPUS LOGIC[A]E* (see *Fig. 3*). The figure presents a hunting allegory with personified Logic as a hunter with horn (*sonus* and *vox*), from which two roses (the *duae praemissae*) emerge. Moreover, all of the following details are clearly labeled: the arms (*argumentae*), the chest (*conclusio*), and the legs (*praedicabilia/praedicamenta*), as well as the hunting knife (*syllogismus*), the bow (*quaestio*), and the dogs (*veritas* and *falsitas*). The hunted prey is a hare (*problema*); the errors (*fallaciae*) are already lying on the ground; weeds (*parva logicalia*) are growing on the right-hand side foreground, and brushwood (*insolubilia*) can be found behind them. The middle ground is occupied by the *silva opinionum*, with the doctrines represented by the thoroughly medieval trees of the *Occamistae*, *Scotistae*, *Thomistae*, and *Albertistae* (cf. Becker 1970: 12; Büttner 2003: 279-280).

A significantly deeper cut is made about 140 years later by Johann Heinrich Alsted and his *Encyclopedia septem tomis distincta* (1630).[33] Here, the inherited spectrum of the seven *artes* is only invoked in order to fill it anew. Furthermore, the layout and the sources of that monumental work of 2,404 narrowly printed folio pages (plus registers) set new standards. The methods developed during the manuscript era have now been perfected into a transparently structured layout in two columns, which can dispense with manuscript structuring elements such as initials (which were still present in Reisch's work) and colored highlights ('rubrics' in the proper sense of the word), since now a differentiated set of instruments such as paragraph marks, enumerations, tables and diagrams (that sometimes extend over many pages), changing font styles (recte vs. italics) and the like emerge in their place. Quotations are emphasized in typesetting and can thus be recognized at first glance, with authors, works, and passages being meticulously named. Original characters are chosen for Greek and Hebrew inserts.[34]

32 On Reisch's layout and image schema between (medieval) tradition and (humanistic) innovation, see Meier 1999: 285f.; Büttner 2003: 281.

33 Alsted 1989 (Vol. 1-2) and 1990 (Vol. 3-4). On the author's vita and his place in the history of ideas, cf. the detailed account in Hotson 2000 (on the structure of the *Encyclopedia*: 163-167).

34 It may be noted, incidentally, that the *Fraktur* typeface for the main text, which was still being used indiscriminately by Reisch, gave way to the *Antiqua* that had since become common for Latin texts. In Alsted's work, the *Fraktur* is reserved only for German-language insertions and terms, e.g., in the glossaries of Book 5 (Alsted 1989: Vol. 1, esp. 138-157) and in the enumeration of the winds in Book 18 (ibid: Vol. 2, 1112-1113).

Scientific accuracy, transparency, and—within the framework of what is medially possible—the easiest accessibility of the material are obviously the guiding principles of the author and his typesetters. No less than four subdivided registers (*Authores*, *Loci Communes*, *Quaestiones*, and *Res et verba*) guide the user's way through Alsted's textualized cosmos.

While all of this points noticeably beyond the pre-modern tradition that Pliny represented for us at the beginning, Alsted's modifications to the genre's design are not entirely without precedent. In their layout, they perfect in the typoscript medium what the manuscript one was almost constitutively characterized by: a both mnemotechnically and didactically optimal linking between text and image. In the internal structuring and interconnecting of the material, the inherited methods are extended, systematized, and combined, but by no means are they 'invented,' as the development outlined above has demonstrated. And even where the new approach is the most striking, e.g., in terms of content, certain constants can still be noted: the mechanical arts had already been integrated into Hugh of St. Victor's *Didascalicon* (whose elaborate systematization of the sciences continues to have a phenotypical effect in Alsted's extensive 'genealogy' of the scientific system, presented in 38 complex tables); historiography had already occupied a prominent position in Vincent of Beauvais's *Speculum maius*; and architecture as an encyclopedic subject stretched all the way back to Pliny. The place, of course, where inherited contents like these meet in Alsted is unconventional: together with magic, alchemy, and gnomology, and with disciplinary curiosities such as tobaccology, charientilogy ('clownery') and many others, they populate a hybrid *Farrago disciplinarum* (Tom. VII), which literally means a 'knowledge barn,' and which brings together everything that does not quite fit into the structures of the other disciplines—namely, because the medieval tradition, at that point in time, still holds too much sway over them. The History, subdivided into mnemonics, historical theory, and chronology, offers another example of this impact: The biblical history of salvation from Creation onwards still forms the foundation, while ancient mythology and profane history are integrated into it or follow it. Chronology is still based on the Augustinian theory of the six ages of the world and the four kingdoms doctrine of the biblical prophet Daniel (Dn. 2), both of them common since late antiquity.[35]

Thus the program behind Alsted's immense extensions of the encyclopedic *orbis*, which simply aims to encompass 'all areas that man is able to learn and teach,' allows the historical starting point and the core of the genre to be recognized, even though the curricular field had been empirically deepened and epis-

35 Cf. the diagram that concludes History in Book 32: Alsted 1990: Vol. 4, 2068.

temically regrouped. Alsted himself was aware of the fact that he was both a continuator and an overcomer of tradition, as he cautiously justified his method with different interpretations of the inherited generic term. Entirely new, however, and far more significant for the further development of the European encyclopedia, is another aspect: the medial switch towards printed books increased the genre's distribution and circulation. The confessional, social, intellectual, and, last but not least, linguistic dynamism also broadened the genre's functional borders. What was largely a monastic or university matter in the Middle Ages now shifted to lay erudition, to emerging journalism and new literary forms. In some respects, the genre was preparing to influence social and cultural developments. With a view to literary fiction, which can only be touched upon in the following final section, a new interaction between encyclopedic knowledge and epic narrative emerged, one which is prototypically represented by certain German Baroque novelists such as Duke Anton Ulrich of Brunswick (*The Syrian Aramena*), Daniel Caspar von Lohenstein (*Arminius*), or Heinrich Anselm von Ziegler und Kliphausen (*Die asiatische Banise*). The Romanticist Joseph von Eichendorff later dismissed their works as "gewissermaßen tollgewordene Realenzyklopädien" ("real encyclopedias that have, to a certain extent, gone mad")—not completely without justice, to be sure, but without any sensitivity for the poetic and aesthetic peculiarity of that type of novel, which in some ways anticipates postmodern narrative fiction.[36]

SOME PROSPECTIVE REFLECTIONS ON ENCYCLOPEDIC NARRATION AND ENCYCLOPEDIC FICTIONS

"But you have had living proof that it [i.e., the kingdom of Prester John] doesn't exist!" This is the objection that Umberto Eco's skeptical logothete Niketas makes to his guest, the Lombard Baudolino, in the 'real-fictive' year 1204 in Constantinople, to which Baudolino replies: "We had proof that we hadn't reached it. That's different." (Eco 2003: 518)

Umberto Eco's medieval picaresque novel (cf. Wyss 2005 and, for a historian's view, Fuhrmann 2003) is a masterpiece of what can be described (without

36 Eichendorff also discredits the aforementioned authors as "authentic savants who are simply interested in a broad exposition of their erudition" and who therefore offer "every conceivable item of knowledge except, regrettably, poetry;" Eichendorff 1990: 905-906.

Eichendorff's negative overtone) as an 'encyclopedic novel.' Just before ending, it offers us an opportunity to extend the space of popular encyclopedic imagination. It would be easy to simply refer to the inventions and fictions within premodern encyclopedic works, which could be readily explained in terms of alteritarian epistemic conceptions and alternate views of reality at that time. But even more interesting is the case of encyclopedically structured narrative fiction under the epistemic-poetological premises of (post-) modernity.

Of course, since the author is the famous Bolognese semiotician Eco, *Baudolino* takes the form of a postmodern satyr play, a hybrid mixture of history, adventure, and detective novel with accessories from travelogues, legends, and encyclopedias. And it is also an essay about the constructedness of everything that can be seen, experienced, and created by men. The protagonist originates from the misty swamplands of 'Italia profonda'; he is a gifted liar who becomes first an adopted son of, then the publicist for, Frederick Barbarossa (reigned 1152-90 CE). His teacher is the (real!) chronicler and bishop Otto of Freising (d. 1158), who teaches his fictive student the questionable, yet nonetheless fruitful and undoubtedly true lesson of his life: "The world condemns liars who do nothing but lie, even about the most trivial things, and it rewards poets, who lie only about the greatest things." (Eco 2003: 43)

Lies and liars—what might sound disrespectful is not at all intended as such by Otto: Historians, chroniclers, and encyclopedists construct their reality, for the empirically factual—Goethe's abovementioned Hydra—is so immeasurable and incomprehensible that it becomes perceptible and comprehensible to human beings only in its reduction to human measure. For the greatest things, only the imagination of the poet may be sufficient. Baudolino's life puts this bold thesis to the test (which is daring only for a chronicler). After Otto's death he accompanies the emperor on the Third Crusade, from which Barbarossa would never return. His death in faraway Syria is a proven fact, but only the novel discloses how this actually happened (since we are concerned here with the encyclopedic aspect and not with the historical or criminological one, nothing more will be said about this). After this early failure of the Crusade, a small, almost apostolic circle (albeit not apostle-like in their characters) centered around Baudolino, most of them former Paris collegemates, does not turn back to Europe. Burdened by the awareness that one of their companions might be a murderer, they penetrate ever more deeply into the sometimes utopian, sometimes dystopian wonders (*mirabilia*) of the Orient. Their journey is accompanied by a multiplicity of naturalistic, ethnographic, geographic, theological, technical, and urban fascinations, yet is overshadowed by tragic deaths, and only knows one goal: the empire of the Prester John, ruler of an enormous Christian realm somewhere in India, which the Christians in the West could open up as a second front in the crusades against Islam. It was Otto of Freising (the real one now) who had made these utopian lands known throughout the Latin West. Peoples and landscapes, streams of rocks, night-black fogs, strange human tribes, animals, and plants roll past the European travelers like a panopticon. But they never attain their actual goal, that literally encyclopedic summa of all desires, hopes, and utopias of civilization.

In the Middle Ages, this goal was no doubt possible to attain, at least in the books of chroniclers and encyclopedists. Ever since its first (and rather concise) mention in Otto of Freising's *Chronicle of the Two States* (book VII, ch. 32f.), this realm became a geopolitical fact, located in the 'real fantastic' space par excellence which actually is India. The kingdom of Prester John was said to be a stronghold of everything good, noble, and true in the world, a place of power and splendor, a land of marvelous things and conditions, of elaborate architectures and sophisticated urban life. In the novels, the heroes usually only touch on this encyclopedic-utopian symbiosis. Only in the Grail narrative and in certain chronicles does it become a real setting and event.

Eco's novel clarifies, in an astonishing manner, the hitherto uncertain origin of the topos—the realm of Prester John—which is not a utopia in the actual sense of the word, because it is topographically guaranteed, and is also recorded on medieval maps of the world. Here, the famous letter that first describes this marvelous realm (a forgery that, during the 12th century, circulated back and forth between the Pope and the Eastern and Western Emperors) is the invention of a bunch of Parisian students with little eagerness to learn but with great talents, rallied around the notorious liar Baudolino. The term 'invention,' of course, once again does not apply, and precisely here lies the novel's significance for postmodern poetics: As in a collage, almost all of Eco's details about Prester John's realm are drawn from medieval sources, which in turn refer back to accepted authorities from antiquity whose names we already know: Pliny, Solinus, Isidore, the epic tradition concerning Alexander the Great, and so on. The setting of the *studiosi*'s compilation, who are inspired by abundant alcohol and drugs, is the venerable library of the Abbey of St. Victor—we already encountered Hugh, the (real) abbot of this abbey, as one of the renowned encyclopedists of that time. The merry academic circle's work merely represents a new arrangement of preexisting material.

As a result, Baudolino and his entourage later set out towards their imaginary goal via routes they have themselves invented, passing through landscapes they have themselves designed according to the ancient sources.

What Eco is staging here in an apparently genuinely postmodernist manner is an eminently medieval procedure: appropriation, explanation, and interpretation of the world through written authorities and narrative fiction. In the later Middle Ages, when the timeless and placeless sceneries of Arthurian epics had lost their attractiveness, and the desire for instruction and orientation had spread to vernacular poetry, a series of novels were written which for good reason can be called 'world books' and which point toward Eco's novel in many ways (not, however, with regard to the comic and ironic style of the latter). Extending from Albrecht's *Jüngerer Titurel* to Hartlieb's *Alexander* and the anonymous *Fortunatus* (to look only at the German side of things), they convey the most diverse kinds of encyclopedic knowledge in a disciplinarily de-contextualized, but narratively re-contextualized, form (cf., e.g., Müller 1984; Ruberg 1995; Herweg 2010: here chap. 3; Herweg 2012). Apart from their accessibility to a lay public that did not read Latin, their advantage in relation to encyclopedias was the capacity to *narrate* the selected knowledge of different disciplines and to project it onto the (virtually replicable) life-itinerary of a realistically designed hero. The (as invoked by Stierle) "Verwilderung des Romans als Ursprung seiner Möglichkeit" (Stierle 1980)—i.e., the exuberance of the late medieval and early modern novel as a source of the very possibility of the genre—becomes manifest most strikingly in its 'encyclopedization.'

Incidentally, Baudolino's past failure, which the logothete Niketas comments on so uncomprehendingly, does not dissuade him from setting out again in search of Prester John's lands. And he acts quite consistently in this regard, since he himself had issued the slogan of highest credibility when he, together with his delirious Parisian friends, penned the fake letter by the potentate which was to serve as a signpost for him for the second attempt (Eco 2003: 141):

> "Someone proposed also mentioning an underground stream rich in precious stones, but he refused to pursue the idea [...]. On the testimony of Pliny and Isidore, they decided instead to place salamanders in those lands, snakes with four legs that live amid flames. 'It only has to be true and we'll include it,' Baudolino said, 'so long as we're not telling fairy tales.'"

The final sentence reads like a program of encyclopedic fiction: Above all, no fairy tales—if something has already been written down someday and somewhere, it must be true. And anything can be found written down somewhere, can't it? A truly breathtaking idea...

REFERENCES

Alsted, Johann Heinrich (1989-1990): Encyclopaedia. New facsimile printing of the Herborn edition of 1630, with a preface by Wilhelm Schmidt-Biggemann and a bibliography by Jörg Jungmayr, Stuttgart: Frommann-Holzboog.

Becker, Udo (1970): Die erste Enzyklopädie aus Freiburg um 1495. Die Bilder der 'Margarita Philosophica' des Gregorius Reisch, Prior der Kartause, Freiburg: Buchhandlung Herder.

Beetz, Manfred (ed.) (1990): Briefwechsel zwischen Schiller und Goethe in den Jahren 1794 bis 1805, Vol. 1, Munich: C. H. Hanser.

Binkley, Peter (ed.) (1997): Pre-Modern Encyclopaedic Texts: Proceedings of the Second COMERS Congress, Groningen, 1-4 July 1996 1996 (Brill's Studies in Intellectual History 79), Leiden: Brill.

Blanck, Horst (1992): Das Buch in der Antike, Munich: C. H. Beck.

Borst, Arno (1994): Das Buch der Naturgeschichte: Plinius und seine Leser im Zeitalter des Pergaments, Heidelberg: Winter.

Brincken, Anna-Dorothee von den (1968): "Mappae mundi und Chronographia. Studien zur 'imago mundi' des abendländischen Mittelalters." In: Deutsches Archiv für Erforschung des Mittelalters (DA) 24, pp. 118-186.

Büttner, Frank (2003): "Die Illustrationen der 'Margarita Philosophica' des Gregor Reisch." In: Frank Büttner/Markus Friedrich/Helmut Zedelmaier (eds.), Sammeln – Ordnen – Veranschaulichen. Zur Wissenskompilatorik in der Frühen Neuzeit, Münster: lit Verlag, pp. 269-300.

Chartier, Roger/Cavallo, Guglielmo (eds.) (1999): Die Welt des Lesens. Von der Schriftrolle zum Bildschirm, Frankfurt am Main: Campus.

Christes, Johannes (1997): "Enkyklios Paideia." In: Der Neue Pauly 3, cols. 1037-1039.

Collison, Robert L. (1964): Encyclopaedias: Their History throughout the Ages: A Bibliographical Guide with Extensive Historical Notes to the General Encyclopedias Issued throughout the World from 350 BC to the Present Day, New York: Hafner Publishing Co.

Darnton, Robert (1979): The Business of Enlightenment: A Publishing History of the 'Encyclopédie' 1775-1800, Cambridge, MA and London: Harvard University Press.

Diderot, Denis/d'Alembert, Jean le Rond (1755): "Encyclopédie ou Dictionnaire raisonné des sciences, des arts et des métiers," Vol. 5, Paris. English transl. Accessed May 16, 2019. https://quod.lib.umich.edu/d/did/.

Diers, Ulrich (1977): Enzyklopädie. Zur Geschichte eines philosophischen und wissenschaftlichen Begriffs, Bonn (Archiv für Begriffsgeschichte, Suppl. 2).

Eco, Umberto (2003): Baudolino. Translated from the Italian by William Weaver, London: Vintage.

Eichendorff, Joseph von (1990): Geschichte der Poesie. Schriften zur Literaturgeschichte, ed. by Hartwig Schultz, Frankfurt am Main: Deutscher Klassiker Verlag.

Englisch, Brigitte (1994): Die Artes liberales im frühen Mittelalter (5.–9. Jh.). Das Quadrivium und der Komputus als Indikatoren für Kontinuität und Erneuerung der exakten Wissenschaften zwischen Antike und Mittelalter, Stuttgart (Sudhoffs Archiv, Beiheft 33).

Feistner, Edith (ed.) (2011): Konrad von Megenberg (1309-1374), Ein spätmittelalterlicher 'Enzyklopädist' im europäischen Kontext (JOWG 18), Wiesbaden: Reichert.

Foucault, Michel (1970): The Order of Things. An Archaeology of the Human Sciences, New York: Vintage Books.

Fowler, Robert L. (1997): "Encyclopedias. Definitions and Theoretical Problems." In: Binkley 1997: 3-29.

Franklin-Brown, Mary (2012): Reading the World. Encyclopedic Writing in the Scholastic Age, Chicago: University of Chicago Press.

Fuchs, Harald (1962): "Enkyklios Paideia." In: Reallexikon für Antike und Christentum (RAC) 5, cols. 365-398.

Fuhrmann, Horst (2003): Das Mittelalter in der Literatur: Umberto Eco und sein Roman 'Baudolino' (Eichstätter Universitätsreden 110), Wolnzach, pp. 4-32.

Grafton, Anthony/Glenn W. Most/Salvatore Settis (eds.) (2010): The Classical Tradition, Cambridge and London: Harvard University Press.

Grenzmann, Ludger/Stackmann, Karl (1984) (eds.): Literatur und Laienbildung im Spätmittelalter und in der Reformationszeit (DFG-Symposion Wolfenbüttel 1981), Stuttgart: J. B. Metzlersche Verlagsbuchhandlung.

Herweg, Mathias (2010): Wege zur Verbindlichkeit. Studien zum deutschen Roman um 1300 (Imagines medii aevi 25), Wiesbaden: Reichert.

Herweg, Mathias (2012): "'Verwilderter Roman' und enzyklopädisches Erzählen als Perspektiven vormoderner Gattungstransformation. Ein Votum." In: Freimut Löser/Robert Steinke/Klaus Vogelgsang/Klaus Wolf (eds.), Die Zukunft des Spätmittelalters, Wiesbaden: Reichert, pp. 77-90.

Hofmann, Heinz (1987): "Artikulationsformen historischen Wissens." In: Hans Ulrich Gumbrecht/Ursula Link-Heer/Peter-Michael Spangenberg (eds.), Grundriß der romanischen Literaturen des Mittelalters XI: La littérature historiographique des origines à 1500, Vol. 1, 2, Heidelberg: Winter, pp. 367-687.

Hotson, Howard (2000): Johann Heinrich Alsted 1588-1638. Between Renaissance, Reformation, and Universal Reform, Oxford: Oxford University Press.

Luff, Robert (2003): "Du compilateur anonyme à l'auteur autonome: Figures de l'écrivain encyclopédique au moyen âge européen." In: Danielle Buschinger (ed.), L' 'effet auteur' au Moyen Age. Actes du Colloque d'Amiens (Mars 2001), Amiens: Université de Picardie-Jules Verne, pp. 55-67.

Meier, Christel (1984): "Grundzüge der mittelalterlichen Enzyklopädik. Zu Inhalten, Formen und Funktionen einer problematischen Gattung." In: Grenzmann/Stackmann 1984: 467-503.

Meier, Christel (1999): "Bilder der Wissenschaft. Die Illustration des 'Speculum maius' von Vinzenz von Beauvais im enzyklopädischen Kontext." In: Frühmittelalterliche Studien 33, pp. 252-286.

Meier, Christel (ed.) (2002): Die Enzyklopädien im Wandel vom Hochmittelalter bis zur frühen Neuzeit, Munich (Münstersche Mittelalter-Schriften 78).

Merton, Robert K. (1985): On the Shoulders of Giants: A Shandean Postscript. San Diego/New York/London: Thomson Learning.

Michel, Paul (2002): "Ordnungen des Wissens. Darbietungsweisen des Materials in Enzyklopädien." In: Ingrid Tomkowiak (ed.), Populäre Enzyklopädien. Von der Auswahl, Ordnung und Vermittlung von Wissen, Zürich: Chronos, pp. 35-83.

Mierke, Gesine/Fasbender, Christoph (eds.) (2013): Wissenspaläste. Räume des Wissens in der Vormoderne (Euros 2), Würzburg: Königshausen & Neumann.

Müller, Jan-Dirk (1984): "*Curiositas* und *erfarung* der Welt im frühen deutschen Prosaroman." In: Grenzmann/Stackmann 1984: 252-273.

Offergeld, Thilo (ed.) (1997): Didascalicon de studio legendi. Studienbuch. Lateinisch-Deutsch (Fontes Christiani 27), Freiburg: Herder. English transl. (1991): The Didascalicon of Hugh of St. Victor: A Medieval Guide to the Arts, translated from the Latin and with introduction and notes by Jerome Taylor, New York: Columbia University Press.

Pfeiffer, Franz (ed.) (1962): Das Buch der Natur von Konrad von Megenberg. Die erste Naturgeschichte in deutscher Sprache (Repr.), Stuttgart.

Pliny (1967): Natural History, Vol. 1: Praefatio, Libri I, II. With an English translation by Harris Rackham (Loeb Classical Library 330), Cambridge and London: Harvard University Press (https://ryanfb.github.io/loebolus-data/L330.pdf).

Reisch, Gregor (1973): Margarita philosophica, with a preface, introduction, and new index by Lutz Geldsetzer (Instrumenta Philosophica III.1), Düsseldorf: Stern-Verlag Janssen.

Ribémont, Bernard (1996): "Enzyclopédisme médiéval et modernité." In: Jean R. Scheidegger/Sabine Girardet/Eric Hicks (eds.), Le Moyen Âge dans la modernité. Mélanges offerts à Roger Dragonetti (Nouvelle bibliothèque du Moyen Âge 39), Paris: Champion, pp. 381-394.

Ribémont, Bernard (1997): "On the Definition of an Encyclopaedic Genre in the Middle Ages." In: Binkley 1997: 47-61.

Ruberg, Uwe (1995): "Zur narrativen Integration enzyklopädischer Texte am Beispiel des Faustbuchs von 1587." In: Franz M. Eybl/Wolfgang Harms/Hans-Henrik Krummacher/Werner Welzig (eds.), Enzyklopädien der Frühen Neuzeit. Beiträge zu ihrer Erforschung, Tübingen: Niemeyer, pp. 64-80.

Schwinges, Rainer Christoph (ed.) (1999): Artisten und Philosophen: Wissenschafts- und Wirkungsgeschichte einer Fakultät vom 13. bis zum 19. Jahrhundert, Basel (Veröffentlichungen der Gesellschaft für Universitäts- und Wissenschaftsgeschichte 1).

Simek, Rudolf (1992): Erde und Kosmos im Mittelalter. Das Weltbild vor Kolumbus, Munich: C.H. Beck'sche Verlagsbuchhandlung.

Stierle, Karlheinz (1980): "Die Verwilderung des Romans als Ursprung seiner Möglichkeit." In: Hans Ulrich Gumbrecht (ed.), Literatur in der Gesellschaft des Spätmittelalters (GRLMA Begleitreihe 1), Heidelberg: Winter, pp. 253-313.

Stolz, Michael (2004): Artes-liberales-Zyklen: Formationen des Wissens im Mittelalter, 2 Vols., Tübingen: A. Francke Verlag (Bibliotheca Germanica 47).

Thraede, Klaus (1962): "Das Lob des Erfinders: Bemerkungen zur Analyse der Heuremata-Kataloge." In: Rheinisches Museum für Philologie N.F. 105, pp. 158-186.

Ventura, Iolanda (2014): "Enzyklopädie." In: Fritz Peter Knapp (ed.), Die Rezeption lateinischer Wissenschaft, Spiritualität, Bildung und Dichtung aus Frankreich (GLMF I), Berlin and Boston: DeGruyter, pp. 161-199.

Wenzel, Horst (2003): "Von der Gotteshand zum Datenhand. Zur Medialität des Begreifens." In: Sybille Krämer/Horst Bredekamp (eds.), Bild–Schrift–Zahl, Munich: Wilhelm Fink, pp. 25-56.

Wiethölter, Waltraud/Bernd, Frauke/Kammer, Stephan (eds.) (2005): Vom Weltbuch bis zum World Wide Web – Enzyklopädische Literaturen, Heidelberg: Winter.

Wyss, Ulrich (2005): "Umberto Eco auf der Suche nach dem Gral." In: Laetitia Rimpau/Peter Ihring (eds.), Raumerfahrung – Raumerfindung. Erzählte Welten des Mittelalters zwischen Orient und Okzident, Berlin: De Gruyter, pp. 313-325.

*Was Cometen eygentlich seyen.** Ways of Imparting Knowledge about the Nature of Comets in Early Modern Ephemeral Literature

Marion Gindhart

ABSTRACT

This article shows how knowledge about the nature of comets materializes in different formats of ephemeral literature between the 1530s and 1619. It examines the strategies used to convey traditional and newly acquired knowledge to different groups of recipients and the specific targets the writers of cometary literature pursued with their publications. The examples considered in the paper include popular media such as printed sermons on comets, which also deal with natural-historical aspects; the wide range of cometary pamphlets, which are quite uneven in quality (with Johannes Schöner's *Coniectur* being a small but instructive text for a broad audience); the format of the annual prognostication (*practica*), which Peter Apian used for presenting learned knowledge (e.g., instructions on cometary observation, and position and orbit determinations), and the new finding of the tails' antisolarity; or the Schreibkalender, which had an important multiplier function for communicating disciplinary (especially astronomical) knowledge, as in the case of Peter Crüger. Besides small formats in the vernacular, the latter also uses a voluminous bilingual treatise (the *Uranodromus cometicus*) to establish the new, astronomy-based knowledge on comets, invalidating the Aristotelian cometary model which he formerly had defended as a student.

* "What comets actually are." This article was translated by Christopher Reid and revised by the author.

INTRODUCTION

In the early modern period, comets were among the natural phenomena which—at least until the end of the 17th century—enjoyed a collective as well as continual interest. They also possessed great affective potential: Embedded in cultural memory primarily as prophets of bad tidings,[1] a comet sighting (or even just the rumor of a cometary appearance) gave rise to a series of communicative actions and interactions, both oral and written. Impressive specimens such as the Winter Comet of 1618/19[2] were true media sensations. As events, they were prominent topics in church sermons, everyday conversations, and in correspondence. They sparked a deluge of printed works that promised to satiate the contemporary hunger for information and the expectations of a wide-ranging audience. Such writings deal with a diverse assortment of cometary knowledge—in Latin, but most of all in the vernacular.

The explanatory and interpretive models presented therein can be assigned to four major discourses, which are characterized by diachronic stability and already directed and structured the statements about comets in antiquity.[3] These include:

- the metaphysical-theological discourse, which interprets comets as signs through which a divine power reveals itself and communicates with mankind;
- the astrological discourse, which assumes that the stars have an influence on the microcosm and provides a set of rules with which statements about the future can be made based on certain cometary parameters, e.g., size, shape and coloring, duration, position, or path;
- the historical-empirical discourse, which reveals the implications of current comets in analogy to historically verified events that occurred after past comet appearances, and which thus updates for the present the relevance of historical evidence;
- the natural-historical discourse, which is dedicated to the explanation of comets as natural, i.e., physical or astronomical phenomena.

1 A detailed synopsis of how comets were seen and explained as signs or as causes of (above all negative) future events is given by Weichenhan 2004: 388-425.

2 Comet C/1618 W1. On the reception of this comet in contemporary literature, cf. van Nouhuys 1998 (for the Netherlands); Gindhart 2006 (for German-speaking countries) and 2017a (summary with additional literature).

3 A paradigmatic study of these discourses for the comet year 1618 can be found in Gindhart 2006.

It is typical of early modern cometary literature that elements of knowledge from all these worlds of meaning can appear polyphonically and synergistically in the context of particular works (e.g., natural-philosophical and astrological knowledge in primarily theological texts, or theological knowledge in astronomical treatises). In each instance, the reading and interpretation of comets can be adapted according to one's needs. For example, a comet with its given parameters could be cleverly employed against religious opponents or—especially in the early Reformation movement—could be interpreted as a herald of the world's imminent demise.[4]

The popular pamphlets, which make up the lion's share of early modern cometary literature, were usually mass-produced for the market and—with the right combination of marketing techniques—could be a lucrative ad hoc source of income. The best market opportunities were offered by prints that responded quickly to a comet's appearance, were inexpensive, and promised to satisfy the interpretive needs of a discursively educated and expectant public.

However, it was not uncommon for a title page to arouse higher expectations than the following, often rapidly produced text could fulfill. In his *Iudicium Astrologicum* (*Fig. 1*) Elias Ehinger, headmaster of the Anna-Gymnasium at Augsburg, for example, had to confess that he could not perform the prominently announced astrological interpretation, stating: "Was nun deß jetzigen Cometae deutung sein werdt/ kan man zur zeit nicht eygentlich wissen" ("What the signification of the present comet will be/ cannot exactly be known at this time")[5] (Ehinger 1618: B1v).[6]

In this article, I would like to highlight some examples from the ephemeral cometary literature—i.e., printed works which responded quickly to actual comets and were intended for immediate sale—and examine how they dealt with the *natural-historical* aspects of comets. How and for what purpose is which specif-

4 On the cultural history of comets, their multi-faceted interpretability, the amalgamation of 'popular' and 'learned' knowledge, and the possibilities of updating traditional patterns of interpretation, cf. Schechner 1997 (with a focus on England). The aim of her study is to show "that the history of cometary science is a tale not of rigid borders, but of the fluid interplay of high and low beliefs about nature and religion" (ibid: 4).

5 Unless otherwise noted, all translations of original sources into English were made by the editorial team and revised by the author.

6 On Ehinger's repeatedly re-printed text, which skillfully manages to address recipients of all levels of education via the title page and text design, cf. Gindhart 2006: 124-130 and 248-249; 2017a: 46-49. Johann Andreas Fabricius aptly summarizes the play with readers' expectations: "Der Titel muß gemeiniglich das Buch verkaufen, und ist auch nicht selten das erste, dadurch die Käufer hintergangen werden [...]" ("The title must generally sell the book, and it is not uncommonly the first item to betray the buyers") (Fabricius 1752: 733).

ic knowledge presented? Are there any particularly strong discursive synergies? How conservative are the ephemeral media and how is new and learned knowledge handled for a broad audience? The chronological framework of the analyzed works will range from the 1530s to 1618/19. A brief outline will first be presented of how the knowledge of the nature of comets developed and expanded during this period.

Figure 1: Elias Ehinger: Iudicium Astrologicum Von dem Newen Cometa [...], *Augsburg: Hans Schultes the Elder 1618 (Title Page)*

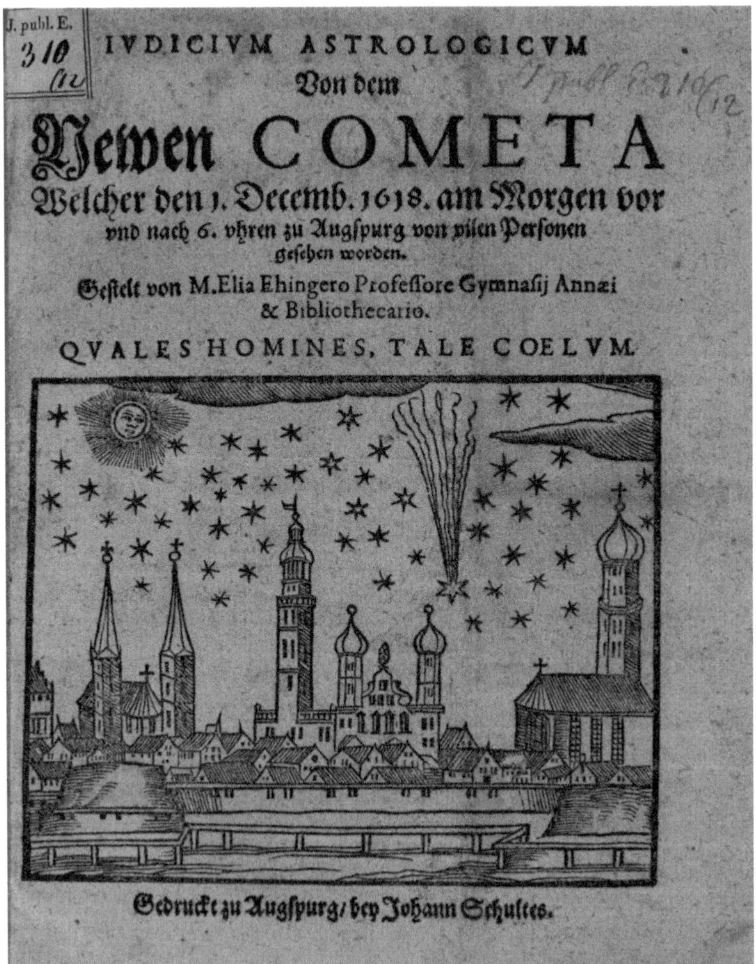
IVDICIVM ASTROLOGICVM
Von dem
Newen COMETA
Welcher den 1. Decemb. 1618. am Morgen vor
vnd nach 6. vhren zu Augspurg von vilen Personen
gesehen worden.
Gestelt von M. Elia Ehingero Professore Gymnasij Annæi
& Bibliothecario.
QVALES HOMINES, TALE COELVM.
Gedruckt zu Augspurg / bey Johann Schultes.

Source: BSB Munich, 4 J.publ.e. 310,12, urn:nbn:de:bvb:12-bsb10513064-6

COMETS— EARTHLY VAPORS OR COSMIC PHENOMENA?

Until the second half of the 16th century, the Aristotelian worldview, with its meteorological theory of comets, dominated cometary literature more or less unquestioningly. The basic model that developed from the commentary tradition was as follows.[7]

At the center of the cosmos is the stationary earth. It is surrounded by the elements water, air, and fire. Extending up to the sphere of fire is the sub-lunar region, i.e., the region below the moon, the planet closest to the earth. This region is characterized by constant change and transience. Above the sub-lunar region is the unchanging super-lunar region, in which the planets and fixed stars move on crystal spheres in eternally uniform, circular, concentric orbits. This 'upper' region consists of a single, fifth element—the aether. Comets are described as hot, dry vapors, which ascend from the earth to the uppermost layer of air through the influence of the sun and other celestial bodies. Here, they are compressed and ignited below the sphere of fire and carried along with the rotation of the spheres, but also have their own, apparently erratic course. They expire when the vapor matter is depleted. The axiom concerning the permanence, immutability, and uniformity of the 'upper' celestial sphere does not allow for the existence of comets in this region, for they are only rarely observed and also appear to show clear variations in their different parameters. Thus, they are assigned to the ever-changing lower world with other fiery *meteora* like shooting stars.

In the High Scholasticism of the 13th century, the Aristotelian worldview was connected with the Christian one and was thus theologically legitimized. It became a dominant model in schools and universities, and was transformed and amplified by various commentary traditions. In the 16th century, Melanchthon after initial reservations against the *Physica* in particular—positioned Aristotelian philosophy (with the exception of metaphysics) and its methods in a supporting role alongside Protestant theology. As a consequence, it remained a fixture in Protestant universities and the authoritative epistemological system.

7 On peripatetic cosmology, cf. Weichenhan 2004: 133-231, with a summary on 229-231. On Fortunio Liceti's interpretative attempt to reconcile the cosmological principles in the Corpus Aristotelicum with astronomical findings, ibid: 233-265. The passage follows Gindhart 2017a: 51-52.

Observations such as the antisolarity of the comets' tails in the 1530s admittedly also gave rise to theories that contradicted the Aristotelian model.[8] Their reception, however, remained peripheral for the time being, as did the adoption of alternative models, which, for instance, were presented and discussed in Seneca's newly appraised *Naturales quaestiones*. The Aristotelian explanation of the comets as *meteora* was supported by distance determinations, which had been carried out in Europe since the second half of the 15th century, but had delivered only faulty results up to the parallax measurements of the Great Comet in 1577/78. Thus, comets were also partly located mathematically in the sub-lunar sphere.

The Aristotelian axiom of the changeable elementary part and unchangeable cosmic part of the world was not seriously challenged until the widely received appearance of a supernova in 1572. This stellar explosion in the constellation Cassiopeia was now conclusively a phenomenon subject to strong changes, which Tycho Brahe's measurements, for example, situated in the super-lunar region. And five years later, when the Great Comet appeared, mathematical proofs successfully demonstrated that comets, as well, could be found far above the moon and were thus to be attributed to the celestial bodies. This finding was at first only accepted by part of the scientific community, however. Brahe published it in 1588 together with his geo-heliocentric system, which relativized the Aristotelian one.

All in all, these developments of the late 16th century represented more of a hairline fissure in than a rupture with Aristotelian cosmology and cometary theory. Even the telescopic observations and discoveries made between 1609 and 1611 did little to bring about a general departure from the Aristotelian worldview for some time. Nonetheless, in the cometary literature of 1618/19, we can observe intensified competition between the traditional, Aristotelian explanation of comets and the new mathematical-astronomical approach based on distance determinations and, along with it, a further interrogation of apparently rigid cosmological assumptions (for instance, via the heliocentric system of Johannes Kepler).

SERMONS ON COMETS—THEOLOGY AND OTHER FIELDS OF KNOWLEDGE

One medium that was responsibly involved in the longue durée of Aristotelian cosmology and cometary theory is the Protestant cometary sermon. In the early

8 Girolamo Cardano, for example, explained comets as celestial globes illuminated by the sun and their tails pointing away from it as refracted solar rays; cf. Kokott 1994: 137-139.

modern Protestant homiletics, contemporary natural events, which showed the omnipotence and immeasurable wisdom of God, were taken up again and again and interpreted in their revelatory character. Since the sermons reached the respective local audiences and their messages were comprehensible in the vernacular, they possessed tremendous opinion-forming potential. Some of these sermons can be found—partly in extended form—in print again shortly after their oral presentation. Thus, their contents were consistently reproducible through individual readings, but also recitals, e.g., in the household.[9] One can also assume that the printed sermons were resold or loaned like other types of ephemeral literature, reaching people who had not heard the oral recitation and/or had limited financial resources.[10] The impact thereby expanded greatly.

The central concern of the cometary sermons is the call for contrition, repentance, and conversion. The comets are interpreted here as signs of warning and admonition sent from God. They refer to his wrath over sinful mankind, which can only be contained through penance. Otherwise, various 'consequences' are unleashed in which unrepentant humans are punished, both during their lifetime and—even more fatally—after death.

The Winter Comet of 1618/19 was accompanied with sermons from the time of its greatest visibility at the beginning of December until its vanishing at the beginning of January. A perfectly timed coincidence was that it had its greatest tail extension on the 2nd Advent Sunday and—as visualized on Ehinger's title page—was to be seen as an impressive manifestation in the morning sky. Indeed, the pericopic text on this Sunday was the synoptic apocalypse in Luke 21:25-36 ("And there will be signs in the sun, the moon, and the stars"). The biblical passage thus proved the significance of the comet in heaven as a divine sign. The passage was verified, in turn, by the comet's striking appearance, perceivable by the listeners themselves. In this way, listening to the book of the Bible and looking into the book

9 On the cometary sermons of the years 1618/19, cf. Gindhart 2006: 25-63. On the potential audiences of ephemeral printed products like broadsheets or pamphlets in the late 16th century and the development of reading abilities among the general population, cf. Schnurr 2009: 202-212. According to Endres (1988), even before the advent of the printing press, the literacy of the middle classes experienced an upswing, which continued to increase with the Reformation. The literacy level in the 16th century is estimated to have been about 5 per cent; cf. Schnurr 2009: 205 with literature.

10 On the costs of ephemeral printed works and their distribution, cf. ibid: 208-209 with literature.

of nature were as one. The same could be said for the Protestant punitive theology and the Aristotelian cometary theory, as will be shown shortly.

More generally, it should be stressed that in the cometary sermons natural-historical knowledge and insights from the other secular discourses can be placed alongside statements from the central theological discourse. The Darmstadt superintendent Heinrich Leuchter, for example, proclaims on the title page of his sermon published in 1619 that he will deal with the following questions:

"1. Was Cometen seyen/ vnd woher sie sich vervrsachen. 2. Was sie bedeuten/ vnd auff was schwere Eventus vnd Hændel sie anzeig thun: Vnd dann zum 3. wie man sich Christlich/ wol vnd heylsam/ bey erscheinung derselbigen/ verhalten soll."[11] (Leuchter 1619: 1)

Following precisely this structure, he begins (1.) by explaining the comets according to Aristotle as meteorological phenomena, but pointing out at the same time that God in his omnipotence was not bound to natural laws and could spontaneously create comets. Next (2.), he lists past comets categorized according to their impacts[12] and specifies their calamitous character using astrological techniques. Thus, he offers his audience insights in various fields of knowledge, whereby the final theological interpretation (3.) remains the most important dimension of the sermon, which is supported by the other discourses.

Leuchter's Swabian colleague, Ulm superintendent Konrad Dieterich, extended his *Ulmische Cometen Predigte* in print to 52 pages. In this sermon, he presents the entire spectrum of contemporary knowledge about comets, refers to theories of cometary superlunarity, and also to positions that deny the hairy stars any significance whatsoever. Thus, unlike Leuchter, he also incorporates knowledge that is not in line with Lutheran orthodoxy. But despite the plurality of viewpoints he offers to the reader, Dieterich unmistakably emphasizes his conformity to the system. He therefore advocates for the vapor model as a natural-philosophical explanatory paradigm. He also simultaneously captures its dramaturgical potential in the context of sin and repentance by interpreting it metaphysically, giving it a higher meaning through this 'translation.' In short, it is the sins of men that effectively ascend as contaminated vapor matter, forming

11 "1. What comets are/ and by what causes they are produced. 2. What they portend/ and what severe events and quarrels they point to: And then, 3., how one should behave/ in a good and wholesome Christian manner/ when they appear."

12 On cometary catalogues and collections of *exempla*, cf. Gindhart 2006: 183-213; Mosley 2013.

the fuel of the comet fire and provoking God's wrath. This, in turn, can only be satisfied by the tears of repentance:

"Vnsere Suͤnde/ vnsere Suͤnde die sind die rechte Fewrige/ hitzige Daͤmpffe/ die dicke/ feiste/ zaͤhe/ Schwefelichte Duͤnste/ welche von der Erden auff empor zu Gott gehen Himmel steigen."[13] (Dieterich 1619: 18)

"Diß Cometen fewr ist Gottes Welt fewr/ darumb hie jedermann Jung vnd Alt/ Mann vnnd Weib/ Reich vnnd Arm/ klein vnd Groß hinzu lauffen/ vnd an demselbigen loͤschen soll. Loͤschen aber sollen wir es nicht mit Wasser/ auß der Donaw oder auß der Blauw/ oder auß S. Peters oder anderen Kaͤsten/ dann das loͤschet hie weniger dan nichts/ Sondern mit vnserm Bitteren herben Hertzwasser vnnd heissen Zaͤheren traͤnen vnserer Augen/ so auß der Brunnquell vnserer Hertzen auffsteigen/ durch die Augen rinnen/ vnnd vber die Backen herunder fliessen/ mit welchen wir vnsere begangene erkante Suͤnd vnnd missethat Hertzlich betrawren vnnd beweinen/ vnnd Gott vmb verzeihung vnnd erlassung derselbigen flehlichen vnd demuͤtiglichen anruffen sollen."[14] (Ibid: 46-47)

This effective metaphor alone made it worthwhile to stick to Aristotelian theory. It was an adherence that would continue, particularly in theological writings, into the late 17th century, defying all new findings from astronomical observations and illustrating the central role that the theologians played as scientific actors with influential powers of interpretation.

13 "Our sins/ our sins, they are the true, fiery,/ hot vapors,/ the thick,/ fat,/ glutinous,/ sulfurous exhalations/ which rise from the earth to God, ascending to Heaven."

14 "This comet's fire is God's fire of conflagration/ to which everyone, young and old,/ man and woman,/ rich and poor,/ humble and eminent, should run,/ and make an effort to extinguish it. But we should not extinguish this fire with water/ from the Danube or from the Blau,/ or from Saint Peter's or other fountains,/ for less than nothing would be quenched with that,/ but instead with the bitter and harsh water of our hearts and the hot tears of our eyes,/ which rise from the wellspring of our hearts,/ run through the eyes,/ and flow down our cheeks;/ with these tears we should sincerely mourn and weep over the sins and misdeeds we have committed and realized,/ and pleadingly and suppliantly call on God to forgive us for and absolve us of these sins."

JOHANNES SCHÖNER'S *CONIECTUR* (1531)— SMALL BUT SUBSTANTIAL SERVINGS OF COMETARY KNOWLEDGE

In the example of Elias Ehinger, we have seen that a popular cometary text often promised more than it could (or wanted to) deliver. Yet there is still something more substantial in this trade: Consider the year 1531, when Halley's Comet appeared, whose periodicity its namesake would discover in 1705. In 1531, there was still no mathematical proof for the superlunarity of comets. On the contrary: So far, all efforts to determine their distance from the earth located them under the moon and thus reinforced the Aristotelian model.[15] There were hardly any systematic attempts to quantify the apparent movement of comets in the heavens; Paolo dal Pozzo Toscanelli's carefully prepared series on six comet sightings between 1433 and 1472 remained unpublished. An important boost for cometary research came in the years 1531-1539, when five conspicuous comets appeared. This offered the opportunity for systematic observations and comparisons in a fairly short period of time. One of the most influential documents in this respect is Peter Apian's monumental work *Astronomicum Caesareum* (1540), which brings together his observations in comprehensive detail in its second part. It was preceded by two German-language publications on the comets of 1531 and 1532 in popular formats by Apian—a topic that will be taken up again in the next chapter.

Another 'professional' who observed Halley's Comet in 1531 was Johannes Schöner, mathematician, astronomer, geographer and cartographer, globe-maker, and horoscope specialist.[16] Since 1526, he was a teacher for mathematical subjects at the Egidien-Gymnasium in Nuremberg. In this position, he was also responsible for the compilation of annual prognostications, blood-letting tables, and calendars,[17] and thus provided the local community with requested astronomical-astrological knowledge concerning everyday life. He had observed the comet on several evenings in August,[18] and his results and explanations were, as we know for instance from the Melanchthon correspondence,[19] eagerly anticipated and even demanded by the scientific community.

---

15 An error analysis of the calculations of Johannes Regiomontan and Johannes Vögelin can be found in Crüger 1619: 86-93.

16 On Schöner, cf. Maruska 2008.

17 Matthäus 1969: 1020-1025; Maruska 2008: 56-58.

18 Kokott (1994: 82) plausibly suggests that Schöner's dates of his evening observations (August 15-25) should be counted back by one day in each case.

19 MBW T 5 (2003): No. 1178, 2 (Letter of August 19, 1531 to Joachim Camerarius the Elder).

He presented his findings in popularized form in a short German cometary text, the six-folio *Coniectur*, which he dedicated to the Council and citizens of Nuremberg in return for a corresponding payment (Schöner 1531a).[20] Since it was "not only difficult but also risky" ("nit allain schwerlich sonder auch geferlich" [ibid: A1v]) to evaluate comets, he preferred to offer some *astronomical* orientation to the general public in light of the many widely circulating opinions and interpretations.[21]

Figure 2: Johannes Schöner: Coniectur odder abnemliche außlegung [...] vber den Cometen so im Augstmonat/ des M.D.XXXj. jars erschinen ist, *Nuremberg: Friedrich Peypus 1531 (Title Page)*

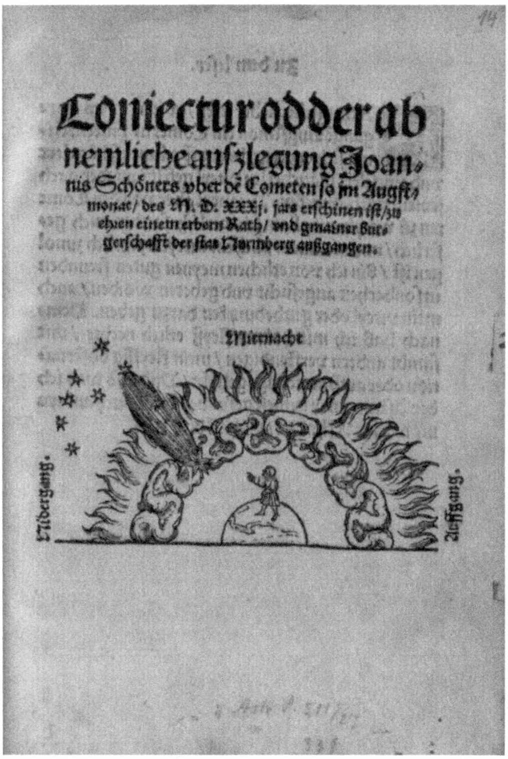

Source: BSB Munich, Res/4 Astr.p. 90 h#Beibd.14, urn:nbn:de:bvb:12-bsb00029070-4

20 On this text, cf. Kokott 1994: 77-83 and 131-133; Meinel 2009: 17-20; Gindhart 2017b: 203-204.

21 A list of contemporary cometary texts is found in Brüning 2000: 18-23.

The *Coniectur* offers the usual elements of cometary pamphlets (observation report, remarks on the genesis and classification of comets, astrological and theological interpretations) and is based on disciplinary know-how, which is 'broken down' for a broad audience and imparted with internal differentiations. The information is presented step-by-step and with some repetition in easily comprehensible, carefully arranged textual units. These are provided with significant headings, introduced with ornamental initials and subdivided with virgules. This mise en page adds value to the print and enables even people of limited literacy to cognitively grasp the content. By also calling upon the practical experience of non-educated recipients, the latter are able to contribute their expertise to the reading and acquire, via transfer, an understanding of natural historical knowledge (or at least parts of it). At the same time, the writing is also interesting for scholars, even for specialists, not only because of the observational data, but also because of the astrological interpretation, which appealed to all educational levels.

The design of the title page, too, is valuable from a commercial and didactic point of view (*Fig. 2*). The woodcut interacts closely with the main text in an intermedial way by visualizing the contents of the first two chapters dealing with the path and genesis of the comet, while at the same time condensing them as a programmatic statement. It refers to Schöner's first observation on the 15th (probably the 14th) of August 1531 at 9 pm: As he describes it, we see the comet above the western horizon ("Nidergang"), its tail forming a straight line with the two rear stars of the Big Dipper, Dubhe, and Merak (α and β UMa). In accordance with Aristotelian theory, it is located in the uppermost atmospheric layer near the fire sphere, where it "print als ain lange grosse kertzen/ so lange biß die sewle gantz verprindt vnd verzert wird durch die flammen" ("burns as a long, big candle,/ until this pillar of fire is utterly consumed and eaten up by the flames") (ibid: A3r). An observer-figure gestures from the earth, which is in the center of the implied cosmos, to the comet, which is depicted as a six-pointed star with tail. In this iconized form, we find comets in the 16th and 17th century on the covers of innumerable publications. It can be assumed that this 'comet pictogram' went a ways toward supporting the promotional impact of the title pages due to the high degree of pattern recognition.[22] A noteworthy detail is that the elongation of the axis of the tail touches the center of the earth. This aspect is not mentioned in the *Coniectur*, but it is referenced in Johannes Regiomontan's

22 On the commodity character of early modern cometary literature, cf. Gindhart 2006: 120-138; 2017a: 44-49.

Problemata, which Schöner published from the latter's literary remains during the time of his comet sighting.[23]

Schöner did not actually apply the rules outlined therein for determining the parallax of comets. As far as we know from the *Coniectur*, his gathered data were insufficient for parallax calculation.[24] He only estimates the distance of the comet and its dimensions (within the framework of the Aristotelian theory), but records all parameters relevant to astrological interpretation in order to provide it—as announced in the preface—with a solid foundation. Wolfgang Kokott examined Schöner's coordinate determinations and stated that they did not fall short of Peter Apian's accuracy. In other words, they show a standard that was at the upper limits of contemporary possibilities.[25]

In his interpretation, Schöner first names the traditional spectrum of disasters that can be expected as 'natural' outcomes following the hairy stars. On the basis of the astronomical data and the determined influence of Mars and Mercury, with which the latest comet was in conjunction at the beginning of its appearance in Leo, he then becomes more precise. For example, he specifies the imminent illnesses or the groups of people, disciplines, cities, and regions particularly affected by the comet's impact. He reports on the interpretative systems of the astrological authorities Albumasar, Haly Abenragel, and Leopold of Austria that can be applied to the cometary parameters, and thus provides his readers with ample material for their own further interpretations.

Schöner's *Coniectur* was extraordinarily successful. It was immediately reprinted in several centers of the Reformation, including Leipzig, Magdeburg, and Zwickau; a Dresden broadsheet offers excerpts from the prognostic chapters. This is probably not least due to the fact that Rome was mentioned as a prominent place threatened by the effects of the comet.[26]

23 Schöner 1531b: C3v (cf. Kokott 1994: 129-131). In the dedication letter (A3r/v, dated August 21, 1531) to Erasmus Ebner, mathematically inclined son of a Nuremberg patrician family, Schöner briefly describes the appearance and substance of the current comet and refers to contradictory interpretations. According to Schöner, humans could never penetrate the depths of divine wisdom to give an exact exegesis of comets; what God determined for them to know, however, on the basis of mathematics, Regiomontan had shown in the *Problemata*.

24 Cf. Kokott 1994: 131-132.

25 Ibid: 10, 80 and 194. Schöner (1531a: A2v-A3r) detects a threefold movement of the comet: the daily movement from east to west; the course from west to east through the zodiac signs Leo to Libra (longitude); and its movement from the north to the zodiac (latitude).

26 Rome is specifically named with other cities (ibid: B1v), and an Albumasar quote at the end of the work is related to "Babylon" Rome: "So ein Comet erscheint erstlich im

PETER APIAN'S *PRACTICA* (1531)— AN ANNUAL PROGNOSTICATION AS A 'STEPPING STONE' FOR LEARNED COMETARY KNOWLEDGE

As mentioned above, Schöner's observation data and position determinations from the popular *Coniectur* were consulted by other experts: For example, Peter Apian, serving as a mathematics lecturer at the University of Ingolstadt since around late 1526,[27] used them as comparative material and an argumentative foundation for his own observations (Apian 1531: K3r). He, too, made his contribution to the year's literary output on comets. But unlike Schöner, Apian employed the format of an annual prognostication, a *practica*, as a 'stepping stone' for, among other things, giving detailed instructions on cometary observation and position and orbit determinations with a wealth of data material. This material provided the basis for the law he introduced in his *Practica* that is still relevant to this day: the antisolarity of the comets' tails.

From the 1520s to the 1540s, Apian regularly produced broadsheet calendars and annual prognostications and established himself as a veritable 'brand name' in this booming sector.[28] His surviving practicas show a range of 7-8 folios with a uniform, thus reliably recurring structure. Apian not only greatly extended the scope of the *Practica* for the year 1532 (now 38 folios), he also intended to give it a significance which exceeded that of a merely temporal commodity.[29] Combining a detailed daily prognosis for the following year (which, as ever, aimed at a broad public)[30] with a comprehensive cometary treatise, the work sought to

---

zaychen des Lewen/ bedeute er krige zwischen den kǒnigen/ vnnd sonderlich den kǒnigen Babilonie/ (vernim wie oben) auch wetagen im pauche" ("If a comet first appears in the sign of Leo,/ it indicates wars among the kings,/ and especially for the kings of Babylon/ [see above] also anguish") (ibid: B2r).

27 On Apian, cf. Röttel 1995 with further literature.

28 On Apian's broadside calendars, cf. Biller 1995; on the format of annual practicas and their market value, cf. Matthäus 1969: 1001-1006 and 1199-1234; Green 2012: esp. 109-139; Kempkens 2014.

29 Explicitly in Apian 1531: A2r.

30 The practica has a distinctly didactic character and also grants the reader clearly formulated methodical insights. At the beginning of the writing, all the used symbols are listed and named and their combined notation is explained, "damit der leßer/ der dißer kunst nicht geu̇bt ist/ die Practica deste leychter versteen mǒcht" ("so that the reader/ who is not practiced in this art/ might more easily understand the practica")

bring not only ephemeral, but also long-term benefit in presenting best-practice models. The unique ambition and quality of the writing are underlined by the dedication to the Bavarian dukes Wilhelm IV and Ludwig X.[31] Apian expressly names his Ingolstadt students (and those of other universities) as addressees who were to profit from both parts of the work for their actual training and future professional work.[32]

In the dedication, Apian explains that in his cometary treatise he wants to go beyond the traditional and unspecific "gemaine ding" ("general things") (ibid: A1v), which are simply repeated in many printed works for lack of observation data. He wants to see "ob etwas mercklichers oder heymlichers darvonn gesagt oder geschriben moͤcht werden" ("whether something more significant or secret can be said or written about this") (ibid). This applies to the genesis and nature of the comet as well as its path. At the same time, the writing is supposed to serve as a guide for third parties to observe future comets (which Apian announces already for 1532 due to planetary constellations) and to calculate their orbits "on sonderliche mühe" ("without much effort") (ibid), thus increasing the pool of data. The detailed mathematical instructions aim at an exclusive circle of readers. Beside the aforementioned students, this probably also included reckoning masters, land surveyors, builders, and other representatives of the educated, non-academic population with an affinity for mathematics and astronomy. For this target audience, Apian wrote a whole series of German-language treatises in which he drew on his didactic expertise and expended considerable intermedial effort to communicate astronomical and mathematical knowledge with practical

(ibid: A3v) In brief chapters, remarks then follow on the solar eclipse forecast for August 30; on the rising and setting, visibility and invisibility of the planets in the course of the year; on the beginning of the seasons with weather prognostications; on the year's ruling planet (Venus); on threatening diseases; and on the fortunes of the social classes. A second part contains the aforementioned annual calendar, which lists all relevant sidereal constellations for each day as well as the weather forecasts derived from them, medical instructions (with a list of the critical days), and propitious dates for haircutting, sowing, and planting.

31 The two dukes are named on the title page. The Wittelsbach lozenges and the coat of arms of Landshut, the residence of Duke Ludwig, are also depicted there. The following dedication letter (ibid: A1v-A2r) dates November 23, 1531.

32 Ibid. On mathematics and astronomy at the University of Ingolstadt and on Apian's work as a mathematics lecturer, cf. Schöner 1994 and 1995. As part of his teaching activity, Apian probably also instructed medical students in iatromathematics. These students and those of the arts faculty should therefore be regarded as addressees of the practica as a kind of teaching model.

relevance and applicability. In this way, he allowed for those who were unfamiliar with Latin to gain access to these branches of knowledge.[33]

Figure 3: Peter Apian: Practica auff dz 1532 Jar [...], *Landshut: Georg Apian 1531 (Title Page)*[34]

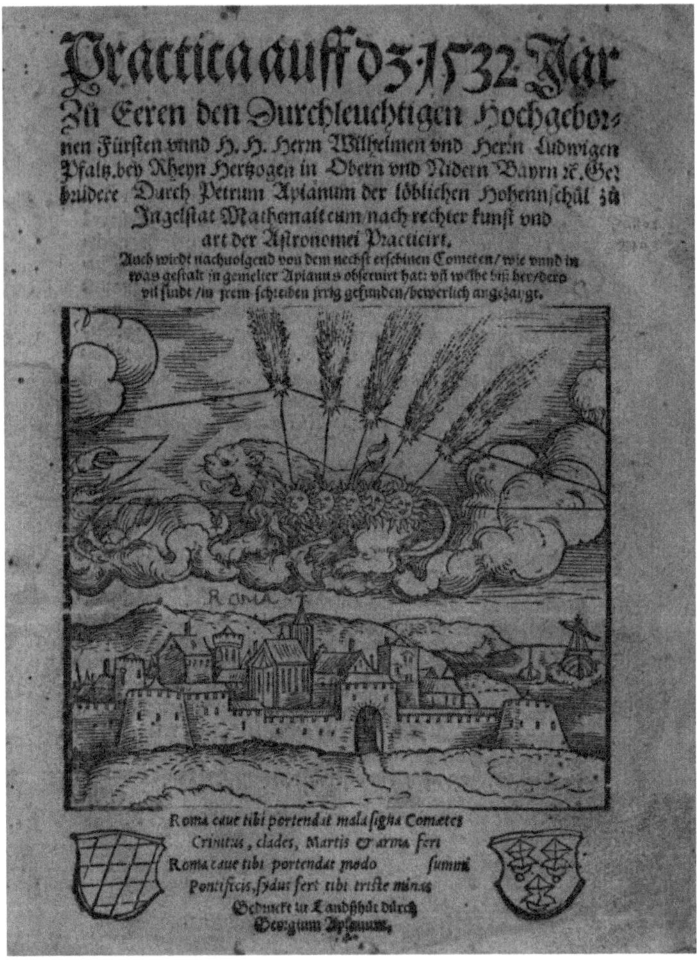

Source: SB Regensburg, 999/Philos.2743, urn:nbn:de:bvb:12-bsb11111496-4

33 Hamel (2008) describes Apian's didactic concern by taking a look at (among other works) his *Instrument Buch* (1533) and the *Kauffmanß Rechnung* (1527). The *Instrument Buch* includes cut-out sheets for gluing onto wooden boards in order to make the instruments also available to customers with limited financial resources.

34 Inspired by the Latin epigram, a contemporary hand added "ROMA" and a cross to the view of Ingolstadt.

The majority of readers who bought the *Practica* to utilize Apian's forecasts as they did every year, and now to also learn more about his cometary observation and interpretation—as advertised on the title page in text and image—, probably did not belong to these target groups. However, when glancing curiously at the cometary part of the writing, they would have at least grasped the framing passages they already knew from cometary pamphlets: the introductory remarks on the formation of comets and the astrological interpretation (the last theorem of 84 in total). Moreover, in leafing through the mathematical chapters, they also would not have failed to take notice of a new piece of knowledge: the antisolarity of the tail.

All the images spread throughout the text visualize this very phenomenon. They accompany, for instance, the reports of every observation day as well as the 56th theorem, in which Apian gives an account of his trigonometrically-verified observation "wie der schwantz sich nach der Sonnen schein alzeyt gewent hat" ("how the tail was always directed in opposition to the sun's light") (ibid: G4v), like a shadow cast in relation to the position of the light source. And, of course, the antisolarity is also illustrated prominently on the title page (*Fig. 3*): Its woodcut shows the first stages of the comet's path in the sign of Leo (to be read from left to right), which reaches towards the ecliptic. Mercury, with which the comet was in conjunction at the beginning of its sighting, is represented by its planetary symbol below the lion's mouth. Connecting lines between the comet's and the sun's positions clearly visualize that the comet's tail was continually directed away from the sun. The whole event is located above the fortress town of Ingolstadt. There, Apian had observed the comet from August 13 to 23 in relation to Arcturus (α Bootis) on the western horizon after sunset, like Johannes Schöner (*Fig. 4*).

It is worth noting another element of the title page: Below the woodcut there is a Latin epigram, which warns Rome of the misfortunes revealed to this city by the comet.[35] What exactly threatens the Pope, however, is left open by a spatium in verse 3. Here, the missing dactylic accusative object may be freely supplemented.[36] A warning like this on the title page of a Catholic print is remarkable and no doubt encouraged Protestant readers everywhere to make the purchase.

35 The twice-cited "Roma cave tibi" (v. 1 and 3) may well be an intertextual reference to the prodigious warning which an ox belonging to Gnaeus Domitius Ahenobarbus is supposed to have uttered during the latter's consulate (cf. Livius, *Ab urbe condita*, 35.21). On the astrological explanation of the threat to Rome and the Pope (and other cities, regions, and countries), cf. Apian 1531: K3v-K4r.

36 A contemporary hand has added the—obviously intended—noun "funera" in one of the copies preserved in the Regional State Library (SB) Regensburg (999/Philos.2092/2098).

Figure 4: Peter Apian: Practica auff dz 1532 Jar [...], *Landshut: Georg Apian 1531 (F3v, Detail)*

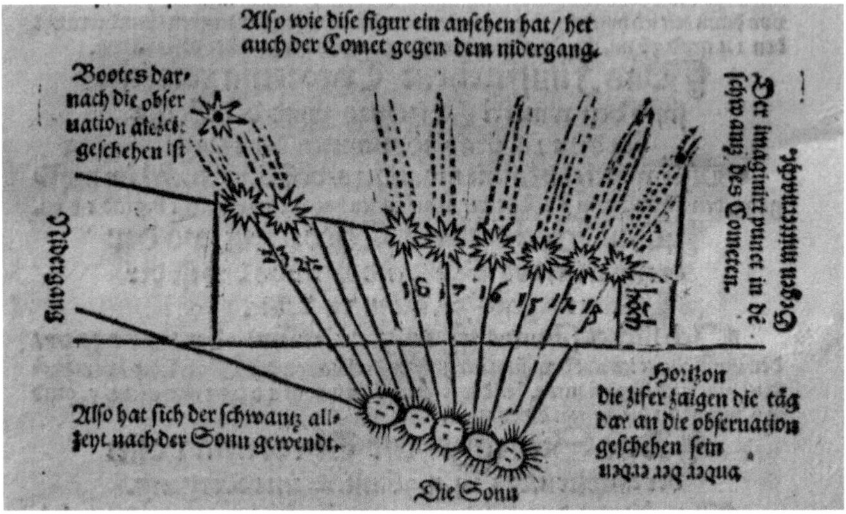

Source: SB Regensburg, 999/Philos.2743, urn:nbn:de:bvb:12-bsb11111496-4

Apian names, visualizes, and mathematically proves the antisolarity of cometary tails, but he does not elucidate the causes of that phenomenon. This is due to the fact that he follows the Aristotelian cometary theory, which is outlined in detail at the beginning of his writing. On this basis, the newly found law simply could not be explained. It required, therefore, an adaptation: As Apian notes in the *Astronomicum Caesareum* of 1540, the tail and its behavior can be explained if one regards the comets not as ignited vapors, but as vapors illuminated by the sun.[37]

Other scholars, such as Girolamo Fracastoro, also recognized the antisolarity on the occasion of the comet of 1531. Apian was the first, however, to establish the law graphically and to make it quasi-official by dedicating it to the Bavarian dukes.[38] Moreover, the accuracy of his observational material was also sufficient for later astronomers to use it as a basis for their own orbit calculations. Edmond Halley, for example, could determine the orbit of the comet of 1531 from Apian's data and identify it as a periodic comet with the appearances of 1607 and 1682. However, he did not receive the data via the German *Practica*, but via the

37 "Sunt tamen qui cometam esse dicunt exalationem subtilem non inflammatam, sed lucentem a lumine Solis, cui sententiae et nos accedimus [...]" ("Some people claim, however, that a comet is a subtle vapor not set on fire but illuminated by the light of the sun, an opinion, which we share too") (Apian 1540: O1v).

38 On early knowledge of the anti-solar tail direction in China, cf. Kokott 1994: 56.

Astronomicum Caesareum. It was in the latter's second part, nine years later, that Apian published observations and evaluation results on the comets of 1531, 1532, 1533, 1538, and 1539 in Latin.[39]

A LARGE AND LONG-LASTING QUARREL—*PHILOSOPHI* VS. *ASTRONOMI*

The discovery of the tails' antisolarity immediately found its way into learned literature.[40] Looking ahead again to the year 1618, one sees that it also had spread in the meantime in popular writings. And not only antisolarity: While quite a few authors (and not just theologians) in 1618/19 still continued to fill the traditional formats with the well-known 'certainties' and to stick to the Aristotelian cometary theory, it can be observed that the astronomical knowledge acquired in the interim was increasingly distributed, propagated, and established via popular texts and entered into competition with the 'old' knowledge. Here, episteme is noticeably in motion. The strategies with which the safeguarding of traditional and the establishment of new knowledge were pursued and undermined, promoted and fought, are particularly intriguing.[41]

The supporters of the Aristotelian model often do not even address superlunarity theories in their writings. They vehemently adhere to their established epistemic system and repeatedly emphasize the coherence of vapor theory. If they mention superlunarity at all, it is in order to critically examine *ancient* models such as the planetary theory of the Pythagoreans, but *not* Brahe's or Kepler's concepts, which for their part do not explain comets as permanent stable bodies, but as transitory condensations of aether. They also deny the methodological suitability of astronomy and mathematics to obtain valid knowledge about com

39 Apian 1540: N2v-O3r. On the documentation of the observations in the *Astronomicum Caesareum*, cf. Kokott 1994: 60-72. The comets are treated here as typical applications for the nomographic reduction method; in the two German cometary writings (Apian 1531 and 1532, cf. Kokott 1994: 58-59), the comets of 1531 and 1532 served as application examples for the sine tables (ibid: 60). The cometary text of 1532 has the same didactic concern as the *Practica* of 1531, namely, to give an introduction to trigonometric astronomy, to guide future comet observations, and thereby to make astrological interpretations more reliable.

40 For example, in the *Norica* (1532) by Joachim Camerarius the Elder, in which the antisolarity of cometary tails is stated, but without any mention of Apian.

41 On the following, cf. Gindhart 2006: 222-267.

ets; or they accept their suitability, while referring to the erroneous distance determinations of the early 16th century as conclusive mathematical proofs for sublunarity. They often claim (despite knowing better) that reliable observation data and parallax determinations of experienced astronomers are *missing*, so that no statements could be made about the distance of the comets. In short, their main concern was a concerted ideological defense of the system.

In return, the advocates of superlunarity make use of their own strategies to assert their supremacy in the field of cometology: In scholarly literature, Aristotelian book learning, with its logical deductions, is argumentatively picked to pieces and confronted with a bulwark of observation data and calculations. Mathematical *veritas* is opposed to peripatetic *probabilitas*, and Aristotelian natural philosophy is, in turn, denied the possibility of being able to make valid statements about cosmic bodies and thus also comets. This discussion is reflected in the popular writings, too. Kepler and Brahe are established as counter-authorities to Aristotle and are panegyrically celebrated; the *Aristotelici* are denounced as deliberate obstructors of true knowledge and slaves to book knowledge. And, although the realization was still a ways off, Aristotelian theory is presented as having already been overcome and being obsolete since the parallax measurements from 1577 onwards.

For example, in a pamphlet entitled *Kurtzer Bericht*, the Coburg city physician Michael Schön declares Kepler to be the greatest living astronomer whose cometary theory (as Kepler asserted himself) even Aristotle would have converted to were he still alive.[42] In his writing, Schön gives a contrasting as well as succinct account of the theories of the *philosophi celebri* and those of the *astronomi exactiores*, thereby offering his readers insight into the models currently in circulation. He decides in favor of superlunarity, since he sees astronomy with its methods as the competent discipline and because the *mathematici* were in agreement that the comets are located in the planetary sphere. In addition, according to Schön, God wanted the true nature of comets to be finally recognized with the help of astronomy. As a veritable gift of His grace, He had now "solche Leut erweckt vnd gegeben/ welche diese Kunst indefesso studio, & exactissimis instrumentis quam diligentissime excolirn" ("brought to life and provided such people/ who most thoroughly cultivate that art with indefatigable eagerness and

42 Schön 1619: A4r/v. The motif of "Aristoteles redivivus," who would adopt the new knowledge and reject his own theory, is widespread. It can be found in the writings of a variety of astronomers such as Tycho Brahe, Galileo Galilei, and Kepler himself, and as early as the year 1576 in the *Responsio* of Thaddaeus Hagecius (documented in Weichenhan 2004: 55-57). Meinel 2009: 42-44 gives a summary of Schön's writing.

the most precise instruments") (Schön 1619: A4r). But whatever their nature—sublunar or superlunar—, comets nevertheless continued to be media of divine communication and messengers of impending (but predictable) disasters for *all* authors. The theological discourse thus remained stable.

Figure 5: Peter Crüger: Kurtzer Bericht Von dem grossen noch zur zeit scheinenden Cometen […], *Gdańsk: Andreas Hünefeldt 1618 (Title Page)*

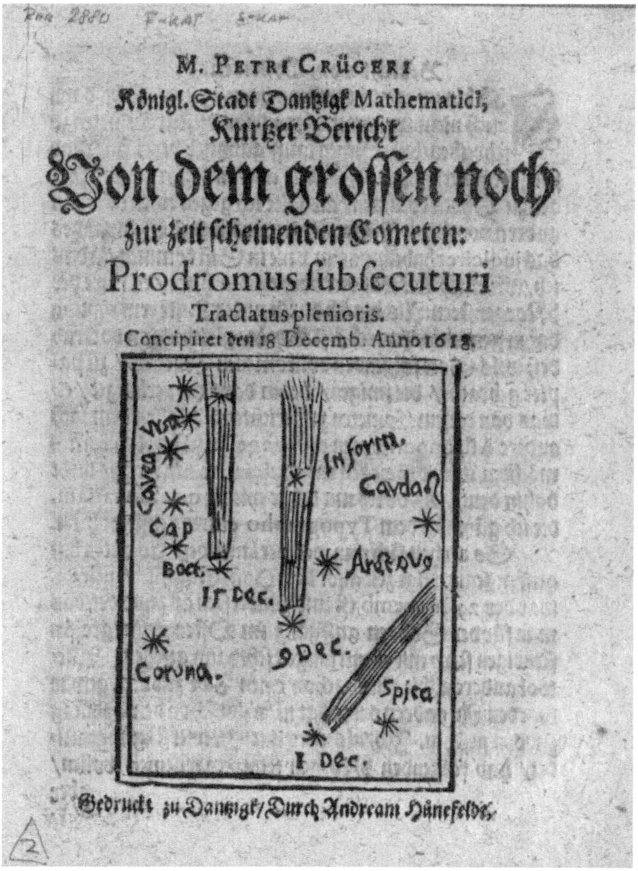
M. PETRI CRÜGERI
Königl. Stadt Dantzigk Mathematici,
Kurtzer Bericht
Von dem grossen noch
zur zeit scheinenden Cometen:
Prodromus subsecuturi
Tractatus plenioris.
Concipiret den 18 Decemb. Anno 1618.

Gedruckt zu Dantzigk/Durch Andream Hünefeldt.

Source: ETH-Library Zurich, Rar 2880, https://doi.org./10.3931/e-rara-443

Peter Crüger,[43] professor of mathematics and poetry at the Gymnasium in Gdańsk, had originally intended to publish one single comprehensive text with all gathered observation data, their evaluations, and astrological interpretations after the last

43 On Crüger, cf. Jensen 2006: esp. 51-126 and 181-186; Herbst 2018.

sighting of the Winter Comet. Nonetheless, as false information about the comet had been circulated under his name, he felt compelled to first draft—before his voluminous *Uranodromus cometicus*—a six-folio pamphlet in the vernacular entitled *Kurtzer Bericht* (like Schön's abovementioned publication).[44] In it, he first vividly describes his observations and position determinations, whereby the text (as in Schöner's *Coniectur*) is closely connected with the title woodcut (*Fig. 5*). Here, the distances of the comet to certain fixed stars are visualized for three dates; the direction and dimensions of the tail are also shown fairly true to scale. The text, serving as a legend, reveals the individual details of the woodcut, which itself brings greater clarity to the observation data. Crüger reconstructs the location of the first appearance and predicts the comet's further path and visibility. He refutes the false theses in circulation and confirms the tail's antisolarity. However, he does not make any statements yet about the distance of the comet, which he reserves for the *Uranodromus*. As with Schöner, the astrological interpretation constitutes a large part of the writing. Crüger, however, denies that the direction of the tail has any astrological significance.

In the following *Uranodromus* (*Fig. 6*), Crüger manages to pull off a remarkable balancing act between popular cometary literature and exclusively learned treatises such as Johannes Kepler's *De cometis libelli tres*.[45] After having reported on the first sightings of third parties, Crüger offers a brief description of the daily cometary positions he observed, which incorporates his statements given in the *Kurtzer Bericht*. The calculations thereafter also begin with German passages in which Crüger explains what astronomical observations entail, what they can contribute to the study of comets, and what kind of instruments he made use of. Almost apologetically, he explains to the reader that sections with Latin calculations will intermittently follow and for what purpose:

"Diß vnd folgende etliche Capitel/ darinnen die Astronomischen Rechnungen enthalten, dienen nicht fuͤr jederman zu lesen/ vnd ich hab sie dennoch vmbstendlich setzen muͤssen fuͤr die Astronomos vnd Philosophos, die da gerne rationem calculi indubitatam haben wollen."[46] (Crüger 1619: 8)

44 Cf. Crüger 1618: A1r/v. On the *Kurtzer Bericht*, cf. also Jensen 2006: 57-59.

45 On the *Uranodromus cometicus*, cf. also Jensen 2006: 59-66; Granada 2016: 287-290.

46 "This and several following chapters,/ which contain the astronomical calculations, are not suitable for everyone to read;/ yet I've had to put them in at full length for the astronomers and philosophers who would like to have an indisputable demonstration of the calculation."

Figure 6: Peter Crüger: Uranodromus cometicus [...], *Gdańsk: Andreas Hünefeldt 1619 (Title Page)*

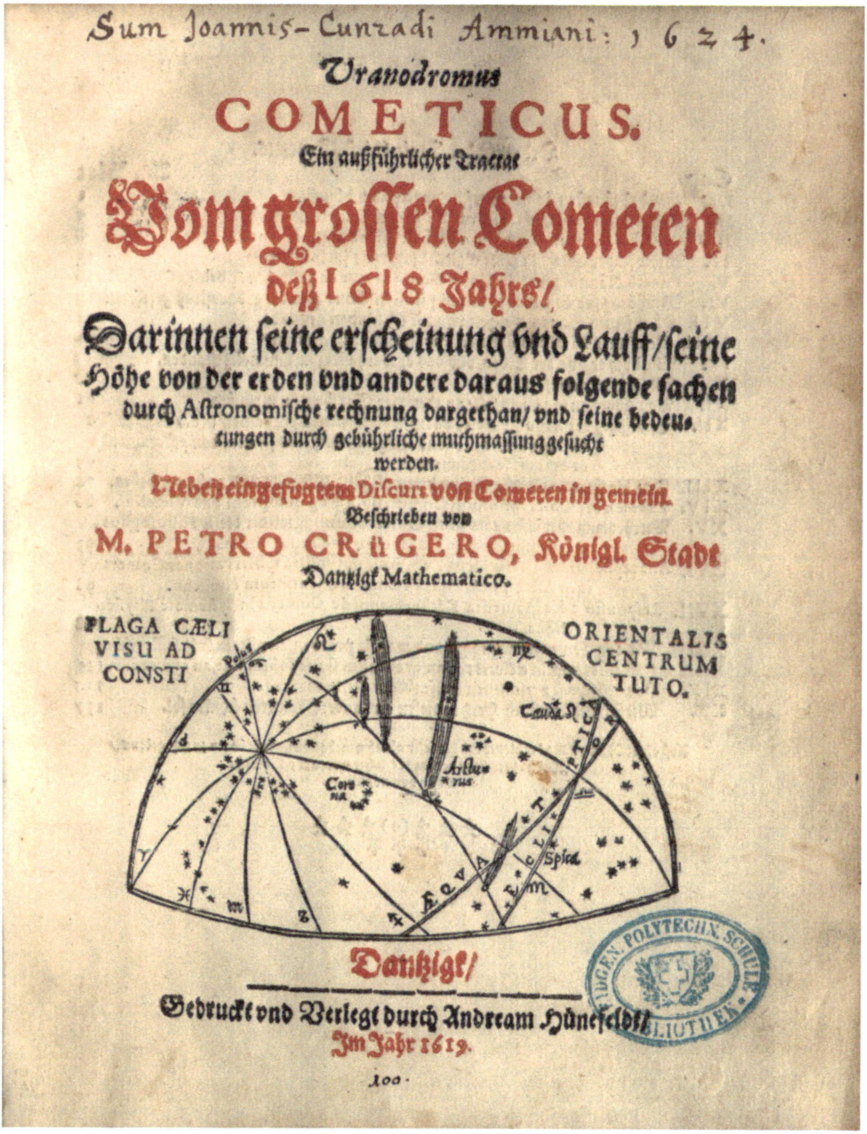

Source: ETH-Library Zurich, Rar 4058, https://doi.org./10.3931/e-rara-314

In these chapters, Crüger formulates and summarizes in German the modus operandi and epistemic goals for the non-specialists and gives basal explanations. For instance, he outlines what a parallax is, which he then determines for the Winter Comet using various methods. His aim is to create a manifesto of 'new' mathematical knowledge and to establish comets "durch Astronomische rechnung" ("through astronomical calculation") (ibid: A1r) as celestial bodies moving in circular orbits. Crüger further argues that because their tails face away from the sun, they must be conglomerates of cosmic matter which the sun acts upon (illumination, tail formation, dissolution).[47] Because he is unable to sufficiently explain the condensation process (ibid: 113-114), Crüger attributes their genesis to the divine: Comets are readable signs sent from God to mankind.[48]

With the *Uranodromus*, Crüger sought to reach educated 'laymen' and scholars. Among the latter, his chief targets were the *Aristotelici.* He thoroughly refutes their cometary model, defends the legitimation and reliability of astronomical observations and calculations against their attacks, and criticizes their stubborn adherence to Aristotelian doctrine.[49] Crüger had already demonstrated that progress could be made here: Fourteen years earlier (on May 14, 1605), he himself acted as respondent in a disputation presided over by the influential Aristotelian Bartholomäus Keckermann at the Gymnasium in Gdańsk. On the basis of co-authored theses, as Crüger confesses,[50] he set forth the sublunar model and its meteorological implications in detail and refuted superlunarity. His mentor Keckermann had realized that the survival of the Aristotelian system was at risk in the wake of observations and calculations following Brahe's. He responded by erecting a veritable bulwark of protection. This was comprised in particular of his disputation and the *Diaskepsis* (published along with it), which emphasized the impossibility of exact astronomical observations and measurements, and his lectures, which were posthumously incorporated in the *Systema physicum* (first printing in 1610). In his *Uranodromus*, Crüger was now trying to

47 A concise summary in German can be found, e.g., in Crüger 1619: 53-55 and 114-115 on the influences of the sun (with an image). Crüger locates the Winter Comet between Venus and Mercury (ibid: 56-57).

48 On the significance of comets in general (with dozens of historical examples), and of the Winter Comet in particular (with remarkably specific interpretations and critical references to contemporary developments), cf. ibid: 115-137. The text closes with an exhortation to repentance and conversion, fully in line with Lutheran orthodoxy.

49 Ibid: 63-113 (with further evidence for the regular course of comets, the antisolarity of their tails, and their scarcely perceptible parallax).

50 Ibid: 98-99. The *disputatio extraordinaria* (Keckermann [pr.]/Crüger [resp.] 1605) also found its way into the influential compendia of Keckermann's writings, such as the *Disputationes philosophicae* (1606 and 1611) and the *Systema systematum* (1613).

tear down this bulwark he once had helped to fortify and dedicated two separate chapters (in Latin) to invalidating Keckermann's positions.[51]

Crüger also communicated the superlunarity of comets in the popular annual Schreibkalender, which he produced for over thirty years during his position as professor of mathematics.[52] In the calendars from 1615 onwards, he treated various topics from the fields of astronomy, astrology, meteorology, and calendar-making in question-answer form in a lucid and didactic way.[53] In 1631, under the title *Cupediae astrosophicae*, the printer Georg Baumann from Wrocław collectively published the questions and answers that had appeared up to that time. In the dedication letter to the Gdańsk secretary Martin Rubach,[54] he underlines the special achievement of Crüger and his vernacular calendars in imparting the fields of knowledge connected with mathematics to a broad public:

"[Gott] hat newlicher Jahren erwecket M[agistrum] Petrum Crügerum [...] welcher [...] die aller fůrnehmbsten vnd nůtzlichsten Lehrpuncten/ sonderlich zur Astronomia/ zur CalenderKunst/ zur Astrologia/ zur Geographia gehörig/ durch Frag vnd Antwort dermassen außfůhrlich/ deutlich/ leicht vnd verständlich in vnser Deutschen MutterSprach erkläret vnd außgeführet/ daß auch gemeine Idioten/ welche sonsten entweder gar nichts studiret/ oder ja von Mathematischen Sachen gantz vnberichtet sein/ dennoch dieselben hohen dinge nunmehr gar leicht fassen vnd begreiffen mögen."[55] (Baumann 1531:):(3v-):(4r)

---

51 Crüger 1619: 98-113. On Keckermann's methods of argumentation and their driving force, cf. Gindhart 2006: esp. 249-253; Granada 2016: 260-266; on Crüger's factual (not personal) positioning against his teacher, who had died in 1609, ibid: 287-290.

52 The calendars cover the years 1609-1640; from 1610 onward, they were printed continuously by Andreas Hünefeldt. On Crüger's calendar production, cf. Jensen 2006: 280-281 and 288; Kremer 2012: 483-488; 2014. On the genre of the Schreibkalender and its functions in imparting knowledge, as well as on its authors and the format's lines of development, cf. Herbst 2010; 2012; Herbst/Greiling 2018.

53 The questions deal, for example, with why farmers still adhere to the old calendar (Baumann 1531: B4r); where the division of the ecliptic and the names of the individual zodiac signs come from (D2r-D3v); whether one can cast a horoscope for buildings and cities (G1v-G2r); whether the world was created in spring (H4r-I1v); whether planets and stars have their own light or are illuminated by the sun (I3v-K1r with reference to the phases of Venus); and much more.

54 Ibid:):(2r-):():(1v. Baumann states the importance of mathematics as a fundamental science. It is therefore a disgrace when scholars who have studied other subjects are outwitted in mathematics by "einfeltige[n] Läyen" ("simple-minded laymen") (ibid:):(2v).

55 "In the recent past, [God] has animated M[agister] Peter Crüger, [...] who [...],/ through questions and answers,/ explained and elucidated the most important and useful

The questions repeatedly deal, among other things, with the predictability of thunderstorms. In this context, comets are also treated. Crüger rejects the notion that they can influence the formation of storms[56] since Aristotle's meteorological model was no longer valid.[57] Referring to his *Uranodromus* again and again, he repeats his main theses on the comets. Namely: Unlike the perfect and eternal heavenly bodies, they are not capable of triggering major developments on earth; they are created by God for another purpose, specifically as signs; they are temporary, loose conglomerations which the sun illuminates and whose substance the sun casts out as a tail. If this tail were to reach the earth, one could expect some cometary impact (or an impact because of the sun's rays descending via the tail). Most comets, however, are too far away for this. They leave no material residue; the power of the sun dissolves them so that they once again become a "subtile himmelsmateria" ("subtle heavenly matter") (ibid: N3r).

The popular Schreibkalender, which had been on the market since the 1550s, had a tremendous multiplier function for communicating disciplinary (especially astronomical) knowledge thanks to additions like Crüger's questions and answers.[58] Due to their unique character (colloquial style, minimal use of technical language, translations or paraphrasing of Latin passages, summaries), they could be profitably read as "astrosophical tidbits" by educated laymen who had no knowledge of Latin, as well as by scholars, who were the target audience for the precisely documented references to the works of ancient and contemporary authorities.[59]

---

teaching points,/ especially in astronomy,/ in calendar-making,/ in astrology,/ and in geography,/ so thoroughly/ and clearly,/ so easily and comprehensibly, in our German mother tongue,/ that even simple laymen,/ who have either studied nothing at all/ or who are completely unacquainted with mathematical matters,/ can now easily grasp and understand these major things anyhow."

56 Baumann 1531: N2r-N3r from the calendar for the year 1621.

57 Cf. ibid: N2r: "Weil aber nun zur zeit gnugsam offenbahr vnd am tage ist/ das die Cometen nicht Elementarischer/ sondern Himlischer materi sind/ vnd im hohen Himmel schweben [...]" ("Because right now it is obvious and clear enough/ that comets do not consist of elementary/ but rather heavenly matter,/ and that they reside high above in heaven").

58 In the Schreibkalender, novel astronomical findings, and even the author's own insights, were frequently debated. For example, in his calendar for the year 1612, Simon Marius published his discovery of Jupiter's satellites, which was independent of Galileo Galilei's, and also refers there to the latter's *Sidereus nuncius* (cf. Kremer 2014: 106).

59 A statistic of the authorities cited by Crüger can be found in Kremer 2014: 109. Among the early modern scholars, Johannes Kepler, with whom Crüger corresponded

Unlike Johannes Kepler, for instance, who regarded the comets as proofs for heliocentricity, Crüger remained true to the Tychonic system. He promoted—also in the calendars—the daily rotation of the earth, but not its annual motion around the sun in a world system, whose spatial dimensions he could not fathom.[60]

CONCLUSION

The aforementioned examples show how various authors decided, out of quite different motivations, what kind of knowledge about the nature of comets they wanted to impart and *how* and which formats of ephemeral literature to use toward this end. We examined the reasons why the Aristotelian meteorological model persisted as the reference model of Protestant sermons well into the 17th century, and indeed why it *needed* to persist. At the same time, we saw—regarding Dieterich's *Ulmische Cometen Predigte*—how alternative natural-historical knowledge could also be presented to the reader. We recognized how an astronomer (Schöner) 'broke down' his exact observation data and his astrological know-how for as wide an audience as possible, achieving great market success with his short cometary text. On the other hand, we learned how his fellow astronomer Apian used the popular medium of an annual prognostication as a 'stepping stone' to present his observations and calculations in extenso (in German, although primarily directed at a learned audience). We also noticed how, in the process, he managed to come up with something new: the antisolarity of cometary tails, which also led him, some time later (in his *Astronomicum Caesareum*), to modify the Aristotelian vapor model. We looked at how an astronomically savvy physician (Schön) informed a broad public about the cometary theories in circulation and tried to convince readers of superlunarity, among other things through a shrewd presentation of Kepler's authority. Finally, we discussed how the Gdańsk city mathematician Crüger intended to establish the new, astronomy-based knowledge on comets in a highly ambitious work (and also in his Schreibkalender) in order to invalidate the Aristotelian cometary model once and for all.

---

from 1515-1526, lead the field by some margin, followed by Tycho Brahe, Joseph Justus Scaliger, David Herlitz, Girolamo Cardano, and Nicolaus Copernicus.

60 Cf. Jensen 2006, 103-110; Kremer 2012: 484; 2014: 113-114 and 120.

REFERENCES

Apian, Peter (1531): Practica auff dz 1532 Jar [...], Landshut: Georg Apian.

Apian, Peter (1532): Ein kurtzer bericht der Observation vnnd vrtels/ des Jüngst erschinnen Cometen [...], Ingolstadt: Peter Apian.

Apian, Peter (1540): Astronomicum Caesareum, Ingolstadt: Georg and Peter Apian.

Baumann, Georg (ed.) (1531): Cupediae Astrosophicae Crügerianae [...], Wrocław: Georg Baumann.

Biller, Josef H. (1995): "Die Wandkalender Peter Apians." In: Röttel 1995, pp. 147-152.

Brüning, Volker Fritz (2000): Bibliographie der Kometenliteratur, Stuttgart: Hiersemann.

Crüger, Peter (1618): Kurtzer Bericht Von dem grossen noch zur zeit scheinenden Cometen [...], Gdańsk: Andreas Hünefeldt.

Crüger, Peter (1619): Uranodromus cometicus [...], Gdańsk: Andreas Hünefeldt.

Dieterich, Konrad (1619): Vlmische Cometen Predigte [...], Ulm: Johann Meder.

Ehinger, Elias (1618): Iudicium Astrologicum Von dem Newen Cometa [...], Augsburg: Hans Schultes the Elder.

Endres, Rudolf (1988): "Die Verbreitung der Schreib- und Lesefähigkeit zur Zeit der Reformation." In: Harald Dickerhof (ed.), Festgabe Heinz Hürten zum 60. Geburtstag, Frankfurt am Main/Bern/New York/Paris: Lang, pp. 213-223.

Fabricius, Johann Andreas (1752): Abriß einer allgemeinen Historie der Gelehrsamkeit. Erster Band, Leipzig: Weidmannische Buchhandlung.

Gindhart, Marion (2006): Das Kometenjahr 1618. Antikes und zeitgenössisches Wissen in der frühneuzeitlichen Kometenliteratur des deutschsprachigen Raumes, Wiesbaden: Reichert.

Gindhart, Marion (2017a): "*Von geschwentzten sternen vnd jhrer wuerckung*. Das Kometenjahr 1618 im Spiegel der zeitgenössischen Druckliteratur." In: Mitteilungen des Instituts für Europäische Kulturgeschichte 24, pp. 31-58.

Gindhart, Marion (2017b): "*De ostentis*. Zur Verhandlung von Vorzeichen in den Werken des Joachim Camerarius." In: Thomas Baier (ed.), Camerarius Polyhistor. Wissensvermittlung im deutschen Humanismus, Tübingen: Narr, pp. 199-220.

Granada, Miguel Á. (2016): "Michael Maestlin and the Comet of 1618." In: Miguel Á. Granada/Patrick J. Boner/Dario Tessicini (eds.), Unifying Heaven and Earth. Essays in the History of Early Modern Cosmology, Barcelona: Edicions de la Universitat de Barcelona, pp. 239-290.

Green, Jonathan (2012): Printing and Prophecy. Prognostication and Media Change 1450-1550. Ann Arbor: The University of Michigan Press (http://oapen.org/download?type=document&docid=625257. Accessed August 10, 2020).

Hamel, Jürgen (2008): "Peter Apian als Popularisator und Didaktiker der Mathematik und der Naturwissenschaften." In: Joseph W. Dauben/Stefan Kirschner/Andreas Kühne/Paul Kunitzsch/Richard P. Lorch (eds.), Mathematics Celestial and Terrestrial. Festschrift für Menso Folkerts zum 65. Geburtstag, Stuttgart: Wissenschaftliche Verlagsgesellschaft, pp. 457-475.

Herbst, Klaus-Dieter (2010): Die Schreibkalender im Kontext der Frühaufklärung, Jena: Verlag HKD.

Herbst, Klaus-Dieter (ed.) (2012): Astronomie – Literatur – Volksaufklärung. Der Schreibkalender der Frühen Neuzeit mit seinen Text- und Bildbeigaben, Bremen and Jena: edition lumière/Verlag HKD.

Herbst, Klaus-Dieter (2018): "Crüger, Peter." In: Biobibliographisches Handbuch der Kalendermacher von 1550 bis 1750 (https://www.presseforschung.uni-bremen.de/dokuwiki/doku.php?id=crueger_peter. Accessed August 10, 2020).

Herbst, Klaus-Dieter/Greiling, Werner (eds.) (2018): Schreibkalender und ihre Autoren in Mittel-, Ost- und Ostmitteleuropa (1540-1850), Bremen: edition lumière.

Jensen, Derek (2006): The Science of the Stars in Danzig from Rheticus to Hevelius, San Diego: eScholarship (https://escholarship.org/uc/item/7n41x7fd. Accessed August 10, 2020).

Keckermann, Bartholomäus (pr.)/Crüger, Peter (resp.) (1605): Theoremata exegetica de cometis in genere et in specie [...], Gdańsk: Wilhelm Guillemoth.

Kempkens, Dieter (2014): "Der Erfolg der Prognostica auf dem Buchmarkt in der Frühen Neuzeit." In: Jahrbuch für Kommunikationsgeschichte 16, pp. 5-27.

Kokott, Wolfgang (1994): Die Kometen der Jahre 1531 bis 1539 und ihre Bedeutung für die spätere Entwicklung der Kometenforschung, Stuttgart: Verlag für Geschichte der Naturwissenschaften und der Technik.

Kremer, Richard L. (2012): "Mathematical Astronomy and Calendar-Making in Gdańsk from 1540 to 1700." In: Herbst 2012, pp. 477-492.

Kremer, Richard L. (2014): "Galileo in Danzig, as Portrayed in Peter Crüger's Schreibkalender." In: Andrea Albrecht/Giovanna Cordibella/Volker R. Remmert (eds.), Tintenfass und Teleskop. Galileo Galilei im Schnittpunkt wissenschaftlicher, literarischer und visueller Kulturen im 17. Jahrhundert, Berlin and Boston: de Gruyter, pp. 103-124.

Leuchter, Heinrich (1619): Cometa, Oder Ein Predigt von Cometen [...], Darmstadt: Balthasar Hofmann.

Maruska, Monika (2008): Johannes Schöner – "Homo est nescio qualis". Leben und Werk eines fränkischen Wissenschaftlers an der Wende vom 15. zum 16. Jahrhundert, Vienna: E-Theses (http://othes.univie.ac.at/1354/. Accessed August 10, 2020).

Matthäus, Klaus (1969): "Zur Geschichte des Nürnberger Kalenderwesens. Die Entwicklung der in Nürnberg gedruckten Jahreskalender in Buchform." In: Archiv für Geschichte des Buchwesens 9, col. 967-1396.

MBW T 5 (2003): Melanchthons Briefwechsel. Kritische und kommentierte Gesamtausgabe, herausgegeben von Heinz Scheible. Band T 5: Texte 1110-1394 (1531-1533), bearbeitet von Walter Thüringer unter Mitwirkung von Christine Mundhenk, Stuttgart-Bad Cannstatt: Frommann-Holzboog.

Meinel, Christoph (ed.) (2009): Grenzgänger zwischen Himmel und Erde. Kometen in der Frühen Neuzeit, Regensburg: Universitätsverlag Regensburg.

Mosley, Adam (2013): "Past Portents Predict: Cometary historiae and Catalogues in the Sixteenth and Seventeenth Centuries." In: Dario Tessicini/Patrick J. Boner (eds.), Celestial Novelties in the Eve of the Scientific Revolution 1540-1630, Firenze: Olschki, pp. 1-32.

Nouhuys, Tabitta van (1998): The Age of Two-faced Janus. The Comets of 1577 and 1618 and the Decline of the Aristotelian World View in the Netherlands, Leiden/Boston/Cologne: Brill.

Röttel, Karl (ed.) (1995): Peter Apian. Astronomie, Kosmographie und Mathematik am Beginn der Neuzeit, Buxheim and Eichstätt: Polygon.

Schechner, Sara J. (1997): Comets, Popular Culture, and the Birth of Modern Cosmology, Princeton: Princeton University Press.

Schnurr, Eva-Maria (2009): Religionskonflikt und Öffentlichkeit. Eine Mediengeschichte des Kölner Krieges (1582-1590), Cologne/Weimar/Vienna: Böhlau.

Schön, Michael (1619): Kurtzer Bericht Von Cometen […], Coburg: Kaspar Bertsch.

Schöner, Christoph (1994): Mathematik und Astronomie an der Universität Ingolstadt im 15. und 16. Jahrhundert, Berlin: Duncker & Humblot.

Schöner, Christoph (1995): "Peter Apian und die Universität Ingolstadt: Aushängeschild oder Außenseiter?" In: Röttel 1995, pp. 39-46.

Schöner, Johannes (1531a): Coniectur odder abnemliche außlegung […] vber den Cometen so im Augstmonat/ des M.D.XXXj. jars erschinen ist, Nuremberg: Friedrich Peypus.

Schöner, Johannes (ed.) (1531b): Ioannis de Monteregio Germani […] de cometae magnitudine, longitudineque ac de loco eius vero, problemata XVI, Nuremberg: Friedrich Peypus.

Weichenhan, Michael (2004): "Ergo perit coelum …". Die Supernova des Jahres 1572 und die Überwindung der aristotelischen Kosmologie, Wiesbaden: Steiner.

More Publicity through Very Short Books. Epitomes in Late Antiquity and the Renaissance[1]

Markus Sehlmeyer

ABSTRACT

From ca. 360 to 410 AD, most published works on Latin history were epitomes, i.e., summaries that condensed the preceding longer works to 10-30 per cent of their original length. One of the main aims of epitomes was to provide necessary basic knowledge of Roman history to new men in politics and administration. Epitomes were also sometimes 'misused' by Christian authors for apologetic aims. As such, they had quite a broad, heterogeneous public. Renaissance epitomes (presented here in a selection from 15th-century Italy) didn't serve any particular political goals. They were written by humanists for learned readers who were often looking for intellectual entertainment—as opposed to the practical and easy-to-read epitomes of late antiquity, which were aimed at inerudite readers. The change in media from manuscripts to printed books increased the impact and circulation of late antique epitomes, but did not have a similar positive effect for Renaissance epitomes.

---

1 I would like to express my warmest thanks to Jesús Muñoz Morcillo and Caroline Y. Robertson-von Trotha for the invitation to contribute to this volume. My colleagues in Osnabrück gave me some advice from the perspectives of their respective fields. And Roman Elenbogen (Bielefeld) also contributed with many fruitful remarks. — Abbreviations used in this essay: DEH (*De excidio Hierosolymitano*); DVI (*De viris illustribus*); OGR (*Origo gentis Romanae*).

INTRODUCTION

In different epochs of world history, information overload caused by too many publications has been criticized. In the Hellenistic period, libraries began to gather information, and encyclopaedic works emerged. The Romans also managed their civilization's knowledge by compiling encyclopaedias. Pliny used over 2000 books in his work *Naturalis Historia*, which itself consisted of only 37 books.[2] The change in media from papyrus and parchment to printed books led to a certain information overload—the age of printing proliferated such multi-volume works.[3] Whether those works actually had an influence on the public[4] is questionable, however. In contrast, short books were more popular, and even though these were often criticized by their contemporaries because of their meagerness, they do not contradict scientific values in any way. Short histories have their own politics.[5] They are seldom cited and often remain in the darkness of writing desks and libraries. An example from the present is the dtv-Atlas *Weltgeschichte* from 1964 (Kinder/Hilgemann 1964). It has had the largest print run in the field of world history in the German language—over 40 editions and more than five million copies. And although it is old-fashioned and in part outdated, many people are still buying it today, over 50 years after its initial publication. Indeed, some short books or epitomes enjoyed a wide audience. In late antiquity, such short histories emerged and were read by many people, as we can see from the number of extant medieval manuscripts. This paper examines whether the same level of publicity can be assumed for the short historical publishing forms of the Renaissance.[6] The writing of abridgments was taught in some of the secondary

2 On Pliny as an early encyclopedist, cf. Carey (2003) and the contribution of Mathias Herweg in this volume (pp. 255-284).

3 The *Encyclopédie ou Dictionnaire raisonné des sciences, des arts et des métiers* (1751-1780) might be an exception, with a press run of more than 23.000: Darnton 1979: 37.

4 The pre-modern 'public' cannot be defined as precisely as in the modern world; 'publicity' is here the possibility of having a wide audience of free citizens.

5 'Politics' here means the social embedding of literature, i.e., the aims of the texts' writers and patrons, cf. Habinek's (1998) book *The Politics of Latin Literature*.

6 Comparisons between the late antique 'Renaissance' of Latin literature during the Theodosian dynasty and the humanist Renaissance are quite common (Treadgold 1984; Celenza 2001; 2012). For condensed literary forms, cf. Horster and Reitz (2010), and especially Mülke (2010).

schools in antiquity,[7] but the exact technique used was a matter of personal preference. In the following, I will call all types of very short historical books 'epitomes' (in Greek the plural is *epitomaí*). Most of these books are abridgements, about 10-30 per cent of the length of a longer work. There are certain ambiguities with regard to terminology that occur from time to time. A breviary can be an epitome, but sometimes it is styled more individually.[8] Authors of epitomes were not ignorant of historical details, but were catering to the influential politicians, generals, or clerics who intended to use the briefer forms of these texts. For the time around 400 AD, the most productive period of Latin literature under the Theodosian dynasty, a quite detailed study on the production and reception contexts of such short works is available.[9]

Epitomes were often "doomed by an aesthetical damnatio memoriae."[10] Since they are known only to specialists, we first have to describe some epitomes in general terms before a more detailed analysis can actually be conducted. The Latin epitomes are known especially to ancient and medieval historians, and the same is true of the neo-Latin breviaria of the Renaissance. To begin with, in late antiquity, from ca. 360 to 410 AD, most of the published Latin histories were epitomes. I counted about 20 epitomes, but only one full-scale history, *Ammianus Marcellinus* in 31 volumes. Epitomes from late antiquity became public commodities through the copying of the original manuscript, i.e., handwritten copies made or allowed by the author.

I will compare this to the situation in the Renaissance, and will have to limit myself to some particularly telling examples, since the Renaissance produced hun-

7 In her detailed work, Blair (2010) focused on the types of abridgment that flourished following the innovation of printing, e.g., commonplace books or tools that were used to find information in long works (lists of authorities, indexes of all kinds).

8 It is easier to say which elements of traditional Roman historiography *cannot* be found in epitomes: speeches and dialogues, long descriptions of places and institutions. Other epitomes also excerpt some passages of the source text in nearly complete form; cf. Zinsli (2017: 209-210) for Xiphilinos (from Cassius Dio).

9 Sehlmeyer (2009). For the definition of epitome, cf. Wheatley (2011: 10). It is extremely difficult to constitute the (historical) epitome as a genre, since its forms are quite multilayered; cf. Hofmann (1987: 371). The only work cited by Hofmann (p. 569) with the title 'Epitoma' is, strangely enough, a biography (*Epitoma vitae Roberti regis*).

10 Wording of the Call for Papers of the Roman conference "Epitome. From Fragmentation to Re-composition (and back again)" (Rome, June 5-6, 2018; http://www.epitome.ugent.be/call-for-papers).

dreds of Latin historical works.[11] Most of my examples will stem from 15th-century Italy. I will present the typical forms of epitomes and then try to analyse whether the works were only *private* tools, or whether they were *public* forms of historiography that intended to address a wide audience—and were, therefore, works of popular science.[12]

LATE ANTIQUITY

Types of Epitomes

The 20 extant epitomes are rocky ground for historiographical approaches. Only rarely do we have any information on the author; most of the works are anonymous. The prooemium (introduction) of the short works has not always survived, and the original epitomized texts themselves are often lost. I will try to characterize the following works[13] in spite of these crucial gaps in transmission:

1. Short extracts from the work of Livy, an Augustan historian on Rome since the foundation, provide orientation in the form of a *detailed table of contents*. From his 142 books, a so-called *periocha*[14] was made on 120 book-pages.[15]
2. *Biographically oriented epitomes* are difficult to assess, since the epitomized texts are missing. We have collections of Republican heroes like *De viris illustribus urbis Romae*[16] and many short narratives about the history of the

11 There is no space here to introduce the newest developments in Renaissance research: Burke (1998), Najemi (2004), and Mazzocco (2006). For historiography, cf. Woolfson (2002) and also note 50 in the present essay.

12 Popular science is not limited to the natural sciences here. In this essay, the term means all popularizations of scientific approaches for a wider audience, and especially historical knowledge. Popular science in history is a part of "public history;" cf. Lücke and Zündorf 2018.

13 An overview of these works can be found in Sehlmeyer (2009: 284-304). I have not included chronicles here.

14 Text: *Periocha* (1984), with the recent contribution by Horster (2016). Florus (2nd cent. AD) was freer in his *Epitoma de Tito Livio*.

15 The work of Iulius Obsequens, who excerpted the *prodigia*, i.e., divine signs, from Livy, has a different intention. It seems to have been aimed at pagans. Text: Giulio Ossequente 2005. A shorter table of contents is available for Trogus' *Historiae Philippicae* (n. 76).

16 Text with commentary: DVI, 2016. Fugmann commented on the first half of the work more extensively (Fugmann, 1990-2004). However, literary-historical works called

emperors. The freest adaption seems to be that of Aurelius Victor, who divided Roman history into four phases and made critical remarks on individual emperors whom he considered inferior.[17]

3. *Adaptions of the Roman history as a whole epoch* were produced by Eutropius[18] and Rufius Festus.[19] Both relied on a book that is no longer extant called *(Enmannsche) Kaisergeschichte*,[20] which was itself an epitome.
4. *Shortened translations* could lend a new intention to a work. For this type of text, we have an example in the Latin translation of Flavius Josephus' *The Jewish War* by a man called Hegesippus.[21]
5. Some more geographical works had *excurses* on history (Solinus),[22] or included lists of renowned persons and had a wider scale (Ampelius).[23]

Relatively speaking, the Greek epitomes of the Theodosian age are not very great in number. We have the Greek version of Eutropius, but a greater number of long works: Magnus of Carrhae, Eutychianos, Eunapius, Olympiodoros, and perhaps Eusebius of Nantes (all only fragmentary). In the Justinian age (middle of the 6th cent. AD) John Malalas wrote in short form early in his life. An overview on this kind of historiography is missing; one has to rely on handbooks of Byzantine historiography (Treadgold 2007; Völkel 2006: 79-96). Latin authors from the same period, such as Cassiodorus (ca. 485-580) and Isidore of Seville (ca. 560-636), sometimes used the short form (Brunhölzl 1975: 28-29 and 86-88), but histories of the Germanic empires[24] were in most cases extended narratives—they were not epitomized at that time.

De viris illustribus by Jerome, Gennadius, and others are often not abbreviations, but rather first presentations of the biographies; cf. Baker 2015: 15-19.

17 Text: Aurelius Victor 1997. For other editions, cf. Sehlmeyer 2009: 284-285.

18 Text: Eutrop 1879 (with the Greek versions and the supplement of Paulus Diaconus). Eutrop 2018.

19 Festus (1967) restructured the history of the empire as a history of the provinces.

20 Ernst Hohl (1911) tried to reconstruct two passages of this work in his dissertation (pp. 35-40). Cf. Schmidt 1989: 196-198; Sehlmeyer 2005; Cameron 2011: 628.

21 Text: DEH (1932). Cf. Sehlmeyer 2009: 293-294.

22 Text: Solinus 2014. The original title of the work in the 2nd edition seems to be *Collectanea rerum memorabilium*, but it mostly contains a geography of the Mediterranean world.

23 Text: Ampelius (1993), but a new critical edition is needed: Holford-Strevens (1995).

24 Cf. the list in McKitterick 2004: 49. Riché 1976: 487-488 points only to extracts of hagiography in liturgy.

Authors—Readers—Manuscripts. Publicity within Historiography

The Authors

The late antique epitomes often stem from persons who might have worked as rhetoric teachers in their younger years. Some were close to political leaders[25]—as were many of the Renaissance authors. But most of the late epitomes remain anonymous works. The technique of abbreviating a text may have come from other literary approaches, but the epitomes differ in their intentions. Such techniques are (Sehlmeyer 2009: 30-31): accentuation; supplementation; new arrangement, for example the change from a chronological approach to a thematic approach; rhetorical devices; religious changes.

These approaches show that the public of a single epitome varied greatly: Some epitomes address a religious group like pagans *or* Christians;[26] but most of the epitomes were religiously neutral—they don't reveal the religion of their authors (Sehlmeyer 2009: 263). This could be one of the possible answers to the mystery of why many epitomes were read hundreds of years later in medieval and modern times.

The Readers

Knowledge of Roman history and the basics of Greek history seem to have been expected of people seeking to rise through the ranks of the Roman bureaucracy. Peter Brown (1971: 29-30) has suggested that the short form was very convenient for those social climbers, the new men in the imperial staff:

"The ancient classical education provided the bridgehead between the two worlds (sc. the imperial servants and the educated upper class). This culture, studiously absorbed, formed a trompe l'œil against which the new man could merge. As one provincial governor professed: *'I came from a poor father in the countryside: now by my love of letters, I have*

25 Eutropius and Aurelius Victor administered Roman provinces; Rufinus of Aquileia, an author of a short Church history complementing the *Ecclesiastical History* by Eusebius of Caesarea, founded a monastery.

26 An anonymous writer excerpted the prodigies by Livy (cf. footnote 15)—these were only interesting for pagans. The collection of painful deaths of emperors seems to address Christians more than pagans (Lactantius, *De mortibus persecutorum*; cf. A. Wlosok, in: Herzog and Schmidt (1989), pp. 394-398).

come to lead the life of a gentleman.'[27] Much of the classical culture of the fourth century was a 'success-culture': its most egregious product was a mere thirty-page 'briefing' – a *breviarium* – of Roman history for the new rulers of the empire."

In my opinion this is only *one* important point: The new man in the Roman bureaucracy needed some basics of Roman history that were easy to find in epitomes. But the reception by other audiences[28] can be assumed through many channels:

- For the elite, the basics of Roman history and *exempla*[29] were necessary to write laudatory speeches on the emperor and other important persons.[30] These exempla are easier to find in epitomes than in the original long histories or the exempla-collection by Valerius Maximus.[31]
- Christian writers were trying to strike the Roman pagans with their own weapons: They used *exempla* in an apologetic manner. Some tried to connect universal history and the history of salvation, which had, for a long time, only ever been presented as a history of persecution.[32]
- School teachers used epitomes to prepare discussions of Roman literature and to instruct the pupils on writing speeches, but they were not schoolbooks (Vendramini 2012: 278-280; Gasti 2015).

On the other hand, it is not easy to determine whether the middle class read such epitomes, although we are well informed on the Theodosian age. The crisis of the 4th century stems from Germanic migration connected with invasions, plunder, and raids. Ongoing Christianization might have induced people unwilling to change their religion—pagans—to give more thought to their Roman identity

27 Brown (1971: 29-30) is here citing Aur. Vict. 20,5; cf. Sehlmeyer 2009: 153f. 284f. Cameron 2011: 384. Vendramini 2012: 306.

28 I agree with Vendramini (2012: 325) ("el público de los breviarios era heterogéneo y no se limitaba a la nueva élite burocrática y military").

29 *Exempla* are the authoritative examples from Roman history that every Roman should know. Often the exemplum is representative of a special virtue (or, in contrast, a vice). For a collection, cf. footnotes 31 and 63.

30 The preserved *Panegyrici Latini* from late antiquity were masterfully edited and translated by B. Müller-Rettig (Panegyrici, 2008-2014).

31 Text: Valerius 1998. This Latin edition includes the epitomes from late antiquity by Iulius Paris and Ianuarius Nepotianius. S. Herzog and Schmidt 1989: 193-195.

32 Sehlmeyer 2009: 197-211 (Jerome, Augustine, Damasus, Ambrose). For Orosius: Orosius 2010 (Text); Van Nuffelen 2012 (Interpretation).

through history (Sehlmeyer 2009: 214-274).[33] We have to assume that Romans searching for their identity occasionally discussed the heroes of ancient Rome on the streets (e.g., on the Forum Augustum with the honorary statues of such exemplary figures), but did they consult epitomes? We won't be able to show this through primary sources, but the three groups mentioned above had an effect as multipliers on the broader public and most certainly drew at least some of their historical knowledge from epitomes:[34]

- In their speeches (*panegyrici*), the elites addressed not only the emperor himself, but also the court and some soldiers.
- Christian writings seem to be diffused quickly through networks of Christians in the whole empire, and so had an impact on those communities. New forms of literature arose, such as universal chronicles and collections of martyrs' lives.
- In the elementary schools, which were attended by many citizen-children, teachers may have used examples from Roman history that were easy to find in epitomes. Grammar and rhetoric schools were only attended by a few children of the upper class—and often reflected on Roman history and exempla.

In particular, the historical knowledge in the anonymous *Kaisergeschichte* (cf. footnote 20 of the present essay) seemed to be very interesting for a broader readership. While the text itself has been lost, half a dozen epitomes (in the widest sense of the term) distributed the basics of Roman imperial history that it contained: Aurelius Victor, Eutropius, Rufius Festus, the Greek version of Eutropius, the *Libellus de vita et moribus imperatorum* (epit. Caes.), and Orosius.

The Manuscripts

For some epitomes, we have over 200 medieval manuscripts: Iustinus, *De viris illustribus*, Solinus, and Orosius (cf. the table on page 327-328).[35] Some epitomes were only passed down in one or two manuscripts. Although it is impossible in most cases to show how many readers a single manuscript had, a higher number

33 Cameron (2011: 213-214, 513, 628-631, 635s., 665, 668, 689, 676, 748) has epitomes in mind, but connects them primarily to historiography (and not to the pagan-Christian discourse).

34 Multipliers are here persons who multiply a certain type of knowledge to a greater group; in our case, they are propagators of historical knowledge.

35 Details in Sehlmeyer (2009: 274-279). Epitomes that were frequently copied were considered useful and sometimes delightful reading material; cf. Haskins 1927, especially p. 226; see citation on page 324 of this essay.

of manuscripts points to greater popularity—no one would have hand-copied a book that nobody wanted to read.

Late Antique Epitomes: Between Politics and Popular Science

Some of the epitomes were not influential at all—the work of Ampelius might have been written for a single pupil (Schmidt 1989: 175). Others had great impact. Aurelius Victor, who was a provincial governor and later an urban prefect, was honored with a bronze statue in 361 (Amm. 21.10.6). Was this for his epitome? Receiving such an honor from the Emperor Julian was remarkable in itself, and shows that Julian was aware of the historical interests of Victor,[36] who had not enjoyed any military success up to this point—a typical reason for honoring someone with a statue. Eutropius enjoyed a greater reception, and he, like Rufius Festus, dedicated his epitome to the Emperor Valens; due to declining knowledge of history and and possible gaps in education, it is quite probable that the emperor commissioned these works—as is stated in their introductions.[37] The Greek version of Eutropius is dated to the beginning of Theodosius' reign (380 AD[38]). Likewise, other epitomes that functioned as works of adult education would not have displeased the emperor, who had to rebuild the army after the disastrous Battle of Adrianople. Modern historians like Peter Brown have postulated that these short Roman histories were meant to supply new men in politics and administration with the necessary basics of Roman history. I assume that these followed policies established by the emperor. The bishop Augustine commissioned an epitome on the failures of Roman dominion to a monk named Orosius (note 32). That monk did more than Augustine intended: In the end, the *Historiae adversus paganos* comprised seven books. In some cases, the authors of the epitomes themselves may have been ambitious and might have been trying to fulfil the wishes of society. Iustinus writes that he wants to inform people about Greek history, but that he also wants to entertain, which is quite typical of popular science.[39]

---

36 The poet Claudian was honored with a statue a little bit later; Sehlmeyer (2009: 166-168 with note 266).

37 Details in Sehlmeyer 2009: 169-174. For the development of education, cf. Vendramini 2012.

38 In my opinion, this demonstrates its usefulness—Latin works were rarely translated into Greek; Sehlmeyer (2009: 176-177 with notes 317-323).

39 Text: Iustinus 1971. For the importance of entertainment through popular works, cf. the essay by Oliver Hochadel in the present volume (pp. 65-92).

Excursus: Historical Epitomes in Medieval Renaissances[40]

In the medieval age, the writing of epitomes partially followed late antique models—Eutropius was supplemented by Paulus Diaconus (late 8th century) and Landolfus Sagax (ca. 1000 AD). In order to provide a short overview on the genre in medieval times, I will focus on the so-called "Renaissances"[41] before the actual Renaissance.[42] In the Carolingian Renaissance, the emperor pushed for educational reforms—not in the modern sense, but rather primarily concentrated on his own court. Educated in the *Hofschule* (the court school), Charlesmagne remained interested in education and drew in renowned scholars (Brunhölzl 1975: 243-315. McKitterick 2004: 40-45). Diaconus christianized his extended version of Eutropius (*Historia Romana*); Haimo of Auxerre (d. ca. 850) wrote *Historiae sacrae epitome sive de christianarum rerum memoria*, which covered the period from Jesus' birth to Theodosius, and which was an epitome of Christian texts (Brunhölzl 1992: 260 and 480).

If it is legitimate to speak of a Renaissance in the 12th century,[43] not only economic or political but also cultural developments must be taken into account. From the Byzantine Empire, many Greek texts came to central Europe again—to say nothing of the Greek scientific texts that survived only in Arabic translation (Watt 1969).[44] Haskins has an entire chapter on historiography (Haskins 1927: 224-302) and stresses the importance of the aforementioned epitomes from late antiquity:

> The Latin historians who really *delighted* the mediaeval, as they did the later Roman, world were the epitomators, Florus, Justin, and Eutropius, whose *popularity* was paralleled by that of the summarizers and excerptors in other fields, like Solinus (Haskins: 1927: 226, my emphasis).

---

40 The focus of my article is the historiography of late antiquity and the Renaissance, but some readers will find it useful to have a short introduction to the 650 years between these two periods: cf. Kaiser (2014: 150-151; "Wissen in handlicher Form").

41 Treadgold 1984. Of course, only in certain areas are the earlier Renaissances comparable to the humanist Renaissance that led to modernity.

42 For medieval historiography, cf. Guenée (1980) or Patze (1987).

43 Haskins 1927. It goes without saying that this use of the term is a controversial one; cf. Benson and Constable 1982.

44 Foot and Robinson (2012) take a broader, more global approach, noting that other cultures (e.g., China) were also familiar with the abbreviation of historical works.

Haskins qualifies this by saying that the medieval writers preferred citations from the Roman poets—epitomes were read, but not cited. Other genres, such as annals and hagiography, tended to be long-form. Books like Hugh of St. Victor's *Didascalicon* (ca. 1125)[45] influenced the Christian schools. Greek (Byzantine) literature also has some examples of historical epitomes. Photios is well-known for his work *Bibliotheke* (9th century), which is compiled from 280 literary works. This work is too long to be an epitome, but similar in that it excerpts necessary basics. In the middle Byzantine period (Völkel 2006: 79-96), historical epitomes played an important role (Brunt 1980: 483-493).[46] In the later medieval age, world chronicles could have aims similar to those of epitomes, but were sometimes longer: cf. the *Flores temporum* (13th century) or Heinrich von Herford's *Liber de rebus memoriabilioribus* (no later than 1355 AD).

In the end, we must admit that the historical epitomes of the medieval age followed the general contours of late antique ones, but contained more Church history. We can only rarely find traces of the politics of epitomes in the medieval age. The Carolingian and 12th-century renaissances seem to have promoted the writing of epitomes for educational purposes in the widest sense, but changes in the educational system meant that epitomes never became schoolbooks. Paulus Diaconus could present his short *Historia Romana* as a work for the noblewoman Adalberga.[47] In the monastery at Lorsch, many epitomes were available (McKitterick 2004: 168-217).[48] They remained there as tools for teachers, i.e., clerics at the schools in the monastery and at the bishop's seat. These epitomes were never of the character of grand narratives, but they indirectly determined the views of Roman and early medieval history: The Franks took their origins from the Trojan and Roman past (McKitterick 2004: 208)—the discourse of the *translatio imperii* followed. Lupus of Ferrières soon (844) presented an epitome to Charles the Bald as a collection of examples of successful Roman emperors.[49]

45 Cf. the introductory remarks to that work by Offergeld 1997.

46 Concerning Xiphilinos, Zinsli (2017: 199) has shown that his epitome contains between 5 and 26 per cent of the source work by Cassius Dio.

47 For the initiatives of Paulus Diaconus, cf. McKitterick 2004: 67f.

48 For the dissemination of Church histories, cf. ibid: 218-244.

49 Lup. *epist.* 93 Dümmeler (MGH): *Maxime autem Traianum et Theodosium suggero contemplandos.* ("I suggest to give attention to Trajan and Theodosius especially") The reference to these emperors may be an indication that he was using the so-called *Epitome de Caesaribus* (Epit. Caes. 48,8).

THE RENAISSANCE

Types of Epitomes

To date, no one has tried to collect all the Renaissance epitomes.[50] This would be a Herculean task, because a great number of epitomes in the broadest sense of the word were written in the neo-Latin world from Petrarch's time onward. I have concentrated here on some examples from Italy, but the genre came to life again in the other European kingdoms too—to say nothing of the vernacular epitomes.

(1) It was quite typical for short histories to include an *epitome of one long or, more frequently, truly voluminous work*. In his time as Pope, Pius II abbreviated the *Decades* by Flavio Biondo. From the first two of Biondo's decades, he made an early summarized edition that was about one third the size of the original (35 per cent).[51] The *Decades* were a short universal history from the so-called Fall of Rome in 410 to Pope John XXIII, 1410-15.[52]

(2) In Book 7 of his *Roma triumphans*, Flavio Biondo paraphrased a Roman history and so produced an *excursus in the style of an epitome*.[53] Three pages refer to the late Republic.[54] A list of Emperors follows this—this list is shorter than in more typical epitomes, which generally dedicate a few sentences to each individual emperor. As such, Biondo's work is more like a *chronicle*.[55]

(3) *Collecting the biographies of renowned persons* under the heading *De viris illustribus*, "on famous men," was an ancient kind of historical narrative. As in antiquity, not all Renaissance examples had the character of a *short* text. In the Renaissance's beginnings in the 14th century, Petrarch (ca. 1360) collected

50 Cochrane (1981) mentions the greater part of the epitomes treated in this article (and is more detailed than Woolf [2012]), but there are many others. For England, cf. Taylor 1987 and Wheatley 2011; some historical epitomes from Germany are touched upon by Sehlmeyer (2009: 279-283). Good introductions to Renaissance historiography are the ones by Völkel (2006: 195-215, 242-244) and Laureys (2014).

51 Biondo has 390 pages with 55 lines per page; Pius II has approximately 150 pages with 50 lines per page.

52 Text: Pius II (1533), who summarized only the first two decades (up to the year 1401).

53 Text: Biondo (1531). Biondo (1392-1463) was secretary at the court of Pope Eugene IV (1431-1447), but he did not write exclusively in the name of the curia.

54 Biondo is himself citing the 119th book of Livy (p. 149, ed. 1531). It is clear that Biondo must have used the *Periocha* of Livy—the last book in full text was Book 45.

55 On the genre of the chronicle, cf. Sehlmeyer 2009: 304-308.

the lives of two dozen Romans (Petrarca 1964)[56] and one dozen biblical or mythical figures. His initial aim was to collect material for his epic on the Second Punic War, *Africa*. Therefore, the life of the Roman commander Cornelius Scipio is extended to 150 pages. But other persons have only two or three pages each, and these short sections have the character of the exempla-biographies of late antiquity. While Roman biographies were readily available in the aforementioned late antique works attributed to Aurelius Victor,[57] Renaissance literates had to write short biographies of their contemporaries themselves. Enea Silvio Piccolomini, who would later become Pope Pius II, compiled (ca. 1444-1450) biographies of bishops and noblemen, and 42 of these biographies are known to us. The work was not published in Pius' lifetime, and its beginning has been lost.[58] And just like in late antiquity, here the historical figure has become an example that is useful for the present moment.

(4) Leonardo Bruni from Florence was a humanist in the sense of a promoter of classical learning.[59] He had a deeper knowledge of Ancient Greek than most other scholars of the time, which allowed him to summarize some important facts on the First Punic War, which he found in the first book of the Greek-language universal history by Polybius (1421-22)—he made a Latin epitome from a Greek summarizing text (*Commentaria primi Punici belli*, Bruni 1537; cf. Ianziti 2012: 61-88). He addressed a large audience: "I wish that these deeds (of the Romans) will be known to all people." (*Res ... gestas ...omnibus esse cognitas quam maxime opto*, Bruni 1928: 123)[60] And also: "I delivered comments on

56 The manuscripts vary: Some have only 24 Romans, others 36. The concept of the work also changed; cf. Kessler 1978: 34 (first title of the work: *Epithoma*).

57 The *Corpus Aurelianum* was compiled from two works whose authors remain unknown, *Origo gentis Romanae* (OGR 2004) and *De viris illustribus urbis Romae* (DVI 2016), as well as from the *Historiae abbreviatae* (Aurelius Victor 1997). All three works are available in the Teubner editions (Aurelius Victor 1970). The epitome (namely, *de Caesaribus*) (2009) was transmitted in a different manner, but is still included in the Teubner editions.

58 One could easily imagine that he was hesitant to judge his contemporaries. The text is approximately 60 pages in length, with an average of about one-and-a-half pages for each man. The shortest biography is four lines; the longest is four pages.

59 Ianziti (2012) is quite detailed and has chapters on the epitomes. Leonardo Bruni (ca. 1370-1444) came from Arezzo near Florence. A member of Coluccio Salutati's circle, he was later secretary to the Pope and, from 1427 onward, Chancellor of Florence.

60 We should bear in mind that authors' prooemia quite frequently imply the desire to reach a wide audience, and we can only assume that Leonardo Bruni did indeed reach such an audience. This is a topical remark.

this war for common utility." (*Commentaria huius belli pro communi utilitate suffeci*, Bruni 1928: 123) Bruni sought to instruct all people about the First Punic War, the account of which was no longer available in the Latin version of Livy. He therefore had to paraphrase passages of the Greek text by Polybius. There later followed an epitome of the late antique Greek historian Procopius, the reception of which was very critical.

Authors—Readers—Manuscripts. Publicity within Historiography

The Authors

Petrarch was an author of poetry, philosophical letters, and one unfinished epic. Other epitomizers like Biondo, Piccolomini, and Bruni were close to the Pope, but also held secretarial positions in various Italian cities. Bruni would later become Chancellor of Florence, while Piccolomini was elected Pope. They applied the abbreviation of historical narratives to a broad sphere: The history of the Roman emperors could be minimized to a raw list of emperors or a very short narrative.[61] Some epitomizers were quite autonomous in terms of their thinking and the analysis of their sources, like the late antique text under the name Aurelius Victor. The *Florentine History* (Bruni 2001-2007; Ianziti 2012: 91-146) by Bruni could also be considered such a work; the *de viris illustribus* literature is often free in its narration as well (i.e., it is more than just an excerpt from one work).

The Readers

The Renaissance works discussed above addressed different kinds of audiences. They were not survey courses for people who had never visited schools, as some of the late antique breviaries were. In most cases, they brought the discussions of humanist circles to a wider learned audience:

- The texts are often dedicated to patrons, bishops, or the Pope himself—the readership consisted, at first glance, of people at the court, but scribes and condottiere may also have been readers.

61 Pius' *Epitome* of the *Decades* by Biondo has reduced the text to 35 per cent of its original length (see footnote 51). For late antique texts, the degree of reduction can only be approximately known. For Justin's widespread *Epitome* of Trogus, the assumption is that the reduction is 10-30 per cent of the original; cf. Sehlmeyer (in print).

- Short historical remarks on the Italian cities and renowned inhabitants—as can be found in Biondo's *Italia illustrata*—suggest that the notables of the cities were readers too.[62]
- However, it is difficult to say when precisely such short histories were used by teachers in the schools. It is often assumed that the advent of book printing in Italy (from ca. 1465 onward) had made books widely available.[63]
- The humanists themselves used short books as tools for writing longer books, such as geographies (cf. note 62) or eulogies (McManamon 1989 [funeral orations]; Plett 1993 [exempla]).

Did artists like Leonardo da Vinci (1452-1519),[64] who was born into a middle-class family,[65] use epitomes as a form of adult education? This is by no means out of the question, since artists were long seen as mere craftsmen. In school, Leonardo learned how to use the abacus, but not much else. Only at the age of 30 did he teach himself Latin, using a grammar and a vocabulary. We have lists of his library with roughly 200 titles: Over the course of his life, he received many lucrative contracts, which allowed him to buy expensive books.[66] He had all three of Livy's *Decades*—only a few of his books touch on history. Leonardo was not particularly interested in history, and he was not professionally driven to write texts, a task for which the *exempla* of ancient Rome or humanists' biographies might have been useful.

More than 100 people from Renaissance Venice listed their books in testaments or other lists.[67] Anselm Fremmer has done detailed research on these lists.

62 Clavuot (1992: 13) emphasizes Biondo's goal of describing the development of the Italian communities, but his primary focus is on Biondo's historical method. It is no surprise that Livy's Books 8 to 10 are of exceptional importance for the *Italia illustrata* (Clavuot 1992: 203). Epitomes, however, played a minor role; cf. Grafton 1997 (concerning humanists as readers).

63 Würgler 2013 (for the change in media); Grendler (1989: 255-263) demonstrates the predominance of classical extended histories as literature useful for pupils, e.g., Livy, Sallust, and Caesar. Iustinus and Valerius Maximus, however, were considered to be collections of examples (p. 203).

64 Cf. the biography by Nichols 2006: 487-491.

65 Admittedly, the term 'middle-class' is problematic. Renaissance society was a corporative system, but Leonardo was neither a member of the nobility nor a cleric. However, he was better-off than the poor in both the city and the countryside.

66 The list is to be found in the Codex Madrid II 2v-3r. The reprint from 1975 provides helpful commentaries by Reti.

67 Fremmer (2001: 323-450) lists Venetian owners of books from 1159 to 1498.

Livy was the most frequently found historian, appearing on nine of the lists.[68] Sallust and Valerius Maximus were popular too. Moreover, these private libraries in Venice also contained the work of Orosius, which due to dictation was written down as Orossio, Oroxius, or something similar. But more epitomes were held in the libraries than Fremmer had anticipated. One person had all three decades by Livy, as well as an additional book called *De gestis Romanorum* (present in the library of Lodovico Gradenigo [1372]; Fremmer 2001: 343), which seems to be the epitome (*Periocha*) of Livy's work. When Flavius Josephus is mentioned in Francesco Dandolo's library from 1339 (Fremmer 2001: 332), it is not improbable that this is the shortened Latin version of *The Jewish War* (DEH, note 22). Only one exemplar each of Justin, DVI (*De viris illustribus*),[69] and Solinus are listed in the Venetian testaments.

Now we would assume that the Renaissance epitomes by Petrarch, Boccaccio, Bruni, or Piccolomini would occasionally appear on these lists. However, we have only Boccaccio's *On Famous Women* (twice) and Bruni's *De Italico bello adversus Gothos gesto* (once; Fremmer 2001: 362 and 430, in testaments from 1401 and 1486). Here we have to remember the nature of transmission: Until the end of the 15th century, books were often disseminated in the form of manuscripts. Short manuscripts were sometimes too small to be bound: they were generally appended to a corresponding longer text; or several short books like the epitomes were bound together in a single anthology. In the abovementioned nine references to Livy, the epitome of Polybius by Bruni might have been included; and epitomes might have been hidden within titles like "una sum‹m›a de virtutibus." (Fremmer 2001: 429, testament of Girolamo da Molin, 1444)

The ambivalent results in these private Venetian libraries must not be generalized. We saw above that the individual authors of epitomes invested a great amount of energy into the dissemination of their works. Leonardo Bruni was a very self-assured epitomizer who thought that epitomes were their own intellectual products, even if the source work (in this case Procopius) was in his eyes insufficient as a historical work. His independent historical thinking places him within the classification not of the mechanical epitomizer, but rather of the innovative epitomizer, which is represented in late antiquity by authors like Aurelius Victor, who was quite critical of the history of the Roman Empire, and whose work was much more than merely a revision book.

68 Fremmer (2001: 240-247) collects the historical works from the lists (cf. ibid: 323-450).

69 The widespread DVI (Ps. Aurelius Victor) is mentioned once (library of Paolo de Bernardo, 1374; Fremmer 2001: 345).

The Manuscripts

Whether an epitome was intended for the private use of a single person or for public use can only be determined through the examination of manuscripts.[70] The author, or friends of the author, would have to initiate the copying of the book, which was an expensive task. Alternatively, the dedication of the book to a renowned person like a bishop or a member of the leading family of the city-state could promote the book's distribution, in cases where the epitome was considered important enough to be copied.

For Leonardo Bruni, a detailed catalogue of manuscripts is available.[71] Today, we have 146 manuscripts of his epitome on the First Punic War and 120 copies of the Italian translation of that work (Hankins 1997: 256-257). Bruni's book was likely one of the only books that provided an impression of the work of Polybius, since most of his contemporaries had no Greek language skills. The number of extant manuscripts indicates that the intended readers were not only specialists of Roman history, but also statesmen and the learned public (Ianziti 2012: 63).[72] Eighty-five of the manuscripts are currently in Italian libraries. We can find the book today in nearly every important city of the Italian Renaissance. It is not clear in every case who the original possessors of the manuscripts in the 15th century were. Although the books are now in central libraries, in the time of Bruni they were mostly in private collections. This explains why some modern libraries have ten or more different copies of the epitome by Bruni: They would have obtained these copies through testaments, the demise of aristocratic houses, or clerical institutions.

Renaissance Epitomes: Between Politics and Popular Science

The Renaissance epitomes presented here were not commissioned by bishops or princes. They harken back to the humanists' own initiatives: They were not works that were easy to read by people without erudition, but instead were aimed at a broader learned audience that was also seeking entertainment. Bruni seems to have been very active in the dissemination of his works. He published several Latin

70 Kallendorf (2017) explains techniques to search such codices; if a critical edition is available, the most important manuscripts are listed. For the transmission of ancient texts, cf. Reynolds/Wilson 2013.

71 Hankins (1997) has listed all codices worldwide that contain works by Bruni.

72 From the introduction, we learn that Bruni intended to fill the gap in the early narration of Romano-Italic history by Livy. Why he called his work a *commentarius* is not clear, and puzzled his contemporaries (Ianziti 2012: 67f.: Letter to Tortelli, note 34).

translations of the Greek Plutarch.[73] Other humanists tried to compile epitomes from all the biographies written by Plutarch (Resta 1962). But even renowned humanists needed some help in understanding Ancient Greek. And so Flavio Biondo was happy to have an epitome of the late antique Greek historian Procopius:[74]

"This is a part of history written in Greek that no Latin author between the times of Justinian and us seems to have read. [...] So Leonardo [Bruni] from Arezzo [...] wrote a history of the same period, which from beginning to end contains nothing more than Procopius." (*Decade* I 4, Biondo 1559: 43C)

Biondo had asked an unknown person to translate Procopius from the Greek; he compared this translation against Bruni's version (completed in 1441 and dedicated to Cardinal Giuliano Cesarini)[75] and considered the latter a kind of plagiarism. For a long time, Bruni withheld the fact that he was using Procopius, and continued listing the work as part of his exclusive personal authorship. Indeed, some of his contemporary humanists recognized that Bruni must have used Procopius, and wrote letters to him about this—these letters are, in turn, a singular source relating to the reception of epitomes at the time.[76] In this 'Procopius affair,' Bruni stressed the fact that Procopius wrote in an unsatisfactory style and that his writing was akin to that of testimonials, and was not at the level of a true historian. In this case, the epitomizer has taken on the role of a creator, not that of an abbreviator. We must admit here that plagiarism was not a theme in the Renaissance, but some critical contemporaries (not only Biondo) would have liked more clarity from Bruni.[77] Moreover, Bruni was an innovative historian of the long form—his widely known *History of Florence* linked the Roman origins of the city with contemporary developments.[78]

73 The introduction of the *Vita Antonii* is printed in Baron (1928: 102-104).

74 Unless otherwise noted, all translations are my own.

75 The prooemium is included in Bruni (1928: 147-149). The whole text is printed in Bruni (1470: 6-132). For details, see Ianziti (2012: 278-300, 400-407).

76 Ianziti (2012: 279. 281-283) cites the letters and provides a very detailed analysis.

77 In our modern view, this work by Bruni was indeed a plagiarism.

78 Text: Hankins 2001. The first book contains a very original version of the founding of Florence through Sullan veterans and Etruscans (Ianziti 2012: 105). Baker (2015: 202) stresses the book's inventiveness.

CHANGE IN MEDIA: PRINTED EPITOMES AS POPULAR HISTORIOGRAPHY

The transformation of knowledge into new forms of media is a mark of respect for the original work. Many older books stopped being distributed if they were not printed. The advent of book printing in 1460 was an important moment in reception history (Grafton 1980; Kallendorf 2017). It significantly increased the diffusion of texts in Italy. The early Renaissance epitomes were written before this time, but we also have epitomes that were only transmitted in print form. Late antique epitomes on Roman history were printed quite early on; in the five years from 1469 to 1474, the most influential texts by Aurelius Victor, Eutropius, and Iustinus were all in print. Astonishingly, we have 20 early prints of Pseudo-Aurelius Victor's *De viris illustribus.*[79] Two abbreviations from Livy were available (from the *Periochae* and the more autonomous *Florus*).

Table 1: Chronological list of prints

Author or title of the epitome	**Pages**[80]	**Manuscripts today**	**First print (year)**
EKG (Enmannsche Kaisergeschichte)	80?	*Lost in antiquity*	—
Rufius Festus	ca. 25	91	1468 with DVI
Periocha Livii	121	> 90	1469
Iustinus, *Epitoma historiarum Philippicarum Pompei Trogi*	302	> 200	1470
DVI (*De viris illustribus urbis Romae*)	46-50	ca. 200	1470 with Ruf. Festus
Eutropius	71	> 60	1471
Orosius, *Historiae adversus paganos* (417/8)	600	249	1471
Bruni, *De bello Italico* (1441)	**127**	**166**	**1471**
Solinus	232	ca. 300	1473
Petrarca, DVI (1360)	**350**	**> 30**	**1496**
Bruni, *De primo bello Punico* (1421)	**89**	**147**	**1498**
Epit. de Caes. (libellus)	43	19	1504

79 The manuscripts from the 15th century can be found in GW M50379-M50408.

80 Pages according to modern editions. Please bear in mind that only a part of the epitomes from Renaissance Italy could be examined in detail in the present essay.

Iulius Obsequens	31	—	1508
DEH (*De excidio Hierosolymitano)*	415	16	1510
Codex-Calendar 354	140	8	1513?
Beccadelli, *Liber rer. gest. Ferdinandi* (1471)	**80**	**19**	**1520**
Pius II, *Epitome Decadum* (after 1463)	**138**	**?**	**(1533)**
Sulpicius Severus, *Chronica*	103	2	1556
Aur. Victor, *Historiae abbreviatae* (361)	53	2	1579
Ampelius, Liber memorialis	62	1 (lost)	1638
Pius II, DVI (ca. 1444-50)	**60**	**?**	**1759?**
Origo Constantini	10	1	1875
Prologi Trogi	20	> 100	with Iustinus?

Not all of the aforementioned Renaissance epitomizers, typed in **bold face** here, found a printer. From Bruni's epitomes, we have in the 15th century three prints of *De bello Italico adversus Gothos gesto*, but only one Latin edition of *De primo bello Punico.*[81] Piccolomini's epitome from Biondo's *Decades* was printed only once in the 15th century.[82] Petrarch's DVI was only available in his *Opera* (Basel 1496).[83]

It is easy to recognize that late antique epitomes played a prominent role in early Italian printings, while Renaissance epitomes were only rarely printed. Whereas in the age of the manuscript the author and his patrons had to prepare copies by hand—a quite costly procedure for private, non-commercial goals—, the printing of an epitome was a sign that the printer expected the book would find buyers. Economical considerations superseded the politics of statesmen or authors. But the late antique epitomes on Roman (and Greek) history seem to have been more in demand than the Renaissance epitomes on Roman, universal, or Italian history.

Some Renaissance epitomes were quite relevant to their times, and took up current issues. Latin epitomes of Polybius or Procopius were no longer necessary once the original source books had been translated into Latin. As in late antiqui-

81 The prints and the preserved samples are collected in GW; for the mentioned works by Bruni, see GW 05600-05602 (in 122 books today) and 05603 (*de bello Punico*, 74 books from this edition today).

82 M33465 from 1481 in 80 preserved books. M33466 was bound with the full text of Biondo's *Decades* (GW 04420).

83 Three hundred and twenty-three pieces have been preserved in public libraries and institutions (GW M31505).

ty, quite specific epitomes (like the one by Ampelius or the *Origo Gentis Romanae*) were no longer being copied. The only texts that became 'popular science' were those with a more polished literary quality, such as the untypical collections of biographies by Boccaccio or Petrarch, or those that were particularly useful for professional writers, such as epitomes of voluminous works:[84] Piccolomini's epitome survived along with the full text of Biondo's *Decades*.

The use of epitomes, or the writing of them, was not a glorious task in the Renaissance. Bruni pretended to have written an outstanding history of the First Punic War—rather than advertising that he had excerpted Polybius. He also denied Procopius was his source on the Italian (Gothic) War, and instead tried to depict Procopius as an eyewitness to those events. Beccadelli reflected on historiography in his book on King Ferrante of Naples, and differentiated between three categories of historians:[85]

1. The best group: Livy, Sallust, Caesar.
2. Mediocre historians: Tacitus, Curtius, Suetonius.
3. Inferior (*humiles*, *infimi*), but still laudable: Orosius, Eutropius, Lampridius.[86]

The epitomes are here located in the third and worst group. They are by no means useless, but have to be considered with regard to their aims. And Beccadelli seems to classify his own work as a history book of the third kind.

Prints of historical (and other) works can today be compiled with the help of the World Wide Web. Patricia Cox Jensen has analysed the first printed ancient historians (to 1600).[87] Eutropius (52 printings) was printed as often, or almost as often, as the Greeks Diodorus (56 printings), Cassius Dio (52), and Polybius (48). Cox Jensen relied on an online database of early printings, the Universal Short Title Catalogue (USTC).[88] I looked up other epitomes in this database. However, one has to keep in mind that the number of printings cannot be the *only* sign of publicity (since the print run is unknown): Iustinus' is the most printed

84 The differentiation between literature (often fiction) and useful books is typical of late antiquity, when epitomes predominantly served the function of useful tools; cf. Cameron 2011: 388.

85 Beccadelli (1968: 66-68), with excellent commentary by the editor Gianvito Resta. Passage first located in the magisterial article by Schirrmeister (2014).

86 Aelius Lampridius is one of the pseudonyms of the supposed authors of *Historia Augusta*. Sehlmeyer 2009: 303-304 (summarizing unpublished results by P.L. Schmidt).

87 Cox Jensen, 2018 updates the paper by Burke (1966), who did not focus on any epitomizers.

88 https://www.ustc.ac.uk/. The catalogue has a larger chronological scope (until 1600) than the GW, but is not as detailed in its descriptions.

epitome (152 editions); Orosius' *Historiae adversus paganos* has 22 printings; but Pius II's *Epitome Biondi* has only two printings, from 1481 and 1533.

Therefore, the most read epitomes of the Italian Renaissance seem to be works with only a brief period of influence. They were no longer part of the public discourse on history. That many *Italian* Renaissance epitomes were only read for a few decades and then quickly forgotten must not be generalized for all countries. For example, beginning in the 16th century, epitomes seem to have been used as schoolbooks in Germany (note 50).

The Renaissance did indeed begin with the recovery of antique Latin and Greeks texts.[89] The occasionally criticized late antique epitomes were also frequently used (but seldom cited), while the fate of the Italian Renaissance epitomes seems to be rather unlucky. They were often written by renowned humanists, but these humanists had attained prominence through other publications or their political influence. Looking up Bruni in Peter Burke's influential books on the Renaissance (Burke 1998, Index), we find him as a discoverer of rare ancient books, as a philologist, and as someone who judged antiquity to be exemplary, useful until modernity. The change in media came too late: The late Renaissance had other problems[90] that could not be solved by *short* books on history. While we are not taking up the perspective of Guglielmo Libri[91] or the *Meistererzählung* of Johan Huizinga[92] in their quite negative views of the late Renaissance in general, it should not be overlooked that 15th-century Italy was facing new and very critical developments: expansionist European kings, and of course the Ottomans, who had taken Constantinople in 1453. Meanwhile, society was also changing, though not as deeply as in late antiquity, when Theodosius established a Christian empire.

---

89 Recent popular interest in this field can be seen in the Pulitzer Prize-winning book by Stephen Greenblatt, 2011. One can reproach this book from a scientific viewpoint as a kind of *Meistererzählung* (Rexroth 2007), but the rediscovery of long-forgotten antique texts (like those of Lucretius) was of course a prerequisite for the Renaissance. Most of the epitomes needed no rediscovery—they were still in use at the time.

90 I have relied on Biondo's parallel between Roman exempla and Hunyadi—a very rare allusion to contemporary history (cf. footnote 67). Bruni's and Pius II's extended works on contemporary history had more of an impact than their shorter texts; cf. Reinhardt 1997: 65-68, 481-485.

91 *Libri*, 1838. For Guillaume (Guglielmo) Libri (1802-69), cf. Giacardi 2005.

92 Huizinga (1872-1945) wrote his renowned *Herfsttij der Middeleeuwen* (*Waning of the Middle Ages*: Huizinga 1919) during the Great War.

Christopher Celenza (2001: 17; 2004: 1)[93] demonstrated parallels between late antiquity and the Florentine Renaissance. It is not as easy to demonstrate that another parallel lay in the emergence of very short histories, i.e., epitomes. A historical comparison is certainly helpful, but I would assume—also in light of our restricted knowledge of neo-Latin epitomes—that historical epitomes played a lesser role in the Renaissance when compared to the role they played in the age of Theodosius. Late antique epitomes remained valuable because of the long-term use of Latin in the medieval age. The "new" epitomes might have been read out of other motives, but only very few of the Renaissance epitomes were able to find new readers in the modern world.

In the end, we have to concede that the form, or rather genre, of the "historical epitome" does not radically change between late antiquity and the Renaissance. Some of the late antique epitomizers had the good fortune to be read later by people interested in ancient history. However, historical epitomes from the Renaissance did not have the same impact.[94]

REFERENCES

Ampelius (1993): Aide-mémoire. Edited and Translated by Marie-Pierre Arnaud-Lindet, Paris: Belles Lettres.

93 We are not so sure that the statement Celenza cites from Burckhardt's *Die Cultur der Renaissance in Italien* ("The rapid progress of humanism after the year 1400 paralyzed native impulses. Henceforth men looked only to antiquity for the solution of every problem, and consequently allowed literature to turn into mere quotation") actually represents Burckhardt's own opinion on the matter. It seems to be an excerpt from Libri, 1838, who is cited in the note of the German edition (Burckhardt 2018: 137 note 2).

94 The reader interested in epitomes will be delighted at the articles by the late Peter Lebrecht Schmidt (1933-2019) in the *Handbuch der Lateinischen Literatur der Antike, Bd. 6. Die Literatur im Zeitalter des Theodosius (374-430 n.Chr.)*, which was still forthcoming during the writing of the present essay (Berger et al., 2020). The handbook includes not only the geography by Solinus (§ 603.9) but also several historical epitomes (Iulius Exuperantius, Iulius Obsequens, Iustinus, and the *Prologi Pompeii Trogi*; the *Libellus de vita et moribus imperatorum* (*Epitome de Caesaribus*) (§ 637). Schmidt's long-awaited interpretation of Eusebius Nanneticus (§ 639.1)—presumably an important figure in the writing of the *Historia Augusta*—is also included in the book.

Aurelius Victor (1970): S. Aurelii Victoris Liber de Caesaribus, praecedunt OGR et DVI, subsequitur Epitome de Caesaribus. Edited by Franz Pichlmayr and Roland Gründel (2nd ed.), Leipzig: Teubner.

Aurelius Victor (1997): Die römischen Kaiser, Lateinisch-deutsch. Edited and Translated by Manfred Fuhrmann. Commented by Kirsten Groß-Albenhausen, Zürich and Düsseldorf: Artemis.

Baker, Patrick (2015): Italian Renaissance Humanism in the Mirror, Cambridge, UK: Cambridge University Press.

Beccadelli, Antonio (1968): Antonii Panhormitae liber rerum gestarum Ferdinandi regis. Edited by Gianvito Resta, Palermo: Centro di Studi Filologici e Linguistici Siciliani.

Benson, Robert L./Constable, Giles (eds.) (1982): Renaissance and Renewal in the Twelfth Century, Oxford: Oxford University Press.

Berger, Jean-Denis/Fontaine, Jacques/Schmidt, Peter Lebrecht (2020): Handbuch der Lateinischen Literatur der Antike, Bd. 6. Die Literatur im Zeitalter des Theodosius (374-430 n.Chr.), Munich: C.H. Beck.

Biondo, Flavio (1531): De Roma trivmphante libri decem, priscorum scriptorum lectoribus utilissimi, ad totiusque Romanae antiquitatis cognitionem pernecessarij. Romae instauratae libri III. Italia illvstrata. Historiarum ab inclinato Rom. imperio decades III. Omnia multo quam ante castigatoria, Basel: Froben.

Biondo, Flavio (2005): Italy Illuminated. Edited and Translated by Jeffrey White, Cambridge, MA: Harvard University Press.

Blair, Ann M. (2010): Too Much to Know. Managing Scholarly Information before the Modem Age, New Haven: Yale University Press.

Brown, Peter (1971): The World of Late Antiquity from Marcus Aurelius to Muhammad, London: Faber.

Brunhölzl, Franz (1975): Geschichte der lateinischen Literatur des Mittelalters 1. Von Cassiodor bis zum Ausklang der karolingischen Erneuerung, Munich: Fink.

Brunhölzl, Franz (1992): Geschichte der lateinischen Literatur des Mittelalters 2. Die Zwischenzeit vom Ausgang des karolingischen Zeitalters bis zur Mitte des elften Jahrhunderts, Munich: Fink.

Bruni, Leonardo (1471): De bello Italico adversus Gothos, Venice: Nicolaus Jenson.

Bruni, Leonardo (1537): De primo bello Punico, Augustae Vindelicorum: Philipp Ulhard.

Bruni, Leonardo (1928): Humanistisch-philosophische Schriften mit einer Chronologie seiner Werke und Briefe. Edited by Hans Baron, Leipzig: Teubner.

Bruni, Leonardo (2001-2007): History of the Florentine People (Vols. 1-3). Edited and Translated by James Hankins, Cambridge, MA: Harvard University Press.
Brunt, Peter (1980): "On Historical Fragments and Epitomes." In: Classical Quarterly 30, pp. 477-494.
Burckhardt, Jacob (2018): Die Cultur der Renaissance in Italien. Ein Versuch (Kritische Gesamtausgabe Band 4). Edited by Mikkel Mangold, Munich: Beck.
Burke, Peter (1966): "A Survey of the Popularity of Ancient Historians 1450-1700." In: History and Theory 5/2, pp. 135-152.
Burke, Peter (1998): The European Renaissance, Oxford: Blackwell.
Cameron, Alan (2011): The Last Pagans of Rome, Oxford: Oxford University Press.
Carey, Sorcha (2003): Pliny's Catalogue of Culture. Art and Empire in the *Natural History*, Oxford: Oxford University Press.
Celenza, Christopher (2001): "Late Antiquity and the Florentine Renaissance: Historiographical Parallels." In: Journal of the History of Ideas 62, pp. 17-35.
Celenza, Christopher (2004): The Lost Italian Renaissance, Baltimore: Johns Hopkins University Press.
Celenza, Christopher (2012): "Late Antiquity and the Italian Renaissance." In: Scott Fitzgerald Johnson (ed.), The Oxford Handbook of Late Antiquity, Oxford: Oxford University Press, pp. 1172-1199.
Clavuot, Ottavio (1992): Biondos "Italia illustrate" – Summa oder Neuschöpfung?, Tübingen: Max Niemeyer.
Cochrane, Eric W. (1981): Historians and Historiography in the Italian Renaissance, Chicago: University of Chicago Press.
Cox Jensen, Freyja (2018): "The Popularity of Ancient Historians, 1450-1600." In: The Historical Journal 61/3, pp. 561-595.
da Vinci, Leonardo (1975). Codices Madrid: Nationalbibliothek Madrid, Vols. 1-5. Edited, Translated, and Commented by Ladislao Reti, Frankfurt am Main: S. Fischer.
Darnton, Robert (1979): The Business of Enlightenment: A Publishing History of the *Encyclopédie* 1775–1800, Cambridge, MA: Belknap Press of Harvard University.
DEH (1932): Hegesippi qui dicitur historiae libri V (De excidio Hierosolymitano). Edited by Vincenzo Ussani. Ps. 1: Textum criticum continens, Vienna: Hoelder-Pichler-Tempsky.
DVI (2016): Ps. Aurelius Victor: De viris illustribus urbis Romae. Edited, Translated, and Commented by Joachim Fugmann, Darmstadt: WBG.
Epitome (2009): "A Booklet about the Style of Life and Manners of the Imperatores." Translated by Thomas Banchich (2nd ed.), Buffalo: Canisius College. Accessed August 21, 2019. http://www.roman-emperors.org/epitome.htm.

Eutrop (1879): Eutropi breviarium ab urbe condita cum versionibus Graecis et Pauli Landolfique additamentis. Edited by Hans Droysen, Berlin: Weidmann.

Eutrop (2018): Breviarium ab urbe condita. Edited, Translated, and Commented by Bruno Bleckmann and Jonathan Gross, Paderborn: Schöningh.

Festus (1967): The Breviarium. Edited by John William Eadie, London: Athlone Press.

Foot, Sarah/Robinson, Chase F. (eds.) (2012): The Oxford History of Historical Writing 2. 400-1400 AD, Oxford: Oxford University Press.

Fremmer, Anselm (2001): Venezianische Buchkultur. Bücher, Buchhändler und Leser, Cologne: Böhlau.

Fugmann, Joachim (1990-2004): "Königszeit und frühe Republik in der Schrift 'De viris illustribus urbis Romae.'" In: Quellenkritisch-historische Untersuchungen. 1-3, Frankfurt: Peter Lang.

Gasti, Fabio (2015): "La forma breve della prosa nella storiografia latina d'età imperiale e tarda." In: Koinonia 39, pp. 345-354.

Giacardi, Livia (2005): "Guglielmo Libri." In: Dizionario Biografico degli Italiani 65. Accessed March 28, 2020. http://www.treccani.it/enciclopedia/guglielmo-libri_(Dizionario-Biografico)/.

Giulio Ossequente (2005): Prodigi. Edited by Paolo Mastandrea. Translated by Massimo Gusso, Milano: Oscar Mondadori.

Grafton, Anthony T. (1980): "The Importance of Being Printed." In: Journal of Interdisciplinary History 11/2, pp. 265-286.

Grafton, Anthony T. (1997): Commerce with the Classics: Ancient Books and Renaissance Readers, Ann Arbor: University of Michigan Press.

Greenblatt, Steven (2011): The Swerve. How the World Became Modern, New York: W.W. Norton.

Grendler, Paul F. (1989): Schooling in Renaissance Italy. Literacy and Learning, 1300-1600, Baltimore: John Hopkins University Press.

Guenée, Bernard (1980): Histoire et culture historique dans l'Occident médiéval, Paris: Aubier.

GW (Gesamtkatalog der Wiegendrucke) (since 1925). Accessed August 21, 2019. http://www.gesamtkatalogderwiegendrucke.de/.

Habinek, Thomas (1998): The Politics of Latin Literature: Writing, Identity, and Empire in Ancient Rome, Princeton: Princeton University Press.

Hankins, James (1997): Repertorium Brunianum. Vol. I. Handlist of Manuscripts, Rome: Istituto storico italiano per il Medio Evo.

Haskins, Charles Homer (1927): The Renaissance of the Twelfth Century, Cambridge, MA: Harvard University Press.

Herzog, Reinhard./Schmidt, Peter Lebrecht (eds.) (1989): Handbuch der lateinischen Literatur, Vol. 5, Munich: Beck.

Hofmann, Heinz (1987): "Artikulationsformen historischen Wissens in der Historiographie des hohen und späten Mittelalters." In: Gumbrecht, Hans-Ulrich /Link-Heer, Ursula/Spangenberg, Peter-Michael (eds.), Grundriß der romanischen Literaturen des Mittelalters, Vol. 11: La littérature historiographique des origines à 1500. Part 2, Heidelberg: Winter, pp. 367-687.

Hohl, Ernst (1911): Vopiscus und die Biographie des Kaisers Tacitus. Diss. Tübingen. In: Klio. Beiträge zur Alten Geschichte 11/1911, pp. 178-229, pp. 284-324.

Holford-Strevens, Leofranc (1995): "Review of L. Ampelius, Aide-mémoire (Liber memorialis) by Marie-Pierre Arnaud-Lindet." In: Gnomon 67, pp. 600-604.

Horster, Marietta (2016): "Livius-Epitome: ein spätantiker Blick auf die (kurzgefasste) römische Republik." In: Dusil, Stephan/Schwedler, Gerard/Schwitter, Raphael (eds.), Exzerpieren–Kompilieren–Tradieren. Transformationen des Wissens zwischen Spätantike und Frühmittelalter, Berlin: de Gruyter.

Horster, Marietta/Reitz, Christiane (eds.) (2010): Condensing Texts – Condensed Texts, Stuttgart: Steiner.

Huizinga, Johan (1919): "Herfsttij der Middeleeuwen. Studie over levens- en gedachtenvormen der veertiende en vijftiende eeuw in Frankrijk en de Nederlanden." Haarlem: Willink. Accessed August 21, 2019. http://www.dbnl.org/tekst/huiz003herf01_01/.

Ianziti, Gary (2012): Writing History in Renaissance Italy: Leonardo Bruni and the Uses of the Past, Cambridge and London: Harvard University Press.

Iustinus (1971): Epitoma historiarum Philippicarum Pompei Trogi. Edited by Otto Seel, Stuttgart: Teubner.

Kaiser, Reinhold (2014): Neue Fischer-Weltgeschichte 3: Die Mittelmeerwelt und Europa in Spätantike und Frühmittelalter, Frankfurt am Main: S. Fischer.

Kallendorf, Craig (2017): "Using Manuscripts and Early Printed Books." In: Victoria Moul (ed.), A Guide to Neo-Latin Literature. Cambridge, UK: Cambridge University Press, pp. 379-393.

Kessler, Eckhard (1978): Petrarca und die Geschichte. Geschichtsschreibung, Rhetorik, Philosophie im Übergang vom Mittelalter zur Neuzeit, Munich: Fink.

Kinder, Hermann/Hilgemann, Werner (1964): dtv-Atlas zur Weltgeschichte: Karten und chronologischer Abriß 1-2, Munich: dtv.

Laureys, Marc (2014): "The Theory and Practice of History in Neo-Latin Literature." In: Brill's Encyclopaedia of the Neo-Latin World. Macropedia, Leiden: Brill, pp. 363-375.

Libri, Guglielmo (1838): Histoire des sciences mathématiques en Italie 2, Paris: Renouard.

Lücke, Martin/Zündorf, Irmgard (2018): Einführung in die Public History, Göttingen: Vandenhoeck & Ruprecht.

Mazzocco, Angelo (ed.) (2006): Interpretations of Renaissance Humanism, Leiden: Brill.

McKitterick, Rosamond (2004): History and Memory in the Carolingian World, Cambridge, UK: Cambridge University Press.

McManamon, John M. (1989): Funeral Oratory and the Cultural Ideals of Italian Humanism, Chapel Hill: University of North Carolina Press.

Moss, Angelo (2014): "Commonplace Books: Major Items in Print." In: Brill's Encyclopaedia of the Neo-Latin World. Micropaedia. Leiden: Brill, pp. 948-953.

Mülke, Markus (2010): "Die Epitome – das bessere Original?" In: Horster/Reitz 2010: 69-89.

Najemy, John M. (ed.) (2004): Italy in the Age of the Renaissance, Oxford: Oxford University Press.

Nicholl, Charles (2006): Leonardo da Vinci. Die Biographie, Frankfurt am Main: S. Fischer.

Offergeld, Thilo (1997): "Einleitung." In: Hugo von Sankt Viktor. Didascalicon de studio legendi: Lateinisch-Deutsch. Edited and Translated by Thilo Offergeld, Freiburg: Herder, pp. 7-102.

OGR (2004): Ps. Aurelius Victor: Origo gentis Romanae. Die Ursprünge des römischen Volkes. Edited, Translated, and Commented by Markus Sehlmeyer, Darmstadt: WBG.

Orosius (2010): Seven Books of History against the Pagans. Translation by Andrew T. Fear, Liverpool: Liverpool University Press.

Panegyrici (2008-2014): Lobreden auf römische Kaiser 1-2. Edited and Translated by Brigitte Müller-Rettig, Darmstadt: WBG.

Patze, Hans (ed.) (1987): Geschichtsschreibung und Geschichtsbewusstsein im späten Mittelalter, Sigmaringen: Thorbecke.

Periocha (1984): Histoire Romaine: T. 34. Abrégés des livres de l'Histoire romaine de Tite-Live. Edited and Translated by Paul Jal, Paris: Les Belles Lettres.

Petrarca, Francesco (1964): Edizione nazionale delle opere di 2: De viris illustribus (1338/1353). Edited by Guido Martellotti, Florence: Sansoni.

Pius II (1533): "Decadum Blondi Epitome, Qua omnis ab inclinato Romanorum imperio historia, per mille et amplius annos, miro compendio (…) complectitur." Basel: Ioannes Bebelius. Accesssed August 21, 2019. http://www.e-rara.ch/bau_1/content/titleinfo/1204958.

Plett, Heinrich F. (1993): Renaissance Rhetoric/Renaissance-Rhetorik, Berlin: de Gruyter.

Reinhardt, Volker (ed.) (1997): Hauptwerke der Geschichtsschreibung, Stuttgart: Kröner.

Resta, Gianvito (1962): Le epitomi di Plutarco nel Quattrocento, Padua: Antenore.

Rexroth, Frank (2007): Meistererzählungen vom Mittelalter: Epochenimaginationen und Verlaufsmuster in der Praxis mediävistischer Disziplinen, Munich: Oldenbourg.

Reynolds, Leighton Durham/Wilson, Nigel Guy (2013): Scribes and Scholars: A Guide to the Transmission of Greek and Latin Literature (4th ed.), Oxford: Oxford University Press.

Riché, Paul (1976): Education and Culture in the Barbarian West: From the Sixth through the Eighth Century. Translation by John J. Contreni, Columbia: University of South Carolina Press.

Schirrmeister, Albert (2014): "Historiographie." In: Der Neue Pauly Supplement 9 Renaissance-Humanismus: Lexikon zur Antikerezeption, pp. 443-454.

Schmidt, Peter Lebrecht (1989): "Historiographie." In: Herzog/Schmidt 1989: 173-211.

Sehlmeyer, Markus (2005): "Ernst Hohl und die Historia-Augusta." In: Sehlmeyer, Markus/Walter, Uwe (eds.), Unberührt von jedem Umbruch? Der Althistoriker Ernst Hohl zwischen Kaiserreich und früher DDR, Frankfurt: Verlag Antike, pp. 69-87.

Sehlmeyer, Markus (2009): Geschichtsbilder für Pagane und Christen. Res Romanae in den spätantiken Breviarien, Berlin: de Gruyter.

Sehlmeyer, Markus (in print): "Pompeius Trogus (§ 309)." In: Albrecht, Michael von/Kißel, Walter/Schmidt, Peter Lebrecht (eds.), Handbuch der lateinischen Literatur 3, Munich: Beck.

Solinus (2014): Wunder der Welt. Edited and Translated by Kai Brodersen, Darmstadt: WBG (Collectanea rerum memorabilium).

Taylor, John (1987): English Historical Literature in the Fourteenth Century, Oxford: Oxford University Press.

Treadgold, Warren T. (ed.) (1984): Renaissances before the Renaissance, Stanford: Stanford University Press.

Treadgold, Warren T. (2007): The Early Byzantine Historians, Basingstoke: Palgrave Macmillan.

Valerius (1998): Facta et dicta memorabilia, I: Libri I-VI; II: Libri VII-IX. Iuli Paridis Epitoma; Ianuari Nepotiani Epitoma. Edited by John Briscoe, Stuttgart and Leipzig: Teubner.

Van Nuffelen, Peter (2012): Orosius and the Rhetoric of History, Oxford: Oxford University Press.

Vendramini, Darío Sánchez (2012): "Los breviarios históricos y la cultura de la nueva élite del Bajo Imperio Romano (260-395 D. C.)." In: Temas Medievales 20, pp. 275-325.

Völkel, Markus (2006): Geschichtsschreibung. Eine Einführung in globaler Perspektive, Cologne: UTB.

Watt, William Montgomery (1969): The Influence of Islam on Medieval Europe, Edinburgh: Edinburgh University Press..

Wheatley, Chloe (2011): Epic, Epitome, and the Early Modern Historical Imagination, Farnham: Ashgate.

Woolf, Daniel (ed.) (2012): Oxford History of Historical Writing 3: 1400-1800, Oxford: Oxford University Press.

Woolfson, Jonathan (ed.) (2002): Palgrave Advances in Renaissance Historiography, London: Palgrave Macmillan.

Würgler, Andreas (2013): Medien in der Frühen Neuzeit, Munich: Oldenbourg.

Zinsli, Samuel Christian (2017): "Beobachtungen zum Epitomatorenhandwerk des Ioannes Xiphilinos." In: Bruno Bleckmann/Hartwin Brandt (eds.), Historiae Augustae Colloquium Dusseldorpiense, Bari: Edipuglia, pp. 197-221.

Pictorial Science and Enlightenment Art: Joseph Wright, William Pether, and the Cognitive Effect of Grayscales

Oliver Jehle

ABSTRACT

When it comes to popularizing recent knowledge—for the Age of Enlightenment—it is necessary to inquire into art forms that could be collected due to large print runs and low prices, and that thus became part of an upscale everyday culture: The medium of print graphics circulated widely in the 18th century, and included a broad selection of Joseph Wright's work. The fact that the subtle technique of the mezzotint was the perfect answer to epistemological questions that Wright developed in his discussion with Robert Boyle and John Locke can be considered a stroke of luck since *techné* and hermeneutics went hand in hand at this moment of art history. The title of the picture already announces what will come into focus: the act of seeing, whose conditions, especially the gradation of black, white, and gray values, become visually tangible. As part of an English visual school that knew how to combine aisthesis and aesthetic experience, the graphic works are here examined in terms of the cognitive added value of grayscale and light-and-dark contrasts. Because these artistic means were the armor of the learned artist, this relates to the popularization of seeing in all its valences.

INTRODUCTION

In English Enlightenment art, a high level of seeing capacity is required, which seems to contradict—at first glance—the idea of the popularization of recent knowledge stocks (Gramaccini 1999: 435-448). However, it is precisely the countless repetitions of chromatically rich paintings, as practiced in the cheaper reproduction graphics and intended for a broader consumer stratum, which can be seen as evidence of the widespread distribution of academic high art. In the medium of the print graphic, which draws its effect solely from the gradation of black, white, and gray values, knowledge is pictorially staged and popularized. The Englishmen Richard Earlom (1743-1822; *Fig. 1*) and Valentine Green (1739-1813; *Fig. 2*) dominated the European market as engravers in this art direction, as did John Boydell (1719-1804) as a publisher.

Figures 1 and 2: Portraits of Richard Earlom (left) and Valentine Green (right), by Thomas Goff Lupton and Lemuel Francis Abbot

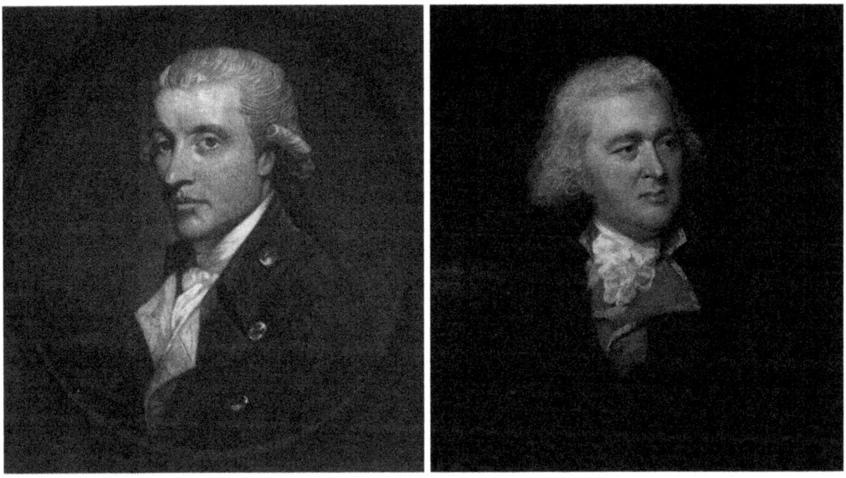

Source: National Library of Wales and National Portrait Gallery, London; Public Domain

The technique of mezzotint, which was considered to be *the* English printing process, was particularly appreciated by collectors in Europe and North America, for the art of scraping was considered to be the "only process that works from the dark to the bright and that separates the visible from the darkness." (Busch 1993: 177)[1] With the mezzotint method, art puts distinct objects in the light and thus becomes an equivalent of the Enlightenment movement itself. Something 'magical' adheres to this technique of forced light-dark contrasts, the effect of

1 Unless otherwise noted all translations are my own.

which is particularly evident in the pages of William Pether (*Figs. 3* and *5*).[2] After Joseph Wright's (1734-1797) large-format oil paintings, these high-contrast leaves show not only hitherto unseen scientific experiments (*Fig. 3*)[3] but also light situations that were understood as 'creation analogies.' Once-smoothed copper plates were roughened with a weighing pan until they were covered with a dense grid of minute dots. A print of such a plate would have led to a printed result, which would be a monochrome and velvety black field on the paper. It is only through the differentiated smoothing of individual parts with the scraper that image areas arise in which the brightness returns: Like an *alter deus*, the copper engraver says it is going to be light.

Figure 3: Valentine Green, A Philosopher Shewing an Experiment on an Air Pump*—After Joseph Wright of Derby, Mezzotint, 44.8 × 58.7 cm*

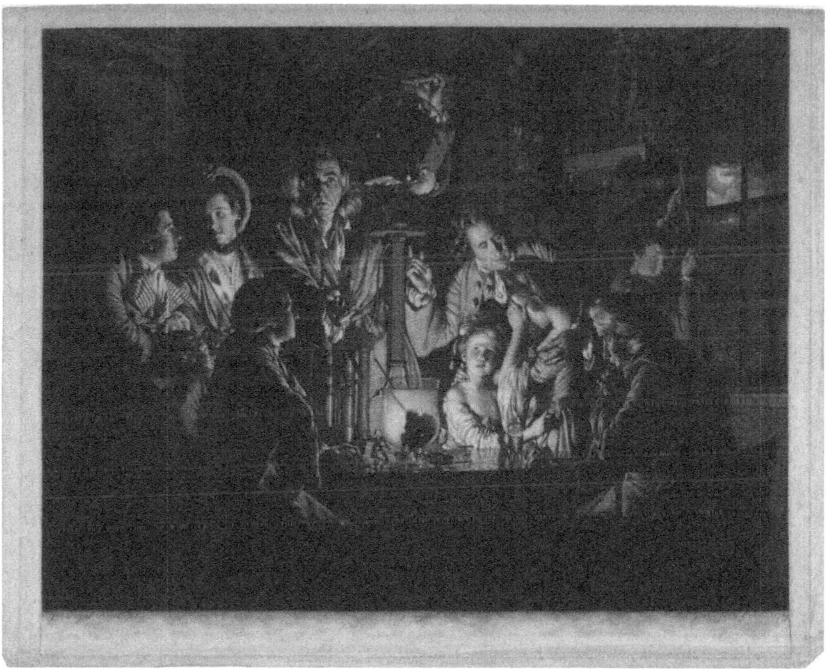

Source: The Elisha Whittelsey Collection; Public Domain

2 See also Joseph Wright, 1765-68 Self-Portrait in a Fur Cap, monochrome pastel on blue-grey laid paper 42.5 x 29.5 cm. Art Institute of Chicago.

3 Valentine Green (1739-1813), A philosopher shewing an experiment on an air pump—after Joseph Wright of Derby, Mezzotint, 44.8 × 58.7 cm. With both artists' names in the plate. Inscribed verso at lower left corner in ink: No. 21, and in pencil. Princeton University Art Museum.

Figures 4 and 5: Joseph Wright, Three Persons Viewing the Gladiator by Candlelight; *William Pether's Mezzotint Version*

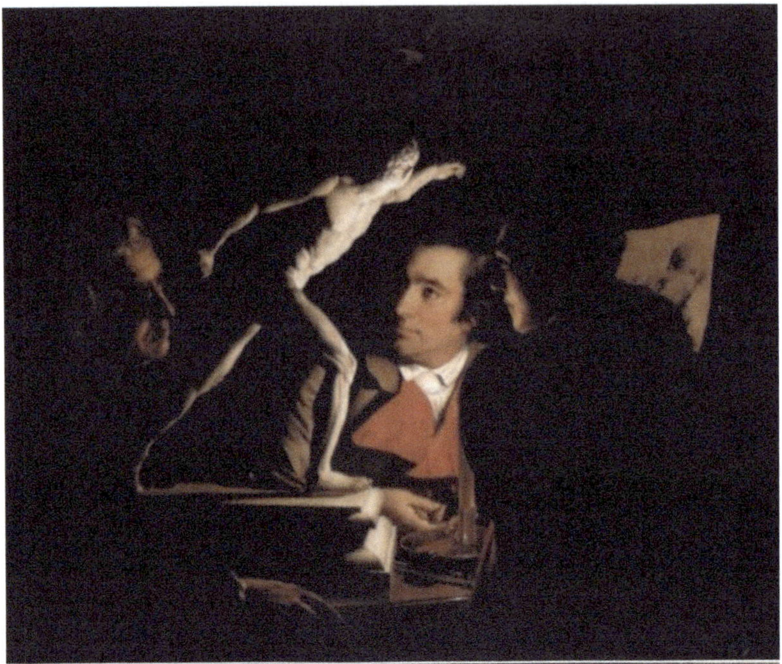

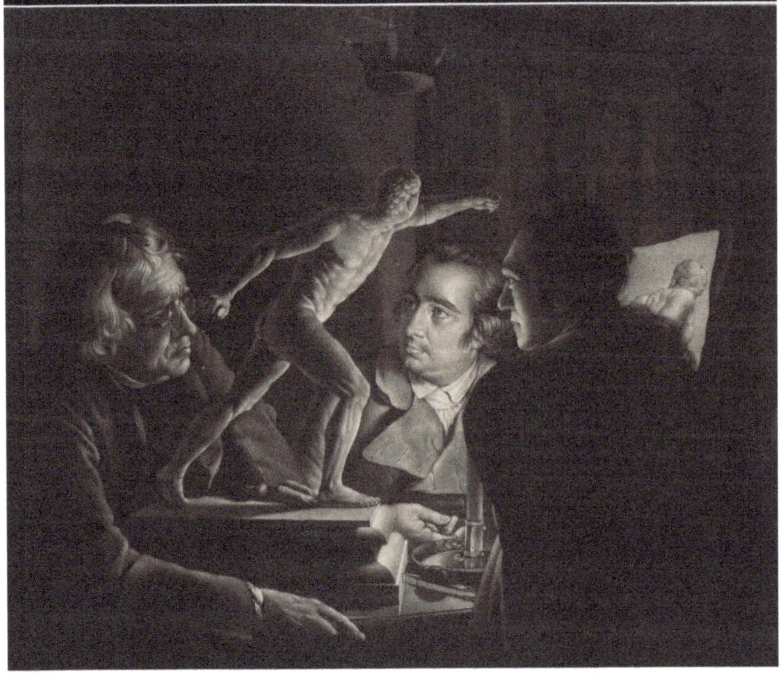

Source: Private Collection and Paul Mellon Fund. 2001.96.13; Photos: Public Domain

The chamber in Joseph Wright's painting, *Three Persons Viewing the Gladiator by Candlelight* (*Fig. 4*)—issued in 1765 for the first time[4]—is filled with a warming light. A light which spreads its rays over the persons and objects and thereby brings them together but also makes their colors and forms visible, classifies them in this way, and gives them outlines. The individual forms are created by the differentiation of the surfaces that are lifted out of the dark by candlelight, the material properties of which become clear in their distinct details—be it the smoothness of the cold copper or the porous surface of the plaster that transforms light into velvet, as in William Pether's mezzotint version (*Fig. 5*).[5]

The image, with its descriptive title, first of all conveys a vivid, enhanced view of an antique sculpture around which three unspecified persons have come together in an intimate space. They seem to communicate with each other, to make a silent discourse about an artifact and the adequate forms of approaching it. The mere fact that sensory perception enables access to the world and forms the basis of all knowledge was inalienably part of the discourse of the "Siécle des lumières." (Foucault 2005: 145; cf. Gearhart 1984: 161-199) Wright, therefore, shows us three observers who, with the light of a hidden source, turn their attention to the cast and the drawing of an antique sculpture: the Borghese Gladiator.[6] In a dark chamber illuminated only by the light of an unseen candle, we take part in a scene of artistic appreciation and in a visual appropriation of the world. Like the act of drawing, the result of which is compared with the given object, so the viewer of the picture visually dissects the selected units of reality in order to create a new, excessively artful reality from the smallest pictorial elements. The play of light and dark, which gives the illumined areas a mimetic intensity, makes seeing the instrument of reason.

"As an operational means of cognition," vision becomes "instrumental in character, it accomplishes intellectual work." (Kanz 1993: 73) Here, the object of knowledge, which is to be transmitted to the thinking eye, requires a suitable

4 In May 1765, a "Lover of the Arts" wrote in *Lloyd's Evening Post and British Chronicle* that it—as a masterpiece—"passed off to some supercilious Connoisseur, at a great price, for an antique picture" (Solkin 1992: 295).

5 William Pether (ca. 1738-1821) after Joseph Wright (1734-1797), Three Persons Viewing the Gladiator by Candlelight, 1769, mezzotint on laid paper, plate: 48.1 x 56.1 cm; sheet: 48.9 x 56.6 cm, Paul Mellon Fund. 2001.96.13.

6 According to Solkin (1992: 216), Wright uses the Borghese Gladiator's imagery to depict "seeing" and "feeling" from the series of the "Five Senses," as Philip Mercier produced in 1740 for the English art market. The representative format that Wright bases on this image resembles the dimensions of a historical painting, both with regard to content and in terms of portraiture and contemporary genre representation (Busch 1986: 21).

form: limited scope, clarity, closed form.[7] This is the only reason why the oldest of the *Three Persons*, who is placed on the left edge of the picture, turns his gaze to a brightly lit piece of paper on which the drawn head of the gladiator and his shoulders and back are visible. This drawing is held up close to the right edge of the picture by a younger man who, as it were, turns his deeply shaded back to the external viewer and therefore, in this position, marks the boundary of the painting.

Thus, the gaze of the eldest passes across the sculpture and traverses the pictorial space. His perspective allows the narrative context of images in the sense that the actors in the picture space compare the casting of the gladiator with a charcoal drawing of the same. Only the eldest man can understand the viewing angle; moreover, the anatomically correct implementation in the drawing can be communicated by him alone. As the picture narrows, the final judgment is up to him. As the bearer of these visual axes and visual references, only the light that brings the central figure—one of the two students—into the immediate vicinity of the antique cast works. Yet not this light carrier, but rather the candlelight, is the feature that constitutes the image. For every object of representation has an active power that can only be articulated from within it (Büttner 1999: 341-351). The elements spread out on the scene determine each other to such an extent that the "impression of space appears to emerge from a play of gradated brightnesses." (Bohlmann et al. 2008: 7) As an active force, light creates space and objects, differentiating the picture elements "according to their different, always lightly expressed assets to affect other picture elements and to be affected by them." (Ibid: 103) The light in Wright's pictorial space is not an accidental element "that shows or illuminates anything," but "transforms the light into a generative and dynamic" element, indeed the "most important image-constitutive element." (Weiss 2008: 8)

In the mezzotint, light creates the space of perception: A dynamic relationship between lighter and darker masses and a play of infinite degrees of light are artfully staged here. The decisive factor is that the question of the purely optical problems of the light intensity and thus of the grayscale is posed and answered solely in the context of print graphics. (Wölfflin 1960: 104) At the same time, Wright does not create the spatial impression through the bundling of aligned lines, but through chiaroscuro and the juxtaposition of light and dark masses. If the impression of space follows from this, the perspective system also undergoes such extensive changes that the viewer's position is also affected. The counterpart of image-space and viewer is complemented by an intra-image light source to form a triangle; the candlelight illumines things, reveals them, and opens up the

7 Thus, according to August Langen, "the tendency is to grasp, to separate, and to disassociate the respective point of apperception, the smallest part of the sharpest and clearest perception" (Langen 1965: 7-8).

self-luminous visual world of the receiver—a pictorial world that primarily tells of the increase in the effect of the chiaroscuro structure (cf. Marin 2003: 205-224).

LIGHT AND DARK: ROBERT BOYLE'S VISUAL EXERCISES

A tradition of differentiated viewing of the world which began with Robert Boyle's[8] visual exercises in 1665 appears here, with the continued dissection of the surface structure aiming to explore the behavior of light on irregular surfaces. The radiant energy of light, according to Boyle, is absorbed or reflected by the surfaces of the objects in different ways. The idea that light can only unfold those shapes and colors when it shines brightly and clearly may have come to Joseph Wright's mind in his struggle with common maladies: He lit a candle in his painting and immediately stopped, wick scissors in hand. The use of candles was commonplace at the time; one would suppose that candles were made of animal fat until the beginning of the 19th century, and that the longer the wick became the more soot and dripping wax there was. The wicks therefore had to be cut regularly. The use of the scissors, which leads to brighter candlelight and improved visual conditions, was known as "light cleaning." The meaning that grows out of the light and its deliberate directing effect, light-dark as a medium which actually creates the visibility of the object's reality, is not simply represented light in this picture, but at the same time "the reflection in the image itself of the possibility of the appearance of the imagery." (Prater 1992: 82) That is why Wright's titling of the painting ends with the phrase "by candlelight." This image-internal reorganization also corresponds to a change in the viewer's position, a change emanating from an insecurity of the viewer's central point of view, if spatial effects are produced solely by light-dark contrasts, thus purely optically, or also if the eye point beside the other focusing points becomes an interchangeable projection point of the image.

By grouping three observers around the classic artifact, Wright achieves a poly-optical perspective that leads to an intensification of the authentic moment of seeing. In the seemingly tranquil gazes, he discovers mobility and a constant search for new perspectives and views, which seem to be designed to turn the astonishment and curiosity of the students into the objectivity of contemplation. What is cognitive about this transformation, however, is not only its internal structure, which allows emotion to become a classifying point of access, but also

8 John Faber Jr., after Unknown artist, Robert Boyle, mezzotint (ca. 1689-1690), 361 x 264 mm paper size, Given by the daughter of compiler William Fleming MD, Mary Elizabeth Stopford (née Fleming), 1931, Reference Collection, London: National Portrait Gallery, D784.

the awareness of a form of reflection of the aesthetic in the aesthetic medium itself—the art of making a mezzotint. The dramatic tonal contrasts that resulted from his attempt to depict the effect of a hidden light source in a darkened room,[9] a kind of *tour de force*, condition the beholder's eye, as it were, to adapt to the brightness conditions. The activity of seeing in all its operations is determined by these external impressions. The precisely crafted print graphics, which cause us to forget the act of making them, along with the accuracy of the artistic rendering, illustrate how Wright, with this sophisticated candlelight picture, focuses on the precision of human perception and seeks to erase the act of producing in the finely executed version.

WRIGHT AND JOHN LOCKE'S *ESSAY CONCERNING HUMAN UNDERSTANDING*

When seeing and understanding are related to each other, the viewer is reminded that the central point of reference of this painting is John Locke's philosophy of perception,[10] which epitomizes an epochal horizon for the 18th century in England and, as an empirical tradition, epitomizes revolutionary impulses in European thinking.[11] It is well known that the aesthetic, thought-provoking, and image-immanent meaning of seeing, and indeed the epistemological approach of the Essay Concerning Human Understanding, was, to explain that

> "[…] external and internal sensation are the only passages I can find of knowledge to the understanding. These alone, as far as I can discover, are the windows by which light is let into this dark room. For, methinks, the understanding is not much unlike a closet wholly shut from light, with only some little openings left, to let in external visible resemblances, or ideas of things without: would the pictures coming into such a dark room but stay there, and lie so orderly as to be found upon occasion, it would very much resemble the understanding of a man, in reference to all objects of sight, and the ideas of them." (Locke 1999: 146 [Vol. II, ch. 11, § 17])

9 The artist's *vite* lists which devices Gottfried Schalcken used to darken and direct the light for his candlelight pictures in his studio. Wright took over this procedure. (Nicholson 1968: 4-7, Vol. I)

10 John Smith, after Sir Godfrey Kneller, Bt, John Locke, mezzotint, 1721. Acquired, 1869, London: National Portrait Gallery.

11 When one considers that Locke's writings "set up the parameters within which the enquiry was conducted for much of the next century and indicated new [...] directions for aesthetic theory" (Pears 1988: 28), Locke's *Essay Concerning Human Understanding* thus establishes the influential tradition of a theory of knowledge based purely on sense perception (Ludwig 1989: 424).

Wright uses the image of the viewer to describe sensory knowledge primarily as a visual act of perception (Langen 1965: 11). In analogy to newly acquired visual conditions of optical instruments, the visual situation is understood and practiced accordingly in order to silence the pre-rational flow of intuitions in the form of a clear and distinct individual image and thus to be able to use it epistemologically. Locke's description of cognitive processes makes use of spatial metaphors, and the dark space and the small opening of which he writes, through which bundled light flows into the cabinet, are not dissimilar to the representation of the painting. In view of Locke's metaphorical use of the camera obscura,[12] a viewer's behavior is illustrated that differs fundamentally from previous descriptions of the visual process (Hüsch 2003: 46-96). According to Richard Rorty, the novelty of this thinking consists in the "conception of the human mind as an inner space" within whose limits "bodily and perceptual sensations [...] were objects of quasi-observation." (Rorty 1979: 49-50)[13]

Locke's baroque spatial philosophy (cf. Leonhard 2006: 11-35), however, not only encompasses the expansion of the scene but also inseparably connects the activity of understanding with the domains of knowledge and vision. Locke uses the metaphor of the dark space since he seeks to design a form of presentation in order to visualize the functioning of the intellect with this means of spatial illustration: "The Understanding, like the Eye, whilst it makes us see, and perceive all other Things, takes no notice of itself." (Locke 1979: I, 1, § 1) Locke refers to the traditional use of philosophical words: "Idea" is etymologically derived from *eidein*, which means "to see." (Mitchell 1987: 5) The primary source of knowledge is experience, which can have two different starting points, *sensation* and *reflection* (Locke 1979: II, 1, § 24). Simple ideas are received in the moment of perception (Locke 1979: II, 1, § 9) and owe primarily to the empirical data; however, the human mind is able to form complex ideas based on them and "by its own power [...], which it never received so united." (Locke 1979: II, 12, § 2) This "action of mind," which presents itself as a free combination of perceptions, leads to the complex ideas whose real content is measured not in relation to things but in their non-contradictory connection, which in turn is the free work of the mind (Locke 1979: II, 22, §§ 1 and 4; II, 23, § 6). Complex ideas arise from the combination of different simple ideas, and these combinations are given names at the same time. This reciprocity has its common root in the sensory experience of seeing; knowledge is acquired and the other is the guarantee of such knowledge. Knowledge results from the connection of received perceptual data, and such a combination gain guarantees the development of cognitive assets.

12 From among the extensive literature on the *camera obscura*, here I will only mention Wheelock (1977) and Snyder (1980: 499-526).

13 John W. Yolton takes the opposite view (1984: 222).

Figure 6: Borghese Gladiator, attributed to Agasias of Ephesus, Copy from the 1st-2nd Century AD, 199 cm

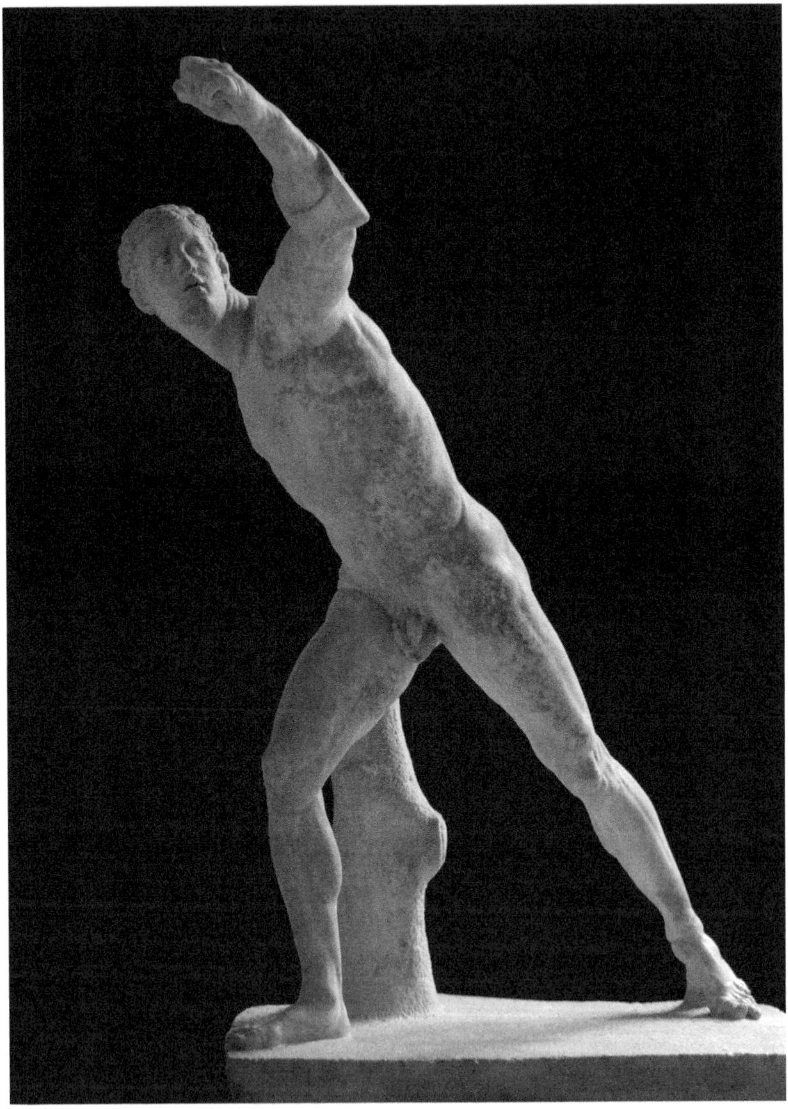

Source: Musée du Louvre, Paris; Photo: Ryan Baumann

If one thus understands the order of images as an aesthetic and epistemologically motivated set, then the painter directs his interest to the act of the classical contemplation of art, without providing a naively immediate identification of essence and appearance. In his process of creating cognitive images, he assures himself of the history of classical art and, in his own practice of art, presents the object-oriented reconstruction of the antique sculpture that has moved into the center of the picture.[14]

The ancient, all-marveling marble statue (*Fig. 6*),[15] which, in Wright's time, was staged as one of the principal works in the Villa Borghese casino and was one of the most famous works of the collection,[16] is described in Winckelmann's *History of the Art of Antiquity* as a "collection of the Beauties of nature." (Winckelmann 1764: 394)

Due to its concentrated and complex plot, the work was considered a bravura piece of antique sculpture, which had advanced to the limits of the technically realizable (Haskell/Penny 1981: 221-223. Bürger 1995: 67).

Sir Joshua Reynolds (*Fig. 7*), director of the academy and guardian of the grand style in England, referred in 1770—in his description of antiquity—to that normative paradigm of a condensed movement figure in whose torsion the lively movement was at the same time "purified and intensified to an ideal one of nature-surpassing invention due to movement." (Krüger 1997: 250)

Moreover, this model of academic discourse is detached, in Wright's pictorial space, from its original presentation structure, selected under mere aesthetic criteria and with a view to its effect in the picture: because there was no other statue that would have been comparable to the gladiator in terms of the expressivity of the movement motif (Reynolds 1997: 151). Thus, Wright's pictorial space becomes a sign of how the recognition of an ancient sculpture can coincide with the intuitive perception of the transience of its plaster cast. Because artifacts are not given to the beholder as things of everyday life, but rather as objects that owe what they are solely to the epistemologically motivated discourse, in this visual-spatial meaning sensory knowledge and rational knowledge are ubiquitous.

---

14 In Pether's mezzotinto, it becomes clear that the knowledge of classical sculptures and artifacts for large circles was made possible precisely by the medium of printmaking—it is only Lessing who reminds us of any autopsy in his media-theoretical examination of the 'Laocoon,' based on printed graphics (cf. Robert/Vollhardt 2013).

15 Wright most likely used the plaster cast of the Borghese Gladiator made by D. Brucciani & Co (London), from original in the Musée du Louvre, 163 cm, London, Collection of the Royal Academy of Arts.

16 The statue was found in May 1609 in Nettuno (Kalveram 1995: 111, n. 99).

Figure 7: Portrait of Sir Joshua Reynolds, by Valentine Green, Mezzotint, 1780

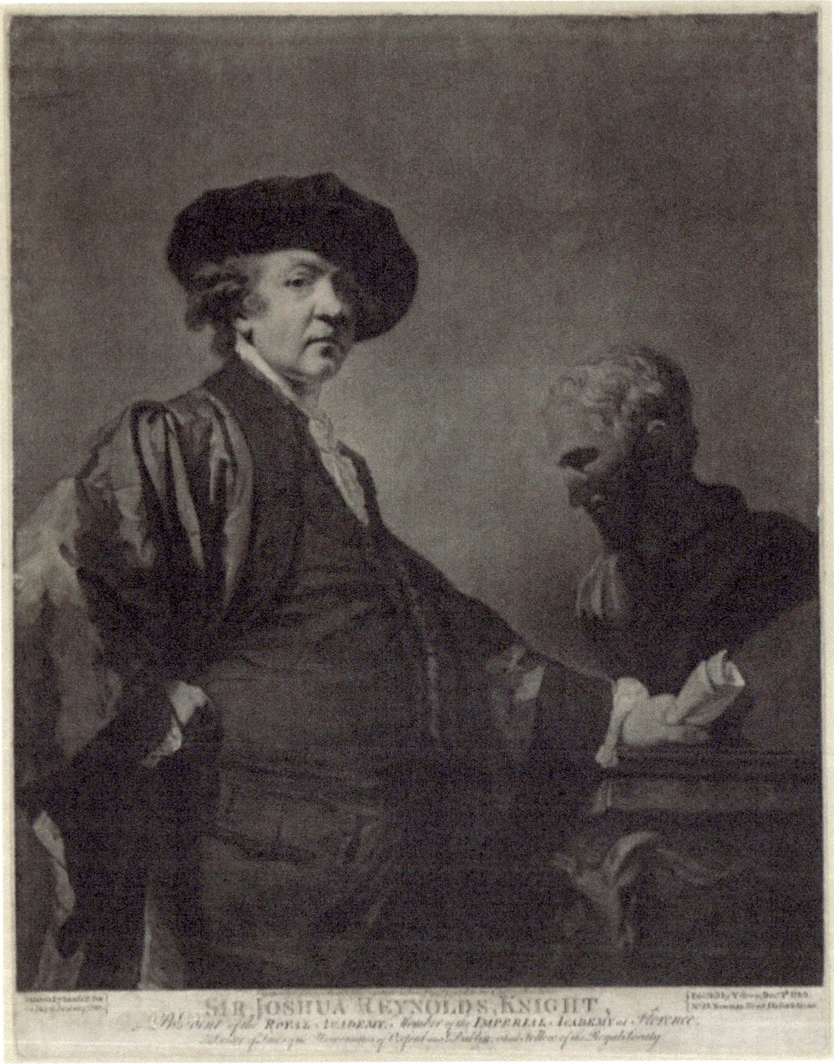

Source: National Portrait Gallery, London, NPG D39705

However, if the repetition of their materialization is not assigned the role of primary meaning, then Wright is concerned with the visualization of the visual act and the subsequent understanding. The sculpture, on the other hand, is marginalized by the insistent gazes of the three men in the painting: The binding force of the classical rule canon, which was derived from the ancient models, disappears to enable a representation of seeing *and* knowing. If one wants to see the empirically guaranteed as possible in its givens, then the forms of seeing must be largely relieved of all iconographic or historical predispositions; furthermore, one must understand the forms of observation only as models that allow variable relations of seeing through their reciprocal combination. The knowledge of observation is thus built up by way of reflection. In order to make this possible, the sculpture that has moved into the center of seeing should not be understood on an ad hoc basis as the embodiment of a classical or canonical ideality, but as an ordering unit for contingent details. This productive insecurity, which results, for the observer, from the fact that ancient sculpture in its dramatic illumination is withdrawn from a semantically evident reading aimed at a clear meaning, is at the same time the condition of an understanding that participates in the constitution of the reality of the pictorial space. If the representational function of the image is undermined, in the sense that Wright does not seek to recognize an already known and iconographically defined subject, the painting aims at the realization of continued invulnerability. In the *Essay Concerning Human Understanding*, Locke explains that the beholder is used to understanding every single substance he encounters as an "entire thing by itself" (Locke 1979: II, Book IV, Chapter VI, § 11) that unites all qualities and is thus independent of all others.

On the other hand, a closer autopsy convinces every observer how much the isolated object is able to develop its specific properties only in the plurality of subjective ways of looking at things. Thus Wright's view of a sculpturally visualized antiquity brings it out of the darkness and places it in the flood of cognitive interest (Iser 1994: 115), which is realized in a threefold act of seeing (Bate 1961: 118). In the staged interplay of complementary seeing and viewing, in the field of tension between contingent perceptions, each observer represents, next to the pole of the feeling and thinking subject, the center of a reflection and observation that obliges them to a particular objectivity. The artist here attempts a mode of explication in which a sense-perceptual experience is combined with the momentary act of seeing—not only temporally and thus consecutively, but also causally. Thus, the figure of reflection picks up the tension of wonder and comprehension, of inside and outside, reflecting the dialogue of observer and object constitutive of a successful receptive-aesthetic relation. The impression of objectivity also supports the observation that the painter's presentation, as likely as it appears to the beholder, is not shown as necessarily compelling, as the only possible interpretation. Perhaps as a reflex to the heightened sensitivity of the philosophical discourse that each instance of knowledge is conditional and depends essentially on the perspective of the observing subject, Wright's pictorial space appears as if he had checked every detail for authenticity and plausibility and tried to make the painterly procedure transparent.

CLASSICAL EN MINIATURE: A CASTING FOR 1 £ AND 11 SHILLINGS

A sculpture that was intended as a classical testimony to a permanent presence in the cultural memory and that was subjected to an ongoing effort of interpretation in order to keep it *readable* (Assmann 1996: 14) makes the claim of the iconographic process evident: Let us forever translate it into transparent meaning. However, the exemplary interpretation as a model of representative reading is adopted in the image space of the gladiator. In the act of perception, there are no transferabilities or substitutions, since viewing the sculpture is considered an experience of the subjective nature itself. Wright uses an ironic refraction of contemporary antiquity with a smaller cast of the Borghese Gladiator, which he acquired for 1 £ and 11 shillings (Solkin 1992: 225). The reading gaze does not appear primarily with a hermeneutic claim; rather, Wright demonstrates the process of seeing, which is considered a reduced copy, as a process of pure observation, which takes place far removed from any kind of dignity of the object that is to be seen. Wright uses this painting to portray the act of pure, unprejudiced perception. The aesthetic tradition of sensibility is concerned with the immediacy of sensory experience. Thus, the mind is defined as a "physical eye" (Vico 1966: 125) and is able to penetrate the phenomenology of the artifact just as it allows the self-determination of the observer in his reflexive consideration. In the title of the picture, when Wright explicitly evokes the concept of seeing[17] through the word "viewing," he joins the scientific debate about the meaning attached to the verb "to view" for the advanced 18th century. "Seeing," as Gehler formulates it in the *Physical Dictionary*, is a "well-known sensation, which the physical objects produce in us through the light and through the eye; a sensation that enables us to judge of the position, shape, size, movement, color, etc. of visible things." (Gehler 1798: cols. 10-29, col. 10) "What is seeing and what one sees in sight is to be demonstrated only through the act of seeing" (Bubner 1989: 55; Barck 1993: 48)—as the interplay of physical processes and various mental actions and reactions whose complexity is to be grasped. Thus, the psychophysical conditions of seeing become a focus of empirical interest.[18]

---

17 At the same time, Bishop Berkeley, as we know, designed the image of an unbodied spirit, to which only the ability to see is given. (Berkeley 1948: Vol. I, p. 233 (= Paragraph 153): "[C]onsider the case of an intelligence, or unbodied spirit, which is supposed to see perfectly well [...] but to have no sense of touch." Cf. Markovits 1984 : 193-282.

18 The new empirical knowledge of the body and its role in the visual process fundamentally changed prevailing epistemological conceptions of the visual act. Confidence in a stable relationship between perceptual content and the perceived object was dwindling, or, as Crary puts it, "the referential illusion is unsparingly laid bare" (Crary 1990: 91; 1998:

The polished spectacle lens condenses all these processes in its optical functioning and thus becomes its symbol (Schaffer 1989: 67-104). The role of the glass in the field of the visible is more than a fortuitous technical instrument that works to optimize perceptual conditions. Like many other prostheses and instruments, it forms an extension of the body and its sensory organs (Fiorentini 2008: 201-221). It can be installed in the image to become a central element in the structure of the visual axes. The technical world relationship promoted by the reassessment of sensory knowledge and mediated by the use of optical instruments and tools not only gave rise to a new subjective sense of self but made that sense tangible as a new sense of sight, as the experience of new visibilities (Barck 1993: 48).

THE POPULAR VIEW OF THE *COGNITIO SENSITIVA*

However, sensory knowledge, which aesthetics is concerned with, cannot be limited to the perception of external objects. Since it is analogous to reason, the *cognitio sensitiva* encompasses more than recognition or apprehension by means of sensory perception (Zelle 1995: 67). It is peculiar to these acts of sensory apprehension that they be determined by what they grasp (cf. Seel 1998: 351-365). If all sensory perception, to use Gadamer's metaphor, can be described as the "articulation" (Gadamer 1970: 85) of an object, then what is thereby recognized is determined no less by the object being recognized than by the practicing observer and his sensual faculties. For he can only perceive what his receptive abilities are. To such a new paradoxical conception of the sensuous as a subjective, active form of experience—which is paradoxical according to the rationalist alternative—, aesthetics comes into play only because it directs its investigation into special images. Every act of reception is a soliloquy of the beholder.

Wright's dark pictorial space, in which light flows in as an articulating agent, is not only bound to the *camera obscura* in terms of its spatial geometry, but also uses the *camera obscura* as an iconic model of knowledge. However, such knowledge also conveys an element of the image that has hitherto been ignored: the image in the image or, to be more precise, the drawing in the image. Positioned at the right edge of the painting and lit up at the same time, a sheet of tinted paper throws back the candlelight and simultaneously bears witness to what the draftsman has seen and held with the help of his hand. In the concentrated vi-

---

475-499). In the wake of sensory-physiological research, new ideas and conceptions of the observer and the visual process arise, whereby the essential difference to the older conceptions lies in the fact that the body was no longer regarded as passive, but as a "productive physiological apparatus." The physiology of the senses proceeds from a self-activity that can be described according to organic laws. (Crary 1988: 46) sion, as the picture suggests, the plaster cast was enclosed and considered as detached until the resulting drawing became the bearer of a further increased visibility. For drawing alone, as *disegno* theory would have it, is predestined to express the painter's cognitive conception; in it alone, the artistic idea gains visibility, even if it is the pervasive copying of a plaster on a small piece of paper: Wright stages the drawn copy of an artifact which is itself only available as a copy. The image-immanent emphasis on modest materials and the ironic access to antiquity may be read as a rhetoric of images that seeks to emphasize the right to insight, to seeing as seeing. Copying a copy touches on a metaphysical prejudice against the mechanical and technical aspects of artistic work. However, drawing is understood in the classical sense as part of a comprehensive humanistic education, that of the *connoisseur*, who previously guided the charcoal pencil and voluntarily practices the perfection of his cognitive abilities.

Wright's drawing within his painting, which is the work of one of the two students, exemplifies this perfection of seeing, which sought to inventory the artifact in its constituent parts in a continued perception and to recreate it through the medium of drawing. The drawing thus links the additively sensual recording with the rational interpenetration of the viewed, and the experience of perception with a successful mimetic imitation. The other two *connoisseurs* of art in this painting approach and reflect on this goal. However, the scene is also charged with symbols: the illumination of the idea and the invention made possible through the medium of drawing; and the paper sheet shining brightly into space, marking the place of understanding and cognition in an early act of transparency, in the darkness of non-knowledge (Roman 1984: 83-84).

In his study of visual acts, Wright reveals the specific nature of sensory perception: a sense of cognition that grasps perception as the interweaving of seeing, knowing, and judging. Thus, this drawing within his painting turns out to be a representation that aims at the act of seeing and knowing, whose reality is primarily constituted by the beholder (Apel 1994: 571).

Wright's pictorial space corresponds to this empirical theory of perception since the process of receptivity of his images demands a nature of perception that begins with sensory pieces of evidence. In place of a traditional canon of modes of representation and inherent meanings, a mode of representation developed from the experience of the artist is set, which addresses the relativity of subjective perception. A traditional hierarchy between the senses and reason is leveled to be able to attribute a real cognitive ability to sensory perception. The visual faculty, like the imagination, is no longer subordinate to reason, but forms with it a unity, a juxtaposition, whereby the suspicion against the senses as carriers of knowledge is to be overcome. Using Locke to describe this perceptual phenomenon—a distinction between sensory impressions and intellectual processing—, Wright's art of black refers, in the epistemological sense, to the sensuous appearance of things without elevating them to complex ideas, and without offering a metaphorical or allegorical interpretation of the perceived.

The obvious parallels with the revaluation of visual perception suggest that Wright's conception of art has taken a decisive position within contemporary theory on the nature of subjective perception (Schofield 1970; Yolton 1983: 3-13, 107-126). If the three unspecified persons are spellbound in the contemplation that seeks knowledge, Wright, with this pictorial representation, refers to the conviction of English empiricism, to which the *cognitio sensitiva* was regarded as a dark but intellectual faculty. Thus, the seeing subject becomes the basis of thought and observation. To claim such a reference, this candlelight painting is a highly condensed image whose fundamental intention is to serve as a stimulus for the beholder and thus to activate his sense of sight. The fact that the observer is so clearly directed to his own seeing, out of which his knowledge feeds, refers to an omnipresent notion in the epistemological discourse of the Enlightenment. The eyes act as organs that mediate between the visible world and human cognition. What proves to be meaningful is only the subjective knowledge of experience. What is seen and recognized in its empirical facticity owes itself to the will not to capture the pictorial offer as a prescription alone, but to let thought succeed contemplation.

REFERENCES

Apel, Friedmar (1994): "Der lebendige Blick. Goethes Kunstanschauung." In: Aust. Kat. Goethe und die Kunst, ed. Sabine Schulze, Frankfurt am Main: Hatje Cantz, pp. 571-578.

Assmann, Aleida (1996): "Metamorphosen der Hermeneutik." In: Aleida Assmann (ed.), Texte und Lektüren. Perspektiven in der Literaturwissenschaft, Frankfurt am Main: Fischer Taschenbuch, pp. 7-29.

Barck, Karlheinz (1993): Poesie und Imagination. Studien zu ihrer Reflexionsgeschichte zwischen Aufklärung und Moderne, Stuttgart and Weimar: J. B. Metzler.

Bate, Walter Jackson (1961): From Classic to Romantic. Premises of Taste in Eighteenth-Century England, New York: Harvard University Press.

Berkeley, George (1948): "A New Theory of Vision," in: The Works of George Berkeley, ed. A. A. Luce and T. E. Jessop, 9 Vols., London.

Bohlmann, Carolin/Fink, Thomas/Weiss, Philip (2008): "Einführung." In: Carolin Bohlmann/Thomas Fink/Philip Weiss (eds.), Lichtgefüge des 17. Jahrhunderts. Rembrandt und Vermeer – Spinoza und Leibniz, Munich: Wilhelm Fink, pp. 7-17.

Bubner, Rüdiger (1989): Ästhetische Erfahrung, Frankfurt am Main: Suhrkamp.

Bürger, Peter (1995): "Wenn der Felsen der Gegenwart zerschellt – Anmerkungen zur Schwierigkeit des Interpretierens heute." In: Anne-Marie Bonnet/Gabriele Kopp-Schmidt (eds.), Kunst ohne Geschichte? Ansichten zur Kunst und Kunstgeschichte heute, Munich, pp. 65-71.

Busch, Werner (1986): Joseph Wright of Derby. Das Experiment mit der Luftpumpe. Eine heilige Allianz zwischen Wissenschaft und Religion, Frankfurt am Main: Fischer Taschenbuch.

Busch, Werner (1993): Das sentimentalische Bild. Die Krise der Kunst im 18. Jahrhundert und die Geburt der Moderne, Munich: C. H. Beck.

Büttner, Frank (1999): "Der Betrachter im Schein des Bildes. Positionen der Wirkungsästhetik im 18. Jahrhundert." In: Aust. Kat. Mehr Licht. Europa um 1770. Die Bildende Kunst der Aufklärung, ed. Herbert Beck, Peter C. Bol, Maraike Bückling, Munich: Klinkhardt & Biermann, pp. 341-351.

Crary, Jonathan (1988): "Modernizing Vision," in: Hal Foster (ed.), Vision and Visuality, Seattle 1988, pp. 29-50.

Crary, Jonathan (1990): Techniques of the Observer. On Vision and Modernity in the Nineteenth Century, Cambridge, Mass: MIT Press.

Crary, Jonathan (1998): "Attention and Modernity." In: Caroline A. Jones/Peter Galison (eds.), Picturing Science, Producing Art, New York, pp. 475-499.

Fiorentini, Erna (2008): "Optical Instruments and Modes of Vision in Early Nineteenth Century." In: W. Busch (ed.), Verfeinertes Sehen. Optik und Farbe im 18. und frühen 19. Jahrhundert, Munich: Oldenbourg, pp. 201-221.

Foucault, Michel (2005): The Order of Things. An Archaeology of the Human Sciences, London and New York: Routledge.

Gadamer, Hans-Georg (1970): Wahrheit und Methode, Tübingen: Mohr Siebeck.

Gearhart, Suzanne (1984): Open Boundary of Fiction and History: A Critical Approach to the French Enlightenment, Princeton: Princeton University Press.

Gehler, Johann Samuel Traugott (1798): "Sehen, Gesicht." In: Johann Samuel Traugott Gehlers Physikalisches Wörterbuch oder Versuch einer Erklärung der vornehmsten Begriffe und Kunstwörter der Naturlehre, here: Vol. 4, Leipzig: Im Schwickertschen Verlage, pp. 10-29.

Gramaccini, Norberto (1999): "Die Druckgraphik im Licht. Der Durchbruch eines populären Mediums." In: Aust.-Kat. Mehr Licht. Europa um 1770. Die bildende Kunst der Aufklärung, ed. Herbert Beck, Peter C. Bol, Maraike Bückling, Munich: Klinkhardt & Biermann, pp. 435-448.

Haskell, Francis/Penny, Nicholas (1981): Taste and the Antique. The Lure of Classical Sculpture 1500-1900, New Haven and London: Yale University Press.

Hüsch, Anette (2003): Der gerahmte Blick – Zu einer Geschichte des Bildschirms am Beispiel der Camera obscura, Karlsruhe: Staatliche Hochschule für Gestaltung.

Iser, Wolfgang (1994): Der implizite Leser. Kommunikationsformen von Bunyan bis Beckett, Munich: Wilhelm Fink.

Kalveram, Katrin (1995): Die Antikensammlung des Kardinals Scipione Borghese, Worms: Wernersche Verlagsgesellschaft.

Kanz, Roland (1993): Dichter und Denker im Porträt. Spurengänge zur deutschen Porträtkultur des 18. Jahrhunderts (Kunstwissenschaftliche Studien; Band 59), Munich: Deutscher Kunstverlag.

Krüger, Klaus (2001): Das Bild als Schleier des Unsichtbaren. Ästhetische Illusion in der Kunst der frühen Neuzeit in Italien, Munich: Wilhelm Fink.

Langen, August (1965 [1934]): Anschauungsformen in der deutschen Dichtung des 18. Jahrhunderts. Rahmenschau und Rationalismus, Darmstadt: Wissenschaftliche Buchgesellschaft.

Leonhard, Karin (2006): "Was ist der Raum im 17. Jahrhundert? Die Raumfrage des Barocks: Von Descartes zu Newton und Leibniz." In: H. Bredekamp/P. Schneider (eds.), Visuelle Argumentationen. Die Mysterien der Repräsentation und die Berechenbarkeit der Welt, Munich: Wilhelm Fink, pp. 11-35.

Locke, John (1979): Essay Concerning Human Understanding, Peter H. Nidditch (ed.), Oxford: Oxford University Press.

Locke, John (1999): An Essay Concerning Human Understanding (1690), The Pennsylvania State University.

Ludwig, Hans-Werner (1989): "Maschine und Baum: Aufklärung und Gegenaufklärung im England der Industriellen Revolution." In: J. Schmidt (ed.), Aufklärung und Gegenaufklärung in der europäischen Literatur, Philosophie und Politik von der Antike bis zur Gegenwart, Darmstadt: Wissenschaftliche Buchgesellschaft, pp. 421-446.

Marin, Louis (2003): Die Malerei zerstören. Berlin: Diaphanes.

Markovits, Françine (1984): "Diderot et l'aveugle." In: J.-B. Mérian (ed.), Sur le problème de Molyneux, Paris, pp. 193-282.

Mitchell, William J. Thomas (1987): Iconology. Image, Text, Ideology, Chicago and London: University of Chicago Press.

Nicholson, Benedict (1968): Joseph Wright of Derby. Painter of Light (Studies in British Art, The Paul Mellon Foundation of British Art), 2 Vols., London and New York: Pantheon Books, here: Vol. I.

Pears, Ian (1988): The Discovery of Paintings. The Growth of Interest in the Arts in England, 1680-1768, New Haven and London: Yale University Press.

Prater, Andreas (1992): Licht und Farbe bei Caravaggio. Studien zur Ästhetik und Ikonologie des Helldunkels, Stuttgart: Steiner.

Reynolds, Joshua (1997): Discourses on Art, edited by Robert R. Wark (Paul Mellon Centre for Studies in British Art), New Haven and London: Yale University Press.

Robert, Jörg/Vollhardt, Friedrich (eds.) (2013): Unordentliche Collectanea. Gotthold Ephraim Lessings ‚Laokoon' zwischen antiquarischer Gelehrsamkeit und ästhetischer Theoriebildung, Berlin and Boston: de Gruyter.

Roman, Cynthia E. (1984): "Academic Ideals of Art Education." In: Children of Mercury. The Education of Artists in the Sixteenth and Seventeenth Centuries. Exhibition Catalogue David Winton Bell Gallery, Providence: Brown University, pp. 81-95.

Rorty, Richard (1979): Philosophy and the Mirror of Nature, Princeton: Princeton University Press.

Schaffer, Simon (1989): "Glass Works: Newton's Prisms and the Uses of Experiment." In: D. Gooding/T. Pinch/S. Schaffer (eds.), The Uses of Experiment: Studies in the Natural Sciences, Cambridge, UK: Cambridge University Press, pp. 67-104.

Schofield, Robert E.: (1970): Mechanism and Materialism. British Natural Philosophy in an Age of Reason, Princeton and New Jersey: Princeton University Press.

Seel, Martin (1998): "Bestimmen und Bestimmenlassen. Anfänge einer medialen Erkenntnistheorie." In: Deutsche Zeitschrift für Philosophie, 46/3, pp. 351-365.

Snyder, Joel (1980): "Picturing Vision." In: Critical Inquiry 6 (Spring), pp. 499-526.

Solkin, David H. (1992): Painting for Money. The Visual Arts and the Public Sphere in Eighteenth-Century England, New Haven and London: Yale University Press.

Vico, Giambattista (1966 [1744]), Die Neue Wissenschaft über die gemeinschaftliche Natur der Völker, Erich Auerbach (ed. and trans.), Reinbek bei Hamburg: Rowohlt.

Weiss, Philipp (2008): "Das Licht des 17. Jahrhunderts." In: C. Bohlmann/T. Fink/P. Weiss (eds.), Lichtgefüge des 17. Jahrhunderts. Rembrandt und Vermeer – Spinoza und Leibniz, Munich: Wilhelm Fink, pp. 17-19.

Wheelock, Arthur K. (1977): Perspective, Optics, and Delft Artists around 1650, New York: Garland Publishing.

Winckelmann, Johann Joachim (1764): Geschichte der Kunst des Alterthums, Dresden: In der Waltherischen Hof-Buchhandlung.

Wölfflin, Heinrich (1960): Kunstgeschichtliche Grundbegriffe, Basel: Schwabe.

Yolton, John W. (1983): Thinking Matter. Materialism in Eighteenth-Century Britain, Oxford: Basil Blackwell.

Yolton, John W. (1984): Perceptual Acquaintance from Descartes to Reid, Minneapolis: University of Minnesota Press.

Zelle, Carsten (1995): Die doppelte Ästhetik der Moderne, Stuttgart: Metzler.

Modern Times:
Arts and Sciences and Media

Popularity Despite Anti-Popularization Thinking of Optical Drawing Devices in the Early 19th Century

Erna Fiorentini[*]

ABSTRACT

Optical aids for drawing were a topic in many artistic manuals of the early 19th century. These publications, however, propagated an academic depreciation of the optical drawing aids and mostly advised against their use. This theoretical skepticism is a negative form of popularization, but it seems not to have affected the popularity of these instruments in the practice of drawing, as their sales figures show. There is, in fact, a glaring gap between the position of the manuals on the one hand and that of the market and drawing practitioners on the other. In this paper, I consider the paths of a rising popularity that was not backed by intentional efforts of popularization. In the artistic practice of drawing, rather, this popularity developed out of its own logic, which belied all theoretical anti-popularization.

* For a systematic repository of the texts and images cited in this paper, see Erna Fiorentini (ed.), Drawing with Optical Instruments – Devices and Concepts of Visuality and Representation (http://tinyurl.com/Optical-Instruments). Where not otherwise stated, all translations in this essay are my own.

SOME WORDS AHEAD

Take a look at *Fig. 1*. An Englishman is painting outside around 1830. He looks inspired and enthusiastic, as one would expect the case to be for an artist. However, this artist is surrounded by a number of instruments, which he apparently needs for his imaging process. There are geometric aids like a ruler, a set square, and a pair of compasses. And there are optical aids as well: the artist is wearing telescopic spectacles, a telescope lays on the ground, and a camera lucida is attached to his drawing board. At some point, the painter will surely switch places with his fellow draughtsman hidden under the tent of a field camera obscura, another well-known optical device used for drawing.[1] These artists combine an individual impetus with some need of quantification, it seems, in accordance with a characteristic trait of the time (for more on this, see Fiorentini 2006a).

Figure 1: Carl Jakob Lindström, "Den engelske konstnären" (The English Painter), 1830, Watercolour, 7,6 x 10,16 in. (19,3 x 25,8 cm)

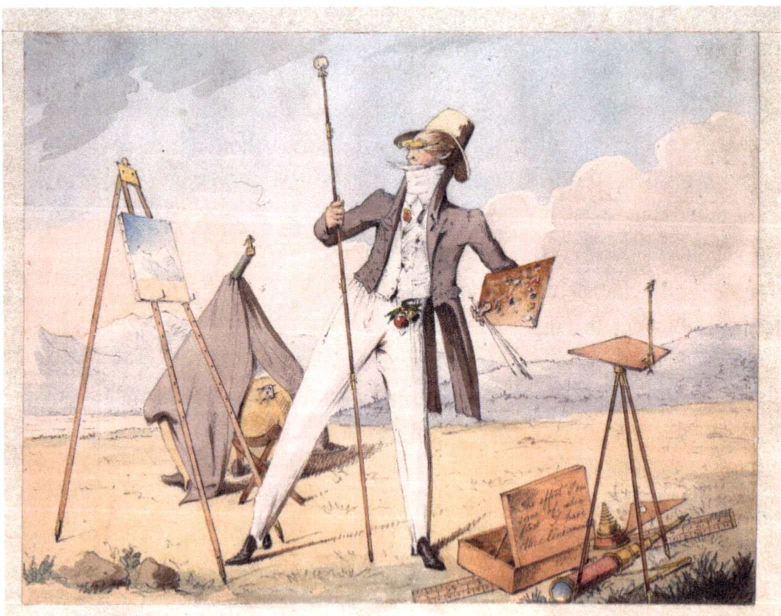

Source: Statens Konstmuseer Stockholm; Photo: Bodil Beckman / Nationalmuseum (CC BY-SA); License Terms: http://collection.nationalmuseum.se/eMP/eMuseumPlus?service=ExternalInterface&module=collection&objectId=64234&viewType=detailView

1 For more about the meaning of Carl Jakob Lindström's 1830 work *The English Painter* in Figure 1, see Fiorentini 2005.

This watercolor is a caricature, ironically pointing to the peculiar behavior of some of the international artists working in Naples in those years. The image ridicules the artists as much as their optical drawing devices, showing that such instrumental practices were already deeply rooted in the collective awareness and had become common—or at least well known—in drawing practice. In other words, what the caricature shows is that optical drawing devices had achieved widespread popularity by this time.

The phenomenon is interesting because here—as we will see—popularity did not result from ad hoc strategies of popularization, although optical aids for drawing had been a relevant topic in drawing technique manuals of the early nineteenth century. Addressing professional drawing practitioners as well as a broader range of drawing amateurs, such manuals in fact mostly did not intend to propagate instrumental practices for drawing. Instead, they depreciated optical drawing aids and advised against their use. In advising drawers against the use of optical devices, these manuals popularized the reasons why such instruments should *not* become popular drawing tools. They pursued a kind of antipopularization, or negative popularization, as I would like to call it in the following.

SOME TRAITS OF ANTI-POPULARIZATION

In order to understand this kind of inverted popularization and the arguments on which the critiques of optical drawing aids are based, we need to consider the nature of these devices. First, they are not "ontological instruments" (Ferraris 2011, 158-159, Tab. 7), for example, spectacles or telescopes, which serve as optical prostheses, as extensions of the visual apparatus thought to enhance seeing and to provide a clear vision of what is out there. Such optical prostheses are not able to produce images, they simply help our perception have a clearer image; in other words, they simply allow us to see better. Furthermore, optical drawing aids are not drawing machines, i.e., devices to which we can delegate seeing and imaging. They are not devices that draw, but devices *for* drawing: They cannot function without a seeing and actively drawing subject. In another words: A person who sees and draws is the *sine qua non* for their functioning.

But how do they function then? One can think of them as visual membranes placed between the looking and drawing person and the world. In analogy to the mechanism of physiological membranes, visual membranes can be seen as a zone allowing the exchange of visual stimuli between the physical world, the perceiving observer, and the picture. As such, these membranes or optical drawing devices moderate, as it were, the process of negotiation between the outer world, the eye, the ideas and notions already present in the mind of the observers, and the image generated by the drawing hand.[2] This is exactly what the most influential drawing manuals of the early 19th century complain about: They point to a lack of 'objective' accuracy and to the draftperson's necessity to fall back on subjective judgment when drawing.

In his manual *Éléments de perspective pratique à l'usage des artistes* (Valenciennes 1973 [1800]), for instance, Pierre Henry de Valenciennes accuses optical devices of inexactness, and even of defectiveness, in their presentation of natural appearances. The picture projected by the lenses of the camera obscura on its screen, Valenciennes contends, presents Nature in a "distinctly wrong" manner (ibid: 208):[3] Lines, he says, are deformed or curved, the margin of the scene is blurry, and its colors falsified (ibid: 296). However, Valenciennes admits, "to see Nature in the Camera Obscura is a very agreeable and very interesting way to study her" (ibid: 295).[4] He therefore prompts art students "to consult it [the Camera Obscura] very often" to train their eyes, since they will "get excellent lessons in the harmony of local color and in the different effects of the sky illuminating terrestrial objects." In this way the student will profit from the optical device and "develop a true style, drawn from Nature" (ibid: 296).[5] In spite of all these sensorial advantages, however, in view of the camera obscura's visual imprecision, Valenciennes "would not advise young artists to copy Nature

2 For this idea, cf. Fiorentini 2004; Fiorentini 2006b. Using 'membrane' in this sense, I distinguish myself from Elkins (1999, pp. 36-42), who understands 'membrane' in an anatomical sense, namely as an impermeable surface mostly connected with the 'skin' as a protective envelope. As a metaphor of permeability, 'membrane' is otherwise used only in the sociological context of identity shaping (Otis 1999).

3 "La chambre noire donne toutes les lignes fausses par une raison bien simple : c'est à cause de la convexité du verre à travers lequel passent les rayons qui réfléchissent la Nature. Plus les lignes droites se rapprochent du foyer et moins elles sont courbes ; mais aussi plus elles s'éloignent, et plus cette courbure est sensible. Ainsi, en dessinant dans la chambre noire et calquant la Nature, elle se trouve visiblement fausse, sur-tout quand elle réflechit de l'Architecture et toutes les lignes droites sans exception."

4 "Il est une manière très-agréable et très-interessante d'étudier la Nature: c'est de la voir dans la Chambre noire..."

5 "Mais nous les exhortons à la consulter très-souvent, [...] ils y prendront d'excellentes leçons d'harmonie de couleur locale et des différents effets du ciel éclairant les objets terrestres; ils s'y formeront une manière vraie, puisée dans la Nature."

in the Camera Obscura, neither when tracing the lines nor when copying the hues [of the camera projection] on their drawing paper" (ibid).[6] A notable piece of negative advice, given the impact of Valenciennes' *Éléments*—as a publication known throughout Europe—on artistic practice (for this, see Denk 2019).

Similarly, Francis Nicholson's *The Practice of Drawing* contends for the same reasons that by 1820 "a limited use of [...] optical contrivances may be allowed to the beginner" due to their merits in the education of the eye (Nicholson 1823 [1820]: 27). Nicholson refers in particular to the so-called Claude glass, the convex black-tinted mirror on which draughtsmen and painters had observed the landscape behind them since at least the 18th century.[7] As Nicholson points out, visual judgement can be trained by comparing the dark tones on the surface of the mirror with the actual appearance of the scene, so this process supports the painter "in discovering what is proper for the pencil" (Nicholson 1823: 21). However, "the student who is desirous to excel will not avail himself of those helps for the indolent" (ibid: 27). The images of the black convex mirror, in fact, deviate so much from natural appearances that they cannot offer cues for an exact representation of nature.

Jacques Paillot de Montabert (*Traité complet de la Peinture*, 1829)[8] asserts on his part that tinted mirrors and camera obscurae do not offer an accurate representation, but rather an "optical translation" of the observed objects.[9] The painter, therefore, is called upon to "carefully appreciate and select the objects,"[10] a process implying "lots of perspicacity."[11] Emphasizing with this the role of the draftsperson in their operating mode, of course, Montabert relativizes the utility of the instruments as devices of objectivity.

Many textbooks for professionals, such as Theodor Thons' *Lehrbuch der Reißkunst* (1832), also deem the use of the camera obscura or camera lucida superfluous for a correct drawing practice. "Such helping instruments will always turn out to be of use only to those who have already gained a sound command of the principles," he writes, but "they will not at all help those who are not ac-

6 "Ainsi, nous ne conseillerons pas aux jeunes artistes de copier la Nature dans la Chambre noire, soit en calquant les lignes sur du papier, soit en copiant les tons des couleurs."

7 For the varied history and significance of the black mirror, cf. Maillet 2004.

8 Montabert 1829: Vol. 9, Ch. 615 ("Des Miroirs, de la chambre obscure etc.").

9 [Cet affaiblissement et cette concentration de la chambre obscure] "sont très propres à faire bien apprécier et bien choisir les objets aperçus ainsi par une traduction optique" (Montabert 1829, Vol. 9, Ch. 615: 635).

10 "[...] bien apprécier et bien choisir les objets" (ibid).

11 "[...] beaucoup de perspicacité" (ibid: 634).

quainted with these principles."[12] August Hertel's *Kleine Academie der zeichnenden Künste und der Malerei* (1844) will then stress that in art "there have been many thoughts about means to support the eye 'in its function,' as, after all, "visual judgment alone is not a sufficient help in drawing from Nature."[13] Optical devices for drawing, however, are still not an option, since they make drawing a mechanical process. At most, the instruments can be allowed for beginners, who "are content with tiny pictures, without knowing about concepts of drawing."[14]

All these manuals, in sum, are not only skeptical about the optical drawing devices' alleged capacity to support drawing as a reliable process of precise reproduction, but also mistrust their suitability for a correct command of pencil and brush. Moreover, they explicitly state that the contrivances—being unable to afford visual reliability by themselves—always need the application of the draftsperson's practical and theoretical knowledge in drawing. An anonymous ironic voice from an 1830 issue of *Athenaeum* puts this in a nutshell:

> "We will conclude by a hint to amateurs, who are sometimes apt to rely too much on the efficacy of mechanical contrivances. It has been said by an able writer on the subject, that such means are like the railing of a road: they may keep the active traveler in the right course, but they cannot make the lame walk. Let not any one imagine that he can learn to draw, merely by purchasing a Camera Lucida; he might as soon learn music, by merely buying a fiddle." (Anonymous 1830: 540-541)

Accordingly, M. H. Shuttleworth's *Remarks on Landscape Painting in Water Colours*, in the first edition of 1831, ignores optical drawing devices altogether. But in the second edition of 1845 it admits the use of "Inventions for sketching objects in correct perspective" for drawers less conversant with the principles of linear perspective "who wish to acquire a thorough and practical knowledge" (Shuttleworth 1845: 41). The devices are thus defined as equipment for begin-

---

12 "[...] so wird man immer finden, daß dieses Mittel nur denjenigen wirklich nützt, welche bereits die Grundsätze der Wissenschaft sich eingeprägt haben, daß es dagegen denjenigen, welche mit diesen nicht bekannt sind, auch so gut als gar nichts hilft." (Thon 1832: 17)

13 "Man hat immer auf Mittel gedacht, dem Auge zur Hülfe zu kommen. Was man gewöhnlich 'ein gutes Augenmaß nennt,' nämlich die möglichst genaue Schätzung wirklicher Naturverhältnisse, ist nicht das, was beim Zeichnen nach der Natur aushilft" (Hertel 1844: § 238, § 237).

14 "Dergleichen Instrumente mögen den Anfängern überlassen bleiben, die sich ein Bildchen schaffen wollen, ohne Begriff vom Zeichnen zu haben." (Ibid: § 241)

ners and amateurs. Among them, Shuttleworth mentions a "Sketchers Guide" and a "New Perspective Delineator" (ibid: 42), which seem to be more portable forms of an older instrument called a *glace à calquer* (a "tracing glass"), used for professional drawing.

SOME TRAITS OF POPULARITY

Portable devices with a glass pane, indeed, had never been contrivances purely aimed at helping unexperienced amateurs in drawing, but belonged all along to the central elements of professional draftpersons' and painters' drawing practices. Take for instance painter Christoph Nathe's 1803 optical contrivance, which he invented for purposes of sketching outside, and which he called a "glass panel" ("Glastafel") or "optical chalkboard" ("optische Tafel")[15] (*Fig. 2*).

Figure 2: Christoph Nathe's "Optische Tafel." Sketch in a letter to Adolf von Gersdorf, Feb. 14, 1803

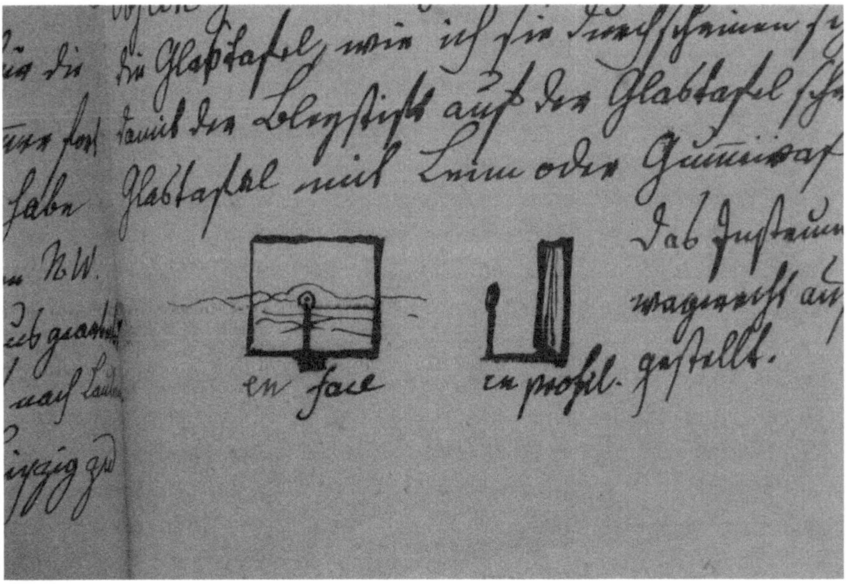

Source: Oberlausitzische Bibliothek der Wissenschaften Görlitz, Sig. ATvG 629; Photo: Courtesy Friedrich Weltzien

15 From a letter from Christoph Nathe to Adolf von Gersdorf, Feb. 14, 1803, Oberlausitzische Bibliothek der Wissenschaften Görlitz, Sig. ATvG 629, 179, unpublished.

As for the procedure, we already know it from John Constable, who also used a glass pane to sketch on the spot.[16] Arthur Parsey describes Constable's method of sketching outdoors with a glass pane in his *The Science of Vision; or, Natural Perspective of 1840*:

"The late Mr. John Constable, R.A., told me, that when he was studying the art of painting in his native place [...] he had attached to the upper part of his easel a frame with a pane of glass in it; this frame was attached by two screw nuts by the two upper corners; to the four corners he attached four strings, which he brought to his mouth in such a manner as to bring the centre of the glass perpendicular to his eye. On this glass [...] he traced with colour the outline of the view he decided on representing." (Parsey 1840: 134)

According to Jo Crook, Constable used oil paint and

"while the paint was still wet, a sheet of paper was pressed onto the glass and the image transferred to the paper. As the transferred image was a mirror image of the scene viewed, the oil paint drawing was traced in pencil on the verso of the paper to correct it. This drawing was then squared for transfer to the canvas." (Crook 1995)

This is a complex procedure, and clearly testifies to the fragility of the idea—perpetuated in the manuals—that working with optical drawing devices must be inescapably mechanical and objective.

In his manual, Shuttleworth also implicitly discourages people from using more recent instruments like the camera lucida, which he deems a "more portable but a very difficult instrument" (Shuttleworth 1845: 41). In spite of this, the camera lucida had also long been part of the equipment of professional painters in many countries. Among its most prominent users we find Caspar David Friedrich (cf. Busch 2003: 55), Jean-Auguste-Dominique Ingres (cf. Fleckner 2001: 160-191), a significant number of Italian and Flamish painters operating in Rome and Naples (cf. Fiorentini 2005: 535-557; 2006c: 195-214), and many English painters (cf. Hammond/Austin 1987: 89-112). One of them was Cornelius Varley (cf. Klonk 1996), who would play a major role in the acceptance of optical drawing devices and whom we shall meet later on.

16 One of these, a preparatory study for his *Flatford Mill* painting (*Scene on a Navigable River*, 1816-17, Tate Gallery N01273) from around 1816, is in the Tate Gallery in London (T05493, https://www.tate.org.uk/art/artworks/constable-study-for-flatford-mill-t05493).

Figures 3a and 3b: Frontispieces of the French original and of the German edition of Charles Chevalier, Conseils aux artistes et aux amateurs, sur l'application de la chambre claire à l'art du dessin *1838 and 1839*

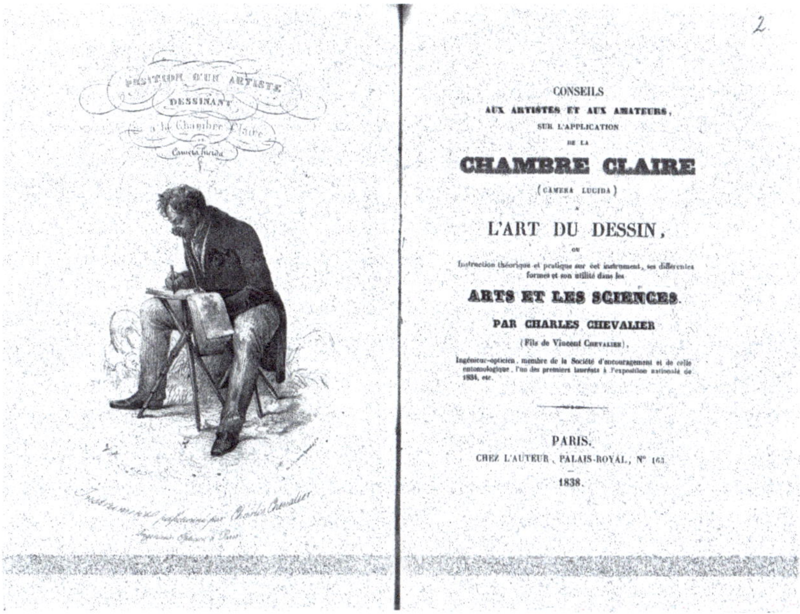

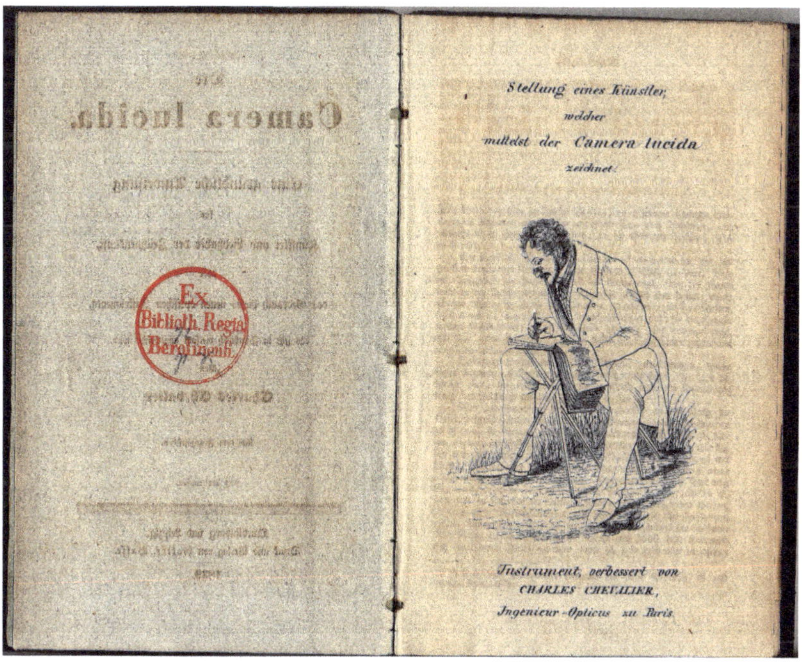

Photo: Erna Fiorentini

Optical drawing devices were always part of professional artistic practices. They were a matter of course among professional painters and not merely an artful stratagem for ambitious drawing neophytes, as the critical painting manuals propagated. Accordingly, despite the negative advice of the different manuals for drawing, optical aids for drawing had been bestsellers at all times, as the lively advertisements in the sales catalogues of instrument makers shows. As early as 1817, the opticians Joseph von Utzschneider and Joseph von Fraunhofer offered camera lucidas for 33 guldens in their glassmaking institute in the Bavarian village of Benediktbeuren (Verzeichnis 1817: 27), which was renowned across Europe for the excellent optical quality of its products; and as late as 1856 artists could still purchase the most up-to-date models of the camera obscura, the camera lucida, and the Claude glass (the latter "in neat embossed Morocco cases") from Pike's *Illustrated Descriptive Catalogue* at prices ranging from 2,50 dollars to 12 dollars (Pike 1856, Vol. 1: 373; Vol. 2: 186-188, 190-193). These are but two of the many examples from the market that testify to the wide distribution of and animated demand for optical drawing devices found throughout the 19th century in Europe and overseas. Such catalogues, which included advertisements for optical drawing devices and detailed instrument makers' descriptions of them, circulated very quickly. Charles Chevalier's instructions and user guide, which accompanied his new camera lucida model released in Paris in 1838 (Chevalier 1838), was printed for the German market as early as in 1839, and was "a thorough instruction for artists and lovers of the art of drawing for the use of this new optical device, which is now broadly utilized in France" (Chevalier 1839; *Fig. 3a* and *3b*).

As this success in the market demonstrates, artists—and amateurs—kept on purchasing and using optical drawing devices. The reaction of the manuals' target audience, that is, was apparently unaffected by the anti-popularization or negative popularization the drawing manuals pursued.

It is remarkable, although not unexpected, that from a certain point onwards practitioners of drawing began to speak out themselves on the utility of optical devices and to make their positions public, just as the theoretical drawing manuals did. They wrote their own books and pieces of advice dealing with the utilization of the instruments in actual drawing practice. In contrast to the way the manuals put forth a negative popularization, the practitioners' handbooks presented a positive estimation of optical drawing aids. *The English Drawing Master*, for instance, was published in 1831 "augmented by Additions and remarks of August Müller, practical drawing teacher" and was thought of as a "nützliches Handbüchlein," as a small and practical portable reference book, as a useful *vade mecum* which wanted "to help both beginners and experienced draftsmen by means of the illustration and description of the many different instruments and machines for drawing currently in use ...," as the book's title reads.[17] This manual emphasizes throughout the utility of optical instruments and encourages readers to use them. Notably, the article on the more modern among the devices described—the camera lucida—laments that such "highly useful instruments are still used too little."[18]

Cornelius Varley, the English landscape painter we met before as a user of optical drawing devices, was an inventor and maker of these instruments as well. He campaigned very early on for the appreciation of optical drawing devices. In 1837, Varley published a "Treatise on Optical Drawing Instruments" in the *Transactions of the Society of Arts*, a highly prestigious and authoritative journal. But he was not content with the inner circle he could reach with his scholarly publication, so he took this one step further and published his treatise as a book for a general audience in 1845 (Varley 1845: 189-224).[19] Declaredly, Varley's intention with this was for broad dissemination of his considerations on optical devices for drawing. Obviously, as a developer and producer of such instruments, Varley must have been interested in awakening the curiosity of potential customers. His treatise in fact considers instruments like the graphic telescope, which Varley had himself invented and which was produced and sold by the opticians Horne, Thorthonwaite, and Woods. Significantly, several advertisements from these opticians and a sales catalogue from their laboratory is attached to Varley's *Treatise*. This notwithstanding, Varley explains his decision to make the *Treatise* more accessible to the public with a cultural motivation, namely, to improve the public's artistic taste by expanding the use of these optical devices:

"The efforts that are being made to draw public attention to the means of improving their taste in the fine arts, and also in the collateral art of design, would alone render a knowledge of such optical instruments as can aid those arts very desirable" (Varley 1845: 3).

However, Varley argues, now that the daguerrotype has added to the instrumental armory for image production and this "splendid discovery [...] only exists by means of the Camera Obscura," the oldest optical device ever, "a correct

17 "[...] nützliches Handbüchlein, das angehenden und geübteren Zeichnern zur Hand gehen will, und zwar mit der Abbildung und Beschreibung der verschiedenen, jetzt gebräuchlichen Instrumente und Maschinen zum Zeichnen," Humphrys 1831. For the history of this manual, cf. Heilmann et al. 2015: 257-259.

18 "Dieses höchst nützlichen und noch zu wenig benutzten Instrumentes," Humphrys 1831: 82.

19 For more on Cornelius Varley in his time, cf. Klonk 1996: 113-126.

knowledge of those instruments becomes a public desideratum" (ibid). Explicitly arguing against the anti-popularization of the drawing manuals, which propagated the inaccuracy and uninspired automatism of the devices, Varley states that his "treatise is offered to the public [...] to make the subject as plain to the users of such instruments as to the maker of them" (ibid). This is not only for the purpose of "explaining the faults and defects incident to them" but also "for the purpose of showing how those faults may be either avoided, lessened, counteracted, or completely cured" (ibid).

With the publication of Varley's treatise, a practitioner of drawing for the very first time exerts himself explicitly for the public critical discussion and the explanation of the nature and use of optical drawing devices. Moreover, different solutions are offered to counteract the defects and difficulties the skeptical drawing manuals complained about in their denigration of these optical aids, so that the devices can be functionally, advantageously, and appropriately used in the practice of drawing. Varley's treatise is a 'theoretical' reaction to a long-lasting skepticism towards optical drawing devices. It gives voice to the expectations and demands of the practical field, where these devices had apparently always been taken for granted. At the same time, Varley's treatise intentionally pursues a strategy of positive popularization, foiling the negative popularization of many drawing manuals.

POPULARITY WITHOUT POPULARIZATION

In sum, the great popularity of optical devices in professional and amateur practices of drawing of the early 19th century doesn't match the skepticism expressed by most drawing manuals and academic treatises of the time. The presence of the devices in the manuals was not the factor that ignited their popularization, or rather, their *becoming popular*, their overwhelming and steady gain in popularity, which could be observed in the market at that time.

But why is it that optical drawing devices were so popular without consistent efforts of popularization behind them? Why, much to the contrary, did they achieve popularity despite the anti-popularization expressed by most of the drawing manuals? And which paths did this process of becoming popular follow?

The case of the so-called Wollaston prism tells us more about these dynamics of becoming popular without strategies of popularization in the early 19th century.[20] This device was patented in December 1806 as an "instrument where-

20 For more on this, cf. Fiorentini 2006a.

by any person may draw in perspective, or may copy or reduce any print or drawing,"[21] and appears in 1807 under the name of "camera lucida" (Wollaston 1807a: 343-347; 1807b: 1-5) (*Fig. 4*).

Figure 4: Illustration of the Camera Lucida in William Hyde Wollaston. "Description of the Camera Lucida," Philosophical Magazine XXVII (1807): 343-347

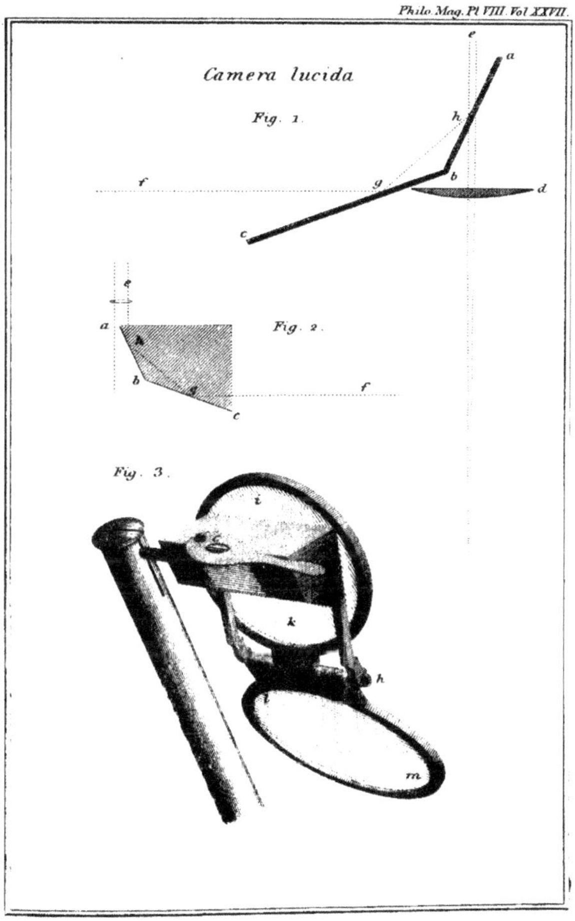

Photo: Erna Fiorentini

21 Patent No. 2993 was published as "Specification of the Patent granted to William Hyde Wollaston [...] for an Instrument whereby any Person may draw in Perspective, or may copy or reduce any Print or Drawing. Dated December 4, 1806," *The Repertory of Arts, Manufactures and Agriculture*, Second Series X (February 1807): LVII, 161-164.

The inventor of the camera lucida, the English polymath William Hyde Wollaston, was a highly appreciated member of the Royal Society in London and a very popular natural philosopher.[22] He was an expert in optics, metallography, metallurgy, and chemistry, among many other sciences. We are indebted to him for epochal findings like the discovery of the metals palladium and rhodium, which remain important to this day. Very early in 1807, Wollaston commissioned George Dollond, an optician in London, with the production, sale, and distribution of his tiny optical device for drawing. In November 1807, Wollaston could jot down in his daybook incoming royalties from camera lucida sales to the amount of £140, which corresponds to about 1000 instruments sold in nine months. Three years later, in December 1810, incoming monthly royalties had leveled off at £100 (cf. Usselman 2015: 149), revealing a brisk and stable demand.

We of course don't know the identity and activities of all those purchasers, but it is remarkable that one of the first registered buyers of Wollaston's camera lucida was the astronomer William Herschel, who paid £2 12s 6d (2 pounds, 12 shillings, and sixpence) for it in February 1807 (cf. Usselman 2015: 150). His son John, later one of the most prominent British scientists and a passionate user of the camera lucida (cf. Fiorentini 2007: 13-34; Schaaf 1990), was a close acquaintance with Wollaston and was one of his most loyal customers. John Herschel was a skilled camera lucida draughtsman. Later, he integrated the device into his complex method of inquiry (Fiorentini 2007).

Coming into existence out of scientific interests in a scholarly context, as one might expect, the camera lucida was immediately adopted as a drawing aid. So, at the beginning, the scholarly and scientific interest in the instrument was one of the major triggers for its spread. A great number of variants and improvements were in fact very quickly attempted by natural philosophers and physicists after Wollaston's first sales. In Vienna, Joseph Schöps published his "newly invented Graphoscope" in 1811; in 1812, August Friedrich Lüdicke, a professor of mathematics in Meissen, gave a "Decription of a modified Camera Lucida"; the mathematician William George Horner found "New and important combinations with the camera lucida" in 1815; and an anonymous scholar from Bavaria developed a "Machine for Drawing" in 1816; and in 1821 Alexander Waddell, an amateur meteorologist and astronomer, wrote up an "Account of a New Optical Instrument, which combines the properties of a Compound Microscope, Camera Obscura, Camera Lucida, and Diagonal Mirror" (cf. Schöps 1812: 338-342; Horner 1815: 281-283; Anonymous 1816: 723-730; Waddell 1821: 143-146).

22 William Wollaston's contemporaries called him "the pope;" cf. Williams 2004.

The variant of the camera lucida that was developed, produced, and marketed by Giovanni Battista Amici, the famous Italian physicist, astronomer, and optician, was the one that gained the largest success. Amici published a paper on his newly designed camera lucida in 1819 (Amici 1819: 25-35), and this was then translated into different languages and commented on by important natural philosophers like Joseph Louis Gay-Lussac, François Arago, and David Brewster (cf. Gay-Lussac and Arago 1823; Brewster 1825: 157).

However, the fame and spread of Amici's camera lucida were initially not increased by his publications alone: Whereas Wollaston had committed production and sales to the laboratory of the optician George Dollond, Giambattista Amici did all production and sales activities by himself. He produced scientific instruments in Modena and fostered their distribution with a clear strategy of his own sales promotion. This strategy included intensive and unique travel activities for the sake of demonstrating his devices. During his travels, Amici presented his optical scientific instruments as innovations and offered them for sale. This strategy of on-the-spot information and selling had a formidable impact, not only in the scholarly circle of scientists for which many of his instruments were intended. Many contemporaries travelled to Amici's laboratory in Modena and Florence to meet him anyway, but his on-site presentations abroad seem to have been of a particularly high quality, and he attracted a broad—and not exclusively scholarly—public interested in the advancements of instrumental and theoretical optics.

Early in 1817, Amici travelled to Florence, Rome, and Naples to promote his devices to the international audience that frequented those towns after the Napoleonic Wars (cf. Grossi 1996: 873-919, in particular 886-897). Politicians, scientists, cartographers, representatives of diverse institutions (scientific and artistic alike), and private citizens literally swarmed to Amici's presentations (his wife reports at least 70 people who came before noon on a single day), and he was even invited to explain his instruments to Pope Pius VII while in Rome (cf. Grossi 1996: 873-919, in particular 886-897). Here, in particular, the camera lucida had become the talk of the town and earned huge success. Amici complains in many letters that he's not able to satisfy the countless incoming orders quickly enough.[23]

The device, of course, immediately attracted the attention of drawing practitioners. "Having heard about it," many came to test and buy the instrument, and

23 "[N]on potete immaginarvi le pressanti ricerche che mi son state fatte di quelle camere lucide per cui mi sono trovato imbarazzato a contentare tutti" (letter from October 19, 1818). See Meschiari 2002: 197-199, letter no. 14.

then passed on the news to their own circles (cf. Grossi 1996: 897). Many orders attest to Amici's connections with the international artistic communities that were active in Rome at that time. The camera lucida had seemingly become a matter of course for painters there. Indeed, shortly after Amici's first visit, the painter Massimo d'Azeglio's father Cesare ordered a camera lucida from Amici, giving the following reason: "having a landscape painter as a son, your Camera Lucida is the subject of my particular attention."[24] Not by chance, d'Azeglio's paintings had been exhibited at the Academy in Turin by the Academy director Giovanni Battista Biscarra, who was a friend of the d'Azeglio family. Biscarra, in turn, had been portrayed by Amici himself with the camera lucida when visiting[25] and had bought one device in 1821, together with the artist Angelo Boucheron (cf. Meschiari 2003: 67). One of Amici's acquaintances from Turin, Nicola Crosa di Vergagni, wrote the following to him: "Under the magic and furious pencil of this draftsman, the device is astonishing the whole town."[26]

The optical drawing device that had come out of the scholarly context, hence, eventually went "into very general use for drawing landscapes" (Brewster 1831: 333) and was "used by all ye artists I find!,"[27] as the painter John Sell Cotman reports in 1817; Cotman used the device himself,[28] as did the sculptors Sir Francis Chantrey (Schaaf 1990: 27-28) and Thomas Campbell (Smailes 1987: 1016, 709-714).

In addition to professional artists, a whole generation of English amateur drawers was fascinated by the camera lucida, representative examples being Henrietta Ann Fortescue (cf. Macinnes 1996: 38-41), a pupil of Francis Nicholson (1823: 28), or the family of Henry Fox Talbot, who would later become one of the pioneers of photography (cf. Kemp 1997: 270-282). The camera lucida also pervaded the international artistic communities and was adopted by international landscape painters, from Wilhelm Huber to Giacinto Gigante (Fiorentini 2005), from Julius Schnorr von Carolsfeld and Franz Horny to Wilhelm Götzloff (ibid).

---

24 "[C]on un figlio paesista, la di lei Camera Lucida è oggetto per me di speciale attenzione," Letter from April 1822 (Biblioteca Universitaria Estense di Modena, Fondo Giovanni Battista Amici, cartella 304, carte 1608-1614). Cesare d'Azeglio would pay for the camera lucida in January 1823; see Meschiari 2003: 69.

25 The portrait is in Modena at the Biblioteca Estense, Fondo Amici-Grossi, Cassetta 2, fasc. 19, disegno no. 27.

26 Crosa di Vergagni 1822. For more on the impact of Amici's camera lucida and its dissemination in international artistic communities, cf. Fiorentini 2006: 195-214.

27 John Sell Cotman, Letter to Dawson Turner, June 12, 1817, quoted from Kay 1925-1926: 94.

28 For instance, for the sketches during his *Picturesque Tour of Normandy* between 1817 and 1820, cf. Pidgley 1972: 4, 781-786.

However, Giovanni Battista Amici's sales records reveal an extremely varied network of camera lucida buyers, exceeding the scientific and artistic fields.[29] The 269 camera lucidas delivered between 1815 and 1832 across Europe and overseas (compared to only 2 camera obscurae sold in that period; Meschiari 2003: 61-62) covered demand in very different fields of application. Such a multifaceted demand is a peculiar trait of the story of the camera lucida and how it become popular. Archaeologists needed the device to draw the views of the monuments they studied, and they put particular emphasis on the use of the instrument. For instance, in the 1819 edition of his *Pompeiana* William Gell deemed it "proper to state, that the original drawings for this work were made with the camera lucida" (Gell/Gandy 1817-1819: xvi; Gell/Gandy 1835: 109). An Orientalist like Edward William Lane decidedly underscored, in the heading of his collection of views named *Description of Egypt*, that "delineations of the monuments, scenery &c. of those countries [...] [were], with few exceptions, made with the camera-lucida."[30] Military institutions of surveying, for instance the Officio Topografico in Naples, were particularly interested in the camera lucida for documentation purposes (Meschiari 2002: 197-199), and the device became a mandatory accessory for the records of voyages in remote countries: Naturalists like William Bullock (1824, 54) and explorers like Captain Basil Hall explicitly propagated the use of the camera lucida. Along these lines, after his travels in North America, Basil Hall recalls the following:

"this valuable instrument ought, perhaps, to be more generally used by travellers than it now is; for it enables a person of ordinary diligence to make correct outlines of many foreign scenes, to which he might not have leisure, or adequate skill, to do justice in the common way." (Hall 1829: ii)

Of course, as may be expected, natural philosophers, physicists, engineers, and anatomists continued to enthusiastically adopt the camera lucida: The Scottish astronomer Charles Piazzi Smyth (Warner 1983), the physicist Marc-Auguste Pictet, the microscopist Friedrich Hoffmann, and the head of the Museum of Physics and Natural History in Florence, Girolamo Bardi, are but a fraction of those who purchased a camera lucida for their scientific studies (cf. Meschiari 2001: 55-114). Furthermore, the device became an indispensable ingredient of microscopic drawing after Amici patented a new catoptric microscope with a camera lucida ocular (Amici 1820: 135-138).

29 Meschiari 2001: 55-114, here 59. For an expanded edition, cf. Meschiari 2003.

30 Cf. Lane's edition by Thompson 2000.

POPULARITY FROM SILENT POPULARIZATION

Hence, although invented in a scientific setting, the camera lucida inundated many other fields of interest and application. Its dissemination did not depend on an intentional strategy of educational popularization. In the practice of artistic drawing of the early 19th century, the popularity of optical devices like the camera lucida seems rather to have resulted from the absorption of the optical instrument as a scientific tool that had permeated art and was independently sustained within the artistic discourse. However, whereas in the sciences optical drawing aids were declared necessary in order to guarantee scientific accuracy, the artistic realm was reluctant to admit their use. In 1830, John Herschel gave a lucid diagnosis of this fact, observing that

"the whole tendency of empirical art, is [...] to place its pride in particular short cuts and mysteries known only to adepts, to surprise and astonish by results, but to conceal processes." (Herschel 1830: 70-72, No. 64)

But the sales volumes of optical drawing devices in the early 19th century allow us to see the picture the artists wanted to hide, namely: the picture of the overwhelming popularity of optical drawing devices in artistic practice. It seems that the course of becoming popular depended here on a kind of silent popularization of the devices among the practitioners of drawing. The use of optical drawing devices propagated itself without programmatic advertising, but rather through the notion that such devices were directed at diffusing a specific technical knowledge as an expertise favoring the improvement of drawing practices.

These devices followed different paths to achieve popularity through such a silent process of becoming popular. These paths were neither a planned, nor a loud, nor a linear strategy of popularization. In the case of the optical drawing devices of the early 19th century, they led to clear popularity in the practice of drawing, despite the active efforts of anti-popularization in the theory.

REFERENCES

Amici, Giovanni Battista (1819): "Sopra le camere lucide." In: Opuscoli Scientifici 3/13: pp. 25-35.

Amici, Giovanni Battista (1820): "Account of an Improved Catoptrical Microscope." In: Edinburgh Philosophical Journal 2, pp. 135-138.

Anonymous (1816): "Maschine zum Zeichnen." In: Anzeiger für Kunst und Gewerbefleiß im Königreiche Baiern 47, November 23, pp. 723-730.
Anonymous (1830): "The Camera Lucida." In: Athenaeum 148, pp. 540-541.
Brewster, David (1825): "Prof. Amici's improved camera lucida." In: Edinburgh Journal of Science 3 (April-October), pp. 157-159.
Brewster, David (1831): A Treatise on Optics, London: Longman/Rees/Orme/ Brown and Green.
Bullock, William (1824): Six Months' Residence and Travels in Mexico, London: John Murray.
Busch, Werner (2003): Caspar David Friedrich. Ästhetik und Religion, Munich: C.H. Beck.
Chevalier, Charles (1838): Conseils aux artistes et aux amateurs, sur l'application de la chambre claire à l'art du dessin, ou Instruction théorique et pratique sur cet instrument, ses différents formes et son utilité dans les arts et les sciences, Paris: chez l'Auteur, Palais-Royal No. 143.
Chevalier, Charles (1839): Die Camera lucida. Eine gründliche Anweisung für Künstler und Liebhaber der Zeichenkunst über den Gebrauch dieses neuen optischen Instruments, das jetzt in Frankreich vielfach angewendet wird, Quedlinburg and Leipzig: Gottfr. Basse.
Crook, Jo (1995): "John Constable – Study for 'Flatford Mill,' Tate Gallery T05493, Technique and Condition." Accessed February 22, 2020. https://www.tate.org.uk/art/artworks/constable-study-for-flatford-mill-t05493.
Crosa di Vergagni, Nicola (1822): Letter from 1822 (Biblioteca Universitaria Estense di Modena, Fondo Giovanni Battista Amici, cartella 304, carte 1608-1614). Biblioteca Universitaria Estense di Moden, Fondo Giovanni Battista Amici, cartella 291.
Denk, Claudia (ed.) (2019): Valenciennes' Ratgeber für den reisenden Landschaftsmaler: zirkulierendes Künstlerwissen um 1800, Munich and Berlin: Deutscher Kunstverlag.
Elkins James (1999): Pictures of the body: pain and metamorphosis, Stanford: Stanford University Press.
Ferraris Maurizio (2011): Il mondo estern, Milano: Bompiani, pp. 158-159.
Fiorentini, Erna (2004): "Subjective Objective. The Camera Lucida and Protomodern Observers." In: Bildwelten des Wissens. Kunsthistorisches Jahrbuch für Bildkritik 2/2, pp. 58-66.
Fiorentini, Erna (2005): "Nuovi punti di vista. Giacinto Gigante e la Camera Lucida a Napoli." In: Martina Hansmann/Max Seidel (eds.), Pittura italiana nell'Ottocento, Venice: Marsilio, pp. 535-557.

Fiorentini, Erna (ed.) (2005-2008), Drawing with Optical Instruments – Devices and Concepts of Visuality and Representation (http://tinyurl.com/Optical-Instruments), accessed February 18, 2020.

Fiorentini, Erna (2006a): Camera Obscura vs. Camera Lucida. Distinguishing Early Nineteenth Century Modes of Seeing, Berlin: Max Planck Institute for the History of Science (Preprint 307, https://www.mpiwg-berlin.mpg.de/sites/default/files/Preprints/P307.pdf), accessed February 18, 2020.

Fiorentini, Erna (2006b): "Instrument des Urteils. Zeichnen mit der Camera Lucida als Komposit." In: Inge Hinterwaldner/Markus Buschhaus (eds.), The Picture's Image. Wissenschaftliche Visualisierung als Komposit, Munich: Wilhelm Fink, pp. 44-58.

Fiorentini, Erna (2006c): "Scambio di vedute. Lo sguardo sulla natura e la Camera Lucida tra i paesaggisti internazionali a Roma intorno al 1820." In: Lorenz Enderlein/Nino Zchomelidse (eds.), Fictions of Isolation: Artistic and Intellectual Exchange in Rome During the First Half of the 19th Century, Rome: L'Erma Di Bretschneider, pp. 195-214.

Fiorentini, Erna (2007): "Practices of Refined Observation. The Conciliation of Experience and Judgement in John Herschel's Discourse and in his Drawings." In: Erna Fiorentini (ed.), The Osmotic Dynamics of Romanticism. Observing Nature – Representing Experience 1800-1850, Berlin: Reimer, pp. 13-34.

Fleckner, Uwe (2001): "Porträt und Vedute. Strategien der Wirklichkeitsaneignung in den römischen Zeichnungen Jean-Auguste-Dominique Ingres." In: Margaret Stuffmann/Werner Busch (eds.), Zeichnen in Rom 1790-1830, Cologne: Walter König, pp. 160-191.

Gay-Lussac, Joseph Louis/François Arago (1923): "Sur la chambre claire (camera lucida). Par le Professeur J.B. Amici, de Modène. Traduit de l'italien." In: Annales de Chimie et Physique 22, pp. 137-155.

Gell, William/Gandy, John Peter (1817-1819): Pompeiana: The Topography, Edifices, and Ornaments of Pompeii, London: Rodwell and Martin.

Gell, William/Gandy, John Peter (1835): Pompeiana: The Topography, Edifices, and Ornaments of Pompeii: the Result of Excavations since 1819, Volumes I and II, London: Lewis A. Lewis.

Grossi, Giovanna Amici (1996): "I Diari dei viaggi e altri documenti della vita e dell'attività di Giovanni Battista Amici." In: Atti della Fondazione Giorgio Ronchi 51/6, pp. 873-919.

Hall, Basil (1829): Forty etchings — from sketches made with the camera lucida in North America in 1827 and 1828, Edinburgh: Cadell & Co.

Hammond, John H./Austin, Jill (1987): The Camera Lucida in Art and Science, Bristol: Adam Hilger.

Heilmann, Maria/Nanobashvili, Nino/Pfisterer, Ulrich/Teutenberg, Tobias (eds.) (2015): Lernt zeichnen! Techniken zwischen Kunst und Wissenschaft, 1525-1925, Passau: Klinger. February 18, 2020. DOI: https://doi.org/10.11588/artdok.00003621, Kat. 37, pp. 257-259.

Herschel, John Friedrich William (1830): Preliminary Discourse on the Study of Natural Philosophy, London: Longman Press, pp. 70-72.

Hertel, August W. (1844): Kleine Academie der zeichnenden Künste und der Malerei, Weimar: B. Fr. Voigt.

Horner, William George (1815): "New and important combinations with the camera lucida." In: Annals of Philosophy 6, October, pp. 281-283.

Humphrys, Charles (1831): Der englische Zeichenmeister, oder die neuesten Methoden, Erfindungen und Verbesserungen im Zeichnen, Tuschieren, Colorieren, Malen und Farbenbereiten, Quedlinburg und Leipzig: Basse.

Kay, H. Isherwood (1925-1926): "John Sell Cotman's Letters from Normandy." In: Walpole Society 14, p. 94.

Kemp, Martin (1997): "Talbot and the Picturesque View. Henry, Caroline and Constance," History of Photography 21/4, pp. 270-282.

Klonk, Charlotte (1996): Science and the Perception of Nature. British Landscape Art in the Late Eighteenth and Early Nineteenth Centuries, New Haven: Yale University Press.

Macinnes, Katherine (1996): "Unconventional Tourist." In: Antique Dealer and Collector's Guide 50/1, pp. 38-41.

Maillet, Arnaud (2004): The Claude Glass. Use and Meaning of the Black Mirror in Western Art, New York: Zone Books.

Meschiari, Alberto (2001): "Il Libro de' conti del laboratorio di Giovanni Battista Amici." In: Atti della Fondazione Giorgio Ronchi 56/1, pp. 55-114.

Meschiari, Alberto (2002): "Giovanni Battista Amici e il Reale Officio Topografico di Napoli: Corrispondenza con i Colonnelli Visconti, De Sauget, Melorio." In: Physis 1, pp. 161-247.

Meschiari, Alberto (ed.) (2003): Il Libro de' conti del laboratorio di Giovanni Battista Amici e altri documenti inediti, Florence: Tassinari.

Montabert, Jacques Nicola Paillot de (1829): Traité complet de la Peinture, Paris: Bossange.

Nathe, Christoph (1803): Letter from Christoph Nathe to Adolf von Gersdorf, 14. Feb. 1803, (179). Oberlausitzische Bibliothek der Wissenschaften Görlitz, Sig. ATvG 629, 179, unpublished.

Nicholson, Francis (1823 [1820]): The practice of Drawing and painting landscape from nature in water colours; exemplified in a series of instructions calculated to facilitate the progress of the learner, London: Murray.

Otis, Laura (1999): Membranes. Metaphors of Invasion in Nineteenth-Century Literature, Science, and Politics, Baltimore: The Johns Hopkins University Press.

Parsey, Arthur (1840): The Science of Vision; or, Natural Perspective, London: Longman & Co.

Pidgley, Michael (1972): "Cornelius Varley, Cotman, and the Graphic Telescope." In: Burlington Magazine 114/4, pp. 781-786.

Pike, Benjamin (1856): Pike's Illustrated Descriptive Catalogue of Optical, Mathematical and Philosophical Instruments, manufactured, imported, and sold by the author, Vol. 1, 2, and 3, New York.

Schaaf, Larry J. (1990): Tracings of Light, San Francisco: The Friends of Photography.

Schöps, Joseph Gotthelf (1811): Beschreibung und Erklärung eines neu erfundenen Graphoskop. das ist: Zeichnungsmaschine, mittels welcher man alle Gegenstände nach der Natur abzeichnen [...] kann, Vienna: Verlegt auf Kosten des Erfinders.

Schöps, Joseph Gotthelf (1812): "Beschreibung einer veränderten Camera Lucida. Von Professor Lüdicke in Meissen," Annalen der Physik 42/3, pp. 338-342.

Shuttleworth, M. H. (1845 [1831]): Remarks on Landscape Painting in Water Colours, London: Houghton & Co, 30, Poultry.

Smailes, Helen (1987): "Thomas Campbell and the Camera Lucida," The Burlington Magazine 129/1016: pp. 709-714.

Thompson, Jason (ed.) (2000): Edward William Lane, Description of Egypt: notes and views in Egypt and Nubia, made during the years 1825, 1826, 1827, and 1828: chiefly consisting of descriptions and delineations of the monuments, scenery &c. of those countries; the views, with few exceptions, made with the camera-lucida, Cairo: American Univ. in Cairo Press.

Thon, Theodor (1832): Lehrbuch der Reißkunst oder der wahren Grundlage der Zeichenwissenschaft, Ilmenau.

Usselman, Melvyn C. (2015): Pure Intelligence: The Life of William Hyde Wollaston, Chicago: University of Chicago Press.

Valenciennes, Pierre Henry de (1973): Éléments de perspective pratique à l'usage des artistes. Suivis de Réflexions et conseils à un elève sur la peinture, et particulierement sur le genre du paysage, Geneva: Minkoff Reprint (Reprint of the original edition, Paris, [1800]).

Varley, Cornelius (1845): A Treatise on Optical Drawing Instruments, London: Horne, Thornthwaite, & Wood [Transaction of the Society of Arts 51 1837, pp.189-224].

Verzeichnis (1817): "Verzeichnis der optischen Werkzeuge welche in dem Optischen Institute zu Benedictbeuren Utzschneider u. Fraunhofer für nachstehende Preise verfertiget und ... besorgt werden." In: Anzeiger für Kunst- und Gewerbfleiß im Königreiche Bayern, Dritter Jahrgang, Monatliche Anzeige 2, p. 27.

Waddell, Alexander (1821): "Account of a New Optical Instrument, which combines the properties of a Compound Microscope, Camera Obscura, Camera Lucida, and Diagonal Mirror." In: The Edinburgh Philosophical Journal 5, April to October, pp. 143-146.

Warner, Brian (1983): Charles Piazzi Smyth: Astronomer-Artist: His Cape Years, 1835-1845, Cape Town: Published for the University of Cape Town by A.A. Balkema.

Williams, Trevor I. (2004 [print], 2010 [online]): "Wollaston; William Hyde (1766-1828)." In: Oxford Dictionary of National Biography (http://www.oxforddnb.com/view/article/29842), accessed February 18, 2020.

Wollaston, William Hyde (1807a): "Description of the Camera Lucida." In: Philosophical Magazine 27, pp. 343-347, and in: Journal of Natural Philosophy, Chemistry and the Arts 17: pp. 1-5.

Wollaston, William Hyde (1807b): "Description de l'appareil appellé Camera Lucida." In: Bibliothèque Britannique 35, pp. 255-264.

Wilhelm Lübke. Art History for Feuilletons

Alexandra Axtmann

ABSTRACT

In addition to important manuals on architecture, sculpture, painting, and the history of art, Wilhelm Lübke (1826-1893) wrote innumerable articles for daily newspapers and popular journals. He was sometimes accused of inaccuracy and a lack of depth by his colleagues. However, this kind of criticism failed to take into account the different types of texts he was writing and the contexts of their reception—from early 'coffee table books' on Raphael and Dürer to articles for feuilletons. Lübke's popular texts on art history were not aimed at reaching art experts but rather a broader target audience. In the early days of art history as an academic discipline, scholars such as Lübke were concerned with arousing the bourgeoisie's interest in works of art and art history while training their own knowledge and aesthetic judgement. Lübke's contribution to the art-historical heritage, the "popularization of art-historical knowledge," as formulated by Lübke's friend Theodor Fontane, remains undisputed to this day. Yet despite the popularity of some of his books, such as the bestseller *Outlines of the History of Art* (1860), there are no studies on his rhetorical approach to art communication and popularization. This essay analyzes Lübke's "animated speech and style of writing" in some of his feuilleton texts with a special focus on the rhetorical means he employed and the differences of his various target audiences (newspaper vs. journal audiences).

INTRODUCTION

After his first lectures at the Academy of Architecture (the Bauakademie) in Berlin, the German architecture and art historian Wilhelm Lübke (1826-1893) worked, from 1857 until his death, as a professor of art history at three polytechnics: in Zurich, in Stuttgart, and finally in Karlsruhe, where he was also director of the local collection of paintings (today the Staatliche Kunsthalle Karlsruhe; Betthausen 2007; Axtmann/Gawlik 2019: 7-44). In addition to numerous monographs, important handbooks on architecture, sculpture, painting, and the history of art in several European countries, some of which are still considered fundamental today, he wrote countless reviews and essays for newspapers and various art history journals, popular science magazines, and other bourgeois journals. During his lifetime, he compiled many of these texts and republished them in four anthologies (Lübke 1869, 1885, 1886, 1891a). Furthermore, there are many more publications than those listed in the aforementioned bibliographical data on Wilhelm Lübke (Kukula 1892; Rohling 1957; Betthausen 2007). A thoroughly researched bibliography proves (Axtmann/Gawlik 2019: 105-143) that it was with good reason that Lübke's contemporaries called him a "frivolously fast writer."[1] For example, about 100 contributions over a span of eight years can be documented for the art journal *Deutsches Kunstblatt* (published from 1850 to 1858), as well as more than 270 for the daily press, of which about 255 were for the *Allgemeine Zeitung* from 1861 until 1893 (Axtmann/Gawlik 2019: 105-143).[2] Until his first teaching position at the Berlin Bauakademie in 1857, it was above all this literary activity that ensured Lübke's livelihood (Oechelhäuser 1906: 531; Axtmann/Gawlik: 8-9). His shorter essays, reviews concerning new engravings and printed works, photographs, and exhibitions, as well as literary works, poetry, musical compositions, and opera performances, all bear witness to Lübke's interest in and knowledge of the contemporary artistic and cultural life and their popular communication. His cultural-historical approach is closely linked to his friend Jacob Burckhardt (1818-1897), whom he succeeded as Professor of Art History in 1861 at the Federal Polytechnical School of Zurich (the Eidgenössisches Polytechnikum).

Like many of his contemporaries in those early days when art history was still being established as an academic discipline, Lübke's concern was to impart knowledge about art history to a broader public. He is thus connected to the popularization tendencies of all disciplines in the 19th century, and above all the natural sciences, which relates to a broader participation in education (Daum

1 "Leichtfertigen Schnellschreiber." (Riegel 1874: 33)

2 There are certainly many more contributions that have yet to be found. For example, it is suspected that he wrote many articles for the Austrian daily press, since Lübke also published in Austria. Furthermore, some contributions are difficult to document since the author's name is absent or merely abbreviated, as was the case with many periodicals of that time.

1995: 2, 25). To say it in the category of Andreas Daum, Lübke was one of the "university popularizers" (Daum 1995: 422-423) who wanted to arouse the interest of a growing educated middle class to train themselves, e.g., in the artistic field and at the same time in their aesthetic judgment (Locher 1999).[3]

His survey books, such as the *Grundriss der Kunstgeschichte* (*Outline of the History of Art*, 1st ed. 1860), which went through eleven editions and was translated into four languages (English, French, Danish, and Swedish), were bestsellers during his lifetime and part of every contemporary private library. Moreover, the quite prompt translations are evidence of the widespread international scholarly and scientific exchanges that were taking place in the 19th century.[4] As one of the best-known art historians of his time, Lübke received many biographical acknowledgments thoughout his life in various yearbooks, magazines, and newspapers, e.g., in *Westermann's illustrierte Monatshefte* of 1862, where he received a life and literature report that even included a contemporary portrait drawing of him just one year after he had begun teaching at the Polytechnikum in Zurich (n.n. 1862). Alternatively, in the *Illustrirtes Unterhaltungs-Blatt* of 1874, he was introduced to the readership as Professor of Art History in Stuttgart, which was called one of the "most respected places" (n.n. 1874) in general art history. It should be mentioned here that the Stuttgart Chair of Art History at the Technical University was not established until 1865 and was one of the first such professorial chairs in Germany, and that such independent chairs were established at technical universities, institutes, and academies much earlier than they were at 'full' universities, primarily for the purpose of educating architects (Beyrodt 1991).[5] He took over this post in 1866 and remained there until his appointment to Karlsruhe in 1885. As with many articles and reports in the daily press, the short reports about his life and work emphasized his rhetoric above all else; thus, he is said to have lectured, in Zurich, about the entire history of the fine arts to immense applause, and his lectures were said to

3 Lübke also wrote several essays on this, as well as on the role of art history as a teaching subject and on its contents; (e.g., Lübke 1866a, 1871a).

4 For the exchange between Germany and Sweden and some translations of texts, see Elin Manker's essay in the present volume. Carl Lemcke was Lübke's successor as Professor of Art History in Stuttgart; he wrote an obituary for Lübke in 1893 (Lemcke 1893) and a lexical biography in 1906 (Lemcke 1906). The translations into Swedish of Lübke's and Lemcke's writings from these years occurred almost immediately after the original texts were published.

5 From 10 to 13 January 2019, a conference on this topic was held at the TU Wien entitled "Art History at Polytechnic Institutes, Technical Academies and Technical Universities. History – Positions – Perspectives." A conference volume is in preparation and will be published by Böhlau-Verlag.

have been enjoyed "by the students of the university and the educated from all stands of a great frequency." (n.n. 1874) Lübke's teachings and his writing style, his emphatic manner of speaking and his enthusiasm, are in every respect connected with his previous training as a school teacher and his experiences as the son of an elementary school teacher. After his state examination in Berlin in 1848, he worked for a year as a teacher trainee at the Friedrich Werdersches Gymnasium (High School), and it was only after this that he embarked on the path to becoming an art historian.

LÜBKE'S STYLE OF WRITING

In the following close-ups, a short look will be taken at the texts which Lübke wrote for feuilletons of the daily press, notably the *Allgemeine Zeitung* from Munich. These are mainly reviews, but also include longer essays, some of which were published in serial form, from 1861 until his death in 1893. This essay explores what Lübke's style of writing expresses, what rhetorical means are used, and whether there are rhetorical differences depending on the specific textual genres and their readership (e.g., newspaper versus journal). Concurrently, this local historical spotlight is also fairly representative of the scholarly production of art-historical texts in the context of art history establishing itself as a discipline at polytechnical institutes. Art history's close relationship to the natural sciences, and above all the function and task of art history as general education subject, like philosophy and history, for all students of the natural sciences at that time at polytechnical institutes—as well as the pressure for the discipline to justify itself in this technical context—played a decisive role in professors' own understanding of the genre, in the correspondingly tailored teachings, and, beyond that, in the published texts. At that time, teaching art history at technical universities meant generalizing historical overviews for architects and engineers, unlike the detailed and more specific art history that was taught to students of the humanities at non-technical universities (Axtmann 2020).[6] Lübke took up the new mechanisms for the generally accessible communication of the complex knowledge contents of the natural sciences in the 19th century (Kretschmann 2003a, 2003b: 194), e.g., by frequently using one textual genre which also played a significant role in natural history texts, as Andreas Daum and Carsten Kretschmann have pointed out (Kretschmann 2003b: 200; Daum 1995: 257), namely, the travel description or travelogue in the tradition of Alexander von

6 With a focus on developments at the Polytechnikum in Karlsruhe from 1868-1919.

Humboldt—which on the one hand was about local landscapes, but on the other hand was also about exotic or undiscovered landscapes (Kretschmann 2003b: 200; Daum 1995: 263-264).

As an example, we turn to an essay from September 1866, where Lübke reports for the *Allgemeine Zeitung* on his entirely unexpected discovery of a *danse macabre* in Badenweiler during a trip:

"Anyone who knows Badenweiler's shady forests, lush green meadows, and lovely vineyards knows that this is one of the most beautiful spots on German soil. Stretching gently over the valley, surrounded by the protective ridges of the Blauen and other smaller mountains, it looks far into the distance, over the magnificent Rhine plain and up to the softly curving chain of the Vosges, this natural western mark of the German country. Here the eye finds the change of a richly graded middle ground, the free distant view over the laughing plain and its splendid mountainous borderline, and finally quiet immersion in the shadowy night of the oak, beech, and fir forests. It is a place where one can live in the quiet peace of nature and forget the restless hustle and bustle of the human world for a while. Yet a surprise was reserved here for the art historian, and a discovery made its way toward him, one that he had not at all thought to find here."[7]

After visiting the known Roman ruins and the ruined castle of Baden, Lübke lets us explore the interior of the church of Badenweiler, which he perceives inconspicuously from the outside as an "unworthy stable," although he, as a self-proclaimed conscientious explorer, is not deterred from entering. Inside, he then discovers individual traces of human figures on a wall, alternating with skele-

7 "Wer Badenweilers schattige Wälder, saftiggrüne Matten und liebliche Weingelände kennt, der weiß daß dieß einer der schönsten Punkte auf deutscher Erde ist. Auf weitvorgestrecktem Arme sanft über den Thalgrund emporgehoben, umfaßt von den schützenden Bergrücken des Blauen und seiner Trabanten, schaut es weit in die herrliche Rheinebene bis zu der weichgeschwungenen Kette der Vogesen, dieser natürlichen Westmark des deutschen Landes. Hier findet das Auge den Wechsel eines reich abgestuften Mittelgrundes, den freien Fernblick über die lachende Ebene und ihre prachtvolle gebirgige Gränzlinie, und endlich stilles Versenken in die schattende Nacht der Eichen-, Buchen- und Tannenwälder. Es ist ein Ort wo man im stillen Frieden der Natur hausen, und das unruhige Treiben der Menschenwelt auf eine Zeitlang vergessen mag. Und doch sollte dem Kunsthistoriker hier eine Ueberraschung aufgespart seyn, und es drängte sich ihm eine Entdeckung entgegen an die er am allerwenigsten hier gedacht hatte." (Lübke 1866b: 4350) All subsequent English quotations have been translated from the German by the author. Therefore, the original German citations are always included in the footnotes.

tons, and is enraptured by this surprising find of a *danse macabre*, one that had not previously been documented. Lübke then conducts a thorough inspection of the church and its walls, and spontaneously decides to stay a few days to examine the wall painting more closely and, as he further reports, to free it from the whitewashed cover that was later attached. This is followed by a detailed description, classification, and valuation of this dance of death (Lübke 1866b).

Lübke's art-historical discovery report allows the reader to embark on an imaginary journey to Badenweiler. It is a re-experienceable adventure. As with many of his essays, in the beginning the focus is on nature, which can be experienced through the senses, and which serves as starting point for the subsequent closer examination of the cultural objects. He writes down his impressions in detail and lets the reader's eye glide over the individual mountain peaks and surfaces. The text thus combines subjective and objective narration and is a mixture of novel-like narration, an imparting of impressions by vivid illustrations, poetic turns, objective descriptions, facts, and technical explanations—in other words, Lübke's text combines the genre of travelogue and a modern variant of ancient *ékphrasis* (see the text by Sara Matías Pérez and Muñoz Morcillo's introduction to this volume, pp. 41-47). Furthermore, the transfer of art-historical knowledge and the classification of newly discovered works of art or architecture are achieved incidentally and subtly during his report. This is what Hermann Pietsch, for example, characterized as Lübke's "flourishing language, the kindling warmth, the sweep of his descriptions, the beautiful, honest capacity for enthusiasm, and on the other hand the masculine incisiveness in his judgment."[8]

In other essays, Lübke takes the reader by his hand by acting explicitly as a travel companion, in a manner comparable to natural scientific travelogues of his time (Daum 1995: 262), and addresses the fictitious reader directly, as an essay about his journey through southern Germany in 1855 illustrates:

"If you, dear friend, have followed me up to this point with the encouraging indulgence which I know and admire in you, I may well ask you now to remain loyal to me on the last stage of my journey, which must quickly come to an end, both on paper and in reality. Since I granted you a

8 "Seine blühende Sprache, die zündende Wärme, der Schwung seiner Schilderungen, die schöne ehrliche Enthusiasmusfähigkeit und andererseits die männliche Schneidigkeit in seinem Urtheil verfehlten nie des Eindrucks in weiten Kreisen, während die auf's Sorglichste und Delicateste ausgearbeiteten kritisch-ästhetischen Essays des feinsinnigen, geistvollen Herausgebers, Friederich Eggers, immer nur von einer ganz kleinen Gemeinde nachdenklicher Gemüther völlig gewürdigt werden konnten." (Pietsch 1877: 272)

rest day in the monastic silence of Maulbronn, you must now permit me to quickly pass a multitude of changing images before your eyes. So be patient a little while longer!"[9]

In the sense of popularization, Lübke works by means of involvement—he asks the reader direct and rhetorical questions, and uses the simple "we" or "you" forms of address.[10] With emphasis and identification, he tries to make past events immediately re-experienceable by integrating subjective feelings and impressions, for example (Imorde/Zeising 2016: 14-17), as well as physical perceptions and the exhaustion of traveling, which the reader could understand very well back then and can still understand today. For example, in a three-part essay from August 1871 about Lübke's journey from "the tropical heat" of Stuttgart towards Donaueschingen, he begins his report with a description of a terribly long train journey—a narrative that almost anyone can understand, even today:

"I had always believed that it was impossible, anywhere in the entire world, to travel more slowly than from Stuttgart to Friedrichshafen by rail. Now I have come back from this delusion; in the race for the maximum slowness possible, the upper Neckar Railway surpasses even the main Württemberg line. If you are particularly lucky, you can make it to Singen in 13 hours."[11]

Here, as in other texts, the "cicerone" Lübke is also able to convey a series of insider tips for tourists, such as special routes and suitable means of transport, inns, and guesthouses with ratings of room and board, and he also mentions his personal

9 "Wenn Du, lieber Freund, mir bis hierher mit der aufmunternden Nachsicht gefolgt bist, die ich an Dir kenne und schätze, so werde ich Dich wohl bitten dürfen, nun auch auf der letzten Station meiner Reise, die sich auf dem Papier wie in der Wirklichkeit rasch den Ende zuneigen muß, mir treulich zur Seite zu bleiben. Habe ich Dir in der klösterlichen Stille von Maulbronn einen Rasttag gegönnt, so wirst du nun mir dafür gestatten, schnell eine Menge wechselnder Bilder an Deinen Augen vorüberzuführen. Also noch eine kurze Weile Geduld!" (Lübke 1855b: 439)

10 For similar scientific travelogues of the time, Andreas Daum (1995: 259-264) cites August Böhner's *Kosmos* (1864).

11 "Ich hatte immer geglaubt: langsamer als von Stuttgart nach Friedrichshafen könne man auf der ganzen Welt mit der Eisenbahn nicht fahren. Jetzt bin ich von diesem Wahn zurückgekommen, im Wettfahren um höchstmögliche Langsamkeit übertrifft die obere Neckarbahn selbst noch die württembergische Hauptlinie. Wenn man besonderes Glück hat, gelangt man in 13 Stunden bis Singen." (Lübke 1871b: 3970) (The distance amounts to approximately 160 km and currently takes between two and three hours by train.)

contacts and conversations with locals (Lübke 1871b: 3971).[12] The large towns are described quickly, as in a short travel guide, with the most exciting sites coming first, in order to then discuss special attractions, which he pays more attention to (Lübke 1871b: 4018).

There are numerous examples of similar texts, but also some more scientific treatises, each dealing with individual works of art, architecture, painting, sculpture, or handicraft, in which Lübke proceeds in a very narrative fashion and with many personal interjections. He always proves to be an accurate and precise describer, presenting the reader with the historical background and context of the objects he is looking at, as well as his own impressions and judgments. In addition to imparting art-historical knowledge, he wants to hone the broader public's own aesthetic perceptions and taste. The art historian thus functions as a kind of trainer for seeing and recognizing art in its various manifestations. In this context, one could take a closer look at Lübke's numerous reviews and critiques, most of which he provided with an essay-like introduction, before discussing the books or essays under review in a second step. A comparative examination of the rhetorical means in these texts would certainly be worthwhile as a supplement to the travelogues presented here, but this is outside the scope of the present essay.

It now has to be asked whether Lübke's particular way of conveying knowledge is only a phenomenon of the texts for feuilletons or if it is also a characteristic of his other writings. Upon examination, we can see that the abovementioned rhetorical means can also be found in his essays and reviews for bourgeoise journals[13] and some art-historical periodicals and yearbooks, such as the *Zeitschrift für Bildende Kunst* or *Deutsches Kunstblatt*. Lübke's handbooks and monographs, on the contrary, do not contain descriptions of his own experiences of nature or his travel impressions. Here he is much more objective, and the focus is on the level of content, a clear presentation of large correlations on the basis of the most important data and objects while at the same time avoiding overly detailed knowledge. Compared to previous published histories of architecture or art—of which, moreover, there were only a few at that time—, the great success of his handbooks was surely based on their overall presentation, which began in ancient times and went all the

12 The Badische Landesbibliothek in Karlsruhe has 17 travel notebooks, which can be used to trace Lübke's travels and discoveries (BLB, K 1279-1295). However, there are no personal impressions or diary notes listed therein; they primarily contain precise sketches, ground plans, detailed drawings, and some scenes or descriptions in note form.

13 E.g., *Die Gegenwart: Wochenzeitschrift für Literatur, Kunst und öffentliches Leben*; *Nord und Süd*; and *Westermann's illustrierte Monatshefte*. For Lübke's publication list, cf. Axtmann/Gawlik 2019: 105-143.

way up to the present day, and which had broad historical and cultural-historical introductions written in a lively, accessible, non-academic, and everyday language that was characterized by both brevity and convenience, as some book reviews described it (Karge 2019; n. n. 1874).[14] These handbooks have an encyclopedic claim and are characterized by the principles of generalization and the systematic order of knowledge. Often, in a single volume of modest size, the material is organized in fairly small chapters: the table of contents of the *Grundriss der Kunstgeschichte* (*Outline of the History of Art*), for instance, comprises about five pages (Lübke 1860a: IX-XIII). As in the early examples of encyclopedias, Lübke equipped his handbooks with several very detailed indexes for better handling and to allow specific information to be easily searched: a city index, which also, in each instance, lists the individual buildings by place; and a list of artists.[15]

Lübke's friend Theodor Fontane (1819-1898) honored his contribution to the popularization of art history several times in his reviews. Both were members of literary circles in Berlin and wrote several critiques for each other, and Lübke even accompanied Fontane, e.g., on one of his journeys through the March of Brandenburg (Karge 1998; Axtmann/Gawlik 2019: 9, 34, 81-87). In Fontane's review of the *Geschichte der italienischen Malerei* (*History of Italian Painting*), he explained the characteristics of Lübke's writing style and personality with the following words:

"To sum up the merits of the work [*History of Italian Painting*, Stuttgart 1878] once again, they are: the brilliant abundance of detail, structure, clarity, and measure everywhere. More than any other writer in his field, not excepting even the most famous, Lübke has contributed to the popularization of art-historical knowledge. And not just in Germany! His happy nature, which is in no way doctrinaire, his lively way of speaking and writing, allowed him to achieve great results, to which the numerous editions of his books bear witness. It was also by happy chance that the teacher and the writer in him were supportive of one another."[16]

---

14 In some of them (n.n. 1874), his textbooks are characterized as "the most dignified works in the whole field of art-historical literature due to their excellent clear presentation, genuine popularity, and philosophical education […]."

15 For more on early encyclopedias, see the essay of Mathias Herweg in this volume.

16 "Fass' ich die Vorzüge des Werkes [Geschichte der italienischen Malerei, Stuttgart 1878] noch einmal zusammen, so lauten sie: bei glänzender Fülle des Details überall Gliederung, Klarheit und Maß. Mehr als irgend ein anderer Schriftsteller seines Fachs, auch die berühmtesten nicht ausgenommen, hat Lübke zur Popularisirung kunstgeschichtlicher Kenntniß beigetragen. Und nicht blos in Deutschland! Seine glückliche, dem Doctrinären abgewandte Natur, seine belebte Sprech- und Schreibweise ließen ihn die großen Resultate, wovon die zahlreichen Auflagen seiner

This was acknowledged in numerous other obituaries and biographies—for example, in a review from 1889 in which the Munich professor and director of local painting galleries Franz Reber (1834-1919) described Lübke's books as "the most prominent popular scientific [*populärwissenschaftlichen*] books not only in Germany, but also, through translations into many languages, in the civilized world."[17]

In the context of popular science communication, it should also be emphasized that Lübke's handbooks with their woodcuts, and above all his first one, the *Geschichte der Architektur* (*History of Architecture* [Lübke 1855a]),[18] are among the first illustrated global histories of architecture, where the pictures are not reproduced as separate panels or even as a separate volume of panels, but within the text itself (Rohling 1957: 155; Karlholm 2004: 15-17; Karge 2019). At the same time, only James Fergusson's *Illustrated Handbook of Architecture* from the same year, 1855, can serve as a point of comparison here (Brouwer 2016; Karge 2019).[19] The possibility for the reader to follow the textual explanations directly beside different illustrative forms such as plans, sketches, scenes, and aerial views into buildings, corresponds to Lübke's pedagogical claim, which is addressed not only to educated experts, but also to the lay public (Engel 2016). In this process, the visual communication, the illustrated textual layout of the handbooks (which we have since become accustomed to), played a very important role for him. This new approach to and presentation of knowledge, which has become increasingly common since then, should not be underestimated in terms of its significance for the establishment of art history as a science or scholarly discipline and its recognition by broad sections of the population (ibd).

CONCLUSION

In contrast to all of this positive feedback, Lübke also had to contend with much criticism by some of his contemporaries in the scientific/scholarly community of art historians, concerning not only his publishing activities but also the scientific claims of his writings. Among other things, he was accused of inaccuracy and

---

Bücher Zeugniß ablegen, erringen. Es traf sich zudem glücklich, daß sich Lehrer und Schriftsteller in ihm gegenseitig unterstützten." (Fontane 1880: 344)

17 "Jedes reiht sich an den glänzenden Erfolg des vorausgegangenen, die zu den verbreitetsten populärwissenschaftlichen Büchern nicht bloß Deutschlands, sondern zum Theil in viele Sprachen übersetzt, der civilisirten Welt zählen." (Reber 1889)

18 The *History of Architecture* is a two-volume work.

19 Wilhelm Lübke is also discussed in Karge 2020.

superficiality on several occasions. However, these critiques, which were often associated with a general criticism of popularization tendencies because of the intense professionalization taking place at that time (Daum 1995: 2-3), failed to take into account the different textual genres and the contexts of their reception. Thus, Lübke's articles in the feuilletons of the daily newspapers and many of his handbooks, as well as his publications on Michelangelo, Titian, and Veronese (all three 1869), or a Raphael (1880), which can be regarded as early forms of so-called coffee table books, were not intended as scholarly works to be read by experts, but as popular works to be read by a large bourgeois circle of people. In his prefaces to the survey books, Lübke explains several times that he has geared his texts to the various reading groups. In his 1891 published memoirs, for example, one can read that after his trips to Italy in 1860 he wanted to publish his field studies in different genres, and that he deliberately distinguished the scientific/scholarly essay from the *popular* scientific essay:

"At the same time, however, I was urged to communicate my art-historical studies to a broader circle, and while I was writing up the results of my architectural research in Italy in an extensive essay illustrated with drawings by myself for the Communications of the Vienna Central Commission [Mittheilungen der Wiener Central-Commission], I decided to present the entire history of art in a popular handbook. [...] My 'Outline of the History of Art' thus came into being."[20]

Moreover, just a few sentences later, he formulated the difficulty of conveying the science/discipline of art history in a popular manner, and the linguistic means he considered necessary for successful writing about art in general:

"People from certain quarters now [in 1891] arrogantly deny the legitimacy of such popular representations and look down on their authors with all the disdain of pedantic stick-scholarship. However, one does not take into account that a good popular book is infinitely more challenging to write than a strictly academic-scientific one. As far as writing about art is concerned, this includes not only erudition, but, above all, a strong touch of

20 "Zugleich aber drängte es mich, einem weiteren Kreise meine kunstgeschichtlichen Studien zu vermitteln, und während ich die Ergebnisse meiner architektonischen Forschungen in Italien in einen umfangreichen mit Zeichnungen von mir illustrirten Aufsatz für die Mittheilungen der Wiener Central-Commission niederlegte, beschloß ich die ganze Kunstgeschichte in einem populären Handbuch darzustellen. [...] Es entstand also mein 'Grundriß der Kunstgeschichte.'" (Lübke 1891b: 322-323) The essay mentioned in the quotation is: Lübke 1860b.

sensitive fantasy, a small piece of poetic talent, and above all the secret of fascinating, lively description, which is possessed by very few."[21]

The texts for feuilletons, which take up the most space in Lübke's extensive oeuvre, were probably the best-suited medium for him to speak about art and make art history popular in an attractive and diverse way, due to the much greater reach that they had when compared to the other textual genres that Lübke made use of. This was obviously in accordance with his nature and his way of lecturing, i.e., through lively, fantastic, poetic narratives, as he describes above, which is why he remained faithful to this popular textual genre from 1861 until the end of his scientific and scholarly life.

REFERENCES

Axtmann, Alexandra (2020): "Die Etablierung der Kunstgeschichte am Karlsruher Polytechnikum." In: Robert Stalla (ed.), Kunstgeschichte an Polytechnischen Instituten, Technischen Hochschulen, Technischen Universitäten. Geschichte – Positionen – Perspektive, Vienna: Böhlau (forthcoming).

Axtmann, Alexandra/Gawlik, Ulrike (eds.) (2019): Wilhelm Lübke (1826–1893). Aspekte seines Lebens und Werkes, Karlsruhe: KIT Publishing.

Betthausen, Peter (2007 [1999]): "Lübke, Wilhelm." In: Peter Betthausen/Peter H. Feist/Christiane Fork (eds.), Metzler Kunsthistoriker Lexikon. 210 Porträts deutschsprachiger Autoren aus vier Jahrhunderten, Stuttgart and Weimar: Metzler, pp. 268-271.

Beyrodt, Wolfgang (1991): "Kunstgeschichte als Universitätsfach." In: Paul Ganz (ed.), Kunst und Kunsttheorie 1400–1900. Wiesbaden: Harrassowitz, pp. 313-333.

Brouwer, Petra (2016): "An illustrated comparison of 'true styles'. James Fergusson's *The illustrated Handbook of Architecture*." In: Matteo Burioni

21 "Nun wird ja von gewissen Seiten solchen populären Darstellungen hochmüthig die Berechtigung abgesprochen und auf deren Autoren mit der ganzen Geringschätzung pedantischer Stockgelehrsamkeit herabgesehen. Allein man bedenkt nicht, daß ein gutes populäres Buch unendlich viel schwerer zu schreiben ist als ein streng fachwissenschaftliches. Was aber vollends das Schreiben über Kunst betrifft, so gehört dazu nicht bloß Gelehrsamkeit, sondern vor Allem ein starker Hauch nachempfindender Phantasie, ein kleines Stück dichterischer Begabung und vor Allem das von sehr Wenigen besessene Geheimnis anziehender, lebensvoller Schilderung." (Lübke 1891b: 323-324)

(ed.), Weltgeschichten der Architektur. Ursprünge, Narrative, Bilder 1700–2016, Passau: Klinger, pp. 42-44.

Daum, Andreas W. (1995): Wissenschaftspopularisierung im 19. Jahrhundert. Bürgerliche Kultur, naturwissenschaftliche Bildung und die deutsche Öffentlichkeit, 1848-1914, Munich: Oldenbourg. (Doctoral thesis)

Engel, Ute (2016): "Popularisierung und Veranschaulichung. Wilhelm Lübke: Geschichte der Architektur von den ältesten Zeiten bis zur Gegenwart." In: Matteo Burioni (ed.), Weltgeschichten der Architektur. Ursprünge, Narrative, Bilder 1700–2016, Passau: Klinger, pp. 40-42.

Fontane, Theodor (1880): "Wilhelm Lübkes Geschichte der italienischen Malerei. Besprochen von Th. Fontane." In: Die Gegenwart. Wochenzeitschrift für Literatur, Kunst und öffentliches Leben 17, pp. 342-344.

Imorde, Joseph/Zeising, Andreas (eds.) (2016): Billige Bilder. Populäre Kunstgeschichte in Monografien und Mappenwerken seit 1900 am Beispiel Albrecht Dürer, Siegen: Universitätsverlag.

Karge, Henrik (1998): "Poesie und Wissenschaft, Fontane und die Kunstgeschichte." In: Claude Keisch (ed.), Fontane und die bildende Kunst. Exhibition Catalogue. Berlin, Staatliche Museen zu Berlin, Nationalgalerie, Berlin: Henschel, pp. 267-278.

Karge, Henrik (2019): "The Formation of Global Architectural History in the Middle of the 19th Century. Wilhelm Lübke's Geschichte der Architektur (1855) in the Context of the Art Historical Works of Karl Schnaase and Franz Kugler." In: Petra Brouwer/Martin Bressani/Christopher Drew Armstrong (eds.), World Histories of Architecture. The Emergence of a New Genre in the Nineteenth Century, Princeton, NJ (forthcoming).

Karge, Henrik (2020): Die Genese der modernen Kunstgeschichte im 19. Jahrhundert. Schnaase – Kugler – Burckhardt – Semper, Hildesheim: Olms.

Karlholm, Dan (2004): Art of Illusion. The Representation of Art History in Nineteenth-Century Germany and Beyond, Bern: Lang.

Kretschmann, Carsten (ed.) (2003a): Wissenspopularisierung. Konzepte der Wissensverbreitung im Wandel, Berlin: Akademie-Verlag.

Kretschmann, Carsten (2003b): "Wissenskanonisierung und -popularisierung in Museen des 19. Jahrhunderts – das Beispiel des Senckenberg-Museums in Frankfurt a. M." In: Lothar Gall/Andreas Schulz (eds.), Wissenskommunikation im 19. Jahrhundert, Stuttgart: Steiner, pp. 171-212.

Kukula, Richard (1892): "Lübke, Wilhelm, Dr., Prof. f. Kunstgeschichte. Techn. Hochsch. Karlsruhe. Geb. 1826." In: Bibliographisches Jahrbuch der Deutschen Hochschulen. Völlig umgearbeitete Neuauflage des Allgemeinen Deutschen Hochschulen-Almanachs, Innsbruck: Wagner, pp. 573-574.

Lemcke, Carl von (1893): "Wilhelm v. Lübke (Nachruf)." In: Allgemeine Zeitung July 19 (Suppl.), pp. 1-6.
Lemcke, Carl von (1906): "Wilhelm Lübke." In: Allgemeine deutsche Biographie. Vol. 52, Nachträge bis 1899, Leipzig: Duncker & Humblot, pp. 106-111.
Locher, Hubert (1999): "Das ‚Handbuch der Kunstgeschichte'. Die Vermittlung kunsthistorischen Wissens als Anleitung zum ästhetischen Urteil." In: Memory & Oblivion. Proceedings of the XXIXth International Congress of the History of Art held in Amsterdam 1996, Dordrecht: Kluwer, pp. 69-87.
Lübke, Wilhelm (1855a): Geschichte der Architektur von den ältesten Zeiten bis auf die Gegenwart, Leipzig: Emil Graul.
Lübke, Wilhelm (1855b): "Eine Fahrt durch Süddeutschland." In: Deutsches Kunstblatt 6, pp. 337-339, 350-352, 357-360, 365-367, 373-374, 409-413, 431-435, 439-441, 448-449, 457-459.
Lübke, Wilhelm (1860a): Grundriss der Kunstgeschichte, Stuttgart: Ebner und Seubert.
Lübke, Wilhelm (1860b): "Reisenotizen über mittelalterliche Kunstwerke in Italien." In: Mitteilungen der K.K. Central-Commission zur Erforschung und Erhaltung der Baudenkmale 5/4-8, pp. 112-120, 134-140, 160-173, 191-203, 222-231.
Lübke, Wilhelm (1866a): "Die heutige Kunst und die Kunstwissenschaft." In: Zeitschrift für Bildende Kunst1, pp. 3-13.
Lübke, Wilhelm (1866b): "Ein Todtentanz in Badenweiler." In: Allgemeine Zeitung September 22, 23 (Suppl.), pp. 4350-4351, pp. 4366-4367.
Lübke, Wilhelm (1869): Kunsthistorische Studien, Stuttgart: Ebner und Seubert.
Lübke, Wilhelm (1871a): "Die Kunstgeschichte und die Universitäten." In: Allgemeine Zeitung June 26 (Suppl.), pp. 3221-3223.
Lübke, Wilhelm (1871b): "Südwestdeutsche Idyllen. I. Donaueschingen, II. Stein am Rhein, III. Heiligenberg." In: Allgemeine Zeitung August 11, 14, 15 (Suppl.), pp. 3970-3972 (I), pp. 4018-4021 (II), pp. 4035-4036 (III).
Lübke, Wilhelm (1885): Bunte Blätter aus Schwaben. 1866 bis 1884, Berlin, Stuttgart: Spemann.
Lübke, Wilhelm (1886): Kunstwerke und Künstler. 3. Sammlung vermischter Aufsätze, Breslau: Schottlaender.
Lübke, Wilhelm (1891a): Altes und Neues – Studien u. Kritiken. 3. Sammlung vermischter Aufsätze, Breslau: Schlesische Buchdruckerei, Kunst- u. Verlagsanstalt.
Lübke, Wilhelm (1891b): Lebenserinnerungen, Berlin: Fontane.
n.n. (1862): "Wilhelm Lübke." In: Westermann's illustrierte Monatshefte 12, pp. 487-489.
n.n. (1874): "Wilhelm Lübke." In: Illustrirtes Unterhaltungs-Blatt, N. F. 7/48, p. 379.

Oechelhäuser, Adolf von (1906): "Wilhelm Lübke." In: Friedrich von Weech/ Albert Krieger (eds.), Badische Biographien, Vol. 5, 1891-1901, Heidelberg: Winter, pp. 527-532.

Pietsch, Ludwig (1877): "Wilhelm Lübke." In: Nord und Süd (October), pp. 268-280.

Reber, Franz (1889): "Rezension zu W. Lübke's Geschichte der deutschen Kunst." In: Allgemeine Zeitung, December 9 (Suppl.), p. 3.

Riegel, Herman (1874): Dem Herrn Wilhelm Lübke, Verfasser mehrerer kunstgeschichtlicher Handbücher und dergleichen mehr in Stuttgart. Offener Brief, Hannover: Rümpler.

Rohling, Ludwig (1957): "Wilhelm Lübke." In: Wilhelm Steffens/Karl Zuhorn (eds.), Westfälische Lebensbilder. Vol. 6, Münster: Aschendorff, pp. 147-165.

Popular Aesthetics of the 19th Century. Ornamental Prints and Pattern Sheets as Actors for Popularization During the 1870s

Elin Manker

ABSTRACT

This article seeks to investigate a branch of popular science that emerged during the second half of the 19th century: popular aesthetics. The primary format for popular aesthetics was the illustrated journal, and especially journals engaging in contemporary manufacturing and handicraft. Ornamental prints and pattern sheets functioned as core actors for the popularization of aesthetic knowledge within these kinds of publications. They also function as core actors for the investigation undertaken in this article.

Popular aesthetics had its roots in German aesthetic philosophy from the turn of the 19th century onward. It aimed for an engagement with aesthetic notions—and even aesthetic practice—in daily life. It thereby differs from other forms of popularization of the period that mainly aimed at conveying a text-based education. Popular aesthetics was, via mediation through internationally circulating illustrated journals, a movement that had implications far outside the borders of Germany. In this article, two ways of approaching popular aesthetics in Sweden are presented: one from a design and engineering journal, and the other from a women's journal. The approach of the women's journal is shown to be the more successful of the two, and to have implications for the sloyd movement well into the 21st century.

INTRODUCTION

When Carl Lemcke wrote *Populäre Aesthetik* in 1865, it was one of several contributions to a Western European movement informing manufacturers and the public of aesthetic knowledge. The intention was to improve emerging industrial businesses and consumer society. It was also an outcome of the contemporaneous undertaking of transferring academic knowledge to a broader audience through the genre of popular science (Papanelopoulou 2009). To carry out this transfer of knowledge, an international and commercial market for publications was crucial. Lemcke's book was soon to be translated into Swedish, as *Populär estetik* in 1868. In the years that followed, popularized aesthetic writing emerged as a growing field in Sweden (as well as in other Nordic countries), not least by journals contributing to the topic. Within the field of Art and Design History, investigations of design reform movements of the 19th century have only recently begun to include the study of these kinds of publications, yet they were crucial in the formation of manufacturing and design businesses during the mid- and late 19th century (Aynsley & Forde 2007; Lees-Maffei 2014).

This article will comment on the popular aesthetic agenda and define our use of this (not generally accepted) concept. We will demonstrate how popular aesthetics migrated from German books and journals into Swedish, and which forms it took in this new environment. We will thereby stress the importance of the commercial book and printing market for popularization as such. Illustrations in the form of ornamental prints and pattern sheets are seen as core actors in this process.

The analytical approach is based on network theories, where the ideas of *heterogenic actors* (Law 1992) and *agency* are the theoretical base for the understanding of *relations* (Latour 2005) and *intra-action* (Barad 2003). The contribution to the genealogy of popular science consists in clarifying how popularization happened within fields outside—but concurrent to—the popularization of the natural sciences, how this was a movement that depended on commercial markets, and how this also occurred in more peripheral parts of Europe, such as Sweden.

A BRIEF THEORETICAL REMARK

When using the concepts of actor and agency, the theoretical grounding in materialism, or "network-approaches" (Olsen 2003: 98), is obvious. We are reluctant to use the concept of the actor-network, since it has very strong connections to the specific approach of ANT (actor-network theory; Latour 1994, 1996, 2005). Bruno Latour's sociological philosophy operates in the background of this research, but our working concepts come from other sources. Our understanding and use of *actor*, *network*, and *heterogeneity* first and foremost derive from the STS (Science and Technology Studies) work of sociologist John Law. Law explains an actor as such according to its heterogenic constitution (Law 1992). What makes an actor is its capacity to establish an intra-acting relation with others. To provide an example that is directly relevant to this article, we would like to begin by describing a journal as a heterogenic actor.

A journal is a heterogenic actor because of what we call *inner actors*. These inner actors are (for instance) covers, articles, images, and advertisements. These inner actors have a relation to *outer actors* such as constitutive ones (writers, illustrators, photographers, advertising agencies, editors, publishers) and consuming ones (resellers, purchasers, readers, libraries). The journal obviously assembles many different actors, and notably all those mentioned above are actors in their own right, as they have other relations too. Heterogeneity thereby alludes to a networking quality. Or, as John Law explains the concept of the actor-network: "Hence the term, actor-network—an actor is also, always, a network." (Law 1992: 384)

Agency, in our understanding, is linked to Karen Barad's concept of *intra-action* (Barad 2003: 815). Barad points out that the commonly used *inter*-action usually refers to entities negotiating with each other; these entities, however, tend to be understood as stable even after the negotiation. By instead using the idea of intra-action, Barad wants to stress the fact that the entities most probably have changed during the negotiation, and that the negotiation constitutes them as actors. Drawing upon Barad, we likewise understand agency as a process of change pertaining to the actors involved. Agency thus implies an intra-acting process.

Methodologically, we work with an unpacking of selected actors. We try to identify the relations between inner and outer actors by tracing where agency, in the intra-action sense of the word, has happened. Latour's suggestion to trace the actors (Latour 2005) is applied throughout the study by following not only the obvious actor (e.g., journals, editors) but also the actors within a particular journal. This could be done in many ways. In this article, *ornamental prints* in design and engineering journals and *pattern sheets* in women's journals are analyzed as inner actors whose relations help reveal the story of popular aesthetics in the 1870s.

CONTEXT: SOCIAL REFORMS

In the 1860s, Sweden was at the beginning of a radical transformation from a poor, rural country to an industrialized nation. For such a transformation, a more educated population as well as higher living standards for more people were seen as crucial. This transformation would take many decades but started in a reform movement that engaged politicians as well as intellectuals and the trading and manufacturing businesses. Two circumstances were particularly important for this development during the 19th century.

Firstly, Sweden wasn't just a rural country. It was also a country of exports of raw material and imports of refined goods. Metal and wood resources were exported as raw materials only to be imported back as more expensive goods. This happened since no industry capable of refining the raw material was present in Sweden. With the main industrial countries in Europe serving as role models, strong arguments for a shift towards a domestic industry were voiced both by politicians and businessmen. A domestic industry would be able to refine the raw materials and reduce imports, something that would create greater profit as well as cater to the needs of the citizen. This last argument was particularly important, since—secondly—more than just raw materials were leaving the country. From the 1860s up until the 1920s, Sweden would see a large part of its population emigrate to America—about a fifth of the population, making Sweden the fourth largest migratory country in Europe. This development lent a sense of urgency to the endeavor of making Sweden a viable country to live in for all of its population (Norman, 2019).

Numerous legislative amendments were made between 1850 and 1900 by the government in Sweden to meet the immediate needs of the migrating population, and in particular the need to earning one's living. New or modernized laws encouraged enterprises, trading, manufacturing, lending institutions, and financial investors to start operating. It was not only private initiatives that were encouraged, however; state institutions were also strengthened. For instance, the governments of the 1850s decided that the railway network and the telegraph network should be operated solely by the state, excluding all competitive private enterprises in these sectors. Furthermore, universities and higher vocational training institutions were successively turned over to the state (Andersson-Skog 2013).

A COMMERCIAL BOOK AND JOURNAL'S MARKET

In this general spirit of reform and enterprise, the market for books and journals changed. Up until the 18th century, printing offices and book binders had control over book production in Sweden. Through the commercialization of the book and journal market, this control instead passed into the hands of publishers and retailers. (Lundblad 2010; Strömquist 2010). This had effects on the publishing process. For instance, new selling strategies emerged. The publishing of booklet versions of a book became common—as was the case with the Swedish version of Lemcke's *Popular Aesthetics (Populär estetik)*, which was published as six booklets before being released as a book. Furthermore, the publishing of books as series of articles before making them into a book became commonplace—as was the case with another German book on popular aesthetics translated to Swedish, Jakob Falke's *Die Kunst im Hause*. This book was translated as a series of articles for the women's journal *Tidskrift för hemmet* (Journal for the home) 1874-75, before the book was released in 1876. Thirdly, new printing techniques made illustrations important both as unique selling points and as a means of provoking new journalistic approaches and layouts (Gretton 2010; Griffiths 2016; Hultzsch 2017; L. Johannesson 1982). These new publication processes affected several genres—popular scientific works as well as fiction and journals (Goldman & Cooke 2012; Lundblad 2010; Van Remoortel 2015).

The commercial market also had an effect on the process of editing journals. Publishers increasingly needed to function as full publishing houses in order to be able to negotiate with the complicated network of artisans, printers, suppliers, and retailers. The successful publisher not only ran an editing business. He (they were virtually all men) also owned a print shop, a binding workshop, and a workshop for illustrations, which had all earlier operated as separate businesses (L. Johannesson 1982). This was also the case for business that published illustrated journals in Sweden. All long-lived journals in the second half of the century were run by publishing houses in the manner described above (E. Johannesson 1980; L. Johannesson 1982).

Popular aesthetics of the 19th century is to be found in this commercial publishing environment. The very idea of spreading knowledge to a broader audience outside the wealthiest class demanded larger editions and far-reaching trading systems. At this point in time, commercialization was a driving force, and was needed to succeed in this undertaking—books and journals were not printed for philanthropic reasons alone.

POPULAR AESTHETICS, OR TO ENNOBLE THE PUBLIC

Popular aesthetics was introduced to a Swedish audience by Carl Lemcke, or more accurately, by the publisher L. Hierta and the translator O. W. Ålund. Prior to Lemcke's book, only two lecturers in philosophy at Uppsala University (F. G. Afzelius and Pontus Wikner) had made any attempts in that direction by presenting recent philosophy (such as Hegel) in a popularized manner. Popular aesthetics, however, was different from the popularization of philosophy. For Lemcke, popular aesthetics meant two things: In essence, it was a popularized (i.e., written for a non-academic audience) description of the philosophy of beauty. More specifically, however, popular aesthetics was meant as a guide for how to approach daily life. In this second sense, popular aesthetics was perceived as vital (and crucial) not only for the person practicing it, but for the creation of modern society as well.

Lemcke highlighted Immanuel Kant as the founder of the kind of aesthetic philosophy he perceived as correct. Nevertheless, Lemcke considered Kant to be too difficult for the general public (Lemcke 1865:5, 24, 294-296; 1868: 2, 15, 224-225). If the main part of *Populäre Aesthetik* is devoted to a review of the doctrines of beauty (divided into three parts: the Concepts of Beauty, the Beauty in Nature, and the Beauty in Art), the underlying argument calls for a use of the knowledge of beauty as an *approach in life*. Popular aesthetics was therefore not simply about conveying a text-based education. Lemcke wanted the reader to change their lifestyle, to be picky about which interests to pursue, and to be meticulous of their own private environment. A popular aesthetic approach was not a matter of trying to achieve a certain ideal, but instead one of constantly trying to *refine* oneself, with aesthetic awareness serving as a beacon (Lemcke 1865: 5-10, 22-27; 1868: 2-6, 14-18). Lemcke's goal, then, was to have the public ennoble—or refine—itself. This, he suggested, was also the aim of society: to constant refine itself by creating better living conditions (Lemcke 1868: 45).[1] In this sense, publications that promoted popular aesthetics differed from other popular science publications, which dealt more specifically with conveying humanistic and scientific knowledge through reading.

Lemcke's most prominent follower in Sweden was the Norwegian art historian Lorentz Dietrichson, who worked in the county between 1860 and 1875. By publishing his own popular aesthetics, *Det skönas verld* (The World of Beauty, subtitled "popular aesthetics," 1869-1873), which largely followed Lemcke's outline and agenda, Dietrichson revealed himself to be a spokesperson for a popular aesthetic movement in Sweden. During the 1870s, Dietrichson turned his interest specifically towards the emerging market of industrially produced goods, arguing that aesthetic knowledge was urgent for manufacturers as well as for consumers—and, in the long run, for society as a whole. In this undertaking, Dietrichson left the book format behind to instead work with articles and journals. In Sweden, from 1870 onwards, journals came to be the most important publication format for popular aesthetics.

1 The Swedish translation was made from the second edition of Populäre Aesthetik. This article uses the Swedish edition as its primary source.

ILLUSTRATING "THE CLAIM OF THE ARTS"

In 1870 a new design and engineering journal entered the Swedish market: *Illustrerad Teknisk Tidning* (Illustrated Technical Journal, hereafter ITT, see *Figs. 1* and *2*). The editor, Wilhelm Hoffstedt, had a clear vision of his journal, of its aims, and of the audience. He wrote: "Prominent journals of several technological sciences are already at hand, but they are not for the greater public, they do not speak a language that everyone can understand." This needed to be changed, he continued, since the sciences occupied more and more of society, and "most of us become workers in the field called industry" (Hoffstedt 1870: 1).[2] Therefore, to provide knowledge of science, ITT would offer "popular treatises" by "the most prominent technologists, scientists, and teachers, as well as engineers, builders, manufacturers, foremen, etc." To contribute to "the claim of the arts" was one of the most important aims according to Hoffstedt, who stated that "designs of beautiful and new models" of craft and manufacturing objects would be presented continuously to fulfill this purpose. The journal would appear once a week, and Hoffstedt stressed that efforts had been made to make it as inexpensive as possible to be affordable for the greater public (Hoffstedt 1870: 2).

Hoffstedt was eager to realize his aim of conveying "the claim of the arts." He did so predominantly by publishing illustrations in the form of ornamental prints, not by publishing articles. For the first sixty-five issues, Hoffstedt presented ornamental prints of new items from manufacturers in Europe on the front page of his technological journal (*Fig. 1*).[3]

The prints (or at least the objects depicted) were mostly retrieved from Swedish, German, or English manufacturers, but also from Austrian, French, Italian, Russian, and Danish manufacturers. Only half as many articles were published discussing aesthetic issues in relation to manufacturing during the same period. Fourteen of these articles were written by the artisan Carl Ahlborn (who was German by birth and lived in Sweden from around 1845 onward) or by the abovementioned Lorentz Dietrichson. Almost as many articles—twelve to be precise—were translations from German journals (mainly *Gewerbehalle,* a German journal that presented popular aesthetic articles and ornamental prints aimed at manufacturing businesses), and three were translations from the English *Art-*

2 All translations from Swedish are my own. For proofreading, I would like to thank Dr. Sara Callahan and Dr. Kareem Abud-Zeid.

3 The layout of ITT changed after 65 issues due to economic difficulties. Illustrations thereafter were inserted as enclosures, a less expensive option for image publication. The case study prior to this article concentrates on the issues of the initial layout of the journal.

Journal. Popular aesthetics in ITT was thereby presented as a blending of international contributions from contemporary manufacturing. There were no national claims or predominant historical style.

Figures 1 and 2: Illustrerad Teknisk Tidning, 1871:2; 1871:50

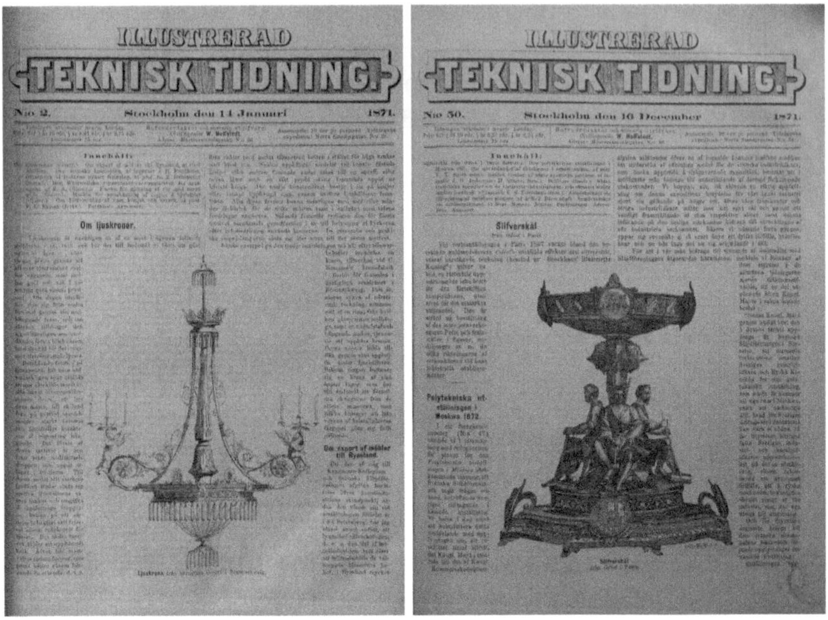
ILLUSTRERAD TEKNISK TIDNING.

N:o 2. Stockholm den 14 Januari 1871.

Om ljuskronor.

ILLUSTRERAD TEKNISK TIDNING.

N:o 50. Stockholm den 16 December 1871.

Silfverskål.

Source: National Library of Sweden; Photo: Elin Manker

Wilhelm Hoffstedt acted on the commercial market for illustrated books and journals. At this point in time, editors of journals with broad target audiences, such as ITT, used a kind of collage technique in their editing work. When international journals arrived in the morning post, interesting articles or news excerpts were selected, translated—often in abbreviated form—, and published in a coming issue, side by side with original articles. Hardly any journalists worked outside the office (E. Johannesson 2001; Nordmark 2001). Journals that had their own print workshop could create illustrations specifically for their articles (Hultzsch 2017; L. Johannesson 1982). In the case of ITT, such a workshop was not available. Instead, Hoffstedt bought illustrations from other publishers. We do not know the exact process of this, but it is likely that he approached illustrations in the same way that he approached texts, i.e., by using a collage technique. Two short examples provide a basis for this suggestion.

The first example is the front page of one of the January 1871 issues of the journal, which presents an ornamental print accompanied by a short article discussing the design of chandeliers. Towards the end of the text, the writer (no author is named) mentions that the illustration was reprinted from *Gewerbehalle*. When looking in *Gewerbehalle*, it becomes obvious that the Swedish article "Om ljuskronor" is a translated and shortened version of "Ueber Kronleuchter" by Constantin Uhde (Uhde 1870). The original article, published in September of the previous year, includes nine ornamental prints—compared to just one in ITT—and three pages of text. A transformation has happened: an intra-action between the original article, the Swedish editor, and the outcome in ITT, manifested through a cut-and-paste performance (Dahlgren 2018).

The second example comes from one of the December 1871 issues of ITT, which presents a silver bowl made by the manufacturer Odiot in Paris, with no article added. A short note informs the reader that this illustration was retrieved from the "Brockhaus' Illustrirter Katalog" (i.e., *Illustrirter Katalog der Pariser Industrie-Ausstellung von 1867*). The Brokhaus Katalog was a publication consisting of 339 pages with interspersed ornamental prints, made as a proceeding for the Paris International Exhibition of 1867.[4] The bowl by Odiot is found on page 123, together with a range of other items made by the same company. An intra-acting transformation is evident, as in the case of the chandelier above. Several actors are employed in this transformation. It is not only the editors, the ornamental prints of two countries, and the outcome in ITT that intra-act with one another; seen in a broader perspective, it can be said that the international exhibitions (world fairs) were contributing actors too.

Further investigation of how and where ornamental prints travelled (Dahlgren 2018) would contribute to a deeper understanding of the design business at this time. In the present contribution, however, the focus must remain on the editing level of popular aesthetics.

The seemingly haphazard manner of choosing illustrations might appear far removed from the popular aesthetic agenda, which demands that people be meticulous in their choices. Hoffstedt does not represent a strong agenda of his own, in his manner of contributing to "the claim of the arts." His editorial work, however, reflects an ongoing and international process to popularize aesthetics, enforced by the commercial market of books and journals. It also reveals ornamental prints to be the most important actors for the transference of popular aesthetics within the design press in Sweden.

However, the presentation of ornamental prints stopped after the first sixty-five issues of ITT. Gradually, popular aesthetics also disappeared from the jour-

4 This was the second of three such catalogues: The first covered the London Exhibition of 1862, and the third contained a selection of illustrations from the first two catalogues.

nal. More than one reason can be suggested for this. It is important though, to point out the fact that if popular aesthetics intra-acted with editorial actors within the design press, it did not seem to intra-act with the reader of the journal. When ITT became more specifically technological and backed off from "claiming the arts," it also became more successful and managed to stay on the market for over eighty years. In any case, there were other journals engaging with popular aesthetics in the 1870s. A look at women's journals tells a different story of how popular aesthetics was implemented in periodicals.

A CULTURAL-HISTORICAL TAKE ON POPULAR AESTHETICS

Tidskrift för hemmet (Journal for the Home, hereafter TFH) entered the market in 1859 and remained the only journal of its kind in Sweden until 1886. The editors were two women known for their emancipatory work: Sophie Adlersparre and Rosalie Olivecrona. Olivecrona left the editorship in 1866, while Adlersparre stayed until 1885, the year the journal was passed on to the women's association Fredrika Bremer Förbundet (a move initiated by Adlersparre). One of the most important agenda items of TFH was women's right to professional work and their right to proper, qualifying education. Particular attention was devoted to craft. Practical knowledge was considered a useful skill in daily life, but was also seen as a potential profession for those women who did not marry.[5]

The first fifteen volumes of TFH consisted solely of texts. The articles focused on political (especially female emancipatory) or popular scientific topics, or had literary content: poems, short stories, and book reviews. In 1873 illustrations entered the journal in the form of pattern sheets. The reason for beginning to publish pattern sheets is not explicitly discussed in the journal. However, a reading of the first fifteen volumes points toward two circumstances as motivating factors: the abovementioned emancipatory agenda of a woman's right to work, with craft being a particularly suitable occupation; and a concurrently developing movement of so-called *home-craft*, visible in society at large.

Towards the end of the 1860s, as a consequence of the aforementioned social reforms, a debate on the waste of labor emerged in the Scandinavian countries. Investigations of folklore lifestyle had resulted in the realization that peasants were

5 This was a relevant concern, since approximately 40 per cent of women older than 15 were unmarried in Sweden at the time. https://www.scb.se/Grupp/Hitta_statistik/Historisk_statistik/_Dokument/Historisk-statistik-for-Sverige-Del-1.pdf, accessed May 31, 2018.

doing very little during the long winters. Several polemical books suggested that people should create handicraft objects during that time, objects that could be sold in those years when the harvest failed. Moreover, according to this debate, crafted objects had a morally beneficial impact on the maker. This debate ended up in a movement arguing for *husflit* or *husslöjd*, here translated as "home-craft."[6]

In its goals and actions, the home-craft movement had strong connections with popular aesthetics. However, the ideal to follow was not the classical philosophy of beauty, but rather old peasant designs and craft knowledge. Home-craft aesthetics sought its roots in regional culture instead of in the (so-called) high cultures of society. This cultural-historical take on popular aesthetics also had German connections, in this case to Cultural History representatives such as the art historian and popular aesthetic writer Jakob von Falke.

Falke argued for a "national home-industry" (*nationale Hausindustrie*) where each region would look specifically towards its local material roots—i.e., for folklore patterns and objects—as inspiration for craft and manufacturing work (Anderson 2009; Houze 2015: 41-70). One of Falke's examples was Sweden. At the 1876 Paris International Exhibition, he had been impressed by the Swedish exposition of folklore dresses and repeatedly reported on the topic in the years to come (Frick 1978). In articles translated in ITT as well as in *Aftonbladet* (a Swedish daily evening post), Falke argued that Sweden was far too dependent on foreign, and especially German, industry. Both Sweden and Norway, he argued, were particularly well positioned to develop a national kind of craft and manufacturing based on folklore patterns and objects. One significant point Falke made was the need to collect patterns and samples for creating a museum, or at least temporary exhibitions, that could spread knowledge of the folklore treasures at hand (Falke 1871a, 1871b).

Notably, the home-craft movement emerged in the Scandinavian countries in the 1860s, which is why Falke was not necessarily a forerunner to Swedish interest in home-craft. Rather, his arguments were connected to a spirit of the time that was

6 As far as we know, there is no generally accepted concept in English for the Scandinavian word husflit (husslöjd is a more specifically Swedish term, but both concepts were used during this time). Another option would be "amateur work," but this concept usually refers to activities carried out for pleasure and as a distraction from professional work (Knott, 2015, pp. xii-xiv). Home-craft is not to be understood as equal to proto-industry. There are similarities, but whereas proto-industry primarily refers to outsourced work, home-craft stems from the work peasants do when agricultural work is not possible. It could be repairing broken objects or creating new items necessary for daily life. The German equivalent would be Hausindustrie.

already present in the Nordic countries. The popular aesthetic agenda Falke argued for nevertheless strengthened the arguments for the home-craft movement.

One person that took Falke's call seriously was Sophie Adlersparre. She paid special attention to the idea of collecting folklore items. Falke might have had manufacturing in mind when suggesting folklore as a source for industry. Adlersparre took the idea a step further by blending her striving for educational and professional possibilities for women with the home-craft agenda and the popular aesthetic notion of ennobling society. She did this by publishing pattern sheets in her journal.

ILLUSTRATIONS AT WORK: THE AGENCY OF PATTERN SHEETS

The first issue of TFH from the year 1873 includes ten embroidery patterns, enclosed as separate sheets (*Fig. 3*). A short article entitled "Förklaring av mönsterplanscherna" (Explanation of the pattern sheets) accompanied the sheets, and in it embroidery techniques and different ways of using the sheets are suggested, as well as an explanation of the origins of the patterns. A certain degree of knowledge of embroidery is needed to understand the suggestions of technique—but even a reader with little background in the subject can understand that no single technique is preferred (TFH 1873: 48). The same goes for the use of the patterns. Several objects found in the home are suggested as suitable: pattern number four could be used on pincushions or padded baskets, and numbers six and seven are suggested as edging for tablecloths, cushions, or padded chairs. They are also said to function well together in the reader's own choice of blending. Pattern number one (top of *Fig. 3*) was considered the most broadly deployable: it was not only suitable for use on different items, but could also be transferred into patchwork, damask, or even woodcarving and sawing.

The next issue of the journal (pp. 100-101) continues to discuss this pattern (pattern number one) by presenting its makers. Several different actors were involved in the making: August Malmström, a professor at the Royal Institute of Art, drew the pattern together with one of his students, "Miss Rysswijk" (Agnes Rijswijk). The "artistic arrangement," i.e., the design and application of the pattern, was made by Hanna Winge. Winge was an artist by profession educated at the Royal Institute of Art (as one of its first female students, 1864-1867), here engaged in textile production. The embroidery was executed by Anna Fleetwood, who had no official education in art or textile but was among the numerous visitors to the public lecture series held in Stockholm by scholars in art and archaeology (Danielson 1991: 93-95). In addition to these directly intra-acting actors, a more distant one also merits attention: the coming international exhibition (world fair) in Vienna, which was the reason for making the pattern in the first place, since the cloth was intended for display there. Finally, TFH and Adlersparre were actors making the pattern part of the periodical network—an act that made it public at the time, and also an act that makes the pattern traceable for us today.

Figure 3: Pattern Sheet, Appendix for Tidskrift för hemmet, *1873: 1*

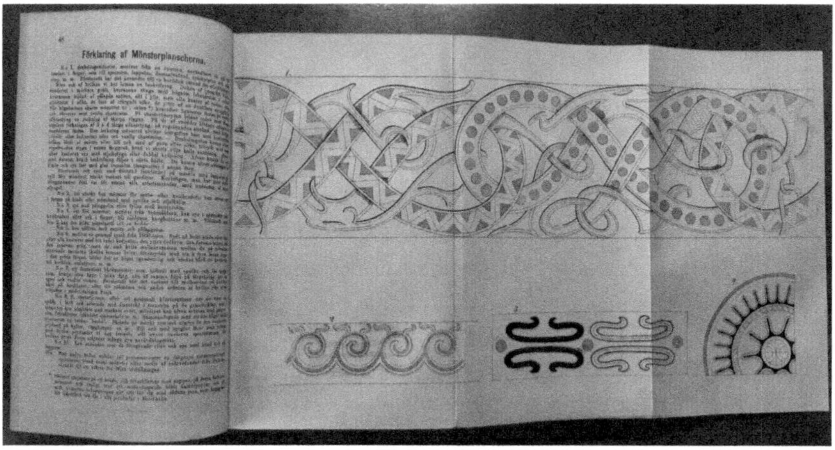

Source: National Library of Sweden; Photo: Elin Manker

When unpacking the actors involved in the making of the pattern, a network of scholars, artists, institutions, and international as well as national exhibitions is revealed. The relation of all these actors could be the object of further study. Here they are left, for now, to simply point toward the network-structure around the pattern-making. When pattern sheets appeared in TFH in 1873, it was not only the first time such illustrations were used in the journal; they were also to be the only kind of illustrations published in it. TFH was therefore not an illustrated journal, in contrast to ITT. Instead, the publishing of pattern sheets has to be understood as a strategic passing of information, necessarily expressed through illustrations. The procedure prior to publication follows a normal design process, with the object-in-the-making intra-acting with different actors along the way: a design brief (an exhibition) starts the process of sketching, followed by design decisions regarding material and technique, the actual making of the item, and the exhibition and potential sale of it. However, in this case, since the pattern was submitted as a rendered pattern sheet in TFH, its agency expanded in other directions, thereby enabling further kinds of usage. Instead of aiming simply to be an exhibited tablecloth, the pattern spread among the reading public with the message of encouraging people to design—or re-design—their own textile objects. Thus, the design process expanded into the public arena.

The decision to include patterns was a successful move for TFH. Patterns were constantly included during the remaining years of the journal's publication. The commentary texts, on the other hand, stopped in 1875. This indicates that the patterns were more important actors than the explanatory texts. The patterns found users among the readers in a way that made them remain as editorial material (in contrast to the corresponding action regarding ornamental prints in ITT).

As early as 1874, Sophie Adlersparre took a step further in her engagement with pattern making by founding the Handarbetets Vänner society (HV, The Friends of Handicraft) together with Hanna Winge and Agnes Rijswijk, among others. Starting as a collector of old textile patterns (especially from the countryside) and as an arranger of the exhibition of those patterns, HV's explicit agenda was to fulfill aims a journal could not. HV's aim with regard to collecting textiles was to convey practical knowledge of old textile techniques. They offered education, and they sold sewing kits as well as embroideries that had already been started (for the customer to finish at home), in addition to finished work; and they furthered their agenda by opening a design bureau in Stockholm (TFH 1874:5: 319). The collecting work and the submitting of patterns in TFH expanded to a significant home-craft business. The collecting work was done by HV but also by the readers of TFH, who were invited to send in samples. The mission of Adlersparre to work for women's rights ended up in a widespread network of popularizing folklore patterns as an aesthetic movement—a work that involved the common female reader and that was most notably mediated by the agency of the pattern sheet.

The pattern sheets in TFH thus intra-acted with the home-craft movement and with the journal as such. The pattern sheets changed the direction of the journal in addition to changing the home-craft movement itself. In TFH (together with HV), the home-craft movement found its most successful branch, showing what home-craft could be "in practice." For the remaining years of the 19th century, textile departments were the most talked-about elements of Swedish participation at the world fairs. Furthermore, the textile home-craft has survived more strongly than other forms of home-craft to the present day. Even today, patterns presented in sloyd magazines continue to echo the kind of folklore patterns that were collected, recorded, and then rendered into new patterns and presented visually by TFH.[7] The practice of hand-craft was transformed into an everyday life practice by Adlersparre and THF through the publication of pattern sheets with a

7 For further reading on home-craft during 20th century in Sweden, cf. Rosenqvist 2007.

popular aesthetic agenda of ennobling the public as a motivating factor, and this practice continues to this day.

CONCLUSION

In this article, our investigation of the popular science has focused on the 19th-century desire to spread aesthetic knowledge outside of academia. This desire had both philanthropic and commercial motives. Even though public lecture series and reading groups were common, having affordable publications was crucial in order to reach a larger number of people. Thus, the emerging mass market of books and magazines was of the utmost importance for 'popular science' as a phenomenon, in the 19th-century sense of the term.

Popularization meant more than merely publishing, however; the texts had to be adapted for a non-academic audience. By looking at the two examples in this article, we can see that the approach taken by the women's journal was the more successful of the two. TFH did more than just present popular aesthetics to its readers: By using pattern sheets, the journal encouraged readers to intra-act with the popular aesthetic movement and thereby take part in it, in a way that ITT did not.

Popular aesthetics meant more than conveying a text-based education, as we learned from Lemcke's definition of the term. It consisted of an undertaking to constantly change one's lifestyle and society as a whole. To contribute to popular aesthetics meant going beyond your present state. Sophie Adlersparre understood this processual approach—popularization had more far-reaching tasks than just to "start spreading the news," as her male editor colleagues were doing.

REFERENCES*

Anderson, Eric (2009): Beyond Historicism: Jakob von Falke and the Reform of the Viennese Interior (diss.), Columbia University.

Andersson-Skog, Lena (2013): "'...att befordra landets förkofvran...' Ekonomiska reformer i brytningstid 1846–1890." In G. Ahlström (ed.), Näringsfri-

* All issues of TFH are accessible at: http://www.ub.gu.se/kvinn/digtid/02/. All issues of ITT are accessible at: http://runeberg.org/tektid/. All issues of Gewerbehalle are accessible at: https://www.deutsche-digitale-bibliothek.de. The Illustrirter Katalog der Pariser Industrie-Ausstellung von 1867 is accessible at: https://digi.ub.uni-heidelberg.de.

hetens tid. Reformer, industriutveckling, kulturklimat, Bromma: Centrum för näringlivshistoria.

Aynsley, Jeremy/Forde, Kate (eds.) (2007): Design and the Modern Magazine, Manchester: Manchester University Press.

Barad, Karen (2003): "Toward an Understanding of How Matter Comes to Matter." In: Signs 28/3, pp. 801-831.

Dahlgren, Anna (2018): Travelling Images. Looking across the borderlands of art, media and visual culture, Manchester: Manchester University Press.

Danielson, Sofia (1991): Den goda smaken och samhällsnyttan. Om Handarbetets Vänner och den svenska hemslöjdsrörelsen (diss.), Stockholm: Nordiska Museet.

Falke, Jacob von (1871a): "Einige Bemerkungen über die Kunstindustrie Schwedens." In: Mittheilungen Des K. K. Österreich. Museums für Kunstgewerbe 6/65, pp. 326-331.

Falke, Jacob von (1871b, March 4): En röst från utlandet i svenska slöjdernas lifsfråga, Aftonbladet.

Frick, Gunilla (1978): Svenska Slöjdföreningen och konstindustrin före 1905 (diss.), Stockholm: Nordiska museet.

Goldman, Paul/Cooke, Simon (eds.) (2012): Reading Victorian Illustrations, 1855-1875. Spoils of the lumber room, Farnham: Ashgate.

Gretton, Tom (2010): "The Pragmatics of Page Design in Nineteenth-Century General-Interest Weekly Illustrated Magazines in London and Paris." In: Art History 33/4, pp. 680-709. DOI: https://doi.org/10.1111/j.1467-8365.2010.00766.x.

Griffiths, Antony (2016): The Print Before Photography. An introduction to European printmaking 1550-1820, London: The British Museum.

Hoffstedt, Wilhelm (1870): "Anmälan." In: Illustrerad Teknisk Tidning 1/1, pp. 1-2.

Houze, Rebecca (2015): Textiles, Fashion, and Design Reform in Austria-Hungary Before the First World War, Farnham and Burlington: Ashgate.

Hultzsch, Anne (2017): "'To the great public': The Architectural Image in the Early London Illustrated News." In: Architectural Histories 5/1: 9, pp. 1-17. DOI: http://dx.doi.org/10.5334/ah.268.

Johannesson, Eric (1980): Den läsande familjen. Familjetidskriften i Sverige 1850-1880 (diss.), Stockholm: Nordiska museet.

Johannesson, Eric (2001): Med det nya på väg (1858-1880). In Den svenska pressens historia II. Åren då allting hände, Stockholm: Ekerlid.

Johannesson, Lena (1982): Xylografi och pressbild. Bidrag till trägravyrens och till den svenska bildjournalistikens historia (diss.), Stockholm: Nordiska museet.

Knott, Stephen (2015): Amateur Craft. History and Theory, London/New York: Bloomsbury Academic.

Latour, Bruno (1994): "On Technical Mediation. Philosophy, Sociology, Genealogy." In: Common Knowledge 3/2, pp. 29-64.

Latour, Bruno (1996): "On actor-network theory. A few clarifications plus more than a few complications." In: Soziale Welt 47, pp 369-381.

Latour, Bruno (2005): Reassembling the social: An introduction to actor-network-theory, Oxford: University Press.

Law, John (1992): "Notes on the Theory of the Actor-Network: Ordering, Strategy and Heterogeneity." In: Systems Practice 5, pp. 379-393.

Lees-Maffei, Grace (2014): Design at Home. Domestic Advice Books in Britan and the USA since 1945, London: Routledge.

Lemcke, Carl (1865): Populäre Aesthetik, Leipzig: E. A. Seeman.

Lemcke, Carl (1868): Populär estetik, Stockholm: L. J. Hiertas förlag.

Lundblad, Kristina (2010): Om betydelsen av böckers utseende. Det svenska förlagsbandets framväxt och etablering under perioden 1840-1914 med särskild hänsyn till dekorerade klotband. En studie av bokbandens formgivning, teknik och relation till frågor om modernitet och materiell kultur (diss.), Lund: Rámus Lunds Universitet.

Nordmark, Dag (2001): Liberalernas segertåg (1830–1858). In Den svenska pressens historia II. Åren då allting hände, Stockholm: Ekerlid.

Norman, H. (2019): "Emigration." In: Nationalencyklopedien. Accessed September 16, 2019. http://www.ne.se/uppslagsverk/encyklopedi/lång/emigration.

Olsen, B. (2003): "Material Culture after Text: Re-Membering Things." In: Norwegian Archaeological Review 36/2, pp. 87-104.

Papanelopoulou, Faidra (ed.) (2009): Popularizing Science and Technology in the European Periphery, 1800-2000, Farnham: Ashgate.

Strömquist, Helena (2010): 'Med coleurt omslag': färgade, dekorerade och tryckta pappersomslag på svensk bokmarknad 1787-1846. En bokhistoria (diss.), Lund: Lunds Universitet.

Uhde, Constant (1870): "Ueber Kronleuchter." In: Gewerbehalle 8/9, pp. 129-132.

Van Remoortel, Marianne (2015): Women, Work and the Victorian Periodical. Living by the Press, Houndmills/Basingstoke/Hampshire: Palgrave Macmillan.

Wassily Kandinsky's Conception of a *Vibration of the Soul*: Art Theory at the Crossroads of Esoteric Literature, Popular Science, and Aesthetics

Beatrice Immelmann

ABSTRACT

Wassily Kandinsky was one of the first artists who produced and exhibited art works with a growing degree of abstraction, from approximately 1910 onwards. His artistic development was accompanied by intense theoretical considerations. The theorizing of a correspondence between artistic production, originating from the artist's *vibration of the soul,* and a resonant aesthetic experience of viewers, was a pivotal aspect of his thinking. While this terminology, in art history, has predominantly been interpreted as a reference to occult and esoteric literature, this article discusses this conception within the wider framework of mutual exchanges between esoteric literature, popular science, art theory, and aesthetics around 1900.

INTRODUCTION

The relation between Wassily Kandinsky's (1866-1944) writings and his artistic oeuvre has always been of major interest to art criticism and art history, dating back to contemporary remarks on the publication of his first book *Über das Geistige in der Kunst* (*On the Spiritual in Art*) in 1911 and that of the *Almanach der Blaue Reiter* (*Blue Rider Almanac*) in 1912.[1] A main focus of research has been the reconstruction of textual sources that influenced Kandinsky's thinking, in order to disentangle the concepts that formed the background of his artistic development. Within this research perspective, scientific, occult, and esoteric literature has been deemed foundational to Kandinsky's art and theory.[2] This interpretation was favored by the statements of the artist himself, who promoted an explanation of his artistic style with references to such sources: In *On the Spiritual,* he referred to well-known personalities from esoteric and occult movements, among them Helena Blavatsky, Rudolf Steiner, Friedrich Zöllner, William Crookes, and Camille Flammarion (Kandinsky 1946: 25-26). In the 1913 autobiographical text *Rückblicke* (*Reminiscences*), Kandinsky took this argumentation one step further and linked the development of his artistic style to the contemporary development of science, and more precisely to atomic research and new insights into the realms of subatomic reality. He stated that, science, given the fact of nuclear disintegration, seemed to be destroyed, hinting at the discovery of radioactivity in 1896. From this, he diagnosed a "dissolution of matter" in contemporary exact and occult science, and, in hindsight, he declared this the starting point for his pursuit of anti-materialist, abstract painting (Kandinsky 2004: 33). Likewise, the "invention of abstract art" from the spirit of esotericism became a popular narrative of art history, starting with Sixten Ringbom's essay *Art in the 'Epoch of the Great Spiritual': Occult Elements in the Early Theory of Abstract Painting*. More specifically, Kandinsky's creation of

1 This is valid for both the positive and negative critiques of Kandinsky's work. In a negative review of the *Blue Rider Almanac*, Tietze (1912) compared Kandinsky's artworks with his written theory and concluded that his theory surpassed his art. In a positive evaluation of Kandinsky's art, Deri (1923: 278-287) explicitly entangles observations on Kandinsky's writings and his art works.

2 Art history research has proposed several influences on Kandinsky's writings. Ringbom (1966; 1972) proposed a strong influence of esoteric and occult literature on Kandinsky; Fineberg (1975) and Weiss (1979) both suggested formative impressions of the art scene in Paris and Munich; Hentschel (2000) argued for the strong influence of Goethe's writings on Kandinsky's work; Schmidt (2002) investigated Kandinsky's scientific references; and Bushart (2016) recently suggested empirical psychology as the root of Kandinsky's aesthetics. And Wünsche (2016), in particular, investigated the effect of psychophysiological factors on the processes of artistic creation and visual perception in the writings of Kandinsky and František Kupka; according to Wünsche, Kandinsky is indebted to Wundt's psychophysical parallelism, since he distinguished between the purely physical effects of color and the psychic effects, with the latter leading to a vibration in the soul.

abstract paintings between 1910 and 1913 has been interpreted to be the result of his knowledge of esoteric and occult literature (Ringbom 1966: 399-405).

In his writings, Kandinsky frequently used the notion *Seelenvibration* (*vibration of the soul*). Accordingly, the origin of this concept has been sought in the research on Kandinsky's textual sources mentioned above. The concept has predominantly been interpreted as a reference to esoteric literature and is central to Ringbom's (1970: 121) reasoning: "Now it is evident that these ideas come straight from Theosophy where the concept of vibration occupies a central place as a cosmic principle."[3] Only recently has research emphasized that the esoteric and occult movements around 1900 did not establish entirely novel programs but drew on science, philosophy, and mythology (cf. Zander 2007; Hanegraaff 2012; Neugebauer-Wölk/Geffarth/Meumann 2013). Art history has therefore criticized one-sided interpretations of Kandinsky's oeuvre and has fostered an understanding of his theory as a representation of "the interrelatedness of many diverse and even contradictory positions" (Short 2010: 1). In this view, Rosenberg (2013: 584) pointed out that Annie Besant and Charles Webster Leadbeater, both active in the Theosophical Society Adyar, employed the notion of vibrations to specifically describe the *effects of colors and lines* in their 1905 book *Thought-Forms*. Rosenberg showed that esoteric literature has drawn on art theory, where the aesthetic effects of basic artistic means have been discussed since the 18th century—for example, in Goethe's color theory (Rosenberg 2007: 309; 2013: 594-600; cf. Brinkmann et al. 2018).

3 Sources that have been discussed in this context are lectures by Rudolf Steiner, which Kandinsky attended in March 1908, and Annie Besant and Charles Webster Leadbeater's book *Thought-Forms* (Long 1980: 32-34). The book also provides important evidence to supplement Ringbom's argumentation, as it contains abstract color and line drawings of the human aura. Kandinsky owned the 1908 German translation *Gedankenformen*, today accessible in the *Fonds Kandinsky*, Centre Pompidou, Paris. The copy is dedicated to "Kandinsky in cordial friendship," but has no further notations. Evelin Priebe relativized Ringbom's interpretation, stating that vibrations have a history as a terminus technicus in the optical and acoustic fields, and that they obtained a psychological connotation around 1900. Priebe (1986), however, held the opinion that Kandinsky alluded to the esoteric belief of a vibratory cosmos with his theory. On the other hand, Georges Roque (2002; 2003) analyzed the notion of vibrations in the wider framework of science and philosophy around 1900 and came to the conclusion that it mostly had no spiritual connotation. However, he specifically excluded Kandinsky from this claim, stating that he used vibrations as a pseudoscientific metaphor to describe the effects of colors (Roque 2002: 185-187).

Such insights challenge the theses of esoteric influences on art and art theory. In the case of Kandinsky in particular, the reconstruction of a one-sided conceptual imprinting on the artist no longer seems to be fruitful. This paper therefore focuses on the wider framework of the concepts of vibrations and on Kandinsky's motives for implementing the terminology in his theory. The first section will briefly discuss the history of the concept of vibrations as a model of aesthetic perception. The second section will discuss this terminology within the framework of shared knowledge and popular science in Kandinsky's network around 1900. The final, summarizing section will discuss the artist's motives for the implementation and dissemination of this terminology and will question the purpose of such references in Kandinsky's writings.

VIBRATIONS AS A MODEL OF AESTHETIC PERCEPTION

The conception of aesthetic perception and aesthetic experience through nervous vibrations or *vibrations of the soul* has a tradition dating back to the beginning of the 18th century. It constitutes a recourse on the scientific model of sensual perception. Physical phenomena, such as light and sound, as well as their perception, were considered derived from the properties of the vibrational motions of matter, as in, for example, Nicolas Malebranche's *De la Recherche de la Vérité* (*The Search after Truth and Elucidations*), from 1675, or Isaac Newton's *Opticks,* from 1704. Since then, this conception has remained a recurring model to explain the physical, physiological, and psychological aspects of the perception of light and sound, as in the writings of Thomas Young, Hermann von Helmholtz, Gustav Theodor Fechner, and Charles Féré. This conception was quickly adopted by aesthetics and art theory, since it provided a model for the *aesthetic* perception of auditory and visual works. Already in 1712, the Swiss philosopher Jean-Pierre de Crousaz, who became acquainted with Malebranche during a stay in Paris in the 1680s, had integrated this model in his music aesthetics in the *Traité du Beau* (*Treatise on Beauty*). Crousaz (1715: 172-173) suggested that sound is propagated through the vibrating particles of the air, whose movement is transmitted to the auditory nerve. Depending on the modalities of these vibrations, sound causes a feeling, which results in aesthetic pleasure or displeasure. In 1725, the French Jesuit Louis Bertrand Castel, who had published on optics and acoustics himself and who was familiar with Malebranche's and Newton's writings, adopted this model for the aesthetic experience of visual stimuli. Castel proposed the construction of a *Clavecin pour les yeux* (*Ocular Harpsichord*), such that when a key was struck, the instrument showed a specific color. By adding succession, harmony, and variety to colors, the instrument would, according to Castel (1725: 2563-2565), add a "musical character" to colors. The concept of vibrations built the scientific foundation for Castel's argument (Rosenberg 2013: 599). Providing a physical and neurological explanation for an analogy of light and sound, or seeing and hearing, the perception—and the appreciation—of visual and auditory stimuli were thus reduced to the same principle. This simplified theory apparently intrigued artists, musicians, and writers, and gave significant impulses to link the aesthetic perception of music and painting, or sound and color, for the next 200 years. In this vein, the terminology became a popular scientific explanation for the processes of perception or perceptual physiology in art theory and aesthetic writings, such as those of Edmund Burke and Charles Baudelaire. This terminology was especially used in theoretical considerations by those artists who produced and exhibited art works with a growing degree of abstraction from approximately 1910 on—not only Kandinsky, but also František Kupka, Robert Delaunay, Georges Vantongerloo, and many others (Roque 1999; 2003; 2007).[4]

Kandinsky in particular theorized a correspondence between artistic production, originating from the artist's vibration of the soul, and a resonant aesthetic experience of the viewer, via an art work. In this process, the artist would compose artistic means, such as colors, forms, and lines, in order to convey the subtle aesthetic experience of a vibration of the soul to the beholder. Kandinsky used this notion of vibrations for the first time in a sketch of his color theory, preserved in a collection of notes titled *Definieren der Farben* (*Definition of Colors*), dated as early as 1904.[5] In these notes, he pleaded for the use of colors according to their psychological effect on a viewer instead of subordinating them to objects. Kandinsky (2007: 253) stated that certain color combinations always have the same effect; they "cause the same psych.[ological] vibration." This manuscript, however, does not discuss any esoteric beliefs. Instead, Kandinsky

4 Unfortunately, the nuances of these conceptions and the complexity of this discourse cannot be discussed in this paper. The continuity of this argumentative pattern is the subject of the author's PhD thesis project, "Visuelle Resonanz. Vibrationen als wahrnehmungsphysiologische Argumentationsfigur in Kunsttheorie und Ästhetik 1725-1925," funded by the University of Vienna, the Vienna Science and Technology Fund (WWTF, as part of the project CS15-036), the Vienna Doctoral School in Cognition, Behavior and Neuroscience, and the German Center for Art History, Paris.

5 These notes are stored at the Gabriele Münter and Johannes Eichner Foundation, Munich. They have been organized and dated by Hans Roethel (Kandinsky 2007: 249-258).

considered the connection of color, in the sense of pigment, to specific psychological qualities, which he ascribed to the pigments through association. These notes thus demonstrate Kandinsky's interest in a correspondence between physical laws and cognitive responses.

VIBRATIONS IN A FRAMEWORK OF ESOTERIC LITERATURE, POPULAR SCIENCE, AND AESTHETICS

Christopher Short (2010) described Kandinsky's writing as a process of adopting and adapting diverse sources, which resulted in the constant development of, and constant changes in, his arguments, leading even to contradictory positions. This means that it is extremely difficult to arrive at clear statements about his theory. Furthermore, due to the abundance of concepts Kandinsky referred to, any assumption of direct influence on his artistic oeuvre or theory must remain speculative. It seems, therefore, to be more fruitful to investigate Kandinsky's theory as part of a complex framework of the mutual exchange of popular knowledge.

The notion of vibrations was indeed commonly used in esoteric literature and was a key concept in the texts to which Kandinsky referred in *On the Spiritual.* Helena Blavatsky, co-founder of the Theosophical Society in 1875, frequently employed the term in her writings to express a continuous motion of atoms permeating matter and space. Blavatsky, however, explicitly endeavored for a thematic continuity of the academic sciences. In her articles, she regularly discussed the insights of contemporary science and occultism and brought up the differences and similarities in both research perspectives. The main angle of her critique is the materialism of science. While science interprets light and sound as different modes of motion of vibratory matter, they are, according to Blavatsky (1891: 14), "*immaterial* perceptive effects." Yet, Blavatsky (ibid: 20) stressed that "whether as substance, matter or effect, sound and light can never be divorced from their modes of manifesting through *vibrations*—as the whole subjective or occult nature is one everlasting perpetual motion of VORTICAL *vibrations.*"[6] As was the case with most authors engaged in this kind of esoteric discourse, she particularly underlined the analogy of sound and color as vibratory phenomena. In the article *Kosmic Mind*, her explanation of the processes of seeing and hearing is identical to those of science. Blavatsky repeats a typical scientific statement on light and sound perception:

6 Emphasis by Blavatsky.

"Vibrations passed along the fibres of our optic nerves to the brain reach our perceptions through our consciousness as sensations of light and color; vibrations affecting our consciousness through our auditory organs strike us as sounds; all our feelings, through whichever of our senses, are due to nothing but motions."

She then adds the following:

"Such are the teachings of physical Science, and such were in their roughest outlines those of Occultism, aeons and millenniums back. The difference, however, and most vital distinction between the two teachings, is this: official science sees in motion simply a blind, unreasoning force or law; Occultism, tracing motion to its origin, identifies it with the Universal Deity, and calls this eternal ceaseless motion—the 'Great Breath.'" (Blavatsky 1890: 93)

Blavatsky obviously did not pretend that there were no differences between her doctrine and science per se on this question; she merely emphasized that occultism knew it first and is able to show a causality of these vibrations in the divine. The chemist William Crookes, who, in 1896, was the president of the Society for Psychical Research, published a *table of vibrations* in which he reduced radiation, heat, electricity, sound, and light to different frequencies of vibrations. In Crookes' view, the different sensual perceptions were a result of the different magnitudes of wavelengths or vibrations (Henderson 2002: 132). The same table was popularized by Camille Flammarion (1900: 16) in his book *L'Inconnu* (*The Unknown*). Additionally, Flammarion reproduced large parts of a discourse on understanding physical phenomena as successive octaves of vibrations and stated that "every sensation corresponds to a vibration in the brain, to a movement of the cerebral molecules, just as every idea does" (ibid: 237). Following from this, it seems that esoteric and occult literature had adopted a common model of sensual perception from the 17th century, but expanded it to include psychic phenomena and the conceptualization of *extra*sensory perception.

In another footnote, Kandinsky refers to the essay *Les lois d'harmonie de la peinture et de la musique sont les mêmes* (*The Laws of Harmony in Painting and Music are the Same*) by the French artillery lieutenant, artist, and composer Henri Rovel (1908). Rovel was probably familiar with Flammarion's popularization of Crookes' table, and he essentially reintroduced it in the aesthetic discourse of sound-color analogies when he suggested that the laws of harmony regarding sound for music and color for painting are basically the same. Rovel examined differing theories of color vision from the preceding years. He examined the Young-Helmholtz theory, which hypothesized three types of photoreceptors in the eye, each of which responds to a particular range of wavelength, and attempted to reconcile this approach with both the seven colors of the spectrum and the empirical application of color pigments. On the other hand, he explained the discovery of rhodopsin (or visual purple) and the finding that this light-sensitive protein reacts differently according to the wavelength of light. This scientific background is, however, merely Rovel's starting point for attempting to transfer the principles of musical order to color composition and to mathematically determine colors in comparison to a musical scale, according to the colors' wavelengths. Rovel saw light and sound as vibratory movements in which the type of sensation depends only on the receiving organ and the intensity of vibrations. Rovel's essay not only suggests there are cosmic principles of harmony; it is, at the same time, a popular scientific synopsis of contemporary optics for artists. Thus, it becomes obvious that science, occult science, esoteric beliefs, art theory, and aesthetics formed a complex, oscillating network that was characterized by mutual ideological takeovers.

By the time Kandinsky was writing his first drafts of a color theory, he was already a close friend of the artists Marianne Werefkin and Alexej Jawlensky, even though their relationship might not yet have been as close as it would later become during their intense collaboration in Murnau in 1908—a collaboration that also included Gabriele Münter. Werefkin had stopped painting during the years 1901 to 1905 due to a personal crisis and was exclusively occupied with theoretical considerations, the *Lettres à un Inconnu* (*Letters to an Unknown Man*), art education, and the establishment of artistic networks in her salon (Fäthke 2001: 61).[7] Werefkin systematically listed the books she had read, including the date of reading, annotations, and commentaries (Folini 1999: 91). From these lists, it is clear that she had excellent knowledge of the literature on physiology, psychology, and aesthetics. These notes have only hesitantly been considered by research to disentangle Werefkin's theory,[8] but they have not at all been reviewed with the aim of disentangling the complex framework of aesthet-

7 Werefkin's influence is visible in the treatise *Grundprobleme der Malerei* (*Fundamental Problems in Painting*) by Rudolf Czapek. Czapek had studied in the private art school of Werefkin's partner Alexej Jawlensky in 1906/07. Werefkin herself was probably responsible for theoretical training at the school, and it is believed that Czapek published his book as an immediate reaction to input from Werefkin (Fäthke 2001: 95). Czapek's book, which is also part of Kandinsky's estate in Paris, shows some resemblances to *On the Spiritual,* yet it is not nearly as complex as Kandinsky's theory and cannot be seen as its role model. Kandinsky wrote a few notes in his copy of Czapek's book.

8 Folini (1999) contrasts only a few quotations by Werefkin with the scientific sources.

ic, scientific, and esoteric content that informed artists in the development of their artistic theories in the early 20th century. Among these notes appear names such as David Sutter, Sully Prudhomme, and Charles Féré, who all have a connection to German psychophysics as established by Hermann von Helmholtz and Gustav Theodor Fechner.[9] All of these authors had tried to develop aesthetic principles according to the contemporary knowledge of physiology and psychology. Fechner (1876: 79) had declared the proportions of light and sound vibrations, propagated through commensurable nervous vibrations, to be foundational for aesthetic pleasure. David Sutter (1870) developed an aesthetic theory of painting in analogy to music in his book *Philosophie des beaux-arts appliquée à la peinture* (*Philosophy of the Fine Arts Applied to Painting*). This included an analogy of the processes of perception. Prudhomme devoted his 1883 book *L'Expression dans les beaux-arts* (*Expression in the Fine Arts*) to his friend Georges Guéroult, who, in 1877, translated Helmholtz's *Die Lehre von den Tonempfindungen als physiologische Grundlage für die Theorie der Musik* (*On the Sensations of Tone as a Physiological Basis for the Theory of Music*) into French. Guéroult himself had theorized specific translation procedures from the feelings of an artist to the visible signs in an artwork and to the feelings elicited in a beholder (Roque 2011: 166). Similarly, Prudhomme (1898: 55, 191) reflected on the communicative processes between perceived objects and perceiving subjects. He considered perception to be a continuous stream of vibrating atoms between the surface of an object and human consciousness. In the cases of seeing and hearing, this communication was conveyed by the vibrating media of air and ether (ibid: 53). Similarly to Fechner, Prudhomme (ibid: 5-6) stated that appreciation is derived from the ratio of vibrations. An artist, in Prudhomme's (ibid: 410) definition, therefore had to have a specific sensibility for harmonious vibrations and had to be able to communicate a sympathetic feeling to the beholder. Féré's book *Sensation et mouvement: Études expérimentales de psycho-mécanique* (*Sensation and Movement: Experimental Studies on Psychomechanics*) also investigated sound and color perception. Using a graphic recording device, Féré was able to measure the physical reactions to stimuli in the fingers of participants under hypnosis and to distinguish their different reactions to different colors. With these experiments, Féré (1887: 42) tried to build a physiological basis for Goethe's theory of cold and warm colors. He discerned that cold colors trigger slow reactions, and warm colors elicit fast reactions or accelerated nervous activity. Féré (1887: 45) stated that vibrations have to be considered a com-

9 Werefkin made these lists between 1883 and 1907. They are stored at the *Fondazione Marianne Werefkin,* Ascona.

mon arousal unit for hearing and seeing. These authors represent just a few examples of esoteric and (empirical) aesthetic literature from this extensive discourse. However, regarding vibrations, it is obvious how deeply intertwined esoteric literature, popular science, and aesthetics were at the turn of the 20th century.

Kandinsky was, from his earliest writings onward, concerned with the reciprocal relationship between artist and beholder and the question of how the artist's intention could be transmitted to the viewer. He verbalized these considerations through a variety of metaphors in several manuscripts, describing the relation as a *channel* (Kandinsky 1998: 150), a *bridge* (Kandinsky 2007: 404), a *path* (ibid: 409), a *sender-receiver relationship* (Kandinsky/Marc 1979: 146), and *telegraphy* (Kandinsky 2007: 292-293). These metaphors all share two commonalities: They all describe a *direct connection between two points*, and that connection is always *employed for transport or transmission*. In this conception, the artist would represent one point, the beholder the other point, with the two being connected through vibratory movements which transmit, via a painting, an abstract message. In this vein, Kandinsky's considerations are closely related to the description of physiological and psychological processes in the aesthetics and art theory of the 19th and early 20th centuries, as discussed above.

Definition of Colors obviously stems from a period during which Kandinsky generated the early notes of his color theory, which mirror the artist's early attempts to define the aesthetic effects of colors on the psyche. Kandinsky postulated that these effects are caused by associations, an explanation that he complemented with the phenomenon of *synesthesia* in subsequent years to describe a more direct impact of an artwork on the beholder (Kandinsky 1946: 41-42). As an example, Kandinsky associated bright yellow with the shrill and loud twittering of canaries and the sour taste of a lemon. It is remarkable that Fechner had suggested exactly the same association in his *Vorschule der Ästhetik* (*Preparatory School of Aesthetics*) (Fechner 1876: 101; Brinkmann et al. 2018: 15-16). Provided that the dating of *Definition of Colors* to approximately 1904 is correct, it is very unlikely that Kandinsky learned of the idea of vibrations from esoteric literature, since such an influence is only assumed from 1906 to 1908 onwards.[10] Around 1900, the conception of vibrations was widely used, not only in esoteric

10 This includes Rovel's text. During a stay in Paris in 1906/1907, Kandinsky started to list and comment on the occult literature he had read. Kandinsky's annotations from the writings of Rudolf Steiner are dated 1908/1909 (Grund 2002: 164). While these annotations show his interest in Steiner's philosophy, no direct relation to Kandinsky's writings is apparent.

literature, but even more prominently in aesthetics and art theory. It was a popular scientific model of sensory perception and of emphasizing a unity of physical and psychological phenomena in the different senses. As such, it has been of great interest to those artists and aesthetic theorists who developed theories of aesthetic experience as *embodied* experience since the early 18th century (cf. Trower 2012). The conception, therefore, is part of a framework characterized by multilayered, interdisciplinary exchanges between science, esoteric belief, aesthetics, and art theory. While Kandinsky's color theory clearly differed from the ideal of color harmony suggested by other authors such as Rovel, he employed a typical argumentative pattern of aesthetic perception, with which he tried to conceptualize the complex relation of artistic creation and aesthetic experience that was no longer bound to recognizable objects in a painting.

Short noted that the placement of references to esoteric literature in the footnotes of *On the Spiritual* is significant as to the role of spiritualism in the book, in that they serve as a supplement to Kandinsky's understanding of *spirit,* without any commitment to one specific understanding of the term (Short 2010: 22). This suggests that Kandinsky used such references as additional post-hoc legitimations of his theorizing and not necessarily as the material of a substantial analysis. But what was the purpose of such conscious references to the given frameworks of popularized knowledge?

KANDINSKY'S MOTIVES FOR THE IMPLEMENTATION AND DISSEMINATION OF POPULAR SCIENCE TERMINOLOGY

Kandinsky's intention was not the interpretation of science for a general audience, which is the content of several other contributions in this book. However, he still might have *had* an intention in his writings. Armin Zweite interpreted Kandinsky's publications as a reaction to the negative criticism of the increasing incomprehensibility of his paintings, from 1909 on. Zweite (1982: 136, 142) suggested that, while the evidence no longer lay in the pictures, the artist's comments became more important.[11] In this vein, Felix Thürlemann (1986: 38-42) suggested that Kandinsky's self-interpretative texts in particular were intended to improve the competence of the critiques of his work as well as the un-

11 Kandinsky probably knew most of the critiques, as he had ordered a service to send press texts to his address (Zweite 1982: 137). In the text *Selbstcharakteristik* (1919) he refers to offensive critiques written between 1908 and 1911 (Thürlemann 1986: 47).

derstanding of the general public. Thürlemann asserted that art criticism, between 1910 and 1913, lacked the discursive means for coping with Kandinsky's emerging abstract style (Thürlemann 1986: 42). Additionally, Barbara Mackert-Riedel (2002: 18) pointed out that Kandinsky's unpublished notes have to be interpreted as a soliloquy on artistic problems. Thus, it seems reasonable to include the conception of vibrations in these considerations.

In this article, it has been shown that the theorizing of aesthetic experience is a key concept of Kandinsky's oeuvre. Considerations of the processes of perception seem to have been a fundamental aspect of coping with the pictorial phenomena he was preoccupied with. In his theorizing of a development towards an abstract style, he was apparently grasping with every concept that would help him in the struggle of clarifying the inner logic of his artistic ideas. Similarly to Kandinsky, Werefkin grappled with the given popular scientific literature and developed a theory of *artistic expression* (Werefkin 1960), although she never published her thoughts on the subject. Simultaneously, references to popular science models offered an argumentative or apologetic backup that, in part, might even have had a strategic purpose. This also explains why Kandinsky's theoretical program is so highly syncretistic. Kandinsky's publications seem to have had an explanatory purpose; he had to explain abstract art to a public that had never seen it before. It can therefore be assumed that Kandinsky tried to push his theory of an abstract pictorial idiom through the general public.[12] With this agenda, the artist employed a language that was rich in metaphors, analogies, comparisons, and specific references to the popular sciences. In this pursuit, he developed a distinct sense of mission. However, even though Kandinsky's theory did not improve the understanding of his artworks and did not bring about more positive critiques, his aesthetic concepts rapidly spread into art criticism and critical literature on art (Thürlemann 1986: 52). Critiques quickly began to describe colors as "vehicles of sensations" that convey a wonderful resonance (Tietze 1912: 547, 549) and to emphasize the goal of dismantling psychological processes (Einstein 1926: 195). Kandinsky's theory, therefore, can also be understood as a medium to communicate accessible metaphors for the effects of his art and to disseminate his ideas into popular discourse.

12 Alexandra Axtmann's article in this volume discusses a similar example of boosting art theory through the awareness of the general public.

REFERENCES

Blavatsky, Helena Petrovna (1890): "Kosmic Mind." In: Lucifer 6/32, pp. 89-100.

Blavatsky, Helena Petrovna (1891): "The Substantial Nature of Magnetism." In: Lucifer 9/49, pp. 8-20.

Brinkmann, Hanna/Boddy, Jane/Immelmann, Beatrice/Specker, Eva/Pelowski, Matthew/Leder, Helmut/Rosenberg, Raphael (2018): "Ferocious Colors and Peaceful Lines: Describing and Measuring Aesthetic Effects." In: Wiener Jahrbuch für Kunstgeschichte 65, pp. 7-26.

Bushart, Magdalena (2016): "Die Expressionisten und die Formfrage." In: Hans Aurenhammer/Regine Prange (eds.), Das Problem der Form: Interferenzen zwischen moderner Kunst und Kunstwissenschaft, Berlin: Gebr. Mann, pp. 238-257.

Castel, Louis-Bertrand (1725): "Clavecin pour les yeux: Avec l'art de Peindre les sons, & toutes sortes de Pieces [*sic*] de Musique." In: Mercure de France 1725, pp. 2552-2576.

Crousaz, Jean-Pierre de (1715 [1712]): Traité du Beau [...] (2nd ed.), Amsterdam: L'Honoré.

Deri, Max (1923 [1919]): Die Malerei im XIX. Jahrhundert: Entwicklungsgeschichtliche Darstellung auf psychologischer Grundlage, Vol. 1 (2nd ed.), Berlin: Paul Cassirer.

Einstein, Carl (1926): Die Kunst des 20. Jahrhunderts, Berlin: Propyläen.

Fäthke, Bernd (2001): Marianne Werefkin, Munich: Hirmer.

Fechner, Gustav Theodor (1876): Vorschule der Ästhetik, 2 Vols., Leipzig: Breitkopf & Härtel.

Féré, Charles (1887): Sensation et mouvement: Études expérimentales de psycho-mécanique, Paris: Alcan.

Fineberg, Jonathan David (1975): Kandinsky in Paris: 1906-1907, Ann Arbor: UMI Research Press.

Flammarion, Camille (1900): L'Inconnu – The Unknown, New York: Harper & Brothers.

Folini, Mara (1999): "Die literarischen Quellen in den Schriften Marianne Werefkins – eine kritische Analyse ihrer Lektüre." In: Margarethe Jochimsen/Barbara Weidle (eds.), Marianne Werefkin: Die Farbe beisst mich ans Herz, Bonn: Verein August-Macke-Haus, pp. 89-99.

Grund, Uta (2002): Zwischen den Künsten: Edward Gordon Craig und das Bildertheater um 1900, Berlin: Akademieverlag.

Hanegraaff, Wouter Jacobus (2012): Esotericism and the Academy, Cambridge, UK: Cambridge University Press.

Henderson, Linda Dalrymple (2002): "Vibratory Modernism: Boccioni, Kupka, and the Ether of Space." In: Bruce Clarke/Linda Dalrymple Henderson (eds.), From Energy to Information: Representation in Science, and Technology, Art, and Literature, Stanford: Stanford University Press, pp. 126-149.

Hentschel, Barbara (2000): Kandinsky und Goethe: Über das Geistige in der Kunst in der Tradition Goethescher Naturwissenschaft, Berlin: Wissenschafts-Verlag.

Kandinsky, Wassily (1946): On the Spiritual in Art (ed. and transl. by Hilla Rebay), New York: Solomon R. Guggenheim Foundation.

Kandinsky, Wassily (1998): Über das Theater – Du Theatre (ed. by Jessica Boissel), Cologne: DuMont.

Kandinsky, Wassily (2004): Autobiographische Schriften (ed. by Hans Konrad Roethel, and Jelena Hahl-Koch), Bern: Benteli.

Kandinsky, Wassily (2007): Gesammelte Schriften 1889-1916: Farbensprache, Kompositionslehre und andere unveröffentlichte Texte (ed. by Helmut Friedel), Munich: Prestel.

Kandinsky, Wassily/Marc, Franz (1979): Der Blaue Reiter (facsimile of the 1912 ed.), Munich: Piper.

Long, Rose-Carol Washton (1980): Kandinsky: The Development of an Abstract Style, Oxford: Clarendon Press.

Mackert-Riedel, Barbara (2002): Wassily Kandinsky über eigene Bilder. Zum Problem der Interpretation moderner Malerei, Weimar: VDG.

Neugebauer-Wölk, Monika/Geffarth, Renko/Meumann, Markus (eds.) (2013): Aufklärung und Esoterik. Wege in die Moderne, Berlin: De Gruyter.

Priebe, Evelin (1986): Angst und Abstraktion. Die Funktion der Kunst in der Kunsttheorie Kandinskys, Frankfurt/Main: Lang.

Prudhomme, Sully (1898): Oeuvres de Sully-Prudhomme. Prose (1883). L'Expression dans les Beaux-Arts, Paris: Alphonse Lemerre.

Ringbom, Sixten (1966): "Art in the 'Epoch of the Great Spiritual': Occult Elements in the Early Theory of Abstract Painting." In: Journal of the Warburg and Courtauld Institutes 29, pp. 386-418.

Ringbom, Sixten (1972): The Sounding Cosmos, Abo: Akademi.

Roque, Georges (1999): "Les vibrations colorées de Delaunay: Une des voies de l'abstraction." In: Centre Georges Pompidou/Pascal Rousseau (eds.), Robert Delaunay 1906-1914: De l'impressionnisme à l'abstraction (ex. cat.), Paris: Centre Georges Pompidou, pp. 53-64.

Roque, Georges (2002): "¿Qué onda? La Abstracción en el arte y la ciencia." In: Peter Krieger (ed.), Arte y Ciencia: XXIV Coloquio internacional de Historia del Arte. Mexico-City: Universidad Nacional Autónoma de México, pp. 169-335.

Roque, Georges (2003): "Ce grand monde des vibrations qui est à la base de l'univers." In: Musée d'Orsay/Serge Lemoine (eds.), Aux origines de l'abstraction. 1800-1914 (ex. cat.), Paris: Réunion des Musées Nationaux, pp. 50-67.

Roque, Georges (2007): "Vantongerloo: Couleurs et vibrations." In: Jean-Etien Grislain (ed.), Vantongerloo. Un pionnier de la sculpture moderne (ex. cat.), Paris: Gallimard, pp. 67-75.

Roque, Georges (2011): "Das Universum der Empfindungen: Eine Parallele zwischen der physiologischen Optik und der Malerei." In: Werner Busch/Carolin Meister (eds.), Nachbilder: Das Gedächtnis des Auges in Kunst und Wissenschaft, Zürich: Diaphanes, pp. 155-170.

Rosenberg, Raphael (2007): Turner – Hugo – Moreau: Entdeckung der Abstraktion (ex. cat.), Munich: Hirmer.

Rosenberg, Raphael (2013): Die Kartographie der Aura aus dem Geist der Wirkungsästhetik: Synästhesie und das Verhältnis von Kunst und Esoterik um 1900. In: Monika Neugebauer-Wölk/Renko Geffarth/Markus Meumann (eds.), Aufklärung und Esoterik. Wege in die Moderne, Berlin: De Gruyter, pp. 583-604.

Rovel, Henri (1908): "Les Lois d'Harmonie de la Peinture et de la Musique sont les mêmes." In: Les Tendances Nouvelles, 3/35 (March), pp. 721-729.

Schmidt, Christiane (2002): "Vassily Kandinsky and Modern Physics: An Alternative Interpretation of the Abstract Art Pioneer's Work Based on an Analysis of the Scientific Books from his Personal Library." In: Peter Krieger (ed.), Arte y Ciencia: XXIV Coloquio internacional de Historia del Arte, Mexico-City: Universidad Nacional Autónoma de México, pp. 313-335

Short, Christopher (2010): The Art Theory of Wassily Kandinsky, 1909-1928: The Quest for Synthesis, Bern: Peter Lang.

Sutter, David (1870): Philosophie des beaux-arts appliquée à la peinture, Paris: Vve. A. Morel et cie.

Thürlemann, Felix (1986): Kandinsky über Kandinsky. Der Künstler als Interpret eigener Werke, Bern: Benteli.

Tietze, Hans (1912): "Der Blaue Reiter." In: Kunst für Alle: Malerei, Plastik, Graphik, Architektur 27/23, pp. 543-550.

Trower, Shelley (2012): Senses of Vibration: A History of the Pleasure and Pain of Sound, New York: Continuum.

Weiss, Peg (1979): "Kandinsky, Wolfskehl und Stefan George." In: Castrum Peregrini 138, pp. 26-51.

Werefkin, Marianne (1960): Lettres à un Inconnu – Briefe an einen Unbekannten 1901-1905 (ed. and transl. by Clemens Weiler), Cologne: DuMont.

Wünsche, Isabel (2016): "Wassily Kandinsky and František Kupka: Between Metaphysics and Psychophysics." In: I. Wünsche and W. Gronemeyer (eds.),

Practices of Abstract Art: Between Anarchism and Appropriation, Cambridge, UK: Cambridge Scholars, pp. 11-29.

Zander, Helmut (2007): Anthroposophie in Deutschland. Theosophische Weltanschauung und gesellschaftliche Praxis 1884-1945, 2 Vols., Göttingen: Vandenhoeck & Ruprecht.

Zweite, Armin (1982): "Kandinsky zwischen Tradition und Innovation: Bemerkungen zur Rezeption seines Werkes in München und zu einem Bildmotiv." In: A. Zweite (ed.), Kandinsky und München: Begegnungen und Wandlungen 1896-1914, Munich: Prestel, pp. 134-177.

Visual Nature Metaphors of Cybernetics in Popular Science and the Arts

Lena Trüper

ABSTRACT

Cybernetics propelled the development of computer science after the Second World War on a technical level. It also triggered a profound paradigm shift in the humanities: By means of the feedback loop, a new concept of the human subject was established, but also a new understanding of "nature" as an all-encompassing ecosystem. This essay explores how this paradigm shift was communicated by (visual) nature metaphors in artworks and films since the 1950s such as *A Communications Primer* (Charles and Ray Eames, 1953), as well as in major popular scientific writings on cybernetics, like Norbert Wiener's *Cybernetics: Or Control and Communication in the Animal and the Machine* (1946), Gregory Bateson's *Steps to an Ecology of Mind* (1968), and Humberto Maturana and Francisco Varela's *The Tree of Knowledge* (1984). As the visual heritage of cybernetics, visual nature metaphors still pervade popular network culture today.

THE VISUAL HERITAGE OF CYBERNETICS

In 2016 Google created a commercial for the "Google Arts & Culture" project called "Cultural Awakening. How Digitalization Opens up New Worlds."[1] Alt-

1 Ger. "Aufbruch Kultur. Wie uns die Digitalisierung neue Welten eröffnet." by Google inc., Mountain View, USA 2018.

hough the commercial promotes the digitalization of artworks, it shows neither a computer nor a display nor any other means of technical support (*Fig. 1*).

Figure 1: Cover Illustration of the Google Arts & Culture Project

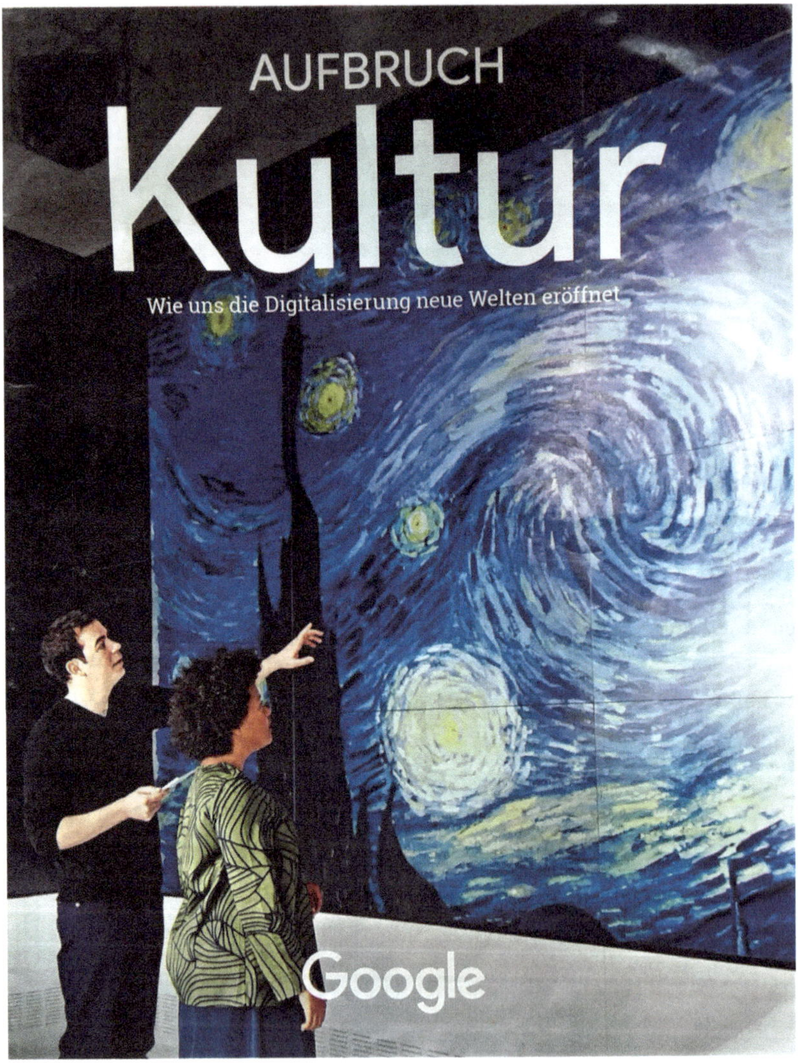

Source: ©2018 Google LLC, used with permission. Google and the Google logo are registered trademarks of Google LLC

Instead, two people are shown looking at the painting "The Starry Night" by Vincent Van Gogh.[2] At first glance, this seems to be a common exhibition situation. However, the photographic imagery contains several subtle allusions to digitalization: Not only is the coarse brushwork of the Van Gogh painting reminiscent of the pixel structure in digital images, but the posture of the two people also indicates access to digital worlds. The person on the left gesticulates, pointing to the middle of a rousing whirlpool wriggling behind two dark cypress trees along the night sky of the painting. Due to the photographic staging, it seems as if the whirlpool was caused by the person—as if the painting was a kind of touch screen the viewer could access and thereby influence the constitution of the painting.

The staging gives the impression that the landscape painting in the commercial has not been primarily deployed to show a typical exhibition situation; instead, the staging focuses on the interactive process between artwork and viewer. Hence in the commercial, the painting works as a visual nature metaphor. It stands for the interactive process of art reception as well as for humankind's access to the digital cosmos.

Such metaphors are omnipresent in contemporary visual culture and have recurred throughout the scientific reception of cybernetics[3] since the 1950s. They are applied whenever an interactive worldview is meant to be demonstrated. Currently, this worldview is negotiated under the term "ecology," for example in the philosophy of Erich Hörl, which characterizes the epistemological precondition in contemporary digital object cultures (Hörl 2013: 122). However, the human is not considered an autonomous subject in this philosophy, but rather one part of a self-organizing system that links human subjects, biological organisms, and electronic artifacts through omnipresent streams of communication.

The aim of this research is to determine what media political visions such visual nature metaphors imply in digital cultures. This essay explores the conceptual change of nature and perception through cybernetics beyond the 1950s, which informs the contemporary ecologic worldview.[4] Through popular scien-

---

2 Vincent Van Gogh: The Starry Night, oil on canvas, 73.7 x 92.1 cm, Museum of Modern Art, New York. The ad shows a reproduction of the painting.

3 Popularly, "cybernetics" is understood as the mathematical theory of self-regulating automated systems developed by the mathematician Norbert Wiener during the Second World War, and is closely connected to the general communications theory established by Claude Shannon (Wiener 1946; Shannon 1949). In this essay, the meaning of "cybernetics" is extended to a broader cultural and scientific context, keeping in mind that, due to its interdisciplinary reception, a single, monololithic "cybernetics" has never existed. For more on the cultural influence of cybernetics, see also Hayles 1999.

4 Orit Halpern described this change as a "midcentury reconfiguration of cognition, perception and sense into algorithm, pattern and process." (Halpern 2014: 80)

tific writings by Norbert Wiener, Gregory Bateson, Humberto Maturana, and Francisco Varela, this shift was communicated via (visual) nature metaphors. These metaphors build the interface between science, the arts, and popular culture. As such, they form the cultural history of cybernetics, which is the foundation of today's visual culture in digital societies.

NORBERT WIENER'S WHIRLPOOL SUBJECT

Cybernetic theory was founded during the Second World War in the context of the US Army (Galison 1994). It is based on the concept that systems are adaptive and that they can react to their environment via feedback mechanisms. As this field of interdisciplinary research was being established in post-war America, cybernetic feedback models became a universal mode of description not only in the natural sciences but also in the humanities. The boundaries between natural and technical systems were thus transgressed on a theoretical level as well as on technological grounds. This transgression caused an epistemic shift resulting in a reformulation of the concept of nature and subject. Nature metaphors were used to communicate this shift between the different disciplines, both among specialists and to the general public.

The popular scientific writings of Norbert Wiener popularized cybernetic ideas of self-regulating systems during the 1960s. In his 1946 book *Cybernetics: Or Control and Communication in the Animal and the Machine*, Wiener used a watery metaphor to describe the new interrelationship of subject and world that was established through the operational logic of cybernetics. He characterized human beings as "patterns of organization" (Hayles 1999: 102) that swim upstream against the entropic tide of the universe. This metaphor was expanded on in *The Human Use of Human Beings: Cybernetics and Society*, his second popular scientific book, where he wrote: "We are but whirlpools in a river of ever-flowing water. We are not stuff that abides, but patterns that perpetuate themselves." (Wiener 1950: 96)

Wiener's watery metaphors elucidate his paradoxical view of nature, which would be recoined during the cultural reception of cybernetics after WWII: "Nature" was figured as an entropic universe in which the subject existed as a "whirlpool" constantly reforming and defining its borders through circulation. Wiener's whirlpool subject asserted a structural alliance between the seemingly incalculable, entropic outside and the homeostatic system of the individuum regulated by —in such a manner that their commonalities could not be refuted by rational argumentation. But Wiener was not alone in resorting to a metaphoric level to explain the principles of cybernetics. The filmmakers Charles and Ray Eames also conceived a host of visual metaphors in their 1953 instructional film *A Communications Primer* which communicated the latest insights of cybernetics and communications theory to the general public.

THE SEA AS A METAPHOR FOR COMPLEX COMMUNICATION SYSTEMS: *A COMMUNICATIONS PRIMER*

In 1953, the 21-minute-long educational film *A Communications Primer* visualized the latest insights of cybernetics and information theory. It contains several animation sequences that show Claude Shannon's communication model (Shannon 1949). In a sophisticated montage technique, these animations are combined with pictures of early calculating machines, scenes of everyday life, as well as land- and cityscapes. A narrator's voice explains what is being shown in the images. Through the montage, the film suggests that communication processes rule all parts of human life. The images of nature are thereby used as visual metaphors for these invisible communication processes. One of these shots shows fields of grain swaying in the wind. The wind is the "medium" of transmission, but remains invisible. It is only through the movement of the grain that the transmission of the signal becomes apparent. The omnidirectional intersection of the vibrating spikes of grain, which seems somewhat chaotic, allow us to picture the interference of signals and noise that are to be eliminated. The chaos of the spikes of grain already indicates what is explained by the narrator soon after: that communication processes, especially those occurring in nature, can reach remarkable degrees of complexity. However, the noise in a communication system may impede deciphering the "code" through which the signals are encrypted while being transferred through the medium, so that the messages might already be transformed by the time they reach the receiver.

Another 30-second sequence of the film shows the surge of ocean waves. The impermanent and ever-changing structure of the surge is accentuated by the frame. The waves foam up and spray, and their peripheral zones entangle with the background. As the position of the camera changes to an aerial shot, it shows how the foam of the surge spreads across the sand in tiny bubbles (*Fig. 2*). The water draws back, and undefinable forms appear. The following commentary accompanies this:

"A wave breaking on a beach brings a world of information about events far out at sea. It can tell of winds and storms, the distance and the intensity; it can locate reefs and islands and many things—if you know the code." (*A Communications Primer*, extract 10'10"-10'40," author's transcription)

According to the narrator, the wave contains an undefinable "world of information" from events far out at sea. However, to be able to understand the wave, its unknown "code" would have to be deciphered. The surge appears harsh and wild and stands metaphorically for the unknown, the uninvestigated, characterizing a form of human relation to the sea (Blumenberg 1979: 10) as well as to communication networks. In the montage, Charles and Ray Eames replace the image of the waves spreading across the sand with the close-up of a flying seagull, followed by a shot of a flock of seagulls flying in an organized confusion across the sky. The recordings of the flock are blended with an aerial shot of a crowd of people on a pedestrian street (*Figs. 3-5*).

Figures 2, 3, 4, 5: Film Stills from A Communications Primer *(1953), by Charles and Ray Eames (16mm, colour, 21' 30", captures at 4:49, 10:41, 11:18, 11:56) (clockwise from top left)*

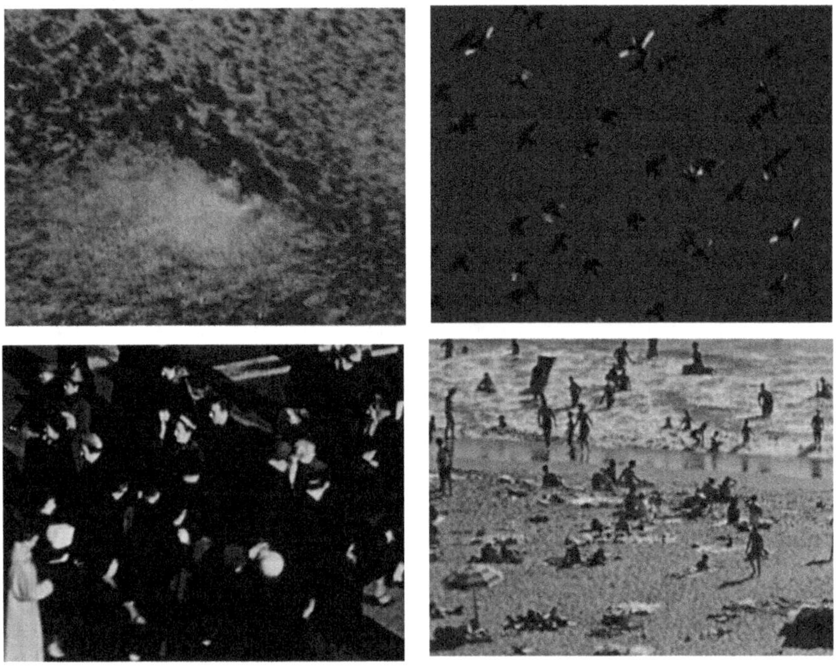

Source: Eames Office LLC, Los Angeles; Screenshots: Lena Trüper

The narrator makes the following comments while all of this is happening:

"When we watch them turning and wheeling, how often have we wondered, what holds such birds together in their flight? Communication is that which links any organism together. It is communication that keeps a society together, and though these people seem unaware of each other's existence, neither looking nor speaking, one group filters through the other with hardly two individuals coming in contact—so constant is the flow of information and so complex the web of communication that keeps them apart and holds them together." (*A Communications Primer*, extract 11'24"-12'02", author's transcription)

The sequence suggests that the members of an anonymous crowd in a big city—although they neither look at each other nor speak to one another—are entangled through an invisible "flow of information." Similar to the seagulls in the flock, the humans in the cityscape seem to be bound by a complex web of information. Likewise, in the heaving sea an invisible communication is processed, as the montaged film images suggest. But while within the flock of birds and the crowd the single members are separable, the picture of the movement of the sea abstracts from the movement of the single components into an overarching motion. As such, the image of the sea is more a metaphor than a model or illustration. It can be read as a visual nature metaphor for the all-encompassing complex communication structures the human beings are involved in. In *A Communications Primer*, this worldview is given particular emphasis in a scene showing people at the beach (*Fig. 5*).

This scene immediately follows the narrator's explanation of the principle of digital encoding. The shot is reminiscent of Norbert Wiener's metaphor of the human "whirlpools" that split off from the ever-flowing universe as separate entities while being structurally affiliated with that universe. Hence, through its pictures, the film further emphasizes the cybernetic ideology, and more specifically the idea that the individual is part of an overarching communication process occurring in "nature." The pictures of the people at the beach give the impression of a harmonic world of leisure. Due to the all-encompassing communication, the human being seems to be returning to the sea as her/his paradisiac evolutionary origin—if only the code could be deciphered!

As with Wiener's popular scientific writings, films like *A Communications Primer* communicated cybernetic thoughts to the public on a visual level. The idea that the human is connected to her/his environment by an all-encompassing communication was radicalized in the 1960s and shaped the environmental movement. One of the protagonists of this epistemological shift was Gregory Bateson. In the 1950s, the Eames anticipated this shift and prepared the public for the new digital worlds.

DOLPHINS, ECOLOGIES, AND *THE TREE OF KNOWLEDGE*

Under the influence of cybernetic models, the concepts of nature and subject changed radically during the 1960s, as well as the concept of perception. The human subject was no longer defined as an observer who had to control and regulate nature from the "outside." She or he was imagined as part of a self-organizing ecosystem that she was integrated into, that she perceived, and that she influenced through her actions. Perception was thereby reconceptualized as an interactive function. Through their continuous transformations, human subjects could interact with their surroundings.

The epistemological shift initiated by cybernetics can be understood through the writings of the anthropologist and cybernetician Gregory Bateson. His diverse scientific career is an extraordinary example of the interdisciplinary nature of the post-war research in the US that disseminated cybernetic models. During the well-known MACY Conferences, Bateson got in touch with the main protagonists of cybernetic thought, including Norbert Wiener, the neurophysiologist Warren McCulloch, the electrical engineer Julian Bigelow, the information scientist John von Neumann, and Bateson's future wife, the anthropologist Margaret Mead (Harries-Jones 1995: 103). Through the new scientific insights he gained at these conferences, he developed a cybernetic understanding of communication, which he then applied in his studies on schizophrenia as well as in his research on ecology and evolution.

One of the basic aspects of Bateson's thinking was his radical rejection of scientific objectivism. According to Bateson, the human subject could never be an objective observer analyzing a system just by watching it from the outside. The subject is always already part of the system he/she observes. In the case of science, the scientist is always part of the systemically structured nature he/she is examining. Bateson (1987b: 246) wrote:

> "The observer must be included within the focus of observation, and what can be studied is always a relationship or an infinite regress of relationships. Never a 'thing.'"

In contrast to the early protagonists of cybernetics such as Norbert Wiener, Bateson did not understand "nature" as an unregulated externality that had to be controlled. For him, "nature" was a complex coupling of homeostatic systems interconnected via cybernetic feedback loops. These systems included individual organisms as well as societies and ecosystems. According to Bateson, the subject participated in those systems via his perception. Bateson thought of perception as a form of communication. He assumed that the subject interacted with his surroundings in a recursive process between acting and perceiving. Throughout this process (Bateson 1978d: 385), the individual set up patterns of recognition structuring subjective consciousness. Driven by a "conscious purpose," those patterns never coincided completely with the complex coupled systems of perceived nature, resulting in a lack of inner complexity of the subject. Bateson thus described the patterning perception as a "semipermeable" connection between the subjective consciousness and the "total mind" of nature (Bateson 1978e: 407-408). According to Bateson, this partial "blindness" of the subject was the reason why humans automatically caused errors in any of their surrounding systems that they did not completely understand, such as biological ecosystems. Bateson aimed to develop a scientific method to approximate the processes of pattern recognition. Through a recursive learning process, he assumed the inner complexity of the subject could be enhanced in order to make his/her actions more suitable for his/her surroundings.

In his research, Bateson was especially interested in communication situations in which the rules of how these communications were processed were not yet fully established. He concluded that during a communicative process, several metacommunicative signals would be established, such as gestures, which in turn would contribute to common patterns of recognition and result in a meaningful understanding between the participants of such a communication process. In search of such acts of nonverbal communication, he analyzed the psychological phenomenon of schizophrenia, which in his opinion was caused by ambiguous communication situations in families, so-called "double binds" (Bateson 1978a: 178-179). He also took a look at aquariums: Between 1963 and 1972, he conducted several studies about the communication processes of octopi and cetaceans together with his colleague John Cunningham Lilly. By studying their sounds and behavior, Bateson hoped to understand how intelligent living creatures that were unable to use verbal language could establish meaningful communication by condensing several metacommunicative levels into their sounds (Harries-Jones 1995: 147-148). Dolphins were especially well-suited to his studies of nonverbal communication because their bodies did not allow them to gesticulate or form facial expressions (Bateson 1978c: 340-341). By studying the patterns of metacommunicative signals among cetaceans, Bateson hoped to develop a method of how to establish a collective consciousness between dolphins and humans that could be transferred to other communication situations. Through his interdisciplinary research, Bateson often chose a metaphoric language that made his studies accessible to several scientific and cultural contexts. Bateson's radical refusal of scientific objectivism formed a basis of postmodern philosophy, prominently in the writings of Gilles Deleuze, who adopted concepts such as "schizophrenia," "double bind," and "plateau" from

Bateson while barely mentioning his name (Shaw 2015). During the 1960s, Bateson's writings also became popular in the American counterculture, which aimed to establish a new form of nonhierarchical society and a new community living in harmony with nature (Turner 2006: 124). Much of the ecological movement in those days had Bateson's view of ecology at its foundation (Harries-Jones 1995: 5).

Figure 6: Vertigo Systems, product video for "Living Floor" (digital, color, 1' 11", capture at 20")

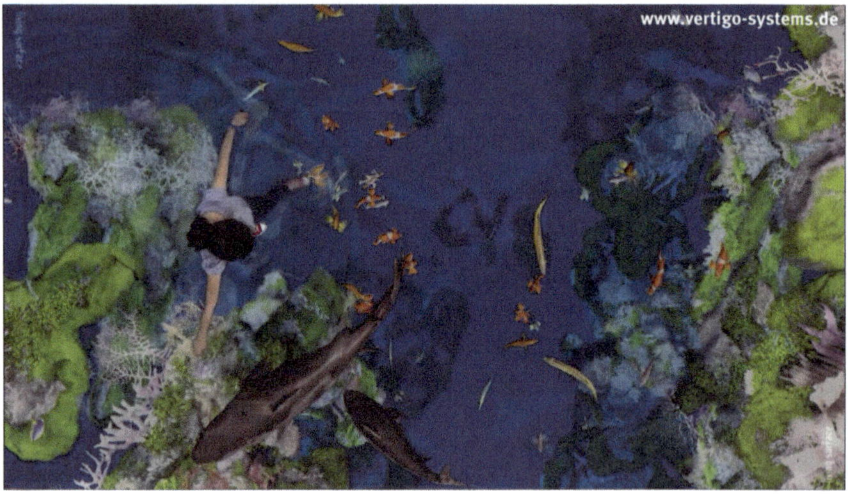

Source: vertigo systems GmbH; Screenshot: Lena Trüper

The video movement of the 1970s in the US received his texts through magazines like *Radical Software*, published by the Raindance Foundation (Bateson 1971, 1972). In popular culture, Bateson's cetacean studies were picked up by the TV series *Flipper* (Jack Cowden 1964), as well as for the films *Flipper* (James B. Clark 1963) and *Day of the Dolphin* (Mike Nichols 1973). Later science fiction movies, such as *Johnny Mnemonic* (Robert Longo 1995) or the recent *Arrival* (Denis Villeneuve 2013), form the visual inheritance of Bateson's view of nature.[5] They can be considered forms of the popular scientific mediation

5 On the metaphorical use of marine creatures as visual nature metaphors of cybernetics in science-fiction films and virtual reality applications, see also my essay (in German) titled "Macht's gut, und danke für den Fisch! Meerestiere als Metaphern der Virtualität. Ein historischer Einblick in das visuelle Erbe der Kybernetik" (Trüper 2019) (So Long, and Thanks for All the Fish. Marine Creatures as Metaphors of Virtuality. A Historical Perspective on the Visual Heritage of Cybernetics).

of cybernetic thought. Right up to our times, cetaceans belong to the main protagonists of augmented reality applications or virtual spaces. Recently, the company Vertigo developed virtual aquariums in which customers interact with fish on so-called "Living Surfaces" (*Fig. 6*).

The underwater world is meant to evoke positive feelings. Marine animals, therefore, belong to visual nature metaphors that accompany the spread of an operational mode of thinking introduced by cybernetics. They convey an ideal picture of nonverbal communication guided by interactive perception and stand for a vision of individual freedom based on nonhierarchical bottom-up organization. In contrast to the whirlpool subjects of Norbert Wiener, the marine animals are concrete embodied metaphors of a concept of enhanced perception. They assign virtual spaces as places of creative interaction that would result in a prolific gain of knowledge.

During the 1980s the biologists Humberto Maturana and Francisco Varela adapted Bateson's view of recursive learning processes while envisioning their own thoughts via nature metaphors. Similar to Bateson, the biologists refused to attribute an objective point of view to the autonomous subject. They too saw the observer as part of the system observed. However, in contrast to Bateson, they thought that the development of an observed system—for example a living organism or an ecosystem—was fundamentally different from the image the observer constructed through his/her own observation. According to Maturana and Varela, all living beings were linked through a "dynamic network of transformations" and so constituted their surroundings ("milieu"), while this milieu in turn influenced the constitution of the organisms. Varela and Maturana called this process "autopoiesis" (Maturana/Varela 1984: 53-54). The organism and its observer had only their similar structure in common, however. Their boundaries were continuously redefined according to their milieu. The biologists described their thesis of autopoiesis with a humorous tone: "The most particular characteristic of an autopoietic system is that it pulls itself up through its own laces as it constitutes itself as different from the milieu through its own dynamic." (Maturana/Varela 1984: 54) This citation comes from the book *The Tree of Knowledge. The Biological Roots of Human Understanding*, which the scientists wrote and designed in 1984. The visual conception of the book in particular reminds one of strategies already applied in *A Communications Primer*. However, unlike in that educational movie, in *The Tree of Knowledge* pictures of natural phenomena were not used to describe technical processes, but rather the process of recognition itself. According to Maturana and Varela, the book is intended to lead the human subject to recognize itself in its process of recognition. This process of getting to "know how we know" was visualized through a circular diagram placed in front of each chapter. It appears like an extended structure of a feedback loop. The boxes of the diagram contain the themes of the chapters. While the reader proceeds from chapter to chapter, the respective box in the diagram is marked in red to emphasize the reader's progress along the path of recognition (*Fig. 7*).

Figures 7, 8, 9, 10: The Tree of Knowledge *(1984), Overview (p. 14) and Illustrations from Chapters 3 (p. 55), 7 (p. 141), and 8 (p. 179) (clockwise from top left)*

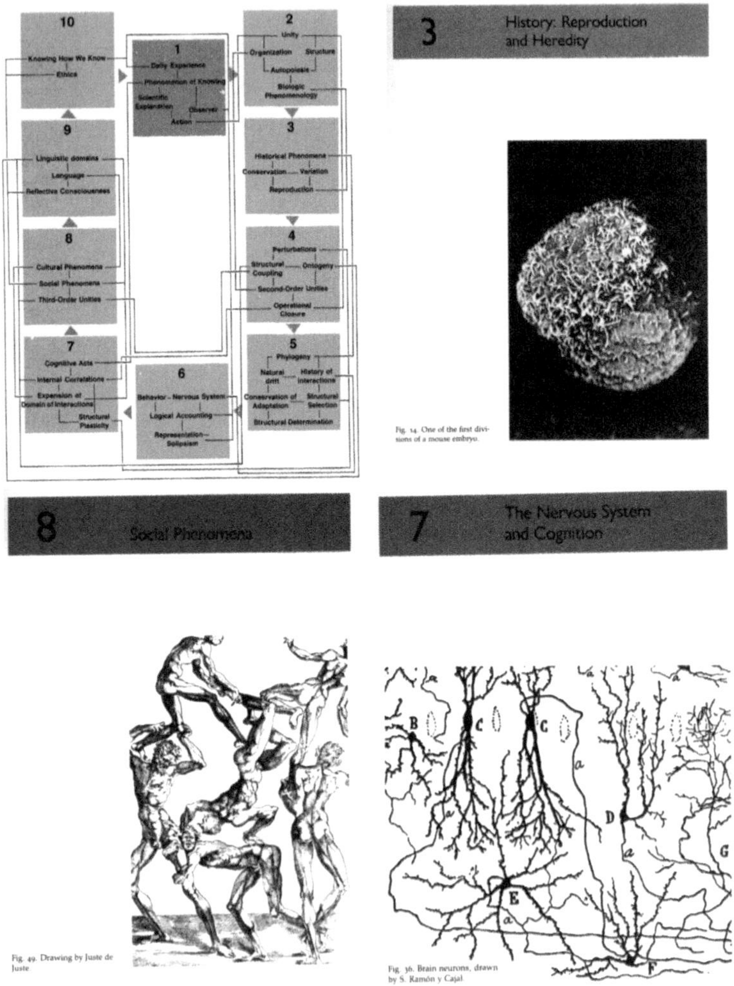

Source: Shambala Publications; Scans: Lena Trüper. From The Tree of Knowledge: The Biological Roots of Human Understanding by HumbertoR. Maturana, Ph. D & Francisco J. Varela, Ph. D., ©1987 by Humberto R. Maturana and Francisco J. Varela. Reprinted by arrangement with The Permissions Company, LLC on behalf of Shambhala Publications Inc., www.shambhala.com

Additionally, each chapter is furnished with a picture. The pictures show galaxies, close-ups of embryonic cells, neuronal networks, and reproductions of works of art using network-like structures (*Figs. 8-10*). The "spiral nebulae NGC 1566 of the Dorado constellation" shown in the book evokes Wiener's whirlpool subject or Van Gogh's "Starry Night" in the aforementioned Google commercial. Thus, the pictures in Maturana and Varela's book not only follow the didactic concept of the book; seen in a broader historical context, they also form part of the visual inheritance of visual nature metaphors established during the cultural reception of cybernetics. They visualize the recognition process that is rooted in cybernetics. Additionally, they reproduce the universalist view, showing that cosmological, biological, and sociological processes all follow the same structure. Thus, they reinforce a pictorial vocabulary that emerged in the 1950s with early cybernetic theory and that has shaped the visual culture of digital societies up to the present day.

EXPANDING THE VIEW: VISUAL NATURE METAPHORS IN THE ARTS AND DIGITAL CULTURE

The examples of popular science writing and film show that the epistemological paradigm shift caused by cybernetic theory during the 1960s was communicated to the public via various visual nature metaphors. Those images formed the interface between scientific knowledge and general public worldviews. A glance at the arts of the time shows that many artists used visual nature metaphors in their compositions while working on the same problems of operational logic as the scientists. One example is the so-called Pepsi Pavilion of the artists' group E.A.T. (Experiments in Art and Technology), which was built for Expo '70 in Osaka (Asendorf, 2008: 118). The spherical pavilion reminds one of a geodesic dome, an architectural form that became popular in the counterculture of the time and broadly reflects the universalist view of cybernetics (Turner 2006: 56-58). Loosely adapting the logic of the observer within the system, the visitor could observe himself while looking into the mirror-faced walls of the pavilion. Due to a kaleidoscopic effect, the boundaries of the room seemed to dissolve. Similarly, the walls on the outside of the pavilion were dissolved optically by a cloud sculpture designed by the Japanese artist Fujiko Nakaya (*Fig. 11*).

Figure 11: Cloud Sculpture by Fujiko Nakaya, 1970

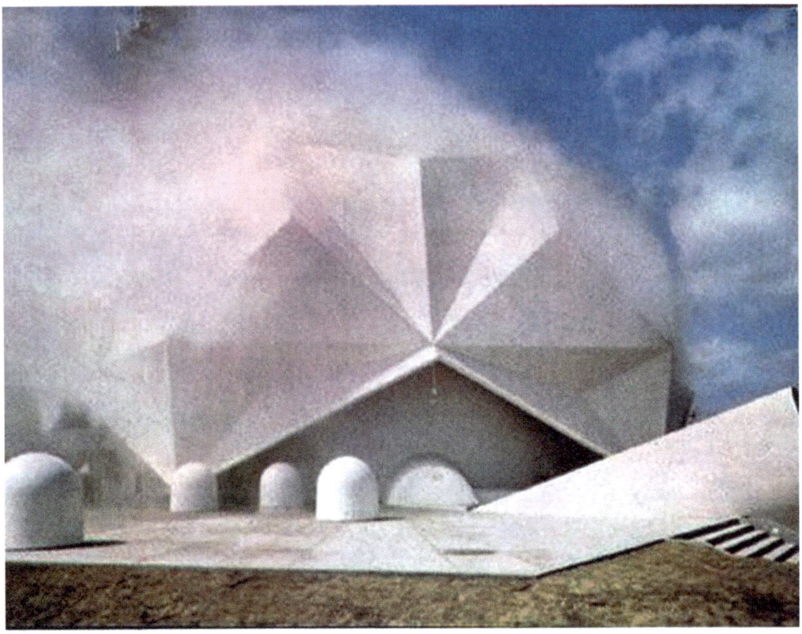

Source: Pepsi-Cola Pavilion, Osaka, Japan (1970); Photo: Shunk-Kender. Daniel Langlois Foundation, Collection of Documents Published by Experiments in Art and Technology (E.A.T.). Courtesy of E.A.T. and Daniel Langlois Foundation

All in all, the visitor was meant to reflect on her/his interactive position in the environment of the pavilion. Together with the self-reflection of the viewer, the triangle building and the cloud of the pavilion served as visual nature metaphors to communicate the new worldview. Hence the engineer Billy Klüver, founder of E.A.T., mentioned the building in the exhibition catalog of the Expo as a "living, active environment" (Asendorf 2008: 118). Complex natural phenomena also became popular in Kinetic Art during the 1960s, as the artist George Rickey stated in 1965 (Rickey 1965: 110, emphasis in the original):

"Nature is rarely still. All the environment is moving, at some pace or other, under laws which are equally a manifestation of nature and a subject for art. The artist finds waiting for him, as subjects, not the trees, not the flowers, not the landscape, but the *waving* of branches and the *trembling* of stems, the piling up or scudding of clouds, the rising and setting and waxing and waning of heavenly bodies, the creeping of spilled water on the floor, the repertory of the sea – from ripple and wavelet to tide and torrent, the antics of people, schools of fish, companies of soldiers, heads of wheat, traffic jams, bees and ants – the 'very many' in motion [...]"

As part of the Center of Advanced Visual Studies (CAVS), Rickey worked on the activation of vision as a means of integrating the viewer and the art object in the same environmental structure. As with the designers Charles and Ray Eames in their educational movie *A Communications Primer* ten years earlier, he described this ever-changing environment by addressing macroscopic natural phenomena like the movement of clouds, the sea, or the trembling of plants, as well as the transformation of technical objects and atoms on a microcosmic level. According to Rickey, these movements constituted the model for the artistic process of Kinetic Art, which simultaneously abandoned the representational paradigm of depicting nature in painting. In general, it can be stated that the reception of cybernetics supported the abandonment of the material art object as well as the growing interest of artists in immaterial concepts and performative practices. The operational logic of cybernetics encouraged the acceptance of real movement as a valuable mode of expression. Artists generated visual nature metaphors in their compositions to deal with the new worldview. The use of visual nature metaphors by artists served to give the artwork a particular sociopolitical relevance.

Furthermore, these metaphors served to endorse the concept of a processual mode of perception that was meant to release certain creative qualities. It culminated in the outline of a concept of "creative interaction" propagated by the so-called "Interactive Art" of the 1990s (Weibel 1993), and furthering theories of swarm intelligence of the time (Kelly 1994). Interaction henceforth no longer served purposes of self-perpetuation but rather the evolution of a system—an idea that is preserved in the present day within the "creative cloud."[6] Once again, visual nature metaphors played a major role in communicating scientific developments and became part of artistic research. Thus, in 1995 the Canadian artist Char Davies developed one of the first artistic virtual spaces in which the so called "immersant" (Jones 1995: 25) could float through a crystallized universe filled with particle swarms to experience a new mode of embodied consciousness.

6 By contrast, the artist Stephan von Huene critically approached such notions of a creative interactivity in a concept he called "intra-action." In his work "Text Tones" (1983), he developed minimalist audiokinetic sculptures that required patient contemplation rather than active audience engagement, while the sculptures' programming transformed unintentional sounds and conversations from the surroundings in a sound collage that was fed back into the exhibition hall (Muñoz Morcillo 2016: 251). Relational ontologies such as Karen Barad's "agential realism" also use the concept of "intra-action" as a critical counterpart of object-based "interactivity" (Barad 2012: 19).

Figure 12: Film Still from Tomorrowland: A World Beyond *(Brad Bird 2015)*

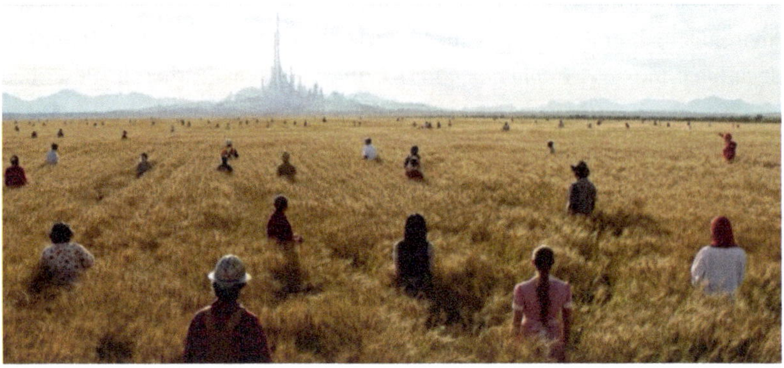

Source: Walt Disney Pictures; Screenshot: Lena Trüper

Figure 13: Film Still from Ex Machina *(Alex Garland 2015)*

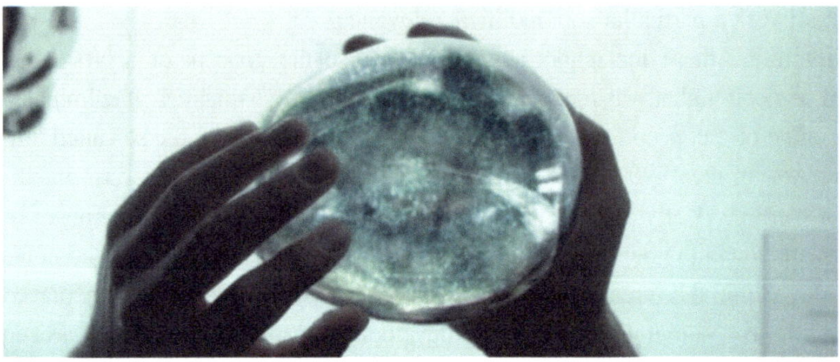

Source: DNA Films; Screenshot: Lena Trüper

Currently, the productive chaos of random structures is frequently addressed in contemporary science fiction films dealing with artificial intelligence. After an apocalyptic scenario in the Disney film *Tomorrowland: A World Beyond* (2015), for example, the technologically liberated humans awake in a lush grain field—a visual nature metaphor for the invisible communication network that sustains and connects them (*Fig. 12*). Their arrangement is reminiscent of the people at the beach in *A Communications Primer*. Unlike the metaphoric sea of the 1950s, the grain field connects the interactions of the networked humans with the notion of fertility and growth. In the film *Ex Machina* (Alex Garland 2016), cloudy structures are used to form an image of the consciousness of the artificial intelligence Ava who—embodied in a female cyborg—rebels against the patriarchal dominance of her maker (*Fig. 13*).

Although cybernetic concepts seem to recall the theories of vibrations in cosmologies of the late 18th and early 19th centuries, their metaphors differ in crucial ways. Instead of metaphors that like "bridge," "channel," or "path," as used for example by Kandinsky, that describe the vibrating connection between two points (cf. Immelmann's article in this volume), visual nature metaphors of cybernetics, like pictures of distributed particle systems and evolutionary growth, emphasize the multidirectional influence of the system's development. The fact that these metaphors became popular among artists and scientists after the Second World War confirms the productivity of an interactive worldview predominant in contemporary popular culture. Accordingly, the photographic scenario of the aforementioned Google ad also shows a creative ecosystem as such: Through their posture and their clothing, the people are integrated into the painting. The black shirt as well as the upright stance of the left person is reminiscent of the vertical position of the dark cypress trees. The choppy pattern on the clothes of the right person echoes the whirling brush stroke of the painting. The image section highlights the act of perception that is envisioned by a gesture. Like a touch screen, the painting is a gateway to the interactive process, which is meant to be creative. Hence, in the context of this photographic scenario, "The Starry Night" by Vincent Van Gogh is not just a painting but also a visual nature metaphor. It highlights the intuitive access to technologies and visualizes the creative participation of humans in the digital ecosystem. Invoking the visions of harmonic ecosystems established in the 1960s, visual nature metaphors facilitate the handling of invisible digital structures up to the present day while hiding the economic and historical conditions of the capitalist world order, under which the development of these technologies proceed. The analysis of such popular scientific and artistic resources helps us reconstruct the historical origins of such metaphors and can provide insights into the ideological implications of these metaphors in contemporary societies.

REFERENCES

Asendorf, Christoph (2008): "Die Künste im technischen Zeitalter und das utopische Potential der Kybernetik." In: Michael Hagner/Erich Hörl (eds.), Die Transformation des Humanen. Beiträge zur Kulturgeschichte der Kybernetik, Frankfurt am Main: Suhrkamp, pp. 107-124.

Barad, Karen (2012): Agentieller Realismus. Berlin: Suhrkamp.

Bateson, Gregory (1971): "Restructuring the Ecology of a Great City." In: RadicalSoftware 1/3, pp. 2-3. Accessed February 23, 2020. http://www.radicalsoftware.org/volume1nr3/pdf/VOLUME1NR3_art02.pdf.

Bateson, Gregory (1972): "Up Against the Environment or Ourselves?" In: RadicalSoftware 1/5, p. 33. Accessed February 23, 2020. http://www.radical software.org/ volume1nr5/pdf/VOLUME1NR5_0035.pdf.

Bateson, Gregory (1978a [1956]): "Toward a Theory of Schizophrenia." In: Douglas 1978, pp. 173-198.

Bateson, Gregory (1978b [1959]): "Minimal Requirements for a Theory of Schizophrenia." In: Douglas 1978, pp. 215-241.

Bateson, Gregory (1978c [1966]): "Problems in Cetacean and Other Mammalian Communication." In: Douglas 1978, pp. 334-348.

Bateson, Gregory (1978d [1967]): "Cybernetic Explanation." In: Douglas 1978, pp. 375-386.

Bateson, Gregory (1978e [1968]): "Effects of Conscious Purpose on Human Adaptation." In: Douglas 1978, pp. 415-422.

Blumenberg, Hans (1979): Schiffbruch mit Zuschauer. Paradigma einer Daseinsmetapher, Frankfurt am Main: Suhrkamp.

Douglas, Mary (ed.) (1978): Steps to an Ecology of Mind, Collected Essays in Anthropology, Psychiatry, Evolution, and Epistemology, London: Granada Publishing Limited.

Galison, Peter (1994): "The Ontology of the Enemy. Norbert Wiener and the Cybernetic Vision." In: Critical Inquiry 21, pp. 228-266.

Halpern, Orit (2014): Beautiful Data. A History of Vision and Reason since 1945, Durham: Duke University Press.

Harries-Jones, Peter (1995): A Recursive Vision. Ecological Understanding and Gregory Bateson, Toronto: University of Toronto Press.

Hayles, Katherine Nancy (1999): How We Became Posthuman. Virtual Bodies in Cybernetics, Literature, and Informatics, Chicago: University of Chicago Press.

Hörl, Erich (2013): "A Thousand Ecologies. The Process of Cyberneticization and General Ecology." In: Dietrich Diederichsen/Anselm Franke (eds.), The Whole Earth. California and the Disappearance of the Outside, Berlin: Haus der Kulturen der Welt, pp. 121-130.

Jones, Mark J. (1995): "Char Davies – VR through Osmosis." In: CyberStage 2/1, pp. 24-28. Accessed April 2, 2020. http://www.immersence.com/ publications/ 1995/1995-MJJones.html.

Kelly, Kevin (1994): Out of Control. The New Biology of Machines, Social Systems and the Economic World, Reading: Addison-Wesley Publishing Company.

Maturana, Humberto/Varela, Francisco (1998 [1984]): The Tree of Knowledge. The Biological Roots of Human Understanding, Boston: Shambhala Publications.

Muñoz Morcillo, Jesús (2016): Elektronik als Schöpfungswerkzeug. Die Kunsttechniken des Stephan von Huene. Bielefeld: transcript.

Rickey, George (1965): "The Morphology of Movement. A Study of Kinetic Art." In: Gyorgy Kepes (ed.), The Nature and Art of Motion, Vision+Value series, New York: George Baziller, pp. 81-114.

Shannon, Claude E./Weaver, Warren (1998 [1949]): A Mathematical Theory of Communication, Urbana: University of Illinois Press.

Shaw, Robert (2015): "Bringing Deleuze and Guattari down to Earth through Gregory Bateson. Plateaus, Rhizomes and Ecosophical Subjectivity." In: SAGE journals 32/7-8, pp. 151-171.

Trüper, Lena (2019): Macht's gut, und danke für den Fisch! Meerestiere als Metaphern der Virtualität. Ein historischer Einblick in das visuelle Erbe der Kybernetik, ffk-journal, Vol. 4, pp. 231-249. Accessed February 23, 2020. http://ffk-journal.de/?journal=ffk-journal&page=article&op=view&path%5B%5D=80.

Weibel, Peter/Grebel, Karl (1993): Genetische Kunst – Künstliches Leben, Ars Electronica, Festival für Kunst, Technologie und Gesellschaft, Linz: PVS Verleger.

Wiener, Norbert (1954 [1950]): The Human Use of Human Beings. Cybernetics and Society, Garden City, NY: Doubleday.

Wiener, Norbert (1965 [1946]): Cybernetics: Or Control and Communication in the Animal and the Machine. Cambridge, MA: MIT Press.

From “The Destroyer of Worlds” to “Atoms for Peace” (and Back?). The Discourse on Nuclear Power in US Popular Science Magazines during the Early Cold War Era

Lars F. Köppen

ABSTRACT

How did the US-American public discuss nuclear power-related topics in the aftermath of World War II? This article examines the coverage by science magazines between the bombs on Hiroshima and Nagasaki in 1945 and the onset of the Cuban Missile Crisis in 1962. Torn between dystopias of a world devastated by atomic bombs and technological euphoria about novel ideas concerning nuclear energy, monthlies like *Popular Science* and *Popular Mechanics* tried to educate their readers about the limitations, risks, and opportunities of a newly-invented technology. The editors themselves proved to be unsure about what to make of the nuclear phenomenon, as they alternated between critical examination and optimistic pieces with more-than-questionable scientific foundations. The ways that this affected the communication of knowledge will be a central point in the present article.

INTRODUCTION

Since the discovery of fire by humankind, or at the very least since the early 20th century, no source of energy has been as controversial as nuclear power. Thus, it is somewhat surprising that while its military and civilian uses have been broadly described and researched by historians (Cooke 2009; Radkau and Hahn 2013;

Radkau 2017), its discussion by the general public, apart from the anti-nuclear-power and anti-war movements, has not. This article aims to fill some of these blanks by analyzing the discourse on nuclear power in long-established popular science magazines such as *Popular Mechanics* and *Popular Science*. As popular science magazines, they act as mediators between professional engineers, researchers, and the results of their work on the one hand and the interests of their scientifically curious layperson audience on the other. They also provide a platform for discourse among members of their audience as well as opinion pieces by prominent members of the scientific community, such as Edward Teller, Wernher von Braun, and Ernest Rutherford. The analysis starts in 1945, when the nuclear attacks on Hiroshima and Nagasaki ended the secrecy surrounding nuclear technology and made it a topic of popular interest, and focusses on the discourse of the 1940s and 1950s, concluding in the late 1950s, just before the Cuban Missile Crisis. The emphasis of this work lies in the ways that these magazines navigate between the expectations of their audience, their self-declared journalistic mission of covering innovation in science and engineering, and the general political and ideological framework of their time. It will fathom the development of the magazines' stances on nuclear power from technological euphoria to doomsday predictions in the 1940s and 50s and explore whether this development contrasted or aligned with the discourse among their audiences and within the scientific communities of the time. Furthermore, this paper will discuss how the magazines in question drew, toed, and in some cases outright ignored the line between scientific sensationalism and up-to-date journalism. Finally, a short outlook will be given on the further development of popular science magazines' views on nuclear power.

This study relies mostly on published issues of magazines with a broader range of scientific topics like *Popular Science* and *Popular Mechanics*. Apart from that, material from diverse sources is also examined, with the aim of contextualizing the views of said magazines within domestic US-American and worldwide debates on nuclear power at the time. *Popular Science* and *Popular Mechanics* have been selected for analysis based primarily on two reasons. First of all, they have a long and continuous publishing history—one that dates back to 1872 in the case of *Popular Science*, and to 1902 in the case of *Popular Mechanics*—and thus allow for long-term diachronic analysis as well as an analysis of future issues. A circulation of at least 100.000 copies per issue for most of the 20th century (Nourie 1990) and low seven-digit numbers recently suggest a broad reader and subscriber base over time. Secondly, as mentioned above, established scientists have contributed to both magazines. Thus, it seems that said scientists viewed the magazines as an appropriate place to publish without tarnishing one's scientific reputation. On the whole, *Popular Science* and *Popular Mechanics* can be assumed to present views that, at the time they were issued, were considered as being generally credible or plausible by the interested public as well as by significant parts of the scientific community.

HORRIFIED FATHERS—NUCLEAR SCIENTISTS AND THE HIROSHIMA ATTACK, 1945

The large-scale use of radioactive material arguably started with the nuclear power programs—and is linked to the nuclear weapons programs—of the main parties involved in World War II. The level which US-American (Hacker 1994), Soviet (Holloway 1994; Josephson 2005), and Nazi German (Walker 2003) research teams had each reached at the end of the war in their quest for nuclear energy production and nuclear weapons has already been broadly documented and discussed. Although the controversy over Werner Heisenberg's involvement in developing the bomb for Germany is less interesting here insofar as the question of whether he would have succeeded had Germany not lost the war is not relevant to the present discussion, it still points to the ideological framework of military nuclear research in the 1940s. Speculation on what course history would have taken had Germany gained the atomic bomb in 1943 or 1944 does not question its potential for devastation, but rather the fact that it might fall in the wrong hands. This line of argumentation dominated the actions and wording of the governments involved in the Second World War. Werner Heisenberg faced criticism from his international colleagues after the war not because his research might have cost thousands of lives, but because he had done it in Hitler's service (Rose 1998: 67). Heisenberg himself, when confronted with the news from Hiroshima, denied that, at the current state of technology, a functional atomic bomb was even possible. Like most of his former colleagues in Germany, he insisted that he had not been tasked with developing any weapons, but merely built reactors for energy production (this assertion has, however, been determined to be utterly untrue, cf. Karlsch 2005).

Because of the amount of secrecy surrounding the nuclear programs, the debate on ways to use nuclear power was confined to the researchers involved. The transcripts from Farm Hall, where most of the German nuclear scientists were held at the end of the war, offer some insight into this debate: As mentioned, Heisenberg flat-out denied the very existence of atomic bombs; Otto Hahn, on the other hand, was horrified by the news from Hiroshima and blamed himself for the theoretical groundwork he had provided. As a consequence, he prominently took part in the public campaign against arming the German forces with nuclear weapons in the 1950s. The same can be said of his American counterpart Robert J. Oppenheimer. In an opinion piece in *Foreign Affairs* magazine (Oppenheimer 1953), the latter opposed the continuation of nuclear arms development and was subsequently removed from all decision-making positions in the national nuclear research program during the McCarthy era (Stern 1969). Even Hahn's relief over the end of the war after Hiroshima was inspired less by technological enthusiasm and more by the notion that dropping the bombs had been a necessary evil to end the war.

THE DEBATE ON NUCLEAR POWER IN *POPULAR SCIENCE* AND *POPULAR MECHANICS* AFTER HIROSHIMA

The bombs on Hiroshima and Nagasaki inevitably ended most of the secrecy surrounding nuclear weapons research and opened up the debate to a broader public. While science magazines like *Popular Science* and *Popular Mechanics* had already occasionally reported on nuclear research as a purely theoretical discipline (Rutherford 1915), the fact that nuclear energy production and warfare was already possible and taking place took them by surprise, as it did the rest of the American public in 1945. Both magazines shared the mission of acting as a mediator between professional scientists and scientifically interested laypeople. While professional scientists took part through occasional contributions as guest authors or interview partners, scientifically interested laypeople utilized the magazines as a platform for peer communication through letters to the editor or paid contact ads that sought out like-minded enthusiasts from various areas of technological interest. This meant that these magazines had to stay abreast of developments in scientific research, as well as provide their audience with suggestions on how scientific findings could be practically applied. In other, more populist, terms, they had to prove the "real"-world applications of results, in contrast with the theoretical and, following the wording, supposedly "unreal" world of academia. This challenge for science communication arguably dates back to antiquity, as Muñoz Morcillo's, Roby's, and Sehlmeyer's contributions to this volume discuss in more depth. As the names of prominent authors and interview partners suggest, both magazines were not just by DIY enthusiasts for DIY enthusiasts with an emphasis on DIY topics, but fulfilled an essential role in bridging a perceived gap between those who were part of the scientific establishment and those who were not. Accordingly, innovations in science and engineering determined the content of the current issues, depending on how the editors estimated the impact of those innovations on the experiences and priorities of their readers in terms of such things as consumption patterns or their sense of security.

Those innovations, as well as the readers' interest, depended largely on global political and economic realities. Thus, during World War II, military topics dominated the front matter of both magazines. This was firstly because the war economy prioritized, in terms of cash flow, technology with military applications over those perceived as purely civilian. This resulted in more military technological breakthroughs, which could then be reported on. Secondly, the war effort affected an increasing number of readers in increasingly extensive ways, in the form of the military draft and additional job openings in the weapons industry. Between January 1942 and September 1945, the front matter of both magazines featured battleships, warplanes, tanks, artillery guns, and soldiers with innovative pieces of equipment almost every month. The bandwidth in *Popular Science* spanned from militaristic technological enthusiasm (*Popular Science*, June 1945), to aggressive anti-Japanese propaganda (*Popular Science*, January 1945), and even critical reflections on the American strategy ("Shall we quit building Battleships?" *Popular Science*, April 1943), to a mixture of different approaches ("Has Precision Bombing failed?" and "Can you recognize the 17 new decorations our fighters are winning?" *Popular Science*, July 1943). *Popular Mechanics* was more restrained and only used illustrations and occasionally small advertisement banners for war bonds (*Popular Mechanics*, April, May, and November 1943), or thanked the women in the weapons industry (*Popular Mechanics*, October 1943). Anyone who wanted to know about the advantages and disadvantages of the M4 Sherman when compared with the Tiger or Panther tank—and who did not have access to classified documents or have the misfortune of finding it out on their own in the European theater—could consult these magazines ("Under the Tiger's Skin," *Popular Science*, June 1945). The same issue provided the reader with information on the bomb sights in US warplanes, until then the allegedly "best kept secret" (Torrey, *Popular Science*, 1945) of the war.

The news of the atomic bomb as a weapon that exceeded any other in terms of its development costs and effort and its destructive potential necessarily raised questions among the readers of *Popular Science* and *Popular Mechanics*. Still in 1945, both magazines issued reports that tried to cover the seemingly most urgent of those questions. In September 1945, *Popular Science* published a statement by the editors on their views about the use of the bomb. In contrast to the German detainees at Farm Hall, they showed no sign of horror; in fact, they stated that their magazine had "...looked forward to such a scientific triumph for many years" (Hammond/Newman/Colby/Walton, *Popular Science*, September 1945: 80). They hoped that this event would hold the key to further scientific progress. Pointing to X-ray technology, they even believed there might be beneficial physical side effects to nuclear explosions, in addition to the possibility of less desirable ones. As they professed their trust in the nuclear science community, they pointed out that the less desireable side effects would surely be eliminated in the future:

"Radiations are known to affect living matter. Radium and X Rays for example, have been used in treating cancer. Hence, some of the biophysical effects of atomic explosions may be beneficial. But others may be gruesome." (Ibid: 230)

It is relevant for further analysis and should be noted that in 1945 the editors should already have been aware that radiation, X-rays included, was in fact harmful to human tissue and to organisms in general, since Hermann Joseph Muller had pointed out, in Science magazine in 1927, that radiation, explicitly and especially X-rays, caused harmful genetic mutations (Muller 1927). This complete lack of critical reflection—the aftermath of the Hiroshima and Nagasaki bombings obviously provided no evidence to support such claims—was mainly due to the magazine's aforementioned militaristic stance of the previous years and the pressure to live up to its mission as a science magazine. *Popular Science* had to report on new developments as soon as they happened, making it virtually impossible not to report on Hiroshima without betraying itself. This can be seen in the article for the first anniversary of the bombing. Not only did editor-in-chief Perry Githens sum up the preceding year as a series of shattered dreams about the civilian use of nuclear power, he also admitted that the conclusion of the September 1945 article "by splitting the atom, man may have united the world" had been disproven by the actual course of history. Consequently, he appealed to his readers to commit themselves in favor of the civilian use of nuclear energy and against nuclear armament (Githens, *Popular Science*, August 1946). *Popular Mechanics*, on the other hand, withstood the pressure to publish quickly rather than soundly. With one additional month to consider the topic, its October 1945 issue still offered a broadly optimistic view on the potential of nuclear power for civilian use, while remaining sensitive to the scale of devastation inherent to nuclear weapons (Atomic Power for War and Peace, in: *Popular Mechanics*, October 1945). Also contrary to the editors at *Popular Science*, the author of the piece asked the political and scientific establishment to promote the civilian use of nuclear power instead of demanding brilliant ideas and miracle breakthroughs from the relevant researchers. However, the following month, *Popular Mechanics* aligned with its counterpart's boundless technological euphoria in the form of an article by William F. McDermott. Although McDermott initially opened with a criticism of the contemporary technological optimism at the dawn of the nuclear age, he then drew a line of tradition from the Hiroshima bomb to German physicist Karl Schlessel's vision of a world where hunger, pain, and strife had been ended by nuclear energy. Thus, McDermott promoted ideas of using atomic bombs to stop hurricanes or wildfires and create ice-free ports (McDermott 1945). Similar to Perry Githens' summary in *Popular Science*, at the end of 1946, disillusionment, if not outright pessimism, was dominant among the editors at *Popular Mechanics*. In the October 1946 issue, writer John L. Kent evaluated the possibility and necessity to move life underground in the face of the nuclear threat, likening this process to the regression of humankind to cave-dwellers (Kent 1946).

The oscillation between euphoria over the developments in nuclear science and resignation in the face of the horrors of nuclear warfare, which both magazines demonstrated, points back to the magazines' function in bridging the aforementioned gap. Until 1945, with most innovations they touched upon, the advantages seemed to outweigh the downsides: Airplanes admittedly provided the military with new means to kill and maim, but they also fulfilled the Icarian dream of flying and opened new and increasingly safe options for travel; electricity, if handled incorrectly or with intent to harm, bore the risk of a gruesome death, yet nobody was ready to give up electric lighting, telephones, and radio broadcasts over those concerns. Additionally, with a little effort, it was possible to fully understand the underlying scientific principles of airplanes and electric devices. Nuclear power, on the other hand, was far more ambiguous: The terminology was more complex, and the processes and outcomes were unimaginable. To demonstrate the dimensions of the Hiroshima bomb, Popular Science referred to the Oppau disaster of 1921, an explosion at a chemical production plant in southwest Germany (Haller 2013; Sanner2015) that not only had not been completely understood even during the official investigation, but was neither nearly as devastating nor, presumably, as widely known among the predominantly American readers. Plain and simple, *Popular Science* and the public in general were at a loss for adequate examples and precedents. This constellation led to uncertainty about nuclear power in public, and, in turn, to indecision among the editors about how to adequately address said uncertainty.

THE DISCURSIVE CIVILIZATION OF NUCLEAR POWER IN THE 1950S

In the late 1940s and early 1950s, both *Popular Science* and *Popular Mechanics* settled on reporting on the implementation of their calls and visions for the civilian use of nuclear power from the utopian-dystopian year of 1945/1946, while maintaining a general air of technological optimism that can be also found in other popular media, such as the short documentary *A is for Atom* (1953) by John Sutherland (see *Fig. 1*).

Figure 1: Film Stills from A is for Atom *(1953), by John Sutherland, General Electric*

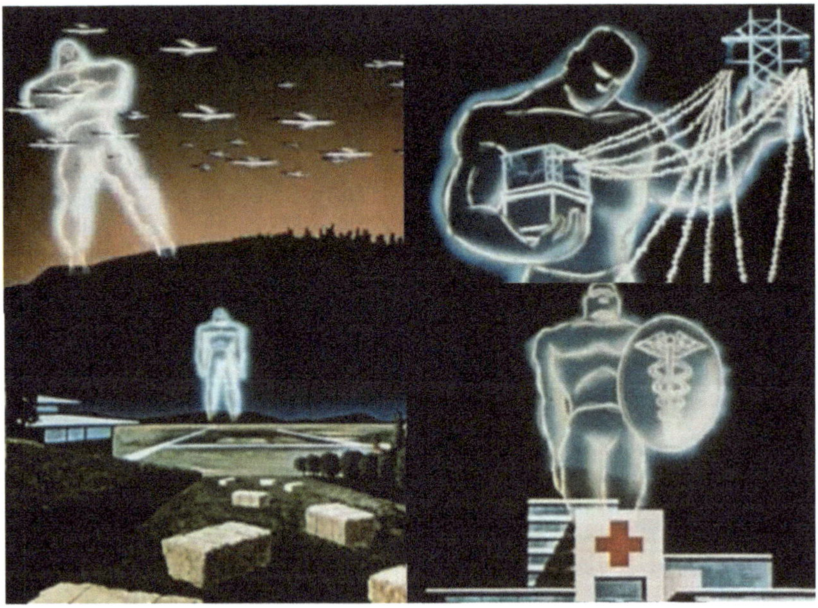

Source: Public Domain; Screenshot: Lars F. Köppen

Since France, the UK, and the USSR had gained nuclear technology and weaponry, the range of reporting expanded beyond US borders. When covering nuclear topics abroad, the majority of articles looked towards Russia, either for technological comparisons or to increase sales through scaremongering. The British made headlines, if very seldomly, with the development of economically feasible power plants ("Britain Leaps Ahead of Us in Race for Nuclear Power," *Popular Science*, August 1954, p. 85 & "Last Minute News and Notes: British Electricity goes Atomic," *Popular Science*, May 1955, p. 95).

After the years 1947 and 1948 had brought forth only very little news on both the civilian and military uses of nuclear power, reporting intensified in 1949, mainly due to the growing confrontation between the opposing blocs of the Cold War. In the April issue of *Popular Science*, Martin Mann investigated the advancements and setbacks in the development of reactors for energy production and vehicle propulsion at the Brookhaven National Laboratory in Upton, NY (Mann 1949). His article once again illustrated the amount of uncertainty surrounding nuclear technology. In his description of the test reactor, he likened it to a steam engine; and when it came to shielding materials, Mann proposed gold as the least expensive option, followed by titanium, rubidium, and cesium, which he admitted were either too rare or were impractical for other reasons.

Editor-in-chief Volta Torrey, on the other hand, was confident when reviewing a book by physicists R.E. Lapp and P.M.S. Blackett. As Torrey summarized, Lapp and Blackett argued that if a nuclear war were to break out between the US and the USSR, the aftermath could not be described in the dimensions and thus the categories of World War II. Despite releasing about 30.000 to 40.000 times the amount of energy that the most powerful conventional bombs of World War II had reached, atomic bombs were not merely that much more valuable to the military. Based on the premise that war was not about causing the biggest explosions but about hitting the enemy at critical points, there was no advantage to be found in choosing nuclear over conventional warheads. Concerning the affected population, the experiences from the surface bombings in London, Coventry, and other European cities had also become irrelevant, because civilian resilience had been rendered obsolete in the face of long-term contamination (Torrey 1949a). The fact that Torrey, presumably citing the United States Strategic Bombing Survey (Gentile 1997) in approval, concludes that "The best way to win a war is to prevent it from occurring," indicates that nuclear weapons had lost their aura as miracle weapons with "controllable" side effects. Just two months later, Torrey turned against nuclear propulsion in ships. Although the content of the article is peripheral—he elaborated on Martin Mann's previous article with a focus on submarines and battleships—the conclusion, again, is noteworthy: Bringing up plans to use submarines for planting naval mines with nuclear warheads, a principle he just had expressed his dismay of, he pleaded to his readers to reflect on the fact that as taxpayers, they were paying for such projects and that they should commit to international controls on nuclear weapons (Torrey 1949b).

The readers' reaction to the amount of coverage that nuclear technology-related topics received in *Popular Science* was somewhat ambivalent, as two letters to the editor from the October 1949 issue show. Mr. Robert Z. eloquently and unambiguously complained that scientific matters, as presumably opposed to the Do-It-Yourself section, took up less and less space in the magazine, and asked for more news from science and particularly from nuclear research. At the other end of the scale, Chester L. would have much preferred to be spared all the technical jargon and given more DIY tips:

> "Sir, I like your magazine very much, but one thing I don't like is stuff about atoms and things that take high geared brains to read. I'm not that smart. Excuse me if I'm blunt. I like things like making things for the darkroom and camera."

The editors answered this with a commonplace reply that despite their best efforts for a balanced publishing agenda, not all readers could be satisfied all the time ("Letters," *Popular Science*, October 1949). Whether or not *Popular Science* was really betraying its scientific claim in favor of DIY tips and household guides, as Robert Z. asserted, is open to discussion. On a descriptive level, regardless, Chester L. was right in pointing out the increase in frequency and scope of coverage on nuclear topics from 1949 onwards. It should be noted however, that despite their names, both magazines were never solely dedicated to scientific knowledge transfer but filled large parts of their issues with paid private and commercial ads.

After an overall silence on the issue in 1950, *Popular Science* opened the following year with two alarmist articles. In February, Alden P. Armagnac warned about something he called "radiological warfare," based on reports from the military establishment (Armagnac 1951). His concerns touched on a tactic that would be best described as a nuclear version of World War I poison gas warfare. The idea of dousing an invasion force in radioactively contaminated sand by airplane is at least strongly reminiscent of German, British, and French troops shooting each other with mustard gas and facing the rebound when the wind blew the gas back into their trenches at Verdun and the Somme. While *Popular Science* did not follow up on these speculations about radioactive sand-spraying airplanes, the article reveals a shift in consciousness among its journalists. They had realized that the actual danger of nuclear weapons lay not primarily in their TNT equivalent, but in the radiation dose and contamination following the blast. It is still somewhat remarkable that *Popular Science* seemed to limit its concerns to warfare, leaving the civilian applications of nuclear energy out of their criticism. Although the editors were aware that civilian reactors still needed proper shielding so that they did not contaminate their surroundings, similar objections to radioactive material as a potential fuel for cars and airplanes, arguably two quite accident-prone means of transportation and thus unfit for nuclear propulsion, cannot be found in the issues from the 1950s.

Besides these somewhat sensationalist and speculative extended features on how the coming war may or may not turn humankind back into cave dwellers and how cars would soon be zooming along with an engine bay full of uranium, the short notice section at the back of every issue provided a certain amount of journalistic balance. The March 1951 issue clarified that the USSR did indeed have the A-bomb, that nuclear energy was currently not fit to replace coal or petrol as a mass power source, and that the only serious research into nuclear propulsion was related to ships (Githens/Torrey/Newman/Howard 1951). However, as another reader's letter demonstrates, the belief in feasible nuclear propulsion outside of ships was persistent at the time. Owen B. of Los Angeles asked the editors about their opinion on an "atomic air train," essentially an assembled train of glider airplanes drawn by a nuclear mothership airplane on intercontinental flights ("Letters: Atomic Air Train," *Popular Science*, January 1952, p. 14). Albeit obviously impossible for economic, technical, and safety reasons, the idea laid out in the letter illustrates the tightrope walk popular science magazines of the time had to perform.

On the one hand, they could not afford to ruin their reputation as serious magazines through unfounded science fiction; on the other hand, their audience expected, and rightfully so, to hear of breakthroughs as soon as possible. Thus, they had to report on innovative and unconventional ideas before they were ripe for actual application, even risking that in the following issue they would have to admit that, once again, the revolutionary idea they had advertised had turned out to be a castle in the air. In this situation, they were almost forced to oscillate between technological euphoria and disillusionment, between nuclear power as a panacea and as a harbinger of the apocalypse. The prime example here is the coverage of nuclear propulsion from 1945 to the late 1950s. Immediately after Hiroshima, this was the first civilian application for nuclear power advertised in *Popular Science*; in 1951 it was ruled out, with the exception of marine vehicles; and then, in April 1954, it was predicted to revolutionize railroads within maybe just two years ("An Atomic Locomotive could be built," *Popular Science*, April 1954); and in October 1957, *Popular Science* expected half of the US Navy to have nuclear propulsion by 1966 ("The Navy goes Nuclear," *Popular Science*, October 1957). Similar sine-line-like changes of mind can be seen in the coverage of atomic bombs. After initially believing that atomic bombs could be "tamed" and put to civilian use, the editors soon came to realize that trying to stop hurricanes this way was not only nonsense from a meteorological perspective, but frankly represented an added danger to the population. Then in June 1954, the magazine interpreted the recently tested hydrogen bomb as the key to a peaceful development of nuclear technology and a promise of "atomic miracles" ("The Good News About the H-Bomb: Atomic Miracles Are Nearer Now," *Popular Science*, June 1954), only to admit one year later that, actually, the hydrogen bomb was just a more fearsome and dangerous version of the atomic bomb ("How the H-Bombs Spread Radioactivity," *Popular Science*, April 1955).

Popular Mechanics did not muster the enthusiasm of its counterpart when playing through scenarios of nuclear warfare. However, it developed a critical stance on the economic and ecological sustainability of nuclear energy production. In April 1955, writer Rafe Gibbs pointed out that a concept for nuclear waste disposal had yet to be found, with storage for said waste driving up the operating costs of power plants (Gibbs 1955). Due to the public's ongoing interest in this topic, *Popular Mechanics* had established a running column called "Keeping up with the Atom" where the editors published the results of a 1953 reactor test accident, during which the researchers had let the test reactor run out of control on purpose and then observed the results. According to them, the fuel overheated and then exploded, launching fragments about 350 feet away while sounding like "a moderate amount of chemical explosive," with no radioactive material leaving the test site and no lethal radiation doses being recorded ("Keeping up with the Atom," *Popular Mechanics*, October 1955). At first glance, the lack of objection to this report by the editors may seem astonishing, but in reality, until the Kyshtym disaster of 1957 (Stone 1999), the only known precedents for comparing nuclear explosions they had were the bombs on Hiroshima and Nagasaki and other bomb explosions at test sites, i.e., military applications. With this lack of background information, it was easy to convince oneself that what was not meant to harm also could not harm. Furthermore, the researchers themselves might have been convinced that their results were sound and subsequently believed that civilian nuclear technology as a whole was safe and controllable.

CONCLUSION

Technological enthusiasm among the readers as well as among the contributing journalists was the raison d'être for science magazines then and has remained so to this day. This dynamic produced outcomes like the ones shown in this study. When the magazines reported on the research from Project Plowshare, the atomic program for civilian and peaceful aims which included the use of atomic bombs for creating sea ports and underground mining (Kirsch 2005), they did not necessarily hold a firm belief that the project would definitely succeed. Instead, they were convinced that it would make an interesting story; and, what's more, a large-scale government-funded project like this one could not go unreported in any case. If the project had succeeded, the magazines could claim to have known about it and to have supported it right from the beginning. If not, they still got away with it and the readers' trust in their expertise did not fade, as letters like the one from Owen N. about "atomic air trains" proved. This phenomenon was, by the way, not confined to nuclear issues. When Wernher von Braun proposed satellite-based solar power as an alternative to nuclear and coal energy, *Popular Science* was ready to spread the word (von Braun, *Popular Science*, 1972), although the research into it has as of today not resulted in any feasible models, and some have dismissed the idea itself as sheer folly (Caton 2015).

In that context, popular science magazines did little actually to translate research results into a readable digest, and instead merely informed on the aims of the research. This can be attributed to a fundamental problem of science journalism: Every simplification bears the risk of distortion, and even moreso when the specific target audience has a broad spectrum of educational backgrounds and interests; however, simplification is necessary if the goal of science journalism is reporting beyond the laboratory. Because it was hard to estimate the average reader's education concerning nuclear science, the magazines did little to introduce the reader to the relevant parameters and theories. Instead, they unified all parameters into a single parameter, namely: "Hiroshima." Hiroshima, with Nagasaki pointedly left out, became a recurring informal parameter for measuring the power of nuclear applications, both civilian and military. Bombs were predicted to kill x times as many people as the Hiroshima bomb; reactors, if they failed, would emit x times as much radiation as the bomb on Hiroshima; the fuel for a reactor provided x times as much energy as the material used in the bomb on Hiroshima; and so forth. The most striking example of this would be found in the report on the 1955 reactor failure experiment ("Keeping up with the Atom," *Popular Mechanics*, October 1955), but the use of Hiroshima as a point of comparison can also be traced back to other articles cited above. Albeit extremely imprecise and distorting, one could argue that "Hiroshima" was the parameter that could safely be assumed to mean something to every reader in the 1950s. It should be noted however, that the material used in this study provides little evidence that the respective editors intentionally tried to establish said parameter or that they had laid out a strategy about how to report on nuclear technology issues beforehand. After all, nuclear technology was only one among many topics they were covering. The unification of parameters was apparently not the result of one executive decision, but rather born out of a situational dynamic between the readers' wishes, the developing understanding of the matter among the magazines' staff, and, as argued above, the need for entertaining stories. To some degree, the magazines arguably repeated the patterns of epitomization that M. Sehlmeyer discusses in his contribution to this volume. Thus, what the magazines did was not so much provide insight into new technologies, but rather help people make sense of such new technologies.

Thus the editorial decisions outlined above beg further investigation. How were those decisions reached? Did later generations of editors reflect upon those decisions, and if so, in what way? Material that documents these processes of the magazines' agenda setting, such as internal memos, editor conference protocols, or even unpublished articles and letters to the editor, has so far proven harder to come by. If finally uncovered—along with, for instance, private correspondence of the editors-in-chief at the time in question—these documents might provide further valuable insights.

REFERENCES

Caton, Jeffrey L. (2015): "Space-Based Solar Power: A Technical, Economic, and Operational Assessment." In: Thompson, C. P. (Ed.): Space-based solar power: feasible idea or folly? New York: Nova Science Publishers, pp. 1-40.

Cooke, Stephanie (2009): In Mortal Hands: A Cautionary History of the Nuclear Age, New York: Bloomsbury.

Gentile, Gian Peri (1997): "Advocacy or Assessment? The United States Strategic Bombing Survey of Germany and Japan." In: Pacific Historical Review 66/1, pp. 53-79.

Hacker, Barton C. (1994): Elements of Controversy: The Atomic Energy Commission and Radiation Safety in Nuclear Weapons Testing, 1947-1974, Berkeley: University of California Press.

Haller, Christian (2013): "Das Explosionsunglück in der BASF vom 21. September 1921. Katastrophenwahrnehmung und -verarbeitung in Presse, Politik und Fachwelt." In: Zeitschrift für die Geschichte des Oberrheins 161, pp. 325-375.

Holloway, David (1994): Stalin and the Bomb. The Soviet Union and Atomic Energy 1939-1956, New Haven: Yale University Press.

Josephson, Paul R. (2005): Red Atom. Russia's Nuclear Power Program from Stalin to Today, Pittsburgh: University of Pittsburgh Press.

Karlsch, Rainer (2005): Hitlers Bombe. Die geheime Geschichte der deutschen Kernwaffenversuche, Munich: Deutsche Verlagsanstalt.

Kirsch, Scott L. (2005): Proving Grounds: Project Plowshare and the Unrealized Dream of Nuclear Earthmoving, New Brunswick: Rutgers University Press.

Muller, Hermann Joseph (1927): "Artificial Transmutation of the Gene." In: Science, July 1927, pp. 84-87.

Radkau, Joachim; Hahn, Lothar (2013): Aufstieg und Fall der deutschen Atomwirtschaft, Munich: Oekom-Verlag.

Radkau, Joachim (2017): Geschichte der Zukunft: Prognosen, Visionen, Irrungen in Deutschland von 1945 bis heute, Munich: Carl Hanser Verlag.

Rose, Paul Lawrence (1998): Heisenberg and the Nazi Atomic Bomb Project – A Study in German Culture, Berkeley: University of California Press.

Sanner, Lisa (2015): "Als wäre das Ende der Welt da." Die Explosionskatastrophen in der BASF 1921 und 1948. In: Veröffentlichungen des Stadtarchivs Ludwigshafen am Rhein 43, Ludwigshafen.

Stern, Philip M. (1969): The Oppenheimer Case: Security on Trial, New York: Harper and Row.

Stone, Richard (1999): "Retracing Mayak's Radioactive Cloud." In: Science. 283, no. 5399, pp. 164.

Walker, Mark (2003): "Otto Hahn – Verantwortung und Verdrängung." In: Ergebnisse. Vorabdrucke aus dem Forschungsprogramm „Geschichte der Kaiser-Wilhelm-Gesellschaft im Nationalsozialismus," Berlin.

Source Material

Armagnac, Alden P. (1951): "What you should know about 'RW.'" In: Popular Science, February 1951, pp. 144-148.

Gibbs, Rafe (1955): "The Worlds Hottest 'Garbage.'" In: Popular Mechanics, April 1955, pp. 124-127 and 250-252.

Githens, Perry (1946): Unhappy Birthday, in: Popular Science, August 1946, p. 7.

Githens, Perry/Torrey, Volta/Newman, F. O./Jensen, H. C.(1951): "P.S. – Last minute news and notes." In: Popular Science, March 1951, p. 280.

Kent, John L. (1946): "Will the Atom drive us Underground." In: Popular Mechanics, October 1946, pp. 148-151 and 240-248.

Mann, Martin (1949): "U.S. Lights New Atomic Pile for Peace." In: Popular Science, April 1949, pp. 121-128.

McDermott, Wm. F. (1945): "Bringing the Atom down to Earth." In: Popular Mechanics, November 1945, pp. 1-6 and 160-168.

Rutherford, Ernest (1915): "The Constitution of Matter." In: The Popular Science Monthly, August 1915, pp. 105-142.

Torrey, Volta (1949a): "So A-Bombs aren't so bad?" In: Popular Science, May 1949, pp. 124-126.

Torrey, Volta (1949b): "Will Atomic Engines be Mobile." In: Popular Science, July 1949, pp. 114-119.

Iconophilia of the Brain, Stage 3? An Epistemic Regime, the Popular Science Magazine *Gehirn & Geist*, and Visual Culture

Dirk Hommrich

ABSTRACT

This article argues that brain research attracts public attention because popular science journalism uses visual means which allow the juxtaposition of seemingly (self-)evident images of brain imaging with everyday images, i.e., pictures which seem to be close to ordinary images and stereotypes of everyday life. The first part of this essay describes the German popular science magazine *Gehirn & Geist* ("Brain & Mind"). The general characteristics of the magazine, its main discursive strategies, and the multimedia presentation of *Gehirn & Geist* are interpreted as means through which proximity to the public and the everyday relevance of brain research are achieved. The second part of the text identifies pivotal types of pictures within the vast and varied visual culture of popular brain research. I conclude that the visual culture of popular brain research is tending toward a kind of new iconophilia—beyond those (functional) brain images we are already used to.

ICONOPHILIA OF THE BRAIN

It is tempting to conceive our makeshift ideas and imaginings of brain research and also journalism of the neurosciences as something exterior to factual assertions regarding neuroscientific findings and theories. However, the popularization of specialized knowledge plays an integral role in constituting brain research and the neurosciences within the contemporary science system and with the general public.

There is certainly more than just one single way of studying those public offerings of knowledg about the brain, the nervous system, etc., and of how such knowledge is presented. This article therefore starts with an epistemic claim that both neuroscientists and scholars of Science and Technology Studies seem to share: For many neuroscientists, the popularity of modern brain research relies on their scientific standards and on powerful instruments such as non-invasive measurement techniques (which, incidentally and for convenience only, produce images). In comparison, Science and Technology Studies argue that the popularity of brain research is primarily linked to the persuasiveness of the visualizations produced by brain imaging technologies in the context of popular science communication. In both cases brain images are prior to other images, scientific or non-scientific. And in both perspectives one might expect science journalism to mainly use brain scans to demonstrate the verisimilitude of modern brain research—and thereby to primarily produce what might be called "popular brain research" (cf. Heinemann/Heinemann 2010; Heinemann 2012; Peters et al. 2013: 328-329).

Putting the enchanting spell of the brain image into a long-term perspective, the historian of science Michael Hagner spoke of the second stage of an iconophilia of the brain which has been (or still is) driven by new functional imaging techniques such as PET and fMRI (Hagner 2006: 219-222). For Hagner, popular visualizations of imaging technologies represent a kind of distorting mirror and projection surface for reductionist and all-too-easy conceptions such as "the mind is the brain," "thoughts are biochemical processes," or "the brain defines personhood." When (functional) brain images are used for the purposes of journalism and public relations, these visual artifacts figure as "labels" and promotions for big neuroscience (Hagner 2008). According to Hagner, brain researchers and neuroscientists agglomerate symbolic and cultural capital in society, which in turn generates beneficial effects within the science system and the competitive conditions of (external) research funding. Concerning the seductiveness and the popularity of brain research, the second stage of the iconophilia of the brain is seen as reminiscent of the first phase in the 18th and 19th centuries, when phrenology claimed to unveil aspects of personality and point to deviant character traits. Hagner therefore speaks of "cyberphrenology" when he refers to imaging technologies which claim to reformulate and localize human traits as brain functions (Hagner 2002).

Hagner's historic insight and the scenario he depicted were connected to the thesis of a new phrenology (cf. Uttal 2001), namely, that brain scans evoke a discursive effect which in turn tends to confuse subjectivity (consciousness, psyche, personality, etc.) with the brain organ, and which can, however, also be transferred to non-scientific images such as advertising graphics or cartoons, which are not created by means of modern technologies such as MRI, PET, and fMRI, and which do not have a (quasi-)argumentative function. The expression "brain image" within Hagner's publications seems to be transferable to morphological, naturalistic-organic, or quasi-anatomical images, and seems to emphasize that subjectivity and personality are being increasingly equated with the human brain among the general public (Hommrich 2019). Hagner's colleague Cornelius Borck describes the effect of being able to see the brain thinking, etc., as a large-scale scientific-historical development that tends to blur the difference between visualization-technological products and neurophysiological processes: "[f]or a long time tools and instruments were simply not sophisticated or complex enough for use as compelling correlates; [...] the arrival of functional imaging promised to discard functional comparisons in general and to conflate the technological visualization with a neurophysiological substrate" (Borck 2012: 114).

I certainly do not want to deny the merits of critical approaches that trace 'science' and scientific assertions in discourse and follow the translation chains of 'true proven facts' and their visualizations in processes of popularizing scientific knowledge or technologies (cf. Hagner 1994, 1996; Dumit 2004). However, such an investigation into the popular pictorial staging of the results of neuroscience seems to have already bought, if not into the scientific claims of the epistemic regime themselves, then at least into the general focus on questions of evidence and truthfulness (cf. Hommrich/Isekenmeier 2016). In other words, in addition to the suggestive effects of scientific data-driven images of the human brain, non-scientific pictures are important for directing public attention towards brain research in the media. Scientific visualizations of the human brain are embedded in a whole variety of 'ordinary' images, and the visual artifacts of brain imaging unfold their rhetorical effects in a comprehensive "visual culture" of brain research.[1] Neuroscience operating with (whole) brain imaging attracts public attention

1 This article refers to my PhD thesis "Theatrum cerebri. Studien zur visuellen Kultur der populären Hirnforschung" (Theatrum cerebri. Studies of the Visual Culture of Popular Brain Research; Hommrich 2019, forthcoming). Within STS research on the neurosciences, the analytical dimensions of the book aim at contributing to several academic fields and transdisciplinary topics: 1) From a *philosophy of science* perspective, my studies analyze which functions the popularization of neurobiological findings, models, and explanations serve for the constitution of the neurosciences as an epistemic endeavor. 2) Drawing on *phenomenological approaches* to both the media and pictorial emanations, the book describes how visual artifacts produce "evident" images that define the contours of the public discourse on neuroscience. 3) Regarding *social philosophy*, my PhD thesis provides insight toin how scientific and popular neuro-talk also enacts the discourse on human enhancement and visions of an optimization of human "potential." 4) Contributing to *metaphorology* and *visual rhetoric*, my studies of the visual culture of popular brain research analyze images of localization, visualization, and causation which structure the discourse on neuro-knowledges and their

because popular brain research uses visual and interpictorial means which allow the juxtaposition of evident images of brain imaging with everyday images, i.e., pictures which seem to be close to ordinary images and stereotypes of everyday life. Brain images refer to specialist scientific codes, whereas images of social life and popular culture refer to commonplace codes. By opening up the pictorial horizon of STS research into the neurosciences, the analysis of, for instance, the popular science magazine *Gehirn & Geist* demonstrates that popular images do indeed have a stake in the communication of cutting-edge research domains.

Thus, the question arises as to what extent the example of *Gehirn & Geist* can be used to speak of a "dominance of the visual" (Hagner 2007: 303), or what kind of iconophilia the magazine promotes. This in turn begs the question of how Hagner conceptualizes the serial arrangement of brain scans and brain images necessary to establish a model function or a centrality of 'the' brain image. If one assumes that similarities between pictures must be present in order to be able to produce a diachronic or synchronous series of brain images, which in turn would establish a scopic regime of 'the' brain image, such cerebrocentric series can result from aesthetic similarities in motif or formal design. The commonalities of brain images, which are shown and cemented in such pictorial series, could thus be reconstructed equally by epistemic and aesthetic qualities (cf. Heßler/Mersch 2009: 42-46; Heßler 2009: 141-146), so that the morphological structure of the brain as visualized in functional brain scans can be just as easily repeated with further brain scans as with advertising or title pages that expose the brain motif. The re-presentation of formal image properties, in turn, can, for example, be produced with colored activations of a brain scan or repeated with contours or lines for the human brain. From this perspective, the question as to whether 'the' brain image should be conceived of as having an epistemological and evidential value or an aesthetic value as an image turns out to be a precarious one.

A SHORT MEDIA PHENOMENOLOGY OF *GEHIRN & GEIST*

For laypersons, popular science is the mouthpiece of science; for scientists, popular science is a seismograph of the social world as well as the researcher's 'delegate' in the question of basic cultural assumptions of scholarly communication.

---

mechanisms of authorization, legitimation, and integration. 5) From a *cultural and communication studies point of view*, as these pertain to popular science, my book shows how popular science combines specialized knowledge and technological imagination both with images of usefulness and with some kind of "pragmethics" regarding normative expectations toward brain research.

Popular science journalism, according to a widespread view, mediates between experts and laypersons by processing and transforming the knowledge of specialists in a layman-friendly way. While scientific communities and whole research domains organize their knowledge acquisition and knowledge production via systematic methods and technical languages that are difficult and very time-consuming for non-experts to understand, experts who speak publicly and journalists both depend on concise forms of presentation as well as simplifications, comprehensibility, and clarity. When popular media inform and report on specialist scientific findings, they transform complex vocabularies into generally comprehensible meanings, as well as into the (presumed pre-)understandings and makeshift ideas of their addressees that can be connected to everyday life and the world in which they live. In this sense, popular brain research—neuroscience as a publicly staged science—is essentially a media phenomenon made possible by the collaboration of science journalists and scientists.

Of course, journalistic techniques from the popular mediation organs of science are by no means trivial. Such attempts to mediate between science and the general public can refer to literary, historical, or completely unrelated examples. They can use helpful suggestions, they can be made of vivid narratives and follow story lines (cf. Viehöver 2011: 180-187), and they can obey certain schemes within a framework (cf. Donati 2011: 151-154). Moreover, science journalism and popular science discourses can present hypothetical thought experiments and use various types of images, such as paintings, comics, photographs, charts, diagrams, and computer graphics. In addition, the significance of plausible comparisons, sound analogies (cf. Hofstadter/Sander 2014), adequately described models (cf. Tetens 2006), and the handling of metaphors and images (cf. Tetens 1999) should not be underestimated regarding the work of science journalists, especially their function as a "formula of meaning" (Geideck/Liebert 2010), as a model (cf. Dierkes et al. 1992), or, as Jäger/Jäger (1997) put it, as a "collective symbol." For example, the "symbolic coding of the human body as a machine [...] can take place in different ways. On the one hand, it can come along as illustration—for instance as caricature. But it can also be staged on film—think of the numerous films in which a human being is transformed into a robot. Finally, the coding can take place in speech images. The decisive moment that makes collective symbolism so effective is its general comprehensibility and the 'effect' of making highly complex and often extremely contradictory facts comprehensible through it." (Jäger et al. 1997: 21).[2] In other words, the representational means used in institutionalized science journalism and in mass mediated expert

2 Unless otherwise noted, all translations in this essay are my own.

knowledge pose a central and permanent professional challenge, which comes to light in the establishment of new branches of science, such as brain research: The public and, to a large extent, journalistic background noise ensures, as it were, that the fanfare procession of brain research is heard above all else. And popular brain research, with its visualizations, establishes performing means and a general visibility that are made for professionals as well as amateurs. They play a central role in media coverage of the neurosciences.

Science journalism has the dual role of demonstrating knowledge and expertise and at the same time addressing 'everyday people' by moderating between different reference systems. Collective orientation patterns and preconceptions, which are effective in the linguistic and visual mediation of scientific knowledge to non-experts, can be found above all in the "interdiscourse" (cf. Link 1986; Jäger 2009: 131-133), in the accompanying "paralanguage" (Janich 2009: 90-135), or in the presentation of knowledge or images in a public "context of persuasion" (cf. Schnettler/Pötzsch 2007: 478). Despite all their differences, these concepts all share one essential feature, namely that they characterize the dissemination of expert knowledge to laypersons through their openness and ability to connect with various vocabularies and subject areas, which are necessary for the rhetorical conviction and public acceptance of statements. Non-fiction books and journals form a kind of podium or arena that is set up in front of an interested public, addressed to the uninformed, and that, occasionally, invites joint debate or reflection. In the moderation between science and society, popular mediation instances amalgamate and mix, in other words, scientific and everyday representations—in both textual and pictorial terms. In this respect, formats of science journalism in particular not only resort to, iterate, and transform general discursive patterns of a socio-cultural nature; they also use and produce popular ideas about the relationship between mind and brain, fields of application of brain research, neuro-technologies, and neuro-focused interventions. The popular staging of brain research thus not only establishes its own knowledge space in textual, literary, and discursive terms; the reinvention of 'science' in the public sphere also involves the organization of a hybrid visual culture, by merging images as media of science communication (Hüppauf/Weingart 2009) and common assumptions and concepts of evidentiality. Before turning to visual culture, I provide an outline of such a public transformation zone of 'science.' As sample material, I refer to the popular science magazine *Gehirn & Geist* ("Brain & Mind"), which can be considered a gatekeeper of popular brain research in German-speaking countries.

Drawing on a formulation by Roger Cooter, it can be said that *Gehirn & Geist* represents an example of a successful attempt "to capitalize on the neuro-turn of our popular culture" (Cooter 2012: 93). Significantly, *Gehirn & Geist* is also involved in turning the current capitalization of the 'Neuro-Turn' into a popular culture of scientific knowledge: The journal offers its readers the prospect of up-to-date information, 'brain literacy,' and psychological competency development. The contextualization and anchoring strategies, which are oriented towards topicality and a proximity to everyday life, create the suggestion that the magazine is a journalistic "mirror of society" and that it reports on scientifically authorized advice for personal problems. In this respect, *Gehirn & Geist* includes both a strong reference to everyday life and a creolization of science with everyday topics and scientific objects. The mixture of everyday language and scientific language thus functions as a "strategic means of gaining acceptance" (Jäger/Jäger 1997: 316).

Founded in 2002, *Gehirn & Geist* is a subsidiary journal of *Spektrum der Wissenschaft*, and the magazine can be regarded as the flagship of popular brain research. Both magazines are published by the Spektrum Verlag (Heidelberg). While other German popular science journals such as *Spektrum der Wissenschaft, Technology Review, Bild der Wissenschaft, P.M., Natur & Kosmos, Hörzu WISSEN*, and *Geo* are multidisciplinary, *Gehirn & Geist* distinguishes itself from such formats primarily through its programmatic competence for, according to its subtitle, "psychology and brain research." Despite this specialization, *Gehirn & Geist* has a characteristic similarity with other general-interest journals that can be assigned to the field of popular "science" in terms of their specialist publications—which, in comparison to daily periodicals on the one hand and journalistic news publications on the other, are hardly geared towards current events. First of all, the publisher Spektrum Verlag uses this specialization in psychology and brain research to complement its range of services and topics. The publishing house makes use of *Gehirn & Geist*'s thematic area by publishing isolated contributions from *Gehirn & Geist* in its own interdisciplinary journal *Spektrum der Wissenschaft* (SdW) (and vice versa), which also edits special issues dealing with the "improved brain" (SdW Highlights: 01/2008), with "brain and consciousness" (SdW Dossier: 05/2009), or with the "fascinating brain" (SdW Dossier: 02/2011). Its unique selling point of being the only journal in the German-speaking countries with a focus on brain research and psychology has given *Gehirn & Geist* a favorable market position and widespread distribution in kiosks and public libraries. Heinemann/Heinemann (2010: 291) stress the importance of the journal in promoting the general publicity and visibility of brain research. The fact that *Gehirn & Geist* survives in the media landscape can be regarded as an indication of the overall popularity of brain research.

In general, it must be pointed out that the journal deals with a wide range of topics and, with 90 to 98 pages (depending on the issue), that it offers a comparatively large amount of space to discourses on brain research and psychology.

As a monothematic, transdisciplinary, and illustrated public journal, G&G differs from daily newspapers, specialist journals such as *Nature Neuroscience* or *Frontiers in Neuroscience*, and also from most popular scientific television formats. The discourse of *Gehirn & Geist* varies and is quite dynamic. On the linguistic level, the journal uses a comprehensive range of journalistic presentation techniques, textual genres, and linguistic styles (vocabulary, allusions, idioms) to translate "sometimes dry research into [...] exciting magazine stories" (editorial, 06/13: p. 3). Published monthly, it is the only journal in the German-speaking world to report on "brain research and psychology" on a monthly basis. The first issues of the early years described and advertised the journal with the statement: "*Gehirn & Geist* is the new magazine for brain research and psychology from Spektrum der Wissenschaft. Leading experts report about the world in our head" (01/2002: 4). *Gehirn & Geist* has not only scientists but also journalists as contact and cooperation partners, sources of information, and authors. Furthermore, it has a scientific advisory board, which vouches for the (scientific) quality of the journal and the articles it contains. The members of the advisory board, researchers who are proven in their field and who have stellar reputations, give *Gehirn & Geist* the nimbus of scientific authority.

The internal structure of *Gehirn & Geist* is characterized by a complex compilation of articles from different text and image genres, and with different journalistic and advertising functions. Reader guidance and entertainment go hand in hand, because in order to make it easy for readers to find their way through the richness and complexity of the content and the visuals, and to maintain an overview of *Gehirn & Geist*'s discourse, the first pages of the journal contain an editorial commenting on the issue and an index of its contents. In addition, for the purpose of greater comprehensibility, the articles are supplemented by short messages, for example keywords, marginal notes, and pull quotes, which are presented in various sizes and colors; furthermore, summary or overview boxes (titled "At a glance"), in which the main theses and core statements of the texts are listed, are built in to the articles. *Gehirn & Geist* also sometimes works with short glossaries, textboxes supplementing or exemplifying matters ("More on the subject," "What is ...?", e.g., 05/2005: 33), and tabular presentations and graphics, all of which serve to explain terms, models, and concepts, to illustrate the text, or to loosen up the visual information density in general. The ease with which *Gehirn & Geist* can be read compared to specialist journals such as *Neuron*, *Nature Neuroscience*, or *Nature Reviews Neuroscience* is further enhanced by high-contrast colored markings or striking word and/or picture marks, which quickly orient the reader and can also establish connections between several articles on a specific title topic, category, or thematic lead.

Each issue of *Gehirn & Geist* is around 92 pages long with an average of 10 to 11 articles per issue, with a standard length of approximately six printed pages per article, and with interviews that are often three to four pages long. The repertoire of the contributions covers glosses, comments, letters to the editor, reports, and reviews, and includes expert debates as well as "laboratory visits" (01-02/2009: 38-39) or curiosities; it ranges from largely factual presentations or reports through interviews that reflect the opinion of an expert or layperson on a certain topic, and on to 'brain teasers.'

If you open the magazine's table of contents, you will first find a short summary of the title topic of the issue. In addition, the articles of *Gehirn & Geist* are arranged according to the relevant domains of knowledge, namely: brain research (10 to 20 pages) and psychology (also 10 to 20 pages). In addition to these two main domains, contributions to other scientific and practical areas such as medicine, pedagogy, artificial intelligence, and philosophy are given around 10 pages; these subject areas, however, are not present in every single issue, but appear from time to time from 2002 to 2015. The magazine's sections also include the editorial, information on (free) online contributions on the homepage of *Gehirn & Geist*, letters to the editor, the "Better Thinking" guide section, and the "On the Air" overview of thematically appropriate and upcoming contributions to various television and radio programs, a calendar of events ("dates") around "mind and brain," book reviews (divided into "Short and Concise," detailed discussions under "Books and More" and "Shop Window – Further New Publications," in the style of short pieces on recent news, which are also summarized under the heading "Brainwave").

Regarding its reading public, *Gehirn & Geist* strives for the broadest possible inclusion, i. e., it strives to expand its target group(s), and also aims for both deliberative and "dialogic" science communication. These claims and efforts to organize consent and dissent among the readership certainly imply implicit knowledge policies of the magazine. Further central characteristics of *Gehirn & Geist*'s editorial institutionalization of popular brain research in the field of science journalism are its suitable media- and communication-technology-based offerings for community-building and identification (on the part of the reader) both with the journal and the topics it deals with. Among its inclusive strategies, *Gehirn & Geist* fosters advantageous conditions to hold and increase the stock of regular readers. The canonization of knowledge, the strategically promoted possibilities for the reader to identify with *Gehirn & Geist*, and the deliberative imperative of this discourse domain legitimize the internal order of the journal and the periodical as such. In this respect, *Gehirn & Geist* grosso modo does not organize value-neutral communication with regard to neuroscience and neurotechnology, but rather formally invents an independent transformation zone of popular brain research that goes far beyond the journalistic tasks of the report and the commentary. *Gehirn & Geist* is thus decisively involved in the generation of public and mass-media attention for the neurosciences and in popularizing brain research in general terms. This is a genuinely *creative* process that, in a certain sense, takes place alongside the science system and that uses particularly affective modes of communication that can be observed in visual communication. Within this transformation zone, the collaboration of (science) journalists, researchers addressing the public, and interested laymen first of all invent "neuro-topics" from scratch. In most cases, these topics are presented as being of societal and personal relevance.

Hagner's claim that scientific images dominate media representations of brain research ignores the variety of brain images within the magazine. Neither the display of scientifically produced (functional) brain images nor the "brain image as label and promotion" within *Gehirn & Geist* should make one forget that the magazine also uses image strategies and visual tactics that on the one hand serve to entertain or make it easy to consume the magazine's content and on the other hand strive for social conditions and contexts in the sense of visually creating proximity to the "meaningful realm of day-to-day experience" (Borck 2012: 118; cf. Sturken/Cartwright 2009: 349-350). Both the linguistic, metaphorical register and that of the visual and graphic manifestations operationalize and functionalize visual phenomena in order to effectively expand the knowledge zone that *Gehirn & Geist* has established.

POPULAR SCIENCE & PICTORIAL VARIETY

On the one hand, one may analyze *Gehirn & Geist* by focusing on the consistent use of scientific and pseudo-scientific visualizations of the brain in the journal's appearance and promotional material. But the creative staging of brain images, on the other hand, needs to be thought of as an aestheticization and as an independent act of producing visual forms whose seductive powers refer to only one of many other forms of visual evidence production. The magazine catches the eyes of its readers by offering a whole variety of different visualizations that is far greater and more comprehensive than the focus on the promise of the objectivity of tomograms and that of "machines as eyewitnesses" (Geimer 2002) suggests. At this point, even the pseudoscientific adaptations and imitations of (functional) brain scans cannot divert from the necessity to also focus on those images that manage to seem evident without a brain motif and without a scientific document function. Visual manifestations, which are conducive to public attention for brain research, or at least to generating attention for *Gehirn & Geist*, to a certain extent stage 'cerebral worlds.' For instance, the use of brain images on the title pages of *Gehirn & Geist* can be roughly divided into two classes: First, the alleged evidence of brain scans serves as presentable 'evidence' for today's possibilities of functional imaging. In addition to this illustration and documentation function regarding neuroscientific knowledge about the brain, both scientific and non-scientific brain images are used in a futuristic and mysterious manner to indicate groundbreaking, promising (technical) developments in brain research and the neurosciences.

Moreover, when *Gehirn & Geist* uses vivid, everyday images and motifs (images with stereotypical subjects referring to the general lifeworld), the primary discourse effect is that, at first glance, they serve as entertainment and illustration. These images evoke aesthetic effect, affects, and reader-specific memories or references to the cultural memory in the register of visuality. Thus, it is not only normalized and normalizing images such as brain scans that are introduced into the discourse of *Gehirn & Geist*, but also seemingly banal images, which prompt the recipients of the journal—subtly or overtly—to reflect on themselves in the light of psychology and brain research. In this respect, references to commonly known visualizations not only serve to trigger associations on the part of the readers (between images of science and everyday perceptions); these visual commonplaces also anchor *Gehirn & Geist* in the visual culture, so that one can speak of a mimesis or adaptation of science journalism to general media culture, which forms a decisive contribution to the production of the magazine's catchy vividness. Therefore, a media phenomenology of popular science and of contemporary science journalism also has to include pictures which are adopting, citing, and transforming elements of popular culture to promote and contextualize science as part of social life and personal lifeworlds.

Assuming that such borrowings, i.e., references to popular culture, reinforce connectivity and the emotional and affective impetus of (visual) communication just as much as visualizations close to everyday life and perception, this is the first of two types of images that should be taken into account for an analysis of visual culture in popular brain research. The usefulness and usability of such illustrative images regarding contextualizing evidentiality add "aesthetic claims to validity" (Schneider 2011) to the epistemic regime of popular brain research as well as to the neurosciences and psychology. Social images and images of popular culture "charge" science communication with passions and emotions one can easily connect with. Such aesthetic seductions are not simply directed at a general audience, but *produce* the readership of science journalism in a decisive way. In this sense, vivid and useful images can be characterized as "rhetorical figures that form the symbolic level of a media system, working on the permanent invisibility and naturalization of its technologies on the one hand and its institutional and ideological integration on the other" (Nohr 2014, p. 26). Moreover, such "normal" images generate a permanent opening for and extend a visual invitation to all readers. These images appeal to affect and vision and thus encourage people to look at them.

If we want to address the visual culture of popular science magazines of today, we therefore should be willing to transform and reconfigure philosophies of evidence (cf. Kelly 2014) and of "pictorial" or "visual arguments" (Mersch 2006; cf. Heßler/Mersch 2009) to accommodate the use of popular images within visual science communication and to take into account the 'atmosphere' and the 'feelings' produced by non-scientific pictures (according to Burri, pictures appeal to their audience as "aesthetic seduction," 2008, p. 350). The pages of *Gehirn & Geist* display visual forms ranging from photographs of scientific objects through reproductions of works of modern art, advertising images, and photographs of playing children, to cartoons and comics. These illustrations remind their viewers of social events, of "scenes" in the sense of something seen elsewhere. These images appeal on the one hand to an individual's personal visual memory by staging a proximity to the everyday and everyday objects; on the other hand, such vivid representations enable readers of *Gehirn & Geist* to both recognize and view the reports on brain science and its results as 'natural' constituents of their lifeworld.

Generally speaking, there are at least two types of "vivid visualizations" which are used in the magazine's arena. First, one can find visual representations that remind 'us' of social contexts and scenes. In contrast, the second type of "vivid visualizations" comprises pictures that have been adopted from the communicative and cultural memory which we might refer to as pop (media) culture. While images that suggest that they refer to social life are only accessible by means of the introspection of a recipient, this second type of image refers to the importation and translation of well-known images as well as to the staging of famous icons.

In *Gehirn & Geist*, there is an almost overwhelming visual plurality of popular images. Lessons about both the state of the art within science journalism and popular (visual) culture can be gleaned from pictures that refer to the sinister cinematic atmospheres of *Alien* (the film poster, 10/2009, 48), of George Lucas' *Star Wars* (06/2004, 49), or to the Ludovico technique scene of Stanley Kubrick's *A Clockwork Orange* (06/2011, 80). Likewise, we encounter familiar figures from TV cartoon series (for instance Vicky the Viking, 9/2008), from advertisements, religion, or spirituality (*Yin Yang*, 03/2003, 14), and from art (for instance, Auguste Rodin's *The Thinker* (10/2010, 51)). Most of the pop images used by the magazine are used within just one issue or an article, but some cultural icons like the image of Sigmund Freud (01-02/2006; 11/2012) or *The Vitruvian Man* are taken up in several issues (11/2005, 59; 06/2009, 14; 05/2010, 31) and for different purposes.

Gehirn & Geist repeatedly adopted the work *The Vitruvian Man* (around 1490) by the Renaissance genius Leonardo da Vinci, which is incorporated into a whole interpictorial series of pictures (cf. Hommrich/Isekenmeier 2016). The picture refers to the anthropometric visualisation and illustration of a text passage by the Roman architect Vitruvius. Today, da Vinci's well-known geometric study of ideal human proportions (*Le proporzioni del corpo umano secondo Vitruvio*) is certainly both a famous piece of art (history) and an inherent part of pop culture: While the original drawing is stored in the Accademia gallery in Venice, the work has taken on its own life as a popular icon. The popularity of da Vinci's drawing depends on its iconic quality. Some people may have come to know the image in school, others may have seen the picture of the Vitruvian Man in everyday life or in the media. The dissemination and circulation of the iconic use of da Vinci's drawing can be tracked when we recognize the Vitruvian Man on the Italian 1 €-coin or on German public health insurance cards; it can be found on the cover of books or on conference posters; someone may associate one of the innumerable examples staging the motif as a product-boosting icon in films like *The Da Vinci Code*, *The Hitchhiker's Guide to the Galaxy*, or *Idiocracy*. The icon was also used for the video game *Deus Ex: Human Revolution*, and for Homer Simpson as a parody of the ideal man. Other examples show the image incorporated into the logo of the local fitness studio or pizza parlor, and some of us may recall seeing it on the packaging of their energy drink in the supermarket.

As a highly mediated picture, the image bears plenty of denotations and connotations, so that adopting the Vitruvian Man may refer to da Vinci as a historical and mythical figure of the Renaissance, or to (occasionally plain) meanings such as "the human body," "the perfection of man" (lat. *homo bene figuratus*), "the plasticity of man," "culture," and also "medicine" or "health and fitness," "geometry," "measurement," "scientific spirit," etc. An analysis of the images that draw on this cultural icon shows that the function of reference varies. Studying the da Vinci drawing within (and as an issue cover of) *Gehirn & Geist* teaches us about the different uses and the semiotic flexibility such pop images have in popular science journalism. The strategies of adopting, stylizing, and contextually embedding this popular image range from affirmative quotations of the image to satiric uses. In *Gehirn & Geist* (11/2005: 59, passim 3/2006, 6), the image is combined with an iconic representation of the brain organ; we find an image of the Vitruvian Man as a figure of puppetry on a parchment-like background: the threads of the puppet are controlled by a hand reaching out of a brain at the top of the image (see *Fig. 1*).

Here, the iconic meaning is part of a pictorial travesty subverting the meaning of the supremacy of (the ideal) man—mastered by the brain as higher actor. Furthermore, the icon of the Vitruvian Man is lacking the geometric lines of da Vinci's drawing, which symbolise (cosmic) order; instead, the puppet master (the brain) holds the strings ("nerve strings," as the commentary says). In short, da Vinci's drawing is used to promote and emotionalize the topics and context of a specific statement or metaphor by means of different functions of reference. Broadly speaking, when using the rhetoric of evidence, popular images, and different functions of reference, *Gehirn & Geist* adopts "the variety of practices of production, dissemination of all kinds of images, as well as concepts which underlie visual perception, practices of looking, and visual representations" (Frank 2008: 473) to both foster and expand its target audience.

Figure 1: "Neuronal Puppet Master"

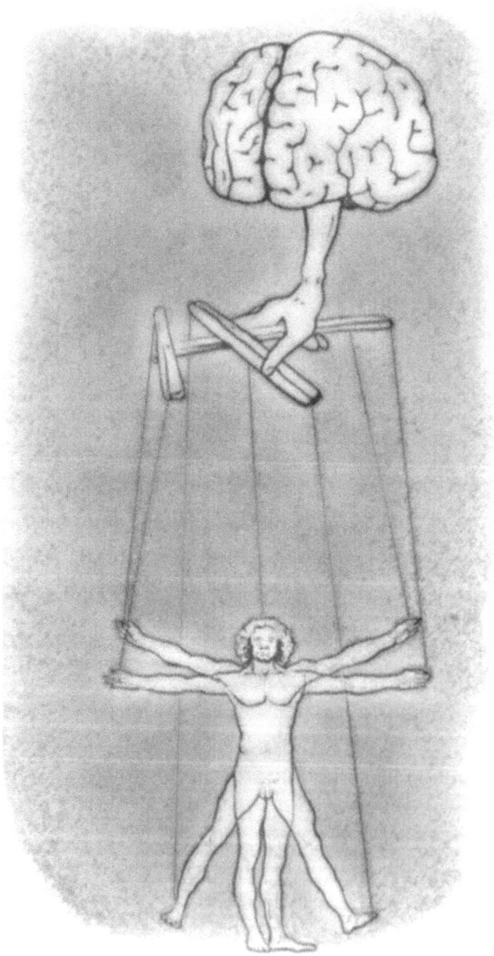

Source: *Gehirn & Geist* 11/2005: 59

In addition to the adaptations of pop and media cultural images found in the magazine, cartoons in particular can be understood as popular "everyday images" that visually stage attitudes toward a certain topic. Peter Weingart, who makes no distinction between cartoons and comics, speaks of "comics as popular culture" (Weingart 2009: 392-394) and states that these "follow the dramaturgical rules of the literary genre, the news values of the media" (Weingart 2009: 403): "As such, they reproduce the stereotypes of science and technology deeply rooted in popular culture" (Weingart 2009, pp. 403-404). From the perspective of media phenomenological observation, it must be added that the clichés of brain researchers, neurobiologists, and psychologists, for example, should be examined in detail. It also should be added that the cartoons in *Gehirn & Geist* explicitly refer to the social, political, economic, etc. contexts. Here, cartoons are quite capable of affirming or counteracting stereotypes of science and technology as well as developing completely new perspectives on clichés, and even of presenting alternatives.

In contrast to the separation of image and caption, *Gehirn & Geist* uses cartoons as combinations of image and text or text in image. Since 2002, these visualizations with an emphatically subjective stance have had their space and place in the magazine. Comparable to glossy and pointed statements, cartoons serve to (mostly) pick up on and evaluate a specific aspect of a topic. Cartoons within *Gehirn & Geist* can therefore be interpreted as a type of image that can be used in a satirical or polemical way, i.e., ironically, mockingly, cynically, etc. In this way, 'particular' and subjective conceptions are encouraged, subverted, reversed, etc. By using these text-image combinations, the cartoons address 'additional' questions of worldview and ideology. In other words, while brain scans make foreign and scientifically certified 'objective' certainties visible and perceptible, cartoons are reserved for calling up and challenging the subjective experiential knowledge and the individual certainties of *Gehirn & Geist*'s readers.

The mostly casual black-and-white or pencil drawings placed as gap fillers seem to be popular with the editorial staff: many quick drawings can be found in almost every issue of the 2002-2015 volumes, quite often also on various topics or articles of an issue. Among the topics are machine learning (03/2002: 53), memory loss (05/2003: 39), disruptive factors and disturbances in functional brain imaging (02/2004: 22), memory training (09/2005: 78), 'mind reading' with brain imaging (03/2006: 40), data interpretation in neuroscience (09/2007: 78), evidence-based medicine (05/2010: 35), brain and IQ doping at the workplace (11/2009: 47), and large-scale research funding in the US and the EU (04/2014: 52).

Another way to combine text and image, information and demonstration, is to present information graphics (as a separate section of *Gehirn & Geist*). In contrast to cartoons and their essayistic commentaries on a topic, infographics aim at catchy messages and visual comprehensibility. For example, the first "G&G-Infographics" demonstrated and explained the basic steps of functional brain imaging (11/2012: 62-63). In any case, cartoons and information graphics, as well as social and popular images, evoke effects and affect among readers.

ICONOPHILIA BEYOND 'THE' BRAIN IMAGE?

In a July 2018 book review, Michael Hagner spoke of the "excessive neurohype of the last 25 years" and stated: "So much has been written about the brain and its relevance for our humanity in the last decades that there is not much to add under the present paradigmatic basic assumptions" (Hagner 2018). The "long summer of the brain," as the science historian also calls the seemingly decaying neurohype, may come to an end, and with it what Hagner described as a second iconophilia of the brain, i.e., the emergence of functional imaging.

In contrast to Hagner's assessment that the general boom of brain research is over, it seems to me, however, that there is a further stage, a third iconophilic stage, in which other visualization technologies, new objects, methodologies, and causal assumptions are emerging, along with characteristic ways to represent and stage them. In issue 11/2012, *Gehirn & Geist* even introduced a new section named "Eyecatcher." The pictures shown there demonstrate progressiveness and suggest that future discoveries and technical possibilities are on the horizon. Most of these "eyecatchers" do not refer to—or only indirectly refer to—human morphology and visualizations of the (morphology of the) human brain, so that one might come to the conclusion that the visual culture of popular brain research and neuroscience on the one hand continues to experiment with different scales, new formats, and an increasing abstraction of its visual artefacts. On the other hand, the diagnosis is that the visual culture (at least of *Gehirn & Geist*) has begun to essentially go with cartoons, information graphics, and other strategies that facilitate quickly consumable, easily adoptable eyecatchers, whether they relate to everyday experiences or to simple messages regarding the spell of brain imaging. To conclude, the images that we sometimes make of ourselves as "homo cerebralis" (Hagner 2000) are fundamentally inspired by popular science communication. At this point, it must remain an open question as to whether the simultaneity of those images that pertain to subjectivity with those images that link to scientific objectification will continue to have anything in common with the visual and pictorial qualities of the brain images we are already used to.

REFERENCES

Borck, Cornelius (2012): "Toys are Us. Models and Metaphors in Brain Research." In: Suparna Choudhury and Jan Slaby (eds.), Critical Neuroscience. A Handbook of the Social and Cultural Contexts of Neuroscience, Chichester: Wiley-Blackwell, pp. 113-133.

Burri, Regula Valérie (2008): "Bilder als soziale Praxis. Grundlegungen einer Soziologie des Visuellen." In: Zeitschrift für Soziologie 37 (4), pp. 342-358.

Cooter, Roger (2012): "Preisgabe der Demokratie. Wie die Geschichts- und Geisteswissenschaften von den Naturwissenschaften absorbiert werden." In: Michael Hagner (ed.), Wissenschaft und Demokratie, Berlin: Suhrkamp, pp. 88-111.

Dierkes, Meinolf/Hoffmann, Ute/Marz, Lutz (1992): Leitbild und Technik. Zur Entstehung und Steuerung technischer Innovationen, Berlin: edition sigma.

Donati, Paolo R. (2011): "Die Rahmenanalyse politischer Diskurse." In: Reiner Keller, Andreas Hirseland/Werner Schneider/Willy Viehöver (eds.), Handbuch sozialwissenschaftliche Diskursanalyse, Wiesbaden: VS, pp. 147-177.

Dumit, Joseph (2004): Picturing Personhood. Brain Scans and Biomedical Identity, Princeton, N.J: Princeton University Press.

Frank, Gustav (2008): "Pictorial und Iconic Turn. Ein Bild von zwei Kontroversen. Nachwort." In: William J. Thomas Mitchell (ed.), Bildtheorie, Frankfurt a. M.: Suhrkamp, pp. 445-487.

Geideck, Susan/Liebert, Wolf-Andreas (2010): Sinnformeln. Linguistische und soziologische Analysen von Leitbildern, Metaphern und anderen kollektiven Orientierungsmustern, Berlin and New York: De Gruyter.

Geimer, Peter (2002): "Sichtbarkeit und Macht. Maschinen als Augenzeugen." In: Peter Geimer (ed.), Ordnungen der Sichtbarkeit. Fotografie in Wissenschaft, Kunst und Technologie, Frankfurt a. M.: Suhrkamp, pp. 171-210.

Hagner, Michael (1994): "Lokalisation, Funktion, Cytoarchitektonik. Wege zur Modellierung des Gehirns." In: Michael Hagner/Hans-Jörg Rheinberger/Bettina Wahrig-Schmidt (eds.), Objekte, Differenzen und Konjunkturen. Experimentalsysteme im historischen Kontext. Berlin: Akademie, pp. 121-150.

Hagner, Michael (1996): "Zwei Anmerkungen zur Repräsentation in der Wissenschaftsgeschichte." In: Hans-Jörg Rheinberger/Michael Hagner/Bettina Wahrig-Schmidt (eds.), Räume des Wissens. Repräsentation, Codierung, Spur, Berlin: Akademie, pp. 339-355.

Hagner, Michael (2000): Homo cerebralis. Der Wandel vom Seelenorgan zum Gehirn, Frankfurt a. M.: Insel.

Hagner, Michael (2002): "Cyber-Phrenologie. Die neue Physiognomik des Geistes und ihre Ursprünge." In: Klaus Peter Dencker (ed.), Die Politik der Maschine. Computer Odysee 2001, Hamburg: Verlag Hans Bredow Institut, pp. 182-197.

Hagner, Michael (2006): "Bilder der Kybernetik. Diagramm und Anthropologie, Schaltung und Nervensystem." In: Michael Hagner: Der Geist bei der Arbeit. Historische Untersuchungen zur Hirnforschung. Ed. Michael Hagner, Göttingen: Wallstein, pp. 195-222.

Hagner, Michael (2007): Geniale Gehirne. Zur Geschichte der Elitegehirnforschung, Munich: dtv.

Hagner, Michael (2008): "Das Hirnbild als Marke." In: Bredekamp Horst/Matthias Bruhn/Gabriele Werner (eds.), Ikonografie des Gehirns. Berlin: Akademie (Bildwelten des Wissens, vol. 6,1), pp. 43-51.

Hagner, Michael (2018): "Rückblick auf den großen Neurohype. Aus eigener Erfahrung: Siri Huvstedt sichtet die Literatur der Hirnforschung." In: Frankfurter Allgemeine Zeitung 2018, 27.04.2018 (No. 172), p. 10.

Heinemann, Linda V./Heinemann, Torsten (2010): "*Optimise Your Brain!* Popular Science and its Social Implications." In: BioSocieties 5 (2), pp. 291-294.

Heinemann, Torsten (2012): Populäre Wissenschaft. Hirnforschung zwischen Labor und Talkshow, Göttingen: Wallstein

Heßler, Martina (2009): "BilderWissen. Bild- und wissenschaftstheoretische Überlegungen." In: Ralf Adelmann, Jan Frercks, Martina Heßler, and Jochen Hennig: Datenbilder. Zur digitalen Bildpraxis in den Naturwissenschaften, Bielefeld: transcript, pp. 133-161.

Heßler, Martina; Mersch, Dieter (2009): "Bildlogik oder Was heißt visuelles Denken?" In: Martina Heßler and Dieter Mersch (eds.), Logik des Bildlichen. Zur Kritik der ikonischen Vernunft, Bielefeld: transcript, pp. 8-62.

Hofstadter, Douglas R./Sander, Emmanuel (2014): Die Analogie. Das Herz des Denkens, Stuttgart: Klett-Cotta.

Hommrich, Dirk (2019): Theatrum cerebri. Studien zur visuellen Kultur der populären Hirnforschung, Cologne: Herbert von Halem (Klagenfurter Beiträge zur Visuellen Kultur).

Hommrich, Dirk/Isekenmeier, Guido (2016): "Visual Communication, Popular Science Journals and the Rhetoric of Evidence." In: JCOM – The Journal of Science Communication 15 (2), pp. 1-8. Accessed May 21, 2019. https://jcom.sissa.it/sites/default/files/documents/JCOM_1502_2016_C04.pdf.

Hüppauf, Bernd; Weingart, Peter (2009): "Wissenschaftsbilder – Bilder der Wissenschaft." In: Bernd Hüppauf and Peter Weingart (eds.), Frosch und Frankenstein. Bilder als Medium der Popularisierung von Wissenschaft, Bielefeld: transcript, pp. 11-43.

Jäger, Margret/Jäger, Siegfried/Ruth, Ina/Schulte-Holtey, Ernst/Wichert, Frank (eds.) (1997): Biomacht und Medien. Wege in die Bio-Gesellschaft, Duisburg: DISS.

Jäger, Siegfried (2009): Kritische Diskursanalyse. Eine Einführung, Münster: Unrast.

Jäger, Siegfried/Jäger, Margret (1997): "Vernetzung biopolitischer Diskurse und ihrer Machteffekte." In: Margret Jäger/Siegfried Jäger/Ina Ruth/Ernst Schulte-Holtey/Frank Wichert (eds.), Biomacht und Medien. Wege in die Bio-Gesellschaft, Duisburg: DISS, pp. 304-354.

Janich, Peter (2009): Kein neues Menschenbild. Zur Sprache der Hirnforschung, Frankfurt a. M.: Suhrkamp.

Kelly, Thomas (2014): "Evidence." In: Edward N. Zalta (ed.), The Stanford Encyclopedia of Philosophy (Fall 2014 Edition), Stanford: The Metaphysics Research Lab. Accessed May 21, 2019. http://plato.stanford.edu/archives/fall2014/entries/evidence/.

Link, Jürgen (1986): "Noch einmal: Diskurs. Interdiskurs. Macht." In: kultuRRevolution (11), pp. 4-7.

Mersch, Dieter (2006): "Visuelle Argumente. Zur Rolle der Bilder in den Naturwissenschaften." In: Sabine Maasen/Torsten Mayerhauser/Cornelia Renggli (eds.), Bilder als Diskurse – Bilddiskurse, Weilerswist: Velbrück, pp. 95-116.

Nohr, Rolf F. (2014): "Nützliche Bilder. Bild, Diskurs, Evidenz." Münster: LIT.

Peters, Hans Peter/Allgaier, Joachim/Dunwoody, Sharon/Lo, Yin-Yueh/Brossard, Dominique/Jung, Arlena (2013): "Medialisierung der Neurowissenschaften. Bedeutung journalistischer Medien für die Wissenschafts-Governance." In: Edgar Grande/Dorothea Jansen/Otfried Jarren,/Arie Rip/Uwe Schimank/Peter Weingart (eds.), Neue Governance der Wissenschaft. Reorganisation – externe Anforderungen – Medialisierung, Bielefeld: transcript, pp. 311-335.

Schneider, Norbert (2011): "Ästhetische Geltungsansprüche." In: Friedrich Jaeger und Burkhard Liebsch (eds.), Handbuch der Kulturwissenschaften, Vol. 1, Stuttgart: Metzler, pp. 266-276.

Schnettler, Bernt/Pötzsch, Frederik S. (2007): "Visuelles Wissen." In: Rainer Schützeichel (ed.), Handbuch Wissenssoziologie und Wissensforschung, Konstanz: UVK.

Sturken, Marita/Cartwright, Lisa (2009): Practices of Looking. An Introduction to Visual Culture, New York: Oxford University Press.

Tetens, Holm (1999): "Die Grenze. Naturwissenschaft lässt sich mit Bildern popularisieren, aber nur mit Mathematik verstehen." Die Zeit Nr. 37, September 9, 1999, p. 55. Accessed May 21, 2019. http://www.zeit.de/1999/37/199937.t-populaerwissen.xml/komplettansicht.

Tetens, Holm (2006): "Wie erlangt der Laie Erkenntnis höherer wissenschaftlicher Welten?" In: John Michael Krois and Norbert Meuter (eds.), Kulturelle

Existenz und symbolische Form. Philosophische Essays zu Kultur und Medien, Berlin: Parerga, pp. 241-254.

Uttal, William R. (2001): The New Phrenology. The Limits of Localizing Cognitive Processes in the Brain, Cambridge, MA: MIT Press.

Viehöver, Willy (2011): "Diskurse als Narrationen." In: Reiner Keller/Andreas Hirseland/ Werner Schneider/Willy Viehöver (eds.), Handbuch sozialwissenschaftliche Diskursanalyse. 3, Wiesbaden: VS, pp. 179-208.

Weingart, Peter (2009): "Frankenstein in Entenhausen? Wahrnehmung von Wissenschaft und Technik in der Öffentlichkeit – kurze Geschichte." In: Bernd Hüppauf and Peter Weingart (eds.), Frosch und Frankenstein. Bilder als Medium der Popularisierung von Wissenschaft, Bielefeld: transcript, pp. 387-406.

Watch and Learn! Image-Based Popularization of Academic Reasoning and Scientific Action in Fictional Movies and Comics

Kathrin Klohs

> Any attempt to understand popular representations of science must begin with a recognition of their complexity.
>
> (Locke 2005: 42)

ABSTRACT

The scientific community typically underestimates fictional representations of science and academia. In contrast, this essay considers movies, comics, and graphic novels as revealing research topics for Science and Technology Studies (STS). Accordingly, we believe that analyzing such visual narrations along the lines of Film Studies und Comics Studies is a means of undertaking Science and Technology Studies. This approach is demonstrated by analyzing a panel from Jim Ottaviani's graphic novel *Feynman* and a screenshot from Louis Leterrier's Hollywood blockbuster *The Incredible Hulk*. We develop the argument that movies and comics do intelligibly translate and directly visualize what it means for their fictional characters to do science or to be a scholar. Storytelling in movies and comics opens up a kind of knowledge that STS are also interested in.

FICTION IS UNDERRATED

Popular fiction[1] about academia and academics[2]—movies like *A Beautiful Mind* (Howard, 2001) or comics like *www.phdcomics.com*—has been underestimated,

1 Concerning the concepts of "fiction" assumed in this text, cf. the overview given by Schaeffer 2012.

especially when it comes to its potential to impart or co-constitute knowledge that is epistemically or epistemologically relevant. In criticism and among scholars and scientists themselves, novels, movies, and comics about academia are often regarded as distortions or oversimplifications, seen either as threatening truth itself or against the backdrop of an assumed scientific monopoly thereof (also discussed in Kirby 2003).[3] Indeed, such accusations against fiction resemble those used in "the dominant view of popularization": In his widely noted paper of the same name, Stephen Hilgartner traces how the popularization of scientific knowledge is usually perceived as a "degradation" or "pollution" (Hilgartner 1990: 519). At most, fiction seems to serve as an icebreaker and eyecatcher at scientific conferences, used to make the audience smile before moving on to the actual topic.

Previous research in Literary Studies and Science and Technology Studies (STS)[4] has focused on archetypes and stereotypes of scientists and scholars in fiction, on the public view on different disciplines, on realism, and on referentiality,[5] asking: Does the depiction of academia in works of fiction popularize how research is actually conducted, or does it just mirror how laypeople imagine this is done?[6] Is the knowledge fiction presents robust or defective? Are the assertions fiction makes right or wrong? This, of course, does not include the way

2 In what follows, the term "academic" embraces both the human and the natural sciences; it is used as an equivalent to the German term *wissenschaftlich*. Thus, "academics" embraces both scholars of the humanities and natural and social scientists.

3 Fictional speech as a threat to an assumed scientific monopoly of truth would be worthy of discussion, keeping in mind the different types of boundary work that Gieryn (1983) has highlighted, as well as the political use of the 'dominant view' according to Hilgartner (1990).

4 For an overview, cf. Haynes (2003) and Weingart (2005). For current developments, cf. Haynes (2016).

5 Mancosu (2011) provides an example of a review that complains about the lack of faithfulness to the history of logic in a graphic novel about Bertrand Russell. Why realism and correctness are not actually the crucial points to discuss with regards to these matters is debated in the third issue of *Public Understanding of Science* 12 (2003), e.g., which is titled *Perception and Representation of Science in Literature and Fiction-Film*. In that issue, cf. Rosenstone (2003) and Rose (2003). Apart from that, the *pragmatic* definition of fictional narration states that fiction abstains from claims of referential truthfulness: "the question of referentiality is irrelevant [...]. [T]he truth claims a text would make if it [...] were a factual text [...] must be bracketed out" (Schaeffer 2012, para. 26, para. 29; as for images, consult para. 6).

6 For the second option—cinema as an indicator of public opinion about science and technology—, consult Elena (1993), for instance.

mad scientists act as characters,[7] what they personify, and to what extent academic staff behave in that manner! Besides, debates in STS about the Public Understanding of Science (PUS) or the Public Engagement with Science and Technology (PEST) are predominantly centered around *factual* speech in educational films or educational comics (concerning comics, cf. Tatalovic 2009; Blank 2010; for film, cf. Verdicchio 2010). Unlike discussions of image practices and imaging processes in Visual Studies and around the technical image, the role of images concerning *fictional* contexts connected with science is hardly ever considered.[8] In sum, commenting seriously on academia in terms of motion-picture images or sequential images seems restricted to those in the nonfiction realm.

This essay proposes a different approach and another viewpoint, by looking at movies and comics as stores of knowledge that reveal insights into both reasoning and acting in academia. It investigates them in order to tap new resources, mainly because, in most cases, fictional storytelling stands out from factual storytelling through its different use of the same medium-specific possibilities and strategies: Characteristic features are utilized more artistically, as means of expression.[9] Methodologically, it is argued that given this *differentia specifica*, applying methods of analysis from Film Studies and Comics Studies to fiction about academia achieves results that are also of interest to STS.

Accordingly, this essay does not call into question the accuracy or correctness of what is presented in fiction: neither in terms of the content of knowledge nor in terms of the processes of accumulating knowledge. Instead, it asks how movies and comics manage to popularize a sense of what it *means* to the characters involved to be a scientist or a scholar, to do brainwork or paperwork in academia, or to solve a specialist problem. The main argument is: When it comes to knowledge about academia derived from fiction, the key to popularization is the fictional characters' *subjectivity*. This refers to their perception, to their consciousness, and to their personal point of view—three constituents of fictional worlds that are expressed through transmedial as well as medium-specific stylistic features (cf. Reinerth & Thon 2017). *Internal focalization* is of particular importance here. This narratological term indicates how a plot has been witnessed, observed, or presented: The narrator or, in our case, the picture reveals "the subjective experience of a character" (Niederhoff 2011: para. 18), disclosing exactly

7 As usual in Literary Studies, the term "character" stands for a personage in a narrative.

8 For exceptions, see the overview of the film genre written by David A. Kirby (2014), or see Lynda Goldstein's (2014) analysis of a comic.

9 This is put into perspective with regard to film production in Verdicchio (2010: especially 67-158).

what the character knows or sees (instead of disclosing more or less than what the character knows or sees).

To elaborate on this, this essay will explore two examples from almost opposite sides of the popular media spectrum—one highbrow and intellectual, the other highly commercial and accessible. One stems from the graphic novel *Feynman* (Ottaviani 2011), and the other from the Hollywood blockbuster *The Incredible Hulk* (Leterrier 2008; cf. Lee 2003). We use these two examples to show how different media's possibilities to make subjectivity intersubjectively accessible play an essential role in the popularization of academic reasoning and scientific action.

FILM STUDIES AND COMICS STUDIES PURSUED AS SCIENCE AND TECHNOLOGY STUDIES

Though an object of study in STS, storytelling in movies and comics must not be investigated the same way it would be in the Social Sciences. This is key to the design of fiction as a piece of art, or at least as entertainment. It does not make sense to stick to, for example, grounded theory or to content analysis in search of its making. As this making reveals the subjectivity behind the work, the analysis must focus on the 'how' instead of the 'what' (for the opposite approach, cf. *fictionmeetsscience.org*). Thus, a movie scene or a comic-book sequence cannot be investigated through a case study that generates data using coding software or a standardized quantitative or qualitative method. Instead, this essay proposes using procedures usually assigned to the humanities and Cultural Studies. Artistic decisions are foregrounded, as are their (prospective) emotional impact on the reader or viewer.

Corresponding analysis criteria are well known in Film Studies, Comics Studies, and Transmedial Narratology, but are less familiar or even unknown in STS. They include the exploration of semiotic systems and storytelling practices of fiction based upon static or moving pictures,[10] such as perspective, view, division of space, verbal-pictorial and audio-visual word-image combinations, lighting, or the use of color. Concerning film, specific means of expression become relevant, such as camera position or angle, background music, or depth of field. For example, in classical narrative cinema, a change in camera movement "may privilege a character's point of view and so invite participatory identification with that character" (Kuhn/Westwell 2012). Concerning comics, equivalents include action lines, soundwords, or the *spatium* between two panels (cf. Mahne

10 Concerning narratology across media and details about movies or comics, cf. Alber & Hansen (2014).

2007; McCloud 1994). The aim and purpose of such an analysis, of course, is a better understanding of popular representations of the sciences and the humanities, which is accomplished by asking the central question: When academia is depicted in movies and comics, how does this affect the audience's insight into this realm, and is there an increase in insight?

TWO EXAMPLES FROM YOUR LEISURE TIME

The following sections will examine two image-based examples from fictional storytelling through close readings. Both of them convert the abstract content of knowledge that was initially restricted to experts into a visible state, making it accessible to laypersons as well while presenting a character's subjective perception or consciousness. Each broaches the issue of cultural techniques in academia: "everyday practices with objects, symbols, instruments and machines," as Krämer & Bredekamp (2003: 24) have summarized. As the first one depicts an object of knowledge that does not exist in physical form—a coherence in mathematics—, it serves as an example for the popularization of academic reasoning. Moreover, as the second depicts an object of knowledge that is invisible in itself—a manipulation of the human genome—, it serves as an example for the popularization of scientific action.

Concerning academic thought, to provide insight within a fictional story seems paradoxical at first sight. Showing and telling audiences how a character thinks and feels might be the core of fiction, but how can we do this if the characters' thoughts and feelings are both motivationally and cognitively built on an enormous specialization? How to communicate a fictional scientist's suitability to or talent for their discipline, the mastery of their field, and the elegance of their solutions to an audience that is in no position to judge these matters? How to present procedures that are difficult to learn, objects that are difficult to identify, problems that are difficult to recognize to laypeople without any practical or background knowledge? As pointed out in previous scholarship about scientific thought at the movies, "there is always a problem in the translation process, of trying to find apt images for an internal process" (Rosenstone 2003: 336) or even "external manifestations of what is often an internal process" (Van Riper 2011: xvi).

The answer I propose for the popularization of academic thought is that image-based media visualize the invisible, thereby transforming what is "*exclusive to the inner realms of a character* [...] into *intersubjectively comprehensible external forms of representation*" (Reinerth/Thon 2017: 3 [emphasis in original]). Movies and comics offer clear images not only for phenomena that turn out to be too big or too small, too slow or too fast for the human eye, but also for reflections that are too complicated or too counterintuitive for the non-expert. In the latter case, exceptional thinking[11] is translated into conventional thinking, and images are used to enhance audiences' capacity to recognize patterns, to create links, or to reason a priori.

Concerning scientific action, popular fiction about academia usually takes for granted that the audience has vague previous knowledge of quotidian practices in research, such as laboratory routines, writing procedures, or viewing habits. A variety of activities that are connected, in the broadest sense, to knowledge-making are shown more-or-less realistically. Apart from the *extent* of the audience's previous knowledge, such movie shots or scenes or panel sequences and their function within the story as a whole can only be understood appropriately when one has a basic understanding of scholarship and technoscience. This text suggests that we should not only consider the practices themselves, as depicted in fiction and carried out in the real world, to be cultural techniques, but also the familiarity of a lay audience with those practices. For example, at the movies, being able to identify a given sequence of characters as a scientific formula, or a glass vial full of bubbling liquids as part of a laboratory, also means being acquainted with "operative processes that enable work with things and symbols" (Krämer/Bredekamp 2003: 27).

POPULARIZATION OF ACADEMIC REASONING—THE MATH BEHIND THE IMAGES

The following example (*Fig. 1*) features a character who is academically seen as a scientific exception and who is socially accepted as an exceptional scientist. This spread from Jim Ottaviani's graphic novel *Feynman*[12] presents the physicist of the

11 I should note here that unlike STS, fiction considers the concept of genius to be much less suspicious, which stems from the dramatic advantages of including geniuses in stories about artists and artistry.

12 Why is *Feynman* subsumed under fictional representations of academia? In general, the extent to which biography as a genre can include fictional elements varies. Notably, the *new biography* movement considers itself "an imaginative art in which [...] techniques borrowed from the [...] novel [are] employed" (Baldick 2008: 38-39). In particular, assuming a *semantic* definition of fictional narration, the ontological status of entities and the truth value status of propositions are sometimes confused. In this regard, two types of fictional modes can be distinguished here, which may or may

same name directly addressing the audience and explaining his intuitive insight into the mathematical aspects of physics. Symbols from the language of mathematical formulae skim across the left panel. It is meaningful that they appear in mirror image, against our direction of reading but in Feynman's, thereby expressing his point of view. Accordingly, the panel immediately to the right fully zooms into his perspective and invites us to share it for a moment, as though we are seeing through Feynman's eyes or his mind's eye. Due to this internal focalization, we realize that Feynman does not differentiate the referentiality of mathematical formulae from their materiality; in his view, they coincide. Although the distinctive language of science represents an abstract or absent object of knowledge, and although it is a much less familiar and much more noticeable sign system than the Roman alphabet (cf. Hoffmann, 2010: 183; Krämer/Bredekamp 2013: 22), it still instantaneously shows something to Feynman that is concrete, visible, beautiful, and expressive.

Figure 1: Jim Ottaviani's Graphic Novel Feynman *(2011: 20), Detail*

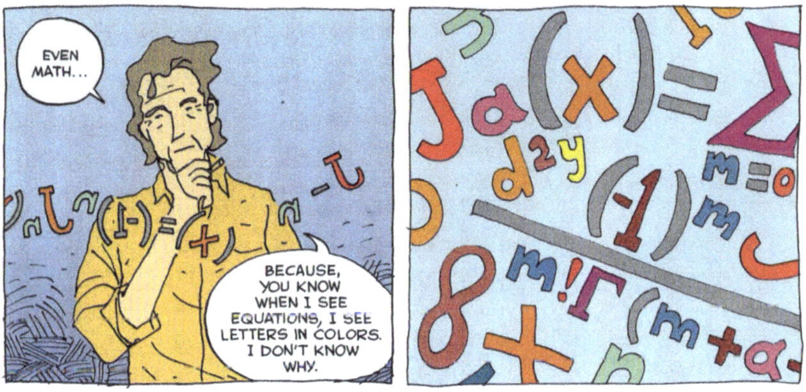

Source: Illustrations Copyright 2011 by Leland Myrick

The next panel (*Fig. 2*) presents Feynman in the same manner, with tri-colored and wavy strings of mathematical formula converging in his hand. Looking at the arrangement and the appearance of the signs themselves—the lettering style (cf. McCloud 1994: Ch. 5)—, it is clear that these strings of formulae are used as symbols, following the work of Peirce, and thus represent something by convention

not be the case with *Feynman* as well: Existing entities can be ascribed fictive properties or actions (e.g., in counterfactual novels), or they can be retained unchanged but act in fictive settings (e.g., in historical novels) (cf. Schaeffer 2012: para. 15). As both cases open up possibilities for the popularization of academic thought, I take Ottaviani's graphic novel for a *fictionalized biography*.

that they do not actually resemble (cf. also Mahne 2007: 49). Vivid and patterned, the spatial arrangement of the written sign on the surface of the paper expresses clarity and lightness, suggesting that the mathematics they represent is distinct and organized, something that Feynman has under control and handles with the greatest of ease. Again, in his view, there is no distance or gap between signified and signifier. On the contrary, he sees through the former (the mathematics) by looking through the latter (the characters). The audience thus learns that Feynman takes possession of his field of research instinctively, through his senses. He is indeed known as a visual thinker, and has become famous for his diagrams.

Figures 2 and 3: Jim Ottaviani's Graphic Novel Feynman *(2011: 21), Details*

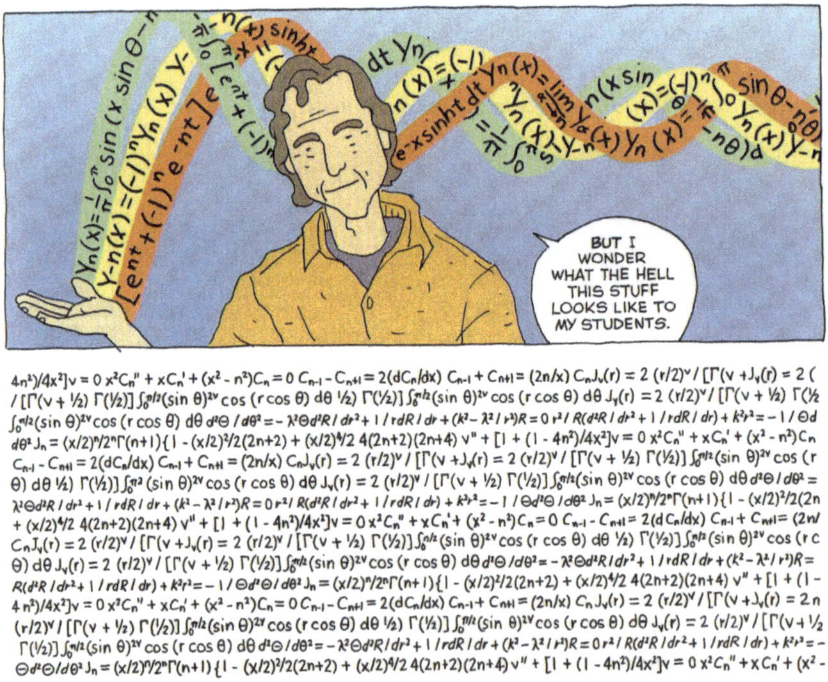

Source: Illustrations Copyright 2011 by Leland Myrick

We see how unique this is in the following panel (*Fig. 3*): The gap between the two panels (i.e., *Fig. 2* and *Fig. 3*) indicates a transition from one aspect to another, a typical narrative strategy in (predominantly Japanese) comics (McCloud 1994: 65-81). *Fig. 3* contrasts Feynman's view with his classes', and through them also the audience's: "I wonder what the hell this stuff looks like to my students," he ponders, revealing that he cannot even imagine how to approach his object of investigation in any other way. However, the graphic novel *can* imagine such a scenario, and we are not surprised by this. By showing the characters in endless grey lines—without any punctuation, breaks, or color, or even a panel frame—, it represents the very same mathematics as monotonous, abstract, and inaccessible.

These two pictures not only confront two views of academic reasoning with one another, one from the inside and the other from the outside. By illustrating that science popularization is by no means a unidirectional top-down journey from expert to layperson, they also envision an advance in knowledge *inside* a scientific discipline: the gap between the Nobel Prize winner's level and the advanced students' level. This reminds us of the question Stephen Hilgartner raises:

> "For example, one could define knowledge as that which is presented by scientific audiences in scientific forums; all the rest, then, would be defined as popularization. Such an approach transforms the problem of locating the boundary between genuine and popularized knowledge into the problem of identifying the experts, audiences, and forums that represent the 'genuinely scientific'. But this manoeuvre produces new ambiguities about how broadly to define these categories. Does the set of speakers who possess genuine scientific knowledge about superconductivity include only a 'core set' of a few world experts [...] or every member of the American Physical Society?" (Hilgartner 1990: 525)

What did we learn from this first example? This graphic novel, as an image-based medium, uses medium-specific features like a specific means of expression to impart a vivid impression and a clear image of the subjective view of the formalized content of knowledge to the audience. We come to know what the character knows and, to a certain extent, how the character feels about his current object of investigation.[13] A hard-to-access sign system that describes a complex mathematical order in particular is transferred into a picture that stands for order in general and that is easy to understand. The audience, though likely mostly consisting of laypeople, learns what it means to be in perfect control of the math behind the images—without being explained the math itself. Suitable instruments to convey meaning like this in the graphic novel are quite simple: firstly, color; secondly, arrangement; thirdly, Western conventions of depicting ease and movement (cf. McCloud 1994: Ch. 4). Mathematical thinking is shown with the aid of a visualization strategy that makes it evident for everyone. Its play with abstraction and concretion, its use of shape and color, and its handling

13 Eder (2008) develops a typology of perception, knowledge, evaluation, wishes, and emotions connected to intentional objects; Reinerth & Thon (2017) summarize the distinctions of this typology.

of text and image all allow it to depict what only the initiate sees behind the language of formulae—paradoxically, in a way that anyone can see.

The image, in sum, opens up an alternative to a "propositional and language-based form of knowledge" here (Krämer/Bredekamp 2013: 23). Fictional image-based storytelling can demonstrate this kind of insight without a hitch because absent or counterfactual aspects can be shown as "acts of pretending, supposing, *or hypothesizing*" (Schaeffer 2012: para. 7 [my emphasis]).[14] The graphic novel draws them; the movie depicts them—if needed, through animation. In fiction, thought experiments can have the same degree of reality as what is presented as real.

POPULARIZATION OF SCIENTIFIC ACTION—THE SCIENCE ALONGSIDE THE IMAGES

The second example deals with the popularization of scientific action. The example chosen stems from Louis Leterrier's film *The Incredible Hulk* and is an example *a fortiori* as it relies on the aforementioned familiarity with research practices and procedures in telling its story: The science explains the plot.

Figure 4: Bruce Banner's Blood Cells in The Incredible Hulk *(2008)*

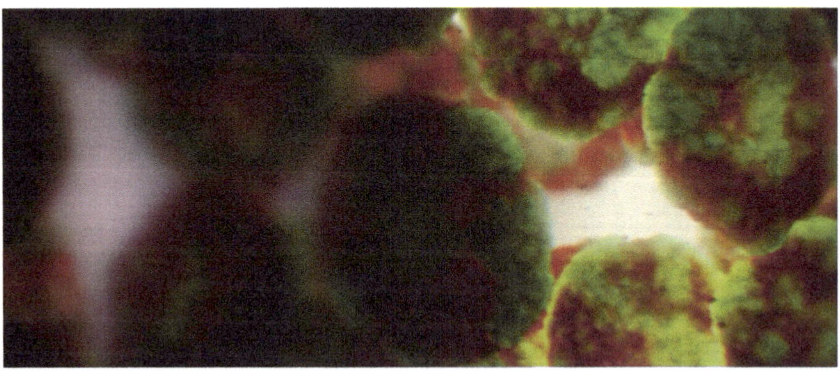

Source: Author's Screenshot from *The Incredible Hulk* (0:11), LEONINE Studios

The shot I will focus on (*Fig. 4*) shows the Hulk's genetically modified blood cells and follows scientific viewing habits and scientific conventions in the use of the image. Just like a large number of movies and comics (concerning comics, cf. Locke 2011), *The Incredible Hulk* refers to conducting a scientific experiment in the labora-

14 As for academic thought in particular, cf. also Schaeffer (2012: para. 10, para. 11).

tory as a stock situation. In doing so, it integrates imagery that was initially developed through scientific imaging procedures (such as microscopy or tomography) with the aid of montage or readjustment. This opens up a new relationship between the technical and the art image. Also, it is a paradigm for internal focalization: With the aid of a subjective shot, the audience is encouraged to see what the scientist sees through the eyepiece of the scientific apparatus; it is not only the gaze of the camera and that of the spectator that merge here, but also that of the scientist.

Figure 5 (a,b,c): Film Stills from Opening Credits of Hulk *(2003)*
Figure 6: Green Fluorescent Protein (GFP) in C. elegans neuron *on the Cover of* Science *(1994, Vol. 263, Number 5148)*

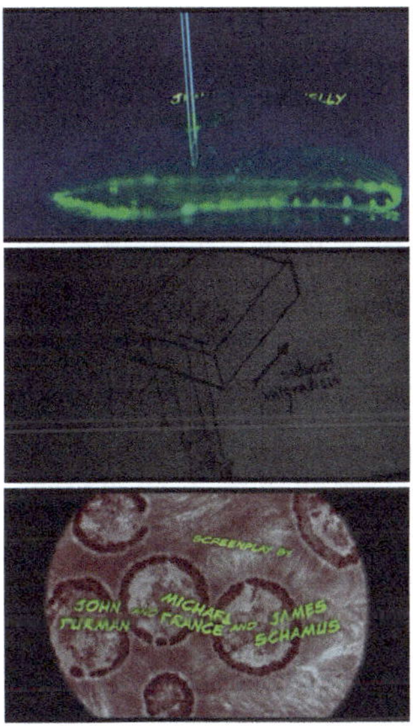

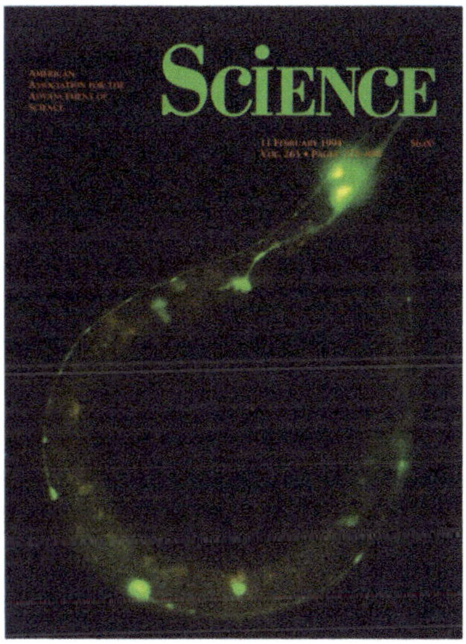

Sources: Figure 5 (a,b,c): Author's screenshots from Lee 2003, 0:01, 0:02, 0:04. Figure 6: American Association for the Advancement of Science

In an earlier take on the Hulk's story (*Fig. 5*), the opening credits of the film *Hulk* (Lee 2003) already interlocked the telescopic, the microscopic, and the endoscopic image in order to tell the origin story of the Hulk/Bruce Banner. The audience becomes familiar with the research of Banner's father, who manipulated his own genes but failed to foresee that he would pass mutations on to his offspring. (In this aspect, Lee's screen adaptation alters the original comic version, but we will discuss this below.) The credits naturally and almost purposefully show him centrifugalizing, dissecting, taking samples, and making laboratory notes. For the audience, it is enough to see how procedures within research activities cause changes of state in their objects amenable to the laws of physics (with regards to factual documentaries, cf. Hellermann 2015: 388). This is even true irrespective of details, causalities, and realism, as also pointed out by Kirby (2011: 79), according to whom they "function as visual cues [...], these images are recognizable only in that they convey the notion of 'science.'"

In the Leterrier film, when the son tries to repair the damage his father's research has done to him and annihilate the Hulk in himself, the course of his scientific experiment turns out to be a metaphor for his identity and an anticipation of his presumed fate. A scientist himself, Bruce Banner examines his blood cells under a microscope, which reveals they are not red but speckled in red and green (*Fig. 4*). Unsurprisingly, it is the same lime green that the Hulk turns when he gets angry, a leitmotif the audience immediately recognizes.[15] During the experiment, Banner applies a remedy he has concocted to his blood sample. As a result, the slide preparation changes to red at first, but then turns completely green.

Inside the fictional world, the movie plausibly explains—at least to the layperson—that the green color stems from the father's research on the *Aequorea victoria* (cf. Kirby 2011: 166): an extraordinarily robust jellyfish from which the so-called Green Fluorescent Protein (GFP) is obtained. Frequently used in cell biology as a contrast medium, GFP (or its glowing representations) is well known to a broad audience from popular scientific images (*Fig. 6*). Thus, the shots showing Banner's blood sample conspicuously resemble award-winning visualizations of scientific origin. The scientific experiment within the plot is based on slide-preparation properties that are visible, or more precisely, that are *made* visible through scientific action.

The audience's understanding of these shots is only developed through their recognition of visual conventions in science and the scientist's everyday practice: Firstly, also in science, there is the use of a contrast medium that dyes particular properties of cell culture. Secondly, in scientific communities, there is the agreement to highlight relevant results afterward, following color schemes and other aesthetic characteristics as well. Thirdly, in scientific film, there is the habit of manipulating time, which means decelerating or accelerating the course of intercellular events (cf. Landecker 2011). Only by knowing this can the audience identify these shots as depicting cells with accentuated properties seen through the microscope and conceive themselves as spectators of life at the movies: The GFP coloring has obviously become popular, accessible, and intelligible to the extent that laypeople recognize the procedure as scientific and the image as an object of knowledge.

15 As McCloud has emphasized with regard to the comic series about Hulk and Batman, colors in comics can gain "*iconic* power" and "symbolize characters in the mind of the reader" (1994: 188 [original in capitals; original emphasis in bold capitals]).

However, there are reasons to suppose that scientific image practices cover only one of several connotations[16] of the Hulk's skin color, while still more exist and can be connected to science. From this perspective, the scientifically plausible reading turns out to be just one of many possible readings in a complex web of references.

Firstly, scientific literacy is also required to understand how the movie expresses collective dreams and fears concerning science. Bernd Hüppauf and Peter Weingart have stressed there is evidence for the assumption that the general image of science is based on archetypical stereotypes consistent with traditions that often reach back to pre-modern beliefs and, equally, on the rules and requirements of the modern popular media; this image must be understood as a surprisingly stable combination of persistent stereotypes and changing patterns (Hüppauf/Weingart 2008: 5; see also Weingart 2005).

By revealing that what Banner's phenotype hides most of the time actually persists in his genotype all the time, the experiment suggests what he *is*. The human genome is presented as evidence, proof, and truth, superseding appearance and even behavior. In terms of how Banner sees himself, i.e., his search for identity and control, this is the worst outcome one could think of. Since the 1990s, some movies (among them, and probably most popularly, *Gattaca* [Niccol 1997; see also Shapshay 2009]) have similarly pointed to the dangers and risks of genetic engineering and genetic determinism: both in terms of health hazards and social dystopias.

Secondly, in popular culture, the lime green color has connotations of the looming aspects of technoscience: See, for instance, the nuclear fuel rod (e.g., in *The Simpsons*[17]) or the mouse pointer and the digital code (e.g., in *The Matrix* trilogy, Wachowski/Wachowski 1999, 2003), both of which refer to public fear, either of nuclear disaster or digitization on screen.[18] Accordingly, whereas in his first comic-book appearance in the 1960s Bruce Banner turned into the Hulk when he was accidentally exposed to gamma radiation, the Hollywood blockbusters from the 21st century alter that origin and instead make it due to genetic engineering and human enhancement. The lime green's potential to express all kinds of technoscientific fear and its function as a warning sign turn out to persist over time. "*The Incredible Hulk* articulates the anxiety over scientific dis-

16 I use this term to describe further associations, as opposed to the more straightforward "denotation."

17 See, for example, https://www.energy.gov/ne/articles/7-things-simpsons-got-wrong-about-nuclear, accessed May 28, 2018.

18 It is worth mentioning that these visual conventions contradict the facts in this case: Gamma radiation is invisible in itself, while fuel rods at most light up the neutralization pond in blue (Tscherenkow radiation). Cf. Vrckovski (2016).

coveries and technological devices through a narrative that literally embodies those discoveries and its devices" (Capitanio 2010: 252).

When we look at the movie to study its ability to express these collective attitudes towards science, we see that in *The Incredible Hulk* the science is not explained at all. Instead, it is presupposed and instructive, and thus fulfills an entirely new role. Science, in other words, is not used as the target of a metaphor, but as its source. This example thus indicates the extent to which the popularization of science is already presupposed in Hollywood.

DISCUSSION AND CONCLUSION—THE STS WITHIN THE IMAGES

Now that we have looked at two images from a graphic novel and a Hollywood blockbuster with the aid of two disciplines unfamiliar to STS, as a kind of short trial run, we should return to our original question: How can graphic novels and movies—as image-based media—popularize academia in accordance with their fictionality and mediality? Moreover, which aspects of academia do they popularize in particular? At first, one might feel tempted to transfer to fictional speech what concepts of popularization say about factual speech: Whereas scientific *outcomes* can be popularized through either appropriate or inappropriate simplification (Hilgartner, 1990, p. 520), the *processes* of knowledge-making in academia—technical or representational procedures, or quotidian practices—might be seen as forming insurmountable obstacles due to their vast prerequisites. However, this is not true for fiction, as a transmedial approach to representations of subjectivity can reveal. Thanks to its various kinds of narrative and artistic freedom concerning realism, its possibilities to process information through stage properties and setting, and its ability to use analogies and focalization, fiction also contributes to the depiction and popularization of knowledge-making.

The example from *Feynman* has shown that comic panels can visualize academic reasoning in a clear way, making the invisible visible, giving shape to psychological and mental processes, and revealing what typically remains closed. The example from *The Incredible Hulk* has shown that film can integrate images of scientific origin and functionalize them for the audience's understanding of the plot. Audiences' existing knowledge about scientific routines and representational techniques can be reactivated at the movies. In both examples, this is mainly due to two narrative strategies: Firstly, the handling of the image; secondly, the use of focalization.

In the portrayal of a community that tends to avoid subjectivity as an epistemological weakness (Daston/Galison 2007), or at least to restrict it to enthusiasm for work and to a thirst for knowledge (Berg & Seeber 2016), storytelling techniques that express a personal point of view seem inappropriate at first glance. But precisely through focalization, both of the examples express non-propositional forms of knowledge (handling images, objects, instruments)—even when, strictly speaking, they are depicting propositional forms (reading formulae, taking laboratory notes) (cf. Krämer/Bredekamp 2013: 23). There is good reason to suppose that the possibilities of fiction to make a personal point of view accessible opens up unexplored but promising possibilities in the visualization and depiction of academia. Thus, it seems helpful, in order to form a more complete understanding of academia, to further study to what extent popular fictional representations of science can contribute to "the sensualisation—the aestheticization—of invisible processes and theoretical objects that are the fuel of scientific change" (Krämer/Bredekamp 2013: 24).

The question remains whether and to what extent image-based fictional storytelling works as the *popularization* of academia. Taking into account the characteristic features of popularization compared to storytelling in previous research, the answer to the first part of that question is: Yes, movies and comics tend to simplify scientific terminology and reformulate it in everyday language; to take an unequivocal stand to scientific dissent; to assign academic topics to characters; to corroborate the audience's assumptions; and to emotionalize their topic.[19] However, the answer to the second part of the question is: "Yes, but…" Image-based fictional storytelling differs significantly from the Public Understanding of Science (PUS) or the Public Engagement with Science and Technology (PEST) and is neither *edutainment* nor *infotainment*. It is noteworthy that most movies and comics do not popularize any scientific content, nor any methodologies or apparatuses. What the Feynman character in the graphic novel explains to the reader is not how to solve the mathematical problem he is working on. What Banner/Hulk demonstrates on screen is not how to dissect an *Aequorea victoria* in order to achieve superhuman powers. Instead, these two narrations emphasize the *importance* of academia to their protagonists. Such fictional characters convey less how the world they investigate is constituted and more how the social system they experience is constituted. Through their perspectives, we do not gain knowledge that is relevant to a single academic discipline, but we do gain knowledge that is relevant to STS.

19 It is important to separate this from the discussion of popularization through mass media. The volumes edited by Daum (2002), Kretschmann (2003), and Blaseio, Pompe, and Ruchatz (2005) point out what storytelling and popularization may have in common.

Establishing the term "scientific *persona*," Lorraine Daston and Otto H. Sibum have pointed out that academics share "a cultural identity that simultaneously shapes the individual in body and mind and creates a collective with a shared and recognizable physiognomy"—an identity that embraces "collective ways of thinking, feeling, judging, perceiving, working" (Daston/Sibum 2003: 2 and 3, respectively). If it is true that fictional characters, too, tend to represent a specimen rather than an individual,[20] they can be compared to *personae*. Accordingly, popular representations of academics might achieve farther-reaching claims than qualitative interviews. Thus, they might not show isolated cases the Social Sciences would refuse as source material. As Rose (2003: 295) underlines:

> "Authenticity is usually interpreted in terms of the depiction of physical reality, meaning the chronology of events, the settings, appearances, etc. However, authenticity can also be meaningfully invoked on other levels, including the personalities and sensibilities of the characters, and the political atmosphere, moral tone, and emotional setting of a story."

It is beside the point here to decide whether fictional storytelling *mirrors* or *teaches* or *develops* any academic knowledge. Instead, it can be concluded that they have the means at hand to do so. It might be of interest for all who are interested in academia as a social system that, though movies and comics do not primarily deal with highly specialized disciplinary knowledge, they should not be called oversimplified, as they certainly co-produce and discuss knowledge about academia. In short: The audience cogitates less on the science behind their images and reflects more on the Science Studies within them.

REFERENCES

Alber, Jan/Hansen, Per Krogh (eds.) (2014): Beyond classical narration. Transmedial and unnatural challenges, Berlin and Boston: de Gruyter.

Baldick, Chris (2008): The Oxford dictionary of literary terms. 3rd ed, Oxford and New York: Oxford University Press.

Berg, Maggi/Seeber, Barbara K. (2016): The slow professor: challenging the culture of speed in the academy, Toronto: University of Toronto Press.

20 The representativeness of fictional characters in general cannot be discussed here. For fictional representations of scientists in particular, also consult Daston & Sibum (2003: 7), as well as the other contributions to the special issue their essay appears in.

Blank, Juliane (2010): "Alles ist zeigbar? Der Comic als Medium der Wissensvermittlung nach dem iconic turn." In: KulturPoetik 10(2), pp. 214-233.

Blaseio, Gereon/Pompe, Hedwig/Ruchatz, Jens (eds.) (2005): Popularisierung und Popularität, Cologne: DuMont.

Capitanio, Adam (2010): "'The Jekyll and Hyde of the Atomic Age': The Incredible Hulk as the Ambiguous Embodiment of Nuclear Power." In: Journal of Popular Culture, 43(2), pp. 249-270.

Daston, Lorraine/Galison, Peter (2007): Objectivity, New York: Zone Books.

Daston, Lorraine/Sibum, Otto H. (2003): "Introduction: Scientific Personae and Their Histories." In: Science in Context 16, pp. 1-8.

Daum, Andreas W. (2002): Wissenschaftspopularisierung im 19. Jahrhundert. Bürgerliche Kultur, naturwissenschaftliche Bildung und die deutsche Öffentlichkeit 1848-1914, Munich: Oldenbourg.

Eder, Jens (2008): Die Figur im Film: Grundlagen der Filmanalyse, Marburg: Schüren.

Elena, Alberto (1993): "Exemplary Lives: Biographies of Scientists on the Screen." In: Public Understanding of Science, 2, pp. 205-223.

Gieryn, Thomas F. (1983): "Boundary Work and the Demarcation of Science from Non-Science." In: American Sociological Review 48, pp. 781-795.

Goldstein, Lynda (2014): "Graphic/Narrative/History. Defining the Essential Experience(s) of 9/11." In: Annessa Ann Babic (ed.), Comics as History, Comics as Literature. Roles of the Comic Book in Scholarship, Society, and Entertainment, Lanham: Rowman & Littlefield, pp. 123-139.

Haynes, Roslynn D. (2003): "From Alchemy to Artificial Intelligence: Stereotypes of the Scientist in Western Literature." In: Public Understanding of Science 12, pp. 243-253.

Haynes, Roslynn D. (2016): "Whatever Happened to the 'Mad, Bad' Scientist? Overturning the Stereotype." In: Public Understanding of Science 25, pp. 31-44.

Hellermann, Michael (2015): Wissenschaft in Film und Fernsehen. Die mediale Morphologie audiovisueller Wissenschaftskommunikation, Münstler: Lit.

Hilgartner, Stephen (1990): "The Dominant View of Popularization: Conceptual Problems, Political Uses." In: Social Studies of Science 20, pp. 519-539.

Hoffmann, Christoph (2010): "Schreiben als Verfahren der Forschung." In: M. Gamper (ed.), Experiment und Literatur. Themen, Methoden, Theorien, Göttingen: Wallstein, pp. 181-207.

Howard, Ron (Director). (2001). A Beautiful Mind [DVD]. Paramount (Universal Pictures).

Hüppauf, Bernd/Weingart, Peter (2008): "Images in and of Science." In: ibid (eds.), Science Images and Popular Images of the Sciences, London: Routledge, pp. 3-31.

Kirby, David A. (2003): "Scientists on the Set: Science Consultants and the Communication of Science in Visual Fiction." In: Public Understanding of Science 12, pp. 261-278.

Kirby, David A. (2011): Lab Coats in Hollywood: Science, Scientists, and Cinema, Cambridge, MA/London: MIT Press.

Kirby, David A. (2014): "Science and Technology in Film. Themes and Representations." In: Massimiano Bucchi/Brian Trench (eds.), Routledge Handbook of Public Communication of Science and Technology, London and New York: Routledge, pp. 97-112.

Krämer, Sibylle/Bredekamp, Horst (2013): "Culture, Technology, Cultural Techniques – Moving Beyond Text." In: Theory, Culture & Society, 30(6), pp. 20-29.

Kretschmann, Carsten (ed.) (2003): Wissenspopularisierung: Konzepte der Wissensverbreitung im Wandel, Berlin: Akademie-Verlag.

Kuhn, Annette/Westwell, Guy (2012): "camera movement (mobile framing)." In: ibid, A Dictionary of Film Studies. Oxford University. Accessed September 17, 2019. http://www.oxfordreference.com/view/10.1093/acref/9780199587261.001.0001/acref-9780199587261-e-087?rskey=ATNXtT&result=86.

Landecker, Hannah (2011): "Creeping, Drinking, Dying: The Cinematic Portal and the Microscopic World of the Twentieth-Century Cell." In: Science in Context 24, pp. 381-416.

Lee, Ang (Director) (2003): Hulk [DVD]. Universal Pictures Germany GmbH.

Leterrier, Louis (Director) (2008): The Incredible Hulk [DVD]. Concorde Video.

Locke, Simon (2005): "Fantastically Reasonable: Ambivalence in the Representation of Science and Technology in Super-Hero Comics." In: Public Understanding of Science 14, pp. 25-46.

Locke, Simon (2011): "Colouring in the "black-box": Alternative renderings of scientific visualisations in two comic book cosmologies." In: Public Understanding of Science 22, pp.304-320.

Mahne, Nicole (2007): Transmediale Erzähltheorie, Göttingen: Vandenhoeck & Ruprecht.

Mancosu, Paolo (2011): "[Book Review: Logicomix]." In: Journal of Humanistic Mathematics 1(1), pp. 137-152.

McCloud, Scott (1994): Understanding Comics, New York: HarperCollins.

Niccol, Andrew (Director) (1997): Gattaca [DVD]. Sony Pictures Home Entertainment.

Niederhoff, Burkhard (2011): "Focalization." In: P. Hühn/J. C. Meister/J. Pier/W. Schmid (eds.), The Living Handbook of Narratology. Accessed September 17, 2019. http://www.lhn.uni-hamburg.de/article/focalization.

Ottaviani, Jim (2011): Feynman, New York: First Second.

Reinerth, Maike Sarah/Thon, Jan-Noël (eds.) (2017): Subjectivity across media. Interdisciplinary and transmedial perspectives, New York: Routledge.

Rose, Christopher (2003): "How to Teach Biology Using the Movie Science of Cloning People, Resurrecting the Dead, and Combining Flies and Humans." In: Public Understanding of Science 12, pp. 289-296.

Rosenstone, Robert A. (2003): "Comments on Science in the Visual Media." In: Public Understanding of Science 12, pp. 335-339.

Schaeffer, Jean-Marie (2012): "Fictional vs. Factual Narration." In: Peter Hühn/Jan Christoph Meister/John Pier/Wolf Schmid (eds.), The Living Handbook of Narratology. Accessed September 17, 2019. http://www.lhn.uni-hamburg.de/article/focalization.

Shapshay, Sandra (ed.) (2009): Bioethics at the Movies. Baltimore: Johns Hopkins University Press.

Tatalovic, Mico (2009): "Science Comics as Tools for Science Education and Communication: A Brief, Exploratory Study." In: Journal of Science Communication 8, pp. 1-17.

Van Riper, A. Bowdoin (2011): "Introduction." In: id. (ed.), A Biographical Encyclopedia of Scientists and Inventors in American Film and TV since 1930, Lanham, Md.: Scarecrow Press, pp. viii-xvii.

Verdicchio, Dirk (2010): Das Publikum des Lebens: zur Soziologie des populären Wissenschaftsfilms, Bielefeld: transcript.

Vrckovski, Simone (2016): "Grüne Fluoreszenz? Voraussetzungen der visuellen Vermittelbarkeit von Radioaktivität." In: Closure. Kieler e-Journal für Comicforschung 3, pp. 112-136. Accessed September 17, 2019. www.closure.uni-kiel.de/closure3/vrckovski.

Wachowski, Lana/Wachowski, Lilly (Directors). (1999): The Matrix [DVD]. Warner Home Video.

Wachowski, Lana/Wachowski, Lilly (Directors) (2003): The Matrix Reloaded [DVD]. Warner Home Video.

Wachowski, Lana/Wachowski, Lilly (Directors) (2003): The Matrix Revolutions [DVD]. Warner Home Video.

Weingart, Peter (2005): "Von Menschenzüchtern, Weltbeherrschern und skrupellosen Genies. Das Bild der Wissenschaft im Spielfilm." In: id. Die Wissenschaft der Öffentlichkeit. Essays zum Verhältnis von Wissenschaft, Medien und Öffentlichkeit, Weilerswist: Velbrück Wissenschaft, pp. 189-206.

Innovative Popular Science Communication? Materiality, Aesthetics, and Gender in Science Slams

Miira Hill

ABSTRACT

This paper is about the communicative construction (Knoblauch 1995, 2017) of science communication in a popular genre called the science slam, and more precisely within the context of materiality, aesthetics, and gender. In science slams, we can observe how popular scientists position themselves. Although seeing the success of female performances (for example Giulia Enders) might lead us to hope for new norms of gender, technical jargon, and visual practices in scientific communication, these hopes are not wholly realized in the science slam genre. Our empirical observations (ethnography, video analysis, interviews) have shown that even if the new genre seems to enable all kinds of revisionist representations of science for men, women remain silent and invisible. Although the technical jargon and visual practices of science slammers are quite different from those used in university lectures, the marginalization of women, among other groups, remains an issue. Drawing on typical tropes of patriarchal societies, successful science slam women are presented as objects of desire, hardworking assistants, or aunts, mothers, and grandmothers. In this way, the science slam can be understood as an expression of still-problematic gender relations in science.

INTRODUCTION

This essay focuses on the communicative construction (Knoblauch 1995, 2017) of science communication in science slams. Its main research question is whether new materiality, aesthetics, and gender politics can be found in the genre of the science slam. The science slam was founded in 2006 in the German town of Darmstadt. Alexander Deppert, who had the idea to create this event, was very inspired by poetry slams. In Germany, there are now more than 30 science slams taking place periodically, and the event continues to expand. Scientists are asked to enter the stage and explain their research topic to a potentially non-academic audience. A science slam is an event in which actors are trying to establish new and improved forms to present and legitimate scientific knowledge. The self-presentation of a scientist at a science slam is expected to be different from conventional modes of science communication. The one strict rule of the science slam is that the presentation has to be short. The genre has a competitive character.

In the first part of this essay, we provide a short historical background on more recent forms of science communication. We develop initial ideas about communication in the context of Habermas' thoughts about publics and Fraser's further developments of subaltern counter-publics. We ask if we can describe the science slam milieu, with Fraser (1992), as one of subaltern counter-publics, or one of a classic public of patriarchal hegemony that is built on the exclusion of gender, class, and alternative publics. Following this, we outline our definition of science communication and the methods we used to study them. Then we introduce the reader to our first encounters with the science slam, and how they raised hopes for changes in science communication. The next part of the text shows other empirical examples from our research and exemplifies ways in which recent science communication can be seen as either changing or remaining the same. In the last part of this essay, we explain why in mediated societies online science communication must be seen as almost equal in influence as situated science communication. In our conclusion, we describe the science slam as a positioning practice of science.

GENDER AND THE HISTORY OF PUBLIC SCIENCE COMMUNICATION

To understand the recent developments in science communication genres, we have to go back and take a look at past relations of social transformation, and at the shifting role of the public and science communication. The history of popularizing science can also be seen as related to the rise and fall of a bourgeois public. In many places, new forms of debating as well as novel lines of cultural differentiation were tested. Habermas's account of *The Structural Transformation of the Public Sphere: An Inquiry into a Category of Bourgeois Society* (*Strukturwandel der Öffentlichkeit* 1990) describes how the bourgeois public sphere emerged in the historical context of French, British, and German developments in the 18th and 19th centuries. At the end of the 18th century, the number of people reading books increased. Before this, reading was something for select groups of scholars, but with the beginning of the age of mass communication (which had its origins in the technical innovation of the printing press in the 15th century) a wide-ranging audience for literature emerged. Habermas provides detailed insights into how a sphere of public communication occurred in connection to these new ways of adapting literature. A small public sphere of townspeople and bourgeoisie was exploring new structures of public communication (Habermas 1990: 13). British coffeehouses, French salons, and German regulars' tables were new spaces for public 'raisonnements' and the consumption of cultural goods. Habermas pointed to the asymmetric concept that the 18th-century bourgeoisie had about culture. The crowd was educated to be part of a sophisticated culture; culture was not reduced to be a culture of the masses (ibid: 254). The idea of educating the uneducated was a top-down concept. It was in the context of this sophisticated public bourgeoisie of the 18th-century that the idea of connecting science to the public in a non-academic environment arose more broadly. At the same time, secular belief systems like those of the natural sciences were spreading. Dominant subjects like theology, medicine, and law were complemented with chemistry and other laboratory sciences.[1] The new men of science had to learn the new scientific methods introduced by Bacon and Descartes, and subsequently had to protect these ideas from thinkers who preferred traditional ways of seeing and knowing the natural world (Sheffield

1 The Copernican turn and Issac Newton's understandings of gravity were starting points for these belief systems (Ebel and Lührs 1988: 15). The pioneers of these new approaches to the study of nature were Francis Bacon (1561-1626) and René Descartes (1596-1650).

2004).[2] (For a more detailed picture of the history of science communication, see the contribution of Oliver Hochadel in this volume.)

In a time of radical changes, in the middle of a revolution and the foundation of the German Reich (1871), it was not just democratic ideas and natural sciences that were on the rise, but also a new bourgeois conception of self. Even though the bourgeois mainly followed a philological-philosophical educational ideal, the notion that humans were capable of understanding the laws of nature extended the more canonical view. Since every member of society hypothetically was understood to have the faculty of reason, there emerged the new idea that every white and able man could take part in the domination of nature (Ebel and Lührs 1988: 15). The popularization of science and the education of the public started to become an ambition of the bourgeoisie after the 18th century (Daum 1998: 4).

For this reason, the era beginning in 1848 is called "the pioneer's period of popular science" (ibid). The impact that associations and clubs had on the transformation of the public sphere and the collectivization of the bourgeoisie was enormous. Scientific academies were comprised of elitist groups of scholars who excluded a broader public, so the bourgeoisie had to create other non-academic spaces to debate science. After 1848, public collections of natural products, chambers of curiosities, museums of natural science, zoos, botanical gardens, observatories, and aquariums were established (ibid: 5). It was a period of public lecturing. By this time, academic scholars had started to favor a new educational ideal for the 19th century and were opening up to a broader public. In addition to talks in associations, there were wandering orators (*Wanderredner*) traveling through Germany with small figures and experimental devices to demonstrate scientific topics. In the same period, the media coverage of popular science also increased.

For Habermas, the fall of the bourgeois public sphere took place in the late 19th century following the change to a more heterogeneous society with fewer class barriers. The first step of commercialization was reducing the economic barrier to education (e.g., cheaper books) while maintaining a psychological bar-

2 The historian Andreas Daum (1998) describes the educational and political dispute between humanists and realists over the relevance of natural science in German schools in the 19th century (ibid: 51). The dominant humanistic ideal with a philological emphasis was contested by realistic explanatory models that aimed to push the relevance of natural science. The worldview of natural science was described as business or utilitarian thinking, as materialism, and as the technicalization and commercialization of society (ibid: 53). Realistic thinkers had to argue against the denunciation of promoting a utilitarian school by using humanistic rhetoric.

rier. The subsequent triumph of cultural industries introduced the second chapter of the commercialization of culture, this time dropping not only the financial but also the psychological barrier. Habermas describes how book clubs after the Second World War reduced the quality of the content of books with the help of their editorial office in order to reach a broader audience. Continuing Adorno and Horkheimer's critique of the culture industry, he describes a change from an acculturating culture to a consuming one. In the emerging pseudo-public sphere of mass media, the deliberative assemblies for 'raisonnements' about cultural topics were replaced by a public of mass media. Habermas describes the ensuing public of mass media as not critical but rather staged, one-sided, and undemocratic. If we believe Habermas, the source for a social class that has been educated to use the intellect in a public sphere has run dry. According to him, the 20th century was a period of incapacitation of the public and the disappearance of non-academic audiences from the academic sphere. After the First World War, an image of the public became dominant in the scientific community, and this image rejected the ability of non-academic people to understand scientific findings.[3]

Habermas' judgments about publics were related to the question of whether communicative genres do or do not empower a public.[4] As we learned from our interviews, science slammers predominantly leave out methodology, statistics,

3 This argument of cultural decay is reflected in contemporary concerns about public science communication. Tufte (2006) argued that PowerPoint has caused a mental enfeeblement of society and is responsible for catastrophes like the crash of the Challenger spacecraft. Moreover, critical voices about newer science communication events say that events like the science slam constitute the public as a "collective of infants" (Klaue 2015).

4 Science slam organizers argue that their audiences are more empowered than those of university lectures. In particular, the audience's ability to participate in voting is described as an empowering act. As part of every science slam, the audience votes for a winner to judge whether the scientific content was delivered in an appropriate form. During the talks, there is mutual presence and immediate emotional response, but no actual dialogue with the audience; the audience is primarily seen as a receiver of information. The arrangement does not question that experts on stage have more knowledge than the audience. Science communication is designed mostly as a monologic and linear process in which the slammer has the monopoly to talk. When listeners suffer while listening to the talk (or if they think a topic was not scientific), they can judge the performance in voting groups or through applause (or the lack thereof). However, the audience's response insofar as it might have implications for research, or the audience's own knowledge, is not the focus of attention.

and theory. These omissions sometimes make the topic more easily comprehensible, but do so perhaps at the cost of not empowering the audience to understand scientific procedures. So one could say that the science slam is not critical, dialogical, democratic, and empowering enough to satisfy Habermas' expectations.[5] In leaving out methodology, statistics, theory, and the diversity of scientific practices, the participants of a science slam do not seem to enable their audience to have a first-hand understanding of science in action. However, since the workplace, the everyday activity of science, and the self of the scientist are central to the science slam, these elements provide listeners second-hand access to scientists' identity formation, which leads us nicely to the scientific critique of Fraser (1992).

Judgments about the public are not infallible. Habermas was criticized for the image he created of a past golden area of public 'raisonnements.' Fraser (1992) argued that Habermas' concept of a bourgeois public sphere has not only falsely presented the idea of one single public sphere, but at the same time uncritically idealized it. She argued that Habermas himself had supported the ideal of a free and equal public discourse with open access. Even if Habermas has pointed to the problem that the bourgeoisie concept of the public was never realized, he did not mention that the bourgeois public was built on the establishing of differences, namely the exclusion of gender, class, and alternative publics. Fraser argues that the bourgeoisie idea of the public is not only an idealized utopia; it is also a male ideological concept that supported and legitimized the leadership of a particular group. Therefore, she argues that Habermas' idea that participants of discourses can argue as if they were equal only by disregarding their backgrounds should be replaced by enabling participants of weaker social groups and removing social disparities. For that reason, Fraser presents an alternative post-bourgeoisie concept of the public spheres. Her concept of subaltern counterpublics is an exemplary form of parallel space in opposition to patriarchal hegemony.

"Public spheres are not only arenas for the formation of discursive opinion; in addition, they are arenas for the formation and enactment of social identities. This means that participation is not simply a matter of being able to state propositional contents that are neu-

---

5 Presentations often include strong asymmetries of speaker and audience. This not only shows in rhetorical remarks like "very primitive approach" (a quote by Johannes Schildgen from a slam) to explain a topic, and in barefaced ways of confronting the audience with context-free facts, but also manifests itself in visual and performative asymmetries.

> tral with respect to form of expression. Rather... participation means being able to speak in one's own voice, and thereby simultaneously to construct and express one's own cultural identity through idiom and style." (Fraser 1992: 166)

For Fraser, identity formation is a part of public participation. In this context, the past enactment of social identities as described by historians such as Shapin must be seen in a more critical context. Shapin (2006) has demonstrated how the man of science in modern times changed his social role from the image of an isolated, unfriendly, confrontational, authoritarian, individualistic, dry, and pedantic character to that of a gentlemanly self-conception. In the gentle culture, the attempt to be civilized, masculine, rational, thinking, distanced, ascetic, and text-oriented was opposed to a more body-oriented approach.[6] The new intellectual order that had been established since the 17th century made modest witnessing a convenient method of creating knowledge in public (Shapin and Schaffer 1985). If we recall Fraser's argument, this modest witnessing must be seen as a specific cultural setting with certain rules that include some people and exclude others. In general, bourgeois culture can be characterized as a way of living where the commitment to a community spirit, the need to construct individuality through socializing, participation in a masculine culture, and practices of self-monitoring were central (Mergel 2001). However, it is important to keep in mind that in an egalitarian and multicultural society there are always multiple coexisting publics that reflect differences in society. Fraser's idea of subaltern counter-publics not only has consequences for common notions of deliberative democracy and its ideas about political equality, but is also a reminder that the communication of science is not open to everybody. Societal inequalities prevent particular groups from becoming part of discursive arenas such as science slams; they exclude parts of the population. Consequently, feminist science studies have diagnosed the body of knowledge of the western scientist as being differentiated from the representations of women. Haraway (1997) has pointed out that the era of the

6 One might wonder about the overlap of the gentle style of communication and the feminine style of communication. Indeed, many researchers who work on the communication of gender relate female communication to indirect and very polite ways of speaking (see Braun and Pasero 1997). However, this depends on the cultural context. Günthner (1997: 139) therefore argues that female speech can by no means be universally set as polite and more indirect, nor is politeness and indirectness always a sign of social subordination and powerlessness. The interpolation of polite/indirect styles depends on the situational context, the overall status of the interacting people, and the linguistic conventions that are ritualized in the cultural context.

modest witness, with its distinction of the knower and the known, supported established inequality. She suggests that leading representations of science might be part of a historically grown balance of power. Haraway is mistrustful about the way western scientists avoid positioning themselves or their responsibility regarding their research. She opposes the "invisible conspiracy of masculine scientists and philosophers" and the "embodied others, who are not allowed not to have a body, a finite point of view" (Haraway 1988: 575). Disembodiment and universal claims are described as part of the western scientist's bag of "God-tricks," which she calls a view from nowhere. For that reason, Haraway describes positioning as a key practice in opposition to the typical visualizing tricks and powers of modern science, which make "various forms of unlocatable, and so irresponsible, knowledge claims" (ibid: 583). These feminist approaches argue for embodiment, partiality, and localization as grounding practices for knowledge claims (objectivity).

SCIENCE COMMUNICATION

Some scholars distinguish science communication and public science (Robertson-von Trotha and Muñoz Morcillo 2018, Bauernschmidt 2018). For them, science communication is the "top-down," institutionalized communication of science at the university (e.g., public relations), while public science is the more voluntary "dialogue-oriented" communication in the tradition of Humboldt. Our use of the term science communication is oriented toward the theoretical concept of communication and the understanding of science that we have developed (Hill 2016). Science communication is not a perfect transmission of scientific information from sender to receiver, but a situated communicative action (Knoblauch 2017). Thus, in our work, the term science communication refers to the situated communicative actions of scientists. Whenever scholars or scientists talk to each other or to a non-academic public, and whenever they refer to the knowledge they characterize as being within their scientific expertise, on these occasions and insofar as this reference is essential to the form of communication, we call this science communication. When analyzing science communication, we look at the validity of scientific knowledge in the communicative genre of the science slam. We refer to the communication (and partly intersubjective validation) of scientific knowledge. The communicative construction of something as scientific happens in the triangle of embodied subjects, their objectivations, and others. In my project, I focus on communication in action. What I call science communication here is therefore beyond (and partly before) the publication and reception of ready-made science qua scientific papers and refers to one-to-many face-to-face situations.

Our work reflects, first of all, the validity of scientific knowledge in a public genre. We argue that, for methodological reasons, we should first empirically study how science is made visible in public today, and show how trust in science is generated. We argue that the genre of the science slam should not be evaluated as being a phenomenon of cultural decay without first being studied. The cultural decay argument pervades many recent studies about the uptake of science into various fields of media; like Habermas, Weingart's (2005) studies of the mass media end up arguing that such phenomena mark a decline of quality in communication about science. Comparing movies, scientific journalism, and literature, Haynes (1994) has shown how the construction of science varies according to the medium of representation. The criticisms of these medial constructions of science were mainly focused on the dominant role of the producing media.[7] Like the idea of cultural decay, mediation is based on the idea that we can have access to the truth. Concerning STS and studies on the sociology of knowledge, we have highlighted how complicated this perspective is (Hill 2016). Taking up a social constructivist standpoint, we will take a perspective of methodological agnosticism (epistemological agnosticism) that is interested in what is acknowledged as true in various contexts. When analyzing science communication, we look at the validity of scientific knowledge in the communicative genre of the science slam.

Secondly, our work focuses on communication processes. Lines of science studies today are focusing on the ways experts are concerned with "boundary work" (Gieryn 1983) to create a difference between the intellectual and nonintellectual. The attention to truth-producing practices points to the importance of the bodily presence of truth in the public sphere (Eyal and Buchholz 2010, Bourdieu 1988). Scholars have recently addressed the problem that past STS research on science communication has been disembodied (Davies 2009, 2013). Some exceptions to this have shown that representations and visibility in science are generated and interpreted by social actors in the context of communicative processes (e.g., Lynch 1988, Amann and Knorr-Cetina 1990, Beaulieu 2002, Alac 2008). Moreover, a handful of scholars have focused on kinesthetic and affective entan-

7 Weingart talks about "mediated science" because the link between science and the public is dominated by the mass media. He focuses on the problems arising from the pressure of mediation. The danger he sees in science that focuses too much on mass media is that the image of science and scientific knowledge is slightly altered through such media.

glement in science communication (Myers 2012) or aspects of body movement and socio-technical arrangements in interaction (Goodwin 1981, Goffman 1981, Knoblauch 2007, Kiesow 2014, Tuma 2012).

The methodology tries to capture the transformation of science communication processes. We address the question of why and how particular communicative actions are taken up in a complex and uncertain relationship between science and the public. Our theoretical and methodological framework is qualitative per the logic of interpretive sociology (*verstehende Soziologie*). Following Schütz, Berger, Luckmann, and Knoblauch, we think that the science slam can best be understood by analyzing the first-order reality constructions and communicative actions of science slam participants. Our general aim is to use a combination of qualitative methods, as is recommended for focused ethnography (Knoblauch 2001). Several data-gathering methods were used to collect different kinds of material. To understand the inner perspective and everyday life of science slam participants, we chose an ethnographic approach. We visited many science slams and tried to capture the spirit and understand the social dynamics of those events. To study the situated performance on stage, we picked the "microscope of interaction studies" (Schnettler et al. 2013), i.e., video analysis. To create multi-perspective validation and to describe different aspects of the science slam phenomenon, we used qualitative interviews as the third central method; with this method, we tried to learn more about the justification of the action. Based on these different methods, we aimed to triangulate features of the communicative genre of science slam.[8]

WHAT IS NEW IN TODAY'S SCIENCE COMMUNICATION?

Let us share a few experiences from one of the first science slams we analyzed. It was February 27, 2012 in a venue called Lido, in Berlin. There were several cameras in the room. Lido was packed with young people. The event was hosted by the science slammer André Lampe. Participants at this evening were Garcia Peters from the University of Hamburg (Meteorologisches Institut), Falko Brinkmann from the University of Münster, Peter Westerhoff from the Charité

8 Akin to the principle of Grounded Theory, my data had to align with other data in a triadic process: With verbal data, visual data, and observer data, the triadic relation of subjective knowledge (legitimations), the relation to others (performance, interaction), and the objectivated world (socio-material arrangement, language, and body) were the centers of interest.

Berlin hospital, and Giulia Enders from Goethe University Frankfurt.[9] After Garcia Peters' presentation, the host introduced Giulia Enders as the second speaker. He introduced her with her name and her academic affiliation and requested the audience welcome her with applause. The audience applauded. In the presence of the moderator, Giulia quickly got her presentation ready. She was dressed in a colorful t-shirt with fast-food images on it, loose jeans, and a blue hoody sweater. She started her talk:

Giulia: Okay, as was said, I am studying medicine.
Voice out of the audience: *woohoo* (cheers).
Giulia: [laughs] (.) yes, that's right. and one thing that medicine is really good for, as you have just seen, is drinking coffee.
(looks to the slide. the slide shows three women with teacups)
or tea with aunts. ah, because while drinking tea you are asked what you are studying,
and while my sister has to spend half an hour explaining what communication design is,
I just say *medicine*.
(looks to the slide. new slide shows smiling women)
Audience: [laughs]

It was clear that Enders felt very comfortable with her discipline, which she presented as a legitimate and widely accepted field of research. In her talk, medicine seemed to be part of common knowledge; she did not have to explain the discipline to her aunts. Her subsequent remarks about visiting her aunts made clear that her research subject—the human gut—is not as popular as medicine in general. After she explained the experience with her aunts to the audience, she said that a talk in the kitchen with her roommate was responsible for her interest in the human gut. One morning, her roommate came into the kitchen and asked her to explain defecation. In her talk, she spoke about things like bowels, farting, sphincter muscles, biomolecular details, and bacteria. Unusual for public science communication, she used a high number of colloquial words like "shit" and "pooping." Her remarks about farting and pooping caused people in the audience to react with giggles and laughter. She explained her topic using cute, comic-style illustrations that her older sister, Jill Enders, had made for her. If the audience had previously thought scientists were boring white men wearing coats in

9 Giulia Enders is writing her dissertation at the Institute of Microbiology at the Goethe University. She wants to become a gastroenterologist. Her doctoral thesis is about a specific bacterium (*Acinetobacter baumannii*). She hopes that there will be more research on the positive influence of specific probiotic bacteria on people's wellbeing.

the laboratory, then Enders' performance must have made an enormous difference to them. As a later newspaper article described it based on a video recording of the event: "The pretty girlish student Giulia Enders explained in her YouTube video not only the last taboos but also the great mysteries of the human body—even going into hard biomolecular details."[10]

In this article, gender is explicitly mentioned as a point of reference and framed as an unconventional attribute of a scientific persona. Her young, authentic, sensitive, casual, and unembarrassed style was unconventional for science communication. With her charming, informal, and easy-going appearance, she easily handled the unpopularity of her topic on stage and told the audience why she thinks the gut is indeed an important field of research.

When we attended the event, we did not know how iconic the talk by Giulia Enders would become and what societal impact her presentation would have. On that night Giulia Enders won the slam with her talk "Darm mit Charme" ("Gut with Charm").[11] Shortly after the video was uploaded on YouTube, her talk became a hit. Although she had attended just four science slams, her videos had more than one million clicks (1.088.800) as of July 2014. Enders received an offer from the publishing house Ullstein Verlag to write a book about her research topic. Shortly after the book's publication in March 2014, it rose to first place on the Spiegel bestseller list in the paperback nonfiction category. It went on to sell more than one million copies. Enders was also invited onto TV talk shows and became prominent in the newspapers. We can observe how in the underground world of science slams a product emerges that is framed both as serious science and as pop culture.

In our later ethnographic experience, many conversations with science slam visitors highlighted that they had seen Enders' slam. Some said that her science slam had motivated them to join a science slam event themselves. In addition to this, various science slammers told us that they were inspired by her style of presentation while preparing their own presentations. Enders' slam could, therefore, be seen as a starting point, as an inspiration for many slammers currently active in science slams. It also served as a starting point for our fieldwork and our research agenda.

This project is not so much about science slammers like Giulia Enders who are successful in the mass media, but more about science slammers who are successful in several situated science slam settings. Nevertheless, we start with Giu-

---

10 http://www.faz.net/aktuell/wissen/darm-mit-charme-von-giulia-enders-12891555.html, accessed September 12, 2020.

11 http://www.youtube.com/watch?v=2qo3ueVlyUY, accessed September 12, 2020.

lia Enders because she had such a massive impact on the public image of science and simultaneously marks our entry into the world of science slam. Of course, her example is exceptional, but it shows how one slam can, hypothetically, establish new communicative practices for science. Her case is a great example that shows how traditional ideas about science are being challenged in science slam presentations. Catalyzed by just one performance, by innovative public practices of sci-ence legitimation, and by unusual styles of self-presentation, these science presentations and their interactional structure can be introduced. Enders proved that scientists could be both professionally and popularly successful (two modes of success that have sometimes been described in the past as being in conflicting with one another), and could even be female. Her example embodies the hope that public science communication can break conventions of science communication regarding, among many norms, gender, technical jargon, and visual practices.

However, is Enders' example truly proof of the beginning of something new in science communication? A science slam is a genre in which, to a great extent, male scientists publicly speak. Since the format's beginning, the gender divide at slams has not been balanced. Only a few women have found it appealing to participate at a science slam. From the successful participants that we observed, only two persons out of 19 were female. Those two female slammers were not successful as touring scientists in various science slam settings (because they only attended a few science slams), but rather on YouTube. In interviews, the slams' organizers recognize and address the problem that this gender imbalance poses. It has even become a ritual part in many events for the moderators to appeal to women in the audience to participate as slammers. In recent years, there have been positive developments in terms of gender, and 50 per cent of the participants in the German championship of 2014 were women.

THE COMMUNICATIVE CONTEXT OF SCIENCE SLAMS

Fraser's idea of subaltern counter-publics is concerned with the question of whether speakers are able to speak in their own voices and to enact their social identities. When focusing on science communication and gender, one could take a look at the way scientists use language. Enders' unusually high number of informal words (such as "shit" and "pooping") could be seen as someone speaking in her own voice. However, this could also be considered with regard to older feminist research on "women's language" (Lakoff 1975) that argued that women tend to use more tag questions ("isn't it?"), polite empty phrases ("thanks"), and hedges ("you know") as a specifically feminine style of communication. Since later feminist research has shown that communication always has to be studied against the background of the specific communicative context of an interactive frame (Günthner 1997: 124), it seems reasonable to highlight not only the language and embodiment of science communication but also the context.[12]

Science slams do not take place in the context of bourgeois culture but rather in the context of a certain communicative milieu. Nevertheless, can we describe the science slam milieu, with Fraser, as one of subaltern counter-publics, or is it one of a classic public of patriarchal hegemony that is built on the exclusion of gender, class, and alternative publics? The milieu of the organizers can be divided into science dropouts who started to go into business for themselves, and science communicators who work for an organization. All of the organizers I met had an academic background. Their original academic training was in the margins, spanning German philology, cultural studies, theater studies, politics, public policy, management, international management, architecture, physics, biology, and technical and environmental support. The gender division of the organizers was fairly balanced. They were mostly between 30 and 40 years old. According to the organizers, the science slam milieu is described as young, sophisticated, and committed. From the faces they observe, organizers get the impression that they have a fairly young audience. Organizers describe their audience as members of academic and creative milieus. It is not only students but also educated people in general who are interested in science who go to science slams. Colleagues and pupils of the research groups of the participating slammers visit the science slam, as well as members from creative industries that live in the neighborhood of the hip event locations. For this reason, one organizer la-

12 Newer perspectives on gender-specific communication include more theoretical reflections about the social construction of gender (Günthner 1997) and other characteristics that are related to differences of power, such as ethnic background, social/economic class, and educational background. Günthner (1997) argues that the production of robust gender identity is connected to permanent acknowledgment in situations of interaction. This perspective makes it necessary to focus on interactions that produce certain genders. The key question then becomes how societal beliefs in so-called "natural" differences of men and women are reproduced by social organization, symbolic order (language), and by everyday interactions (Günthner 1997: 136). In societies, specific genres are related to certain genders. This reproduces cultural beliefs and stereotypes about the nature of women and men.

beled the audience as a "gentrification audience" (MK#45).[13] In the milieu description of Schulze (1992), this would be the 'self-realization milieu,' which is characterized by a higher degree of education and a younger average age, and whose central focus is the pursuit of self-realization.

Empowerment

Science slams can be empowering events for the slammers, who find less recognition in the dominant discourse of science, and who can develop a new form of science through the new format of the science slam.

COMMUNICATION IN SCIENCE SLAMS

When looking at science slams, we are looking at humorous interactions. Science slam lectures are mostly infused with humor. Although organizers emphasize that joy is not the only emotion to be triggered by the performances, the reference to fun is an essential reference. The expression of the audience also finds its place in the interactions. The emotions of the audience are expressed through laughter or interjections. Speakers plan their talks with emotional reactions of the audience in mind, and generally leave short pauses for those reactions.

The Rough-Diamond Style of Science Communication

The organizers of science slams enforce a certain "in-the-making" aesthetic. One typical feature of the genre is that the audience can see the preparation of the presentations and the event as a whole. Not only do slammers have to prepare their presentations while being filmed; the event's staging and preparation are also highly visible. In this way, the science slam promotes a rough-diamond style of science communication. The coaching offered by the organizers is not designed to produce uniformity. In interviews, organizers characterized good science slammers as communicative scientists who present their own research and have creative and artistic skills (Hill 2017). Bad slammers were described as uncreative scientists presenting textbook knowledge they did not produce themselves, and who copy the style of others or create uniformity. Science slammers mostly try to avoid using professional jargon and instead use an everyday lin-

13 Third expert interview. The interviews carried out within the science slam project were numbered chronologically starting with the number 42.

guistic register. There is a slight tendency to use slang and Anglicisms. Several of the slammers address the audience informally with 'du/ihr' in German, while others speak in a much more formal tone. At science slams, it is not considered a lack of decorum to speak in dialect or with a strong accent. Most successful slammers have a slight dialect and even emphasize their dialect in their presentations. Most of the speakers talk rather quickly and have a continuous flow of speech; only a few speak rather slowly.

Most of them have a somewhat professional style of communicating. Sometimes their talk is similar to a moderator or a standup comedian. Mostly we find eloquent professional language, and only a few speakers have issues expressing themselves. A few of the successful participants use speech that is close to high rhetoric. However, even if some follow a quite formal presentation style, they are informal in their content. Some are closer to "talking heads" in their speech orientation; others are more body- or interface-oriented. A typical semantic characteristic is the code-switching of science slam presenters. Science slammers switch from technical jargon to everyday language and vice versa regularly. To provide a specific example: One speaker outlined in his talk that he is especially interested in "bats that feed on fruits." He then referred to this kind of bat in scientific, technical jargon as "*Artibeus jamaicensis*." He goes on to say that he calls this type of bat "AJ."[14] His pronunciation then suddenly switched to English. The everyday formulation "Bats that feed on fruits" turned into a scientific formulation "*Artibeus jamaicensis*," and was then converted into the common Americanized name "AJ."

Visualization

In science slams, the use of presentation software like PowerPoint seems obligatory. As a rule, almost no science slam can do without PowerPoint or similar programs. This technology is therefore likely to be an essential prerequisite for the growing importance of visualizations in science communication. Visuals are just as ubiquitous as presentational software. It is not only a matter of serious, exact, and scientifically founded representations. In addition to this, the repertoire comprises a range of different forms of visuals. Within the scope of their pictorial repertoire or visual style, scientific representations such as statistical visuals form only one part of the overall visuals. Corresponding to the deliberate avoidance of professional jargon and the emphasis on everyday language, we

14 AJ is also, for example, the name of one of the members of the boy band the Backstreet Boys from the 1990s.

observed that vernacular images predominated over scientific images. Science slammers often use pictures from Google Images—i.e., generic visual topoi—to introduce their approach. The vernacular images often have a relation to popular topics like movies (e.g., Star Wars, Batman) or TV shows (e.g., Pimp my Ride,), to cartoons (e.g., the Simpsons), to sports (e.g., soccer), to food (e.g., beer, sausages), to politicians (e.g., Karl-Theodor zu Guttenberg), to childhood images (e.g., the Norwegian children's novel Karius and Bactus), to internet images (e.g., screenshots from Wikipedia or Amazon), to animals (e.g., mice), and to the church (e.g., images of the Pope). They also often include representations of minority groups (overweight crooners, a black boxer with a small brain, women (wearing bras), a working-class lady in a chip shop, old people without teeth). If science slammers produce images for their talks, they often use illustrations or representations in a comic style. Translational slides are often used to bridge the gap between specialized scientific knowledge and common sense by showing scientific visuals and their metaphoric counterparts from everyday life together. Scientific images are also part of the talks. Visuals of experimental setups, genetic codes, magnetic beads, blood samples, nerve cells, glial cells, bones, iron oxide, the brain, and statistical visuals such as diagrams or heat maps are also frequently used. Science slams also often include images from the scientist's workplace (e.g., research buildings, a test person in the hospital, surgeons at OP, students in the cafeteria), and pictures of the researcher or research team in their work environment (e.g., their own team, picture of their student assistant). Furthermore, presenters also like to show the tools they work with, such as a scientific apparatus (e.g., images of MRT machines), or a device (e.g., syringes, petri dishes), and visuals of the results they obtain (e.g., measurements).

Physical Scientific Objects

Presenting physical scientific objects is also central in science slams. These objects become an important feature in the triadic structure between presenter and audience. As objectivations, scientific objects play a special role in science slams. These are not merely images on the slides; sometimes the presenters bring the actual physical objects on the stage so that they can handle them and show them to the audience. An example of this is when one slammer, during his presentation, said that a valuable book was handed over to him by an American library when he asked for it. He implied that he was able to take the valuable book with him. He then held up a shabby book which in his talk he had previously introduced as a mysterious, old book that people do not know much about. He opened the book that he had introduced as the *Voynich Manuscript*.[15] He looked inside it and flipped through a few pages. He unfolded a page and held up the book and turned it with the unfolded side to the audience. He not only mentioned that it was easy for him as researcher to gain access to the desired object, he also brought along the mysterious object and showed it to the audience. He handled it without gloves and gave the audience the possibility to take a look inside themselves. In this way, he was either bringing the mystery of science to the stage, or was disenchanting the scientific object. The speaker was presenting this research object as a valuable commodity.

Orchestration and Timing

Presentations are co-processed in a triadic structure, which unfolds between the presenter, her or his audience, and the objectivations used to communicate. As the objectivations in presentations rely partly on technological infrastructure, the presenter-audience interaction is necessarily co-produced with acts of presenter-machine interactions. These moments are typically well-orchestrated: A click on the laptop or the remote control, often accompanied with a sideways glance toward the projected slide or the screen of the laptop, adds an objectivated time sequencing to the presentation. This technological rhythm pattern (beat) underlies the entire presentation, and its meaning is clear to the presenters and the audience alike. Presenters use the beat of the changing slides as a resource for sequencing the talk (e.g., pre-sequence, general introduction, main argument, excursus, results, conclusion) accompanied by prosodic changes and variations in pace. The audiences use these turning points (tropes) for interjections, applause, and comments. Timing the interaction in the context of the triadic human-human-machine relationship (presenter, audience, hardware and software of the presentation technology) is decisive in the sequencing of science slams.

Another typical characteristic is the significance of timing. Science slammers have well-orchestrated PowerPoint presentations, and they mostly know exactly what appears next on their slides. The orchestration of showing, telling, and leaving pauses for audience processing and reactions seems well prepared.

15 The Voynich manuscript is a book that was discovered in 1912. It is a valuable book from the 15th century which has been an unsolved mystery for scientists and pseudo-scientists. The book is written in an unknown language and includes many illustrations. Today the book is held in one of Yale University's library collections.

Giving Body to Science

Science slammers give body to science and in this way embody knowledge. They present themselves as representatives of science and as producers of knowledge. They offer their body, their body language, their voice, their clothing, their attitude, or their gestures. The embodiment of science in science slams includes, above all, everyday non-perfect bodies. The scientists at the slams can be overweight, can wear T-shirts and pants that are too tight or too loose, and their hair can be poorly done, if they are behaving authentically.[16] Several examples have now shown how science slammers perform on stage. In one case we summarized the bodily performance of a slammer as follows: "He enters the stage wearing casual clothes. He is wearing a green printed T-shirt and black jeans. He has dark short hair and square glasses. His whole style of presentation is loose and laid-back. He is continuously smiling in a whimsical manner. He does not stand straight, nor does he have a striking attitude. Sometimes it is possible to see that he is standing on one leg and loosely shaking off his other leg. He makes many gestures. He uses his arms for his gestures in an unconstrained manner. At first, he seems not to have a very self-confident attitude. He gives the impression of being a shy man. Sometimes it seems as if he is insecure. His prosody is not like that of an actor. When he speaks, we hear an accent (from the Eifel region). Sometimes his voice sounds nervous. From time to time his voice even cracks. The melody typically drops at the end of a sentence. It seems as if he has no fixed text, as if he is acting spontaneously and in an untrained manner. Furthermore, his gestures seem to be out of place relative to the expectations of the stage format. He gives the impression of being very authentic. He does not give the impression that he is playing a character that is different from his real self."

In science slams, insecure behaviors are welcome on stage. Some slammers even flirt with the demarcation from the perfect medial representation of the scientist. We now select a harmless (not insulting) example to illustrate how this is

16 In our ethnography, we once had the feeling that a moderator had a critical eye on a "handsome" and "silky" presenter who was presenting without making any mistakes. In this case, we had the feeling that the moderator was uncomfortable with something the man was presenting or representing. We had the strange impression that this was not the stage for the perfect son-in-law, but rather for people with a human stain. This was only a one-time impression. The man did not win the slam. We will never find out the reasons behind this, but we are sure that the moderators have a lot of power to influence the audience's mood.

so; all the same, the unequal representation of gender is part of science slams. In the presentations of Boris Lemmer, he presents himself as a real scientist:

(holds his right forefinger to his chin)
we will look at a short exerpt. why? much is not real, but one
(keeping the right hand on the chest, thumb and forefinger forming a circle)
part is.
(glances back at the screen, points with his left arm)
THIS **(k)** *is a physicist who is waiting for the first collision of two protons, very excitedly, as you can see. she is not real, she wears a WHITE COAT. If she had ACTually* **(l)** (grabs the fabric of his open check shirt with his left hand) *worn a PLAID shirt, then she would be real.*
(pointing behind him to the screen)
[Audience laughs and applauds]
(smiles briefly)

Here, the media representation of a scientist is confronted with that of a real scientist. A woman wearing a lab coat from a movie is labeled the fake scientist by the man with the plaid shirt on stage, who presents himself as real. The lab coat that represents physicists in movies is replaced by a plaid shirt that represents the physicist in this performance. The speaker points to the media representation behind him on the slide to point out that the female image is a false one; then he touches his plaid shirt while mentioning that *he* is the real scientist. This is a rather subtle example to show how the body of knowledge today is still enacted as male.[17] A real male body of science replaces a fictional female body of science.

In this way, female scientists face two challenges. The first challenge is becoming an equal part of the male-dominated science system and establishing a stiff female representation of women in science (sometimes by wearing a lab coat). The second challenge is becoming an equal part of male-dominated science slam stages and perhaps loosening up the image of the stiff female representation of women in science.

17 Typically, the "othering" of women in science slams is much more explicit. I have chosen this rather harmless example because I did not want to name and shame a specific science slammer.

Gender in Science Slams

In the context of feminist STS, it seems useful to study the inequality of gender in science communication empirically. The study of the language, performance, and communicative context can help us understand why female scientists are still underrepresented in science and science communication. In science slams, we can observe how popular scientists position themselves. One of the best-known performances is that of Giulia Enders, a female scientist from the discipline of medicine who became a successful science slammer and later a bestselling author. She had a large impact on the public image of science. Her case is a great example to show how traditional ideas about science are challenged in science slam presentations. Her success was just as much a personal achievement as it was the success of new public legitimation practices of science and unusual styles of self-presentation, science presentations, and interactional organization. Her example stirs up hope that such public science communication can be more impactful in breaking down conventions of science communication. Although seeing the success of presentations like that of Giulia Enders might lead us to hope for new norms of gender, technical jargon, and visual practices in scientific communication, these hopes are not wholly realized in the science slam genre as we have observed it. Although the technical jargon and visual practices of science slammers are quite different from those used in university lectures, the marginalization of gender, among other groups, remains an issue. Our empirical observations (ethnography, video analysis, interviews) have shown that, even if the new genre of the science slam seems to enable all kinds of revisionist representations of science for men, women remain silent and invisible. As we have shown in this paper, even cinematic representations of female scientists are questioned by 'real' male scientists on stage. New forms of popular scientific personas are presented to the public. The gentleman from early modern times, who was formerly the individual who was trusted to speak the truth, today is replaced by new types of public scientists who are familiar with artistic visual practices, vernacular images, rough-diamond styles of communication, popular culture, and informal and everyday language. In successful science slams, which draw on typical tropes of patriarchal societies, women are presented as objects of desire, hardworking assistants, or aunts, mothers, and grandmothers. In this way, the science slam could be understood as an expression of still-problematic gender relations in science (especially in natural, applied, formal, and health sciences). Furthermore, the slams often include offensive and marginalizing representations of minority groups. Although it is extremely problematic to display such discourses on stage, scientists make themselves and their discursive moves available for critique, both by situated audiences and by other scholars, policymakers, and members of the media.

Situated Communication versus Mediated Communication

On the one hand, we have observed the situated performance of a male researcher who replaces the cinematic imaginary of a female physicist; on the other hand, we have heard of the mediated performance of Enders, who publicly enacted a female scientist. There is still the possibility that Giulia Enders' performance had a more significant impact on the public image of science than the performance of Boris Lemmer. In our research, the relationship between social networks and social movements and the intermeshing of online and offline practices has become central. As our research shows, social networks have become mobilization platforms for the science slam movement. The social dissemination of the science slam in Germany can only be understood if we include the organizers' practice of uploading the videos of the events to YouTube. The diffusion of the genre was powered by Haus der Wissenschaften's practice of uploading science slam videos. Since at least 2009, YouTube has been a disseminator of the science slam genre. Today, one important part of the genre remains that the videos of the most successful slammers are regularly uploaded to YouTube, and organizers often have their own YouTube account. In this way, the performances can have additional success online. Through the uploading of the videos, an archive of successful presentations was created. The science slam is not only a situated event but also a digital phenomenon. Organizers occasionally watch YouTube videos of promising slammers before they invite them. Since organizers include the digital archive in their mode of production, they coach slammers not only according to stage appearance but also to media appearance. The mode of production—with its quick changes and diverse media—is often compared to YouTube.[18]

Our research illustrates diverse informal learning and communication cultures, as well as offline and online practices that mutually reinforce one another. In science slams, we can observe various practices of adoption. It is not only images, videos, and jokes that are the objects of appropriation, but also scientific personas, references, and ways of translating. As we observed, references and analogies in particular become viral and extend through time and space (for example, the reference to beer is a golden oldie in the science slam). In this way,

---

18 "Others then, using completely different media, they appear more YouTubeish. I think that's pretty popular right now." MS#48.

the translocation of social space (Knoblauch 2017: 341) becomes an important topic. The face-to-face situation becomes transcendent[19] and mediated at the moment the performance becomes part of the technical knowledge archive. Communicative actions are permanently available to people who have access to the internet (ibid). Many people can visit this archive, including those who are planning to perform.

"In fact, we have had few substantive questions where someone sent in their lecture and said, 'Is the dramaturgy somehow understandable?' But... that is just because YouTube is such a perfect coaching implement that they do not need us, since most people tell us, 'Before I wrote my speech, I first surfed for five hours on YouTube and looked at different lectures, lectures of others, and pulled out what best is for me.'" MK#45[20]

"One problem might be that this creates uniformity, and then you realize that the new Slammer XY, the new Slammer XY, is a bit like that one from one or two years ago. That's probably going to happen anyway, because people are watching YouTube and learning from each other. " MS#42[21]

People informally learn about science and the genre as they watch YouTube videos. With these forms of self-education, past performances of the genre become part of the further developments of the genre. Science slammers uptake features of successful associates without citation, use images from the internet without paying attention to copyright, and reproduce digital media content in their new context. In this way, the practice of a science slammer could be seen as similar to that of a DJ who makes a remix, or to a social media user who shares a meme (Fischer/Grünewald 2018).

---

19 Because the audience often showed up in great numbers, several organizers also established a public viewing for the event, where people can track the event sitting in a separate room. So the transcendence of the situation not only happens through the digital archive of YouTube, but also through live broadcasting.

20 Expert interview number 3. The interviews carried out within the science slam project were numbered chronologically starting with the number 42.

21 Expert interview number 1. The interviews carried out within the science slam project were numbered chronologically starting with the number 42.

CONCLUSION

We were interested in what happens in scientists' communication to the general public, and therefore, in this project, we considered how the boundary between science and the public was enacted and performed in science slam presentations. Thinking about new genres like the science slam, we asked in which way this communication contains an asymmetry that produces or perpetuates dissimilarity. Thus, boundaries between science and the public can be enacted by socio-technical arrangements, the order of communication, or visual representations. In public science communication, communicative processes shape the questions of what is public and what is science. We paid attention to the enactment of contemporary science communication and highlighted current socio-material arrangements and interaction orders to identify the new genre further. Focusing on the particular situation of the science slam, our project aimed to contribute to the question of how publics are communicatively constructed (Knoblauch 1995). In public science communication, the question of what is the public and what is science is designed into the communicative process. Science slammers imagine their audience (e.g., their level of intelligence, their short attention span), as well as the location and the effect of their presentation. It makes a significant difference if scientists think of their audience as 'not scientific' and 'easily bored,' or if they think of them as 'informed.' So, before the performance, there is an experience-based and/or imagined science slam audience and an imagined performance on stage. The practice of first learning about the genre through the digital knowledge archive YouTube changes this imagination. The texts slammers conceptualize have diverse relations to expectations about the genre. However, the situated performance on stage will always create some contradictions. For example, the prior idea of a nonscientific audience can be revised through a negotiation that unfolds during the performance. In this way, the communicative process on stage is the moment when imagined and situated realities are brought together.

Finally, we would argue against the most general concerns about whether science slams infantilize the public. Similar to the response to the critique on the deficit of PowerPoint communication (Tufte 2006), we would be inclined to argue that science slams probably do not make audiences dumb (c.f. Knoblauch 2008). No matter how much will be said about pooping, sex, or soccer in association with science, this will not be the cause for cultural decay. As with general PowerPoint presentations (Schnettler et al. 2007: 19), with science slams knowledge is becoming more physical. This is even reinforced by societal practices of uploading presentations onto the internet and producing them as media products. Whether for the good or the bad, the science slam is a positioning practice of science. Because of the way they are situated vis-à-vis a public, slammers do not focus so much on the abstract truth of science but embody and localize knowledge claims. This embodiment could be described as an expression and enactment of social identities. In opposition to the usual visualizing tricks and unlocatable knowledge claims of modern science, slammers get naked in public, figuratively speaking.

REFERENCES

Alac, Morana (2008): "Working with Brain Scans. Digital Images and Gestural Interaction in fMRI Laboratory." In: Social Studies of Science 38/4, pp. 483-508.

Amann, Klaus/Knorr-Cetina, Karin (1990): "The fixation of (visual) evidence." In: Michael Lynch/Steve Woolgar (eds.), Representations in Scientific Practice, Cambridge, MA: MIT Press, pp. 86-121.

Bauernschmidt, Stefan (2018): Öffentliche Wissenschaft, Wissenschaftskommunikation & Co.: Zur Kartierung zentraler Begriffe in der Wissenschaftskommunikationswissenschaft. In: Stefan Selke/Annette Treibel (eds.), Öffentliche Gesellschaftswissenschaften zwischen Kommunikation und Dialog, Wiesbaden:Springer.

Beaulieu, Anne (2002): "Images Are Not the (Only) Truth: Brain Mapping, Visual Knowledge, and Iconoclasm." In: Science, Technology & Human Values 27/1, pp. 53-83.

Bourdieu, Pierre (1988): Homo Academicus, Frankfurt a.M.: Suhrkamp.

Braun, Friederike/Pasero, Ursula (eds.) (1997): Kommunikation von Geschlecht, Paffenweiler: Centaurus.

Daum, Andreas (1998): Wissenschaftspopularisierung im 19. Jahrhundert. Bürgerliche Kultur, naturwissenschaftliche Bildung und die deutsche Öffentlichkeit 1848-1914, Munich: Oldenbourg.

Davies, Sarah R. (2009): "Doing Dialogue: Genre and Flexibility in Public Engagement with Science." In: Science as Culture 18/4, pp. 397-416.

Davies, Sarah R. (2013): "Constituting Public Engagement Meanings and Genealogies of PEST in Two U.K. Studies." In: Science Communication 35/6, pp. 687-707.

Ebel, Gerhard/Lührs, Otto (1988): "Urania: eine Idee, eine Bewegung, eine Institution wird 100 Jahre alt!" In: 100 Jahre Urania Berlin. Festschrift Wissenschaft heute für Morgen, Berlin: Urania, pp. 15-74.

Eyal, Gil/Buchholz, Larissa (2010): "From the Sociology of Intellectuals to the Sociology of Interventions." In: Annual Review of Sociology 36, pp. 117-137.

Fischer, Georg/Grünewald-Schukalla, Lorenz (2018): Originalität und Viralität von (Internet) Memes. In: Sonderausgabe kommunikation@gesellschaft 19.

Fraser, Nancy (1992): Rethinking the Public Sphere: A Contribution to a Critique of Actually Existing Democracy. In: Craig Calhoun (ed.), Habermas and the Public Sphere, Cambridge and London: MIT Press, pp. 109-142.

Gieryn, Thomas F. (1983): "Boundary-work and the demarcation of science from non-science: Strains and interests in professional ideologies of scientists." In: American Sociological Review 48/6, pp. 781-95.

Goffman, Erving (1981): Forms of Talk. Pennsylvania: University of Pennsylvania Press.

Goodwin, Charles (1981): Conversational Organization. Interaction between Speakers and Hearers, New York: Academic.

Günthner, Susanne (1997): Zur kommunikativen Konstruktion von Geschlechterdifferenzen im Gespräch. In: Friederike Braun/Ursula Pasero (eds.), Kommunikation von Geschlecht, Paffenweiler: Centaurus, pp. 122-146.

Habermas, Jürgen (1990): Strukturwandel der Öffentlichkeit. Frankfurt a.M.: Suhrkamp.

Haraway, Donna (1988): "Situated Knowledges. The Science Question in Feminism and the Privilege of Partial Perspective." In: Feminist Studies 14/3, pp. 575-599.

Haraway, Donna (1997): Modest_Witness@Second_Millennium. FemaleMan©_Meets_Onco Mouse™: Feminism and Technoscience, London: Routledge.

Haynes, Roslynn D. (1994): From Faust to Strangelove: Representations of the Scientist in Western Literature, Baltimore: Johns Hopkins University Press.

Hill, Miira (2016): Slamming Science. The New Art of Old Public Science Communication, Berlin. (PhD Thesis)

Hill, Miira (2017): "Die Versinnbildlichung von Gesellschaftswissenschaft – Herausforderung Science Slam." In: Stefan Selke/Annette Treibel (eds.), Öffentliche Gesellschaftswissenschaften: Grundlagen, Anwendungsfelder und neue Perspektiven, Wiesbaden: Springer VS, pp. 169-186.

Kiesow, Christian (2014): Exotische Sphären – eine wissenssoziologische Studie zu Kommunikation, Interaktion und Visualität in der mathematischen Forschung. (PhD Thesis)

Klaue, Magnus (2015): "Science Slams. Contra." In: Forschung & Lehre 7/5, pp. 543.

Knoblauch, Hubert (1995): Kommunikationskultur. Die kommunikative Konstruktion kultureller Kontexte, Berlin: de Gruyter.

Knoblauch, Hubert (2001): "Fokussierte Ethnographie." In: Sozialer Sinn 2/1, pp. 123-41.

Knoblauch, Hubert (2007): "Die Performanz des Wissens. Zeigen und Wissen in der Powerpoint-Präsentation." In: Bernt Schnettler/Hubert Knoblauch (eds.): Powerpoint-Präsentationen. Neue Formen der gesellschaftlichen Kommunikation von Wissen, Konstanz: UVK, pp. 117-138.

Knoblauch, Hubert (2008): "The Performance of Knowledge: Pointing and Knowledge in Powerpoint Presentations." In: Cultural Sociology 2/1, pp. 75-97.

Knoblauch, Hubert (2017): Die kommunikative Konstruktion der Wirklichkeit, Wiesbaden: Springer VS.

Lakoff, Robin (1975): Language and Women's Place, New York: Harper and Row.

Lynch, Michael (1988): "The externalized retina: Selection and mathematization in the visual documentation of objects in the life sciences." In: Human Studies 11/2-3, pp. 201-34.

Mergel, Thomas (2001): "Die Bürgertumsforschung nach fünfzehn Jahren." In: Archiv für Sozialgeschichte 41/1, pp. 515-38.

Myers, Natasha (2012): "Dance Your PhD: Embodied Animations, Body Experiments and the Affective Entanglements of Life Science Research." In: Body & Society 18/1, pp. 151-89.

Robertson-von Trotha, Caroline Y./Muñoz Morcillo, Jesús (eds.) (2012): Öffentliche Wissenschaft und Neue Medien: die Rolle der Web 2.0-Kultur in der Wissenschaftsvermittlung, Karlsruhe: KIT Scientific Publishing.

Robertson-von Trotha, Caroline Y./Muñoz Morcillo, Jesús (2018): "Öffentliche Wissenschaft. Von 'Scientific Literacy' zu 'Participatory Culture.'" In: Stefan Selke/Annette Treibel (eds.), Öffentliche Gesellschaftswissenschaften zwischen Kommunikation und Dialog, Wiesbaden: Springer.

Schnettler, Bernt/Hubert Knoblauch (eds.) (2007): Powerpoint-Präsentationen. Neue Formen der gesellschaftlichen Kommunikation von Wissen, Konstanz: UVK Verlag.

Schulze, Gerhard (1992): Die Erlebnisgesellschaft. Kultursoziologie der Gegenwart, Frankfurt a.M.: Campus.

Shapin, Steven (2006): "The Man of Science." In: Lorraine Daston/Katharine Park (eds.), The Cambridge History of Science Vol. III: Early Modern Science, Cambridge, UK: Cambridge University Press, pp. 179-191.

Shapin, Steven/Schaffer, Simon (1985): Leviathan and the Air-Pump: Hobbes, Boyle, and the Experimental Life, Princeton: Princeton University Press.

Tufte, Edward R. (2006): The Cognitive Style of Power Point. Pitching Out Corrupts Within. Cheshire, CT: Graphics Press.

Tuma, René (2012): "The (Re)Construction of Human Conduct: 'Vernacular Video Analysis.'" In: Qualitative Sociology Review 8/2, pp. 152-63.

Weingart, Peter (2005): Die Wissenschaft der Öffentlichkeit. Essays zum Verhältnis von Wissenschaft, Medien und Öffentlichkeit. Weilerswist: Velbrück Wissenschaft.

Epilogue

On Honey, VR Goggles, and Real Medicine

While a genealogical overview sheds much light on the phenomenon of 'popular science,' it also raises many questions about it. At the beginning of this volume, we talked about Lucretius' honey metaphor as indicating an essential aspect of popular science. His didactic poem *On the Nature of Things* embodies the idea of sweet honey that makes the more bitter 'Epicurean medicine' palatable, and even appealing. Indeed, his poem makes it easy to understand the complicated but healing philosophy of Epicurus. Poetry and, by extension, rhetoric and aesthetics, are the 'honey,' i.e., the medium for popularizing science. All societies need to break down intricate scientific knowledge so that the public can enjoy the curative and liberating power of science—i.e., knowledge about the natural world in a broader sense, which includes the human world as well. In this age of fake news, science may very well be the only thing that the majority of us still believe in. For example, for environmentalists science provides the diagnosis of the planet's ailments and prescribes the therapy for healing the Earth or, at the very least, for keeping the planet habitable; and the COVID-19 pandemic is showing us that people and governments largely turn to science in the face of global health crises. But the messages of and about science need further elaboration in order to develop societal agency beyond the confines of academia.

Correspondences with ancient didactic writings can be found even in popular new media, from the protreptics[1] of web videos and science slams to the immersive force of virtual and augmented reality. Virtual reality systems allow immersive, playful, and intuitive experiences of science and of cultural events that serve as sensory evidence in the same way that ancient *ékphrasis* and autoptic narratives provided immersive experiences. The very act of putting on VR-

1 A "protreptic speech" is a rhetorical exhortation which served to attract new students to schools of rhetoric or philosophy. For the significance of the protreptic in antiquity, see the introduction to this volume, pp. 36-38.

goggles has something metaphysical about it that goes beyond the classical honey metaphor. Today, we can literally taste Lucretius' honey by touching and feeling the entertaining realm of scientific knowledge. Indeed, popular science is no longer encapsulated exclusively in vivid descriptions or narratives that spark images in the reader's or listener's mind by building on previous knowledge. The immersion is now supposed to be almost real—an 'as if' experience, at least for a while. This development is not only a matter of virtual reality. Whether you are watching a science slam, a YouTube video, or you are experiencing a virtual environment, popular science has become an act of mystical embodiment, a systemic map that is more spectacular than ordinary rhetorical approaches, but not necessarily more effective than other communications strategies, such as immersive descriptions, literary dialogues, or graphic novels. It does not matter whether or not the science behind the literary or technological 'honey' has a clear moralizing goal. Scientific research that aims at a higher moral purpose seems to be the norm. Even in esoteric societies, the connection between knowledge and salvation is usually accompanied by some kind of initiation. Therefore, popular science can 'cure' in a systemic sense, i.e., it can 'stabilize' a society by providing general knowledge, combating hoaxes, explaining the agency of living beings (including humans) and things, and persuading people about the gifts and usefulness of scientific knowledge in a broader sense. As long as science is part of society, public science does necessarily exist, regardless of the specific, historically contingent vocabulary it uses to define itself. Public understanding of science, scientific literacy, public engagement in science, science communication, or just public communication of science—the explosion of different terminologies and theoretical models of popular science in recent decades corroborates the existence of a more extensive base and the possibilities of the continuation and transformation of its main features in interdependence with current societal and political needs.

But, as I have already indicated, the genealogical overview of this phenomenon raises as many questions as it answers, and these new questions should be addressed in future works. Let us comment on a few of them—the ones that seem most crucial to me.

There are similarities between how ancient ecphrastic techniques sensualized content to spark the listeners' imagination and how fictional movies activate the audience's previous knowledge about scientific routines to popularize academic concepts and processes. Can we then infer that simplification, concision, and sensualization are universal categories of popular science? We are certainly still using vivid descriptions (*ekphráseis*) as sources of knowledge, even if the impact, firstly, of the printed, illustrated book and, then, of the 'pictorial turn' has taken away visibility from a rhetorical device that continued to be taught at school almost until the 18th century. Therefore, ecphrastic studies should go beyond their traditional focus on literature, and should expand their scope so as to include the current rhetorical practices of science communicators. On the other hand, the emergence of contradictions as a result of the encounter between imagined audiences and the real effect of popular science on them is a recent research topic suitable for investigation in modern, performative forms of popular science, but also one that is highly interesting for former historical periods. This would probably imply the development of new methodological approaches to analyze similar phenomena in the past, such as the contradictions between the intention of ancient didactic poetry and its actual effect on the corresponding societies.

If we look at the use of images of the brain in popular media, as discussed by Hommrich in this volume, some of them pertain to subjectivity, and others link to scientific objectivity, in a kind of feedback loop that considers the expectations of the readership, which raises the question of whether this is a general phenomenon that goes far beyond the single case study included here. Indeed, the popularization of science does not always rely on real scientific images. We might therefore ask: What is the motivation for using surrogates for or abstractions of scientific images?

Another big topic is the relation between public science and ideology. There are important and complex reasons behind editorial decisions favoring some topics over others or even for coming up with positive or neutral information policies about new technologies and research projects in general. In this context, the ideological force of visual metaphors can affect society's idea of art, technology, and nature. A good example would be the visualization of nature as a harmonic system where capitalism is naturalized in such a way that it is not even perceived. If the vision of nature as a harmonic ecosystem is ideologically loaded, then we can also see in this example a massive degree of similarity to the visual metaphors that popularized specific philosophical schools in antiquity or theological ideas in the pre-modern period. What is, for example, the point of imparting knowledge to a broad audience about natural phenomena defending or invalidating Aristotelian theories? Is this about the search for truth and its dissemination in society, or are there religious, ideological, or even economic factors at play? In this field of research, we need to think big and be humble at the same time. We are not in a more privileged moment in history than other societies were in terms of popularizing cultural and scientific knowledge. Moreover, popular science is not ideologically neutral. It is an anthropological mirror. Pictorial representations of Enlightenment scientists refer to the conviction of empiricism, as Jehle has shown in his contribution to this volume. In a similar sense, Sehlmeyer argues in this book that the uses of epitomes in late antiquity and in the Renaissance were quite similar, but epitomes produced during the Renaissance were not as successful in later times. This raises the question of whether those short texts became less influential because of modernity itself. Did the idealization of ancient times, in combination with the paragon culture, lead to a seemingly reduced use of those otherwise beneficial texts? Or were Renaissance humanists already confronted with some ideologically shaped awareness of authorship that prevented them from using or recognizing the value of these 'minor' pieces of literature if they were not written by ancient authors? And what are the myths at a modern school if not part of a shared ideology? The values passed down at school are the base of cultural and national identities. The popularization of history, for example, through teaching at school, implies an institutionalized reading of the past, especially regarding the recent history of each country and the continuous accommodation of its myths. Highlighting specific historical episodes seems to be a way of selecting some values over others, as we can read in Fernández Delgado's contribution in this volume. This applies not only to ancient times. Ancient progymnasmatic[2] sketches reveal many societal aspects of popular science; maybe it is time to take a look at school dictations and essays from modern times to understand the interaction of knowledge popularization at school and its sociopolitical implications.

Sometimes it seems that popular and scientific literature only interact with each other in the humanities, and especially in literature and art. Artists also influence society, even if their work is often perceived as elitist. In antiquity, optical and technical knowledge was essential for artists. The rendering of applied science as an artistic aid was thematized in popular anecdotes so that a broad educated audience had access to basic knowledge about the tricks of perspective, the power of imitation, and the boundaries of realistic representation. In light of this, how significant is the impact of the artistic popularization of scientific theories in modern times? Maybe it is time to measure how the avant-garde shaped the general education of 20th-century society, e.g., in relation to the persistence of holistic points of view or faith in technology. There is no doubt about the achievements of popular aesthetics, which, as Manker demonstrates in this volume, have since become part of everyday life. But can art, as a vehicle of knowledge dissemination, transform society? Can we compare its influence to the impact of scientific disciplines such as biology or physics, or to popular

2 *Progymnásmata* are preliminary rhetorical exercises. For a detailed analysis of them, cf. the introduction (pp. 39-43) and Fernández Delgado's article.

technological achievements such as the invention of the bicycle, the car, or the cell phone? I think that, especially in the context of popular science, we should differentiate between art as a source of aesthetic knowledge and art as a systemic communication tool that deals with scientific, cultural, and ideological issues. In this sense, the popularization of art not only creates a new bourgeoisie addicted to classy patterns and collectibles. The popularization of art can also put scientific or cultural knowledge into a broad societal context through communication strategies similar to criticism and activism.

Another less investigated topic related to art is the embodiment of knowledge popularization in cultural artifacts. Greek amphorae, as Streicher claims in this book, show us how history and mythology traveled through the Mediterranean area along with cultural values that held Hellenistic society together. What kind of objects embody popular science today? Surely not only science kits with neutral values and an exclusive focus on scientific experiments, but also movies and narratives that reproduce values, dominant historical or scientific interpretations, and stereotypes like the perpetuation of the gap between the sciences and the humanities. Hollywood is, indeed, responsible for most of the popular scientific ideas we have in our minds.

Popular science also reveals the coexistence of different models of nature, especially in antiquity, where traditional cosmology collides with imaginaries of the void, and harmonic nature encounters violent, unpredictable events, all of this reflected in epic and didactic poems that were intended for a broader audience and composed with educational purposes in mind. Popular science can even be prohibited or censored if the ideas reflected by it represent a threat to the established order. The history of popular science and censorship is still to be written.

And what about the origins of interdisciplinarity in technical literature? Hero of Alexandria, for example, makes mechanics understandable for practitioners and interested lay readers, bridging the gap between the esoteric disciplines and everyday experience, as we learned from Roby's contribution to this volume. Is it possible that this kind of interdisciplinary approach was the key to Hero's success many centuries later, in the Renaissance? Hero could be considered the grandfather of bricolage tutorials and the do-it-yourself culture. If we expand this insight to other historical figures, we could argue, for example, that the father of art history, Giorgio Vasari, is a similar example of interdisciplinary public science.

Under special circumstances, the popularization of science also serves as a kind of pseudo-erudition, i.e., it promotes such things as the knowledge of curiosities and *mirabilia* for educated persons. Simplifications of scientific ideas and name-dropping are recurrent methods used to create a popular science argument based more on previous authority than on scientific reasoning or facts. Even Aristotelian doxography, as Martín-Velasco points out in this volume, is not free of a certain persuasive strength that builds on famous statements and broken syllogisms in order to create the sense that a specific natural phenomenon was already fully understood by a prior authority, thereby encouraging us to believe an abbreviated version of things... This sounds more familiar to us than we would probably like to admit. However, the big question here is why societies select specific topics over others and how they go on to popularize them. These mechanisms are often very carefully hidden, but, as we have seen in former periods, selecting and shaping scientific knowledge is always instrumental for shaping the identity and the political nature of a society. I firmly belive that we should keep exploring continuities, disruptions, and transformations of popular science over large periods of time and across different regions and cultures. Maybe then we will start to understand popular science as a universal phenomenon, similar to religion or politics. Whether we are wearing VR-goggles or not, we, postmodern people that we are, are not as special as we think. Every society has the popular science it deserves.

Jesús Muñoz Morcillo

About the Authors

Jesús Muñoz Morcillo, classicist (PhD) and art historian (PhD), is research fellow at the ZAK | Centre for Cultural and General Studies and the Institute of Art and Architecture History (IKB) at the Karlsruhe Institute of Technology (KIT). In September 2019, the KIT dispatched him for nine months to the United States as the Volkswagen Foundation Fellow at the Getty Research Institute (GRI) in Los Angeles. Currently, his research focuses on the reception of ancient descriptions and pre-modern ecology.

Caroline Y. Robertson-von Trotha, sociologist (PhD, Professor), is founding director of the ZAK | Centre for Cultural and General Studies at the Karlsruhe Institute of Technology (KIT), coordinator of the German network of the Anna Lindh Foundation, member of the Culture Committee of the German UNESCO Commission, and chairlady of the Academic Council for Culture and Foreign Policy (WIKA) at the Institute for Foreign Cultural Relations (ifa).

Oliver Hochadel is a historian of science (PhD) and a tenured researcher at the Institución Milá y Fontanals de Investigación en Humanidades (CSIC, Barcelona). His research focuses on the urban history of science as well as the relationship of science and the public in historical perspective. This includes case studies on electricity in the German Enlightenment, the zoological garden in the 19th century, and human origins research in the 20th century.

Martin Streicher, classical archaeologist (PhD), is the academic coordinator of the Leibniz ScienceCampus – Postdigital Participation – Braunschweig (LWC PdP) at the Georg Eckert Institute for International Textbook (GEI) Research – Member of the Leibniz Association (WGL).

José Antonio Fernández Delgado, classicist (PhD), is Professor Emeritus of the University of Salamanca and former Alexander von Humboldt Scholar at the Freie Universität Berlin. His research interests include Ancient Greek poetry, Plutarchan studies, and the relationship between schooling and literature in antiquity, fields about which he has published ten books and more than 150 papers for peer-reviewed journals and edited volumes.

María José Martín-Velasco, is a classicist (PhD), hispanist, translator, and editor. She spezialized in Greek philology at the University of Santiago de Compostela (USC), where she worked as a lecturer and is currently a reseach associate. In her research, she focuses on Aristotle's theories applied to different literary genres such as epic, theater, and poetry. She has been invited to give talks at different institutions, such as the ICS London, the University of Thessaloniki, the European Council, and the Universites of Lisbon and Coimbra.

Sara Matías Pérez, PhD candidate at the University of Murcia. She studied Classical philology (BA) at the University of Santiago de Compostela and at the University of Salamanca (MA). She is currently conducting research on the Ancient Greek novel.

Maurice Parussel, classicist (MA), is scientific assistant in Greek literature at the Department of Classics of the Ruhr-University Bochum. He is currently writing his PhD dissertation on the reception of the late antique tragedy Christus patiens in the literature of the Renaissance and the early modern Europe.

Dorit Engster, ancient historian (PhD), is lecturer (Lehrkraft für besondere Aufgaben) at the Seminar for Ancient History at the Georg-August University Göttingen. Her research interests include the history of ancient science and technology, historiography and historical theory, women and gender, and the society, religion, and culture of the Roman Empire.

Matteo Rossetti, classical philologist (MA Pavia 2015, PhD Milan 2019), is lecturer in Latin literature at the University of Milan and teaches Greek and Latin language and literature at Italian high schools. He is currently working on Manilius and his relationship with the Aratean tradition, and is also author of articles on Aratus, Manilius, and the reception of Plutarch in Kepler's Somnium.

Courtney Ann Roby is Associate Professor in the Department of Classics at Cornell University, New York. Her research interests focus on the literary aspects of scientific and technical texts from the ancient world, the interaction of verbal and visual elements in those texts, and the definition and dissemination of scientific work. Her first book was *Technical Ekphrasis in Greek and Roman Science and Literature* (2016). Her current projects address how early printed editions of ancient technical treatises rework ancient authors' materials for a new context of production and propagation.

Mathias Herweg (PhD) is full professor of Medieval and Early Modern German Literature at the Karlsruhe Institute of Technology (KIT). His publications include essays on German and European Literature from the Carolingian era to the 16th century, editions of medieval texts such as 'Herzog Ernst' and 'Kaiserchronik,' and ongoing DFG research projects on Encyclopedic Writing and on the medieval Buddha legend 'Barlaam and Josaphat.'

Marion Gindhart, classicist and neo-latinist (PhD, Professor), focuses her studies on the communication, transmission, and transformation of knowledge in antiquity and the early modern period. She is founding member of the DFG-Research Training Group "Early Concepts of Humans and Nature" (Mainz University) and part of its core faculty. Currently, she is working as a researcher in the SNF-Project "Sebastian Brant's Writings in Relation to Early Modern Textual Cultures" (Bamberg University).

Markus Sehlmeyer, ancient historian (PhD, Pivatdozent), was assistant professor at the University of Rostock, research fellow for late antiquity in Jena, and university lecturer in Bielefeld, Osnabrück, Hildesheim, and Marburg. His main interests are ancient historiography, political culture, and the comparison of ancient and early modern migration regimes. During the 2019-2020 academic year, he is serving as Acting Chair of Ancient History in Rostock.

Oliver Jehle, art historian (PhD and Professor), heads the Art History Department (IKB) at the Karlsruhe Institute of Technology (KIT) and currently focuses his studies on the "techniques of naturalness" in the long 18th century. Together with curators from the State Art Gallery Karlsruhe, he recently organized an international symposium on religious and profane art of the early modern period—co-financed by the Volkswagen Foundation. The research results presented there aligned the conception of the well-regarded Grand State Exhibition of Baden-Württemberg 2019 on Hans Baldung Grien.

Erna Fiorentini, art historian (PhD) and geochemist (PhD), is professor for Art History at the Karlsruhe Institute of Technology (KIT). Against a training and research background in both the natural sciences and the humanities, her interests are the ideas, practices, and technologies of image-making that challenge vision and our understanding of visibility and visualization. Her most recent monograph on these themes, co-authored with James Elkins, is *Visual Worlds. Looking, Images, Visual Discipline* (Oxford University Press, 2020). She is the editor of *On Visualization. A multicentric Critique beyond Infographics*, forthcoming from LIT Verlag Berlin 2020.

Alexandra Axtmann, art historian (PhD), is research associate at the Research Data Management Department of the KIT-Library at the Karlsruhe Institute of Technology (KIT). Her current research interests include the history of art historical and architectural history with a focus on their establishment as academic disciplines and the networks of their protagonists, such as Wilhelm Lübke. In addition, she is presently researching the history of the religious parchment and paper cutting.

Elin Manker, art historian (PhD), is a senior lecturer at Umeå University at the Department of Culture and Media Studies. Her research interests include design, decorative arts, and the manufacturing of everyday objects from the 18th century until today. More specifically, she explores the process of designing, the meaning of serial objects in everyday life, and the effects and responsibility that come along with producing and consuming at a local and a global level.

Beatrice Immelmann is lecturer at the Department of Art History at Georg-August-Universität Göttingen (since April 2020), where she is focussing on visual culture studies. She is pursuing a PhD in art history at the University of Vienna. In her thesis on "Visual Resonance" she investigates the conception of aesthetic experience as a resonant relation between artists, artworks, and beholders in art theoretical discourses.

Lena Sophie Trüper is a research assistant at Berlin University of the Arts in the DFG Research Training Group "Knowledge of the Arts." In her PhD project under the working title "Natural Metaphors of Cybernetics: Visual Expressions of Technoecologies from Modernity to the Present" she examines the visual reception of cybernetics in art and popular culture. Her research interests include the transdisciplinary fields of cybernetics, media archeology, film and media studies and postmodern literature, and philosophy.

Lars Frederik Köppen studied history, Islamic studies, and agricultural science at Martin-Luther-Universität Halle-Wittenberg from 2009 to 2017, earning an MA in history with a study on the role of socialist brigades (Brigaden der sozialistischen Arbeit) as executive institutions for social control and workplace discipline in GDR factories. He is currently an archivist and project assistant at the Stiftung Historische Museen Hamburg (SHMH).

Dirk Hommrich (PhD) is board member of the Institute for the Study of Culture Heidelberg, Germany. He has been a research fellow at the Faculty of Translation Studies, Linguistics and Cultural Studies of the University of Mainz as well as senior researcher at the Institute for Technology Assessment and Systems Analysis at the Karlsruhe Institute of Technology. Currently, he is research and policy officer of the German Council for Scientific Information Infrastructures. His publications focus on science studies of neuroscience, philosophy of technology, and post-phenomenology.

Kathrin Klohs, sociologist of science (PhD), has focused her research at the Universities of Freiburg i.Br., Basel, and Lucerne on science studies and fiction, popular media in the 20th and 21st century, narration and governmentality, guidebooks, and Swiss contemporary prose. Currently, she works as a Programme Coordinator at the University of Applied Sciences and Arts Northwestern Switzerland, School of Education (PH FHNW).

Miira Hill (PhD) is a sociologist and post-doctoral researcher at the college of Leuphana University Lüneburg. For many years she has been teaching in fields like emotion sociology, history of science and of sexuality, and visual communication. She is currently researching the relationship between science and the public in the context of digital societies.

Index of Names and Terms